BEITRÄGE ZUR REGIONALEN GEOLOGIE DER ERDE BAND 15

The volcanoes of Agua and Fuego (left), Guatemala. (From VON SEEBACH 1892, Table VII).

BEITRÄGE ZUR REGIONALEN GEOLOGIE DER ERDE

Herausgegeben von Prof. Dr. F. Bender, Hannover,
Prof. Dr. V. Jacobshagen, Berlin, Prof. Dr. J. D. de Jong,
Heemstede/Niederlande, und Prof. Dr. G. Lüttig, Hannover
Band 15

Richard Weyl

GEOLOGY OF CENTRAL AMERICA

Second, completely revised edition

1980

GEBRÜDER BORNTRAEGER · BERLIN · STUTTGART

GEOLOGY OF CENTRAL AMERICA

Second, completely revised edition

by

Richard Weyl

Geologisch-Paläontologisches Institut
der Universität Gießen

With 202 figures and 13 tables in the text
and on 8 folders

1980

GEBRÜDER BORNTRAEGER · BERLIN · STUTTGART

Address of the author:

Professor Dr. Richard Weyl,
Geologisch-Paläontologisches Institut der Universität
Senckenbergstraße 3, D-6300 Gießen

Printed in Germany by Konrad Triltsch, Graphischer Betrieb, Würzburg
ISBN 3 443 11015 0

Preface

The first edition of this book appeared in German in 1961 under the title "Die Geologie Mittelamerikas". Since that time geological research in this part of the world has advanced to such an extent that I felt a revision was necessary. I am grateful to the publisher, Dr. E. NÄGELE, for agreeing to this project and for supporting it in every possible way. I am also grateful to him for the suggestion to bring out a new edition in English in order to reach a broader readership. In doing so, I have complied with a wish expressed by many colleagues in America; however, I do at the same time regret that I have not been able to use my own language to express myself.

The translation was prepared by DEREK and INGE JORDAN in Ottawa, Canada, and I wish to thank them here for their efforts. The German manuscript was completed by the end of 1978, and the translation was ready for printing in July 1979.

I would not have been able to prepare this new edition without the help of many institutions and colleagues in Germany and America. I am grateful in particular to the Deutsche Forschungsgemeinschaft for generously subsidizing many trips to undertake studies or attend congresses.

My work was greatly assisted by the maps and unpublished reports prepared by the Bundesanstalt für Geowissenschaften und Rohstoffe in Hannover, which were kindly made available to me, in particular by my colleagues NICOLAUS and WIESEMANN. I am also indebted to the documentation centre of the above Institute for the bibliographical help which was provided. The University Library of the Justus Liebig-Universität in Gießen always very willingly helped me to obtain the literature material. Nevertheless there are still some gaps in the literature quoted and, in particular, unpublished reports have been missed.

Of my German colleagues I wish to thank H. PICHLER and his co-worker STENGELIN for calculating the norms for many chemical rock analyses.

I am also grateful to various colleagues for critically reviewing certain chapters, namely Messrs. PICHLER (Volcanism), SCHMIDT-EFFING (Costa Rica), STIBANE (Costa Rica) and WIESEMANN (El Salvador).

From Central America I wish to thank my colleagues S. BONIS, O. H. BOHNENBERGER and G. DENGO in Guatemala, R. CASTILLO, C. DÓNDOLI, G. ESCALANTE, R. FISCHER, C. GALLI and M. SÁNDOVAL in Costa Rica, D. DEL GIUDICE and G. RECCHI in Panama, as well as LÓPEZ RAMOS in Mexico for providing me with literature and much helpful advice. US geologists supported my work by letting me have sometimes still unpublished papers; I would like to mention here the names of W. A. VAN DEN BOLD, M. J. CARR, J. E. CASE, R. C. FINCH, S. E. KESLER, R. D. KRUSHENSKY, J. W. LADD, A. R. MCBIRNEY, W. I. ROSE

Jr., R. E. STOIBER, H. WILLIAMS and W. P. WOODRING. I wish to express my gratitude to them and to many unnamed colleagues for their assistance.

I am also deeply indebted to the Instituto Geográfico Nacional of Costa Rica and to its directors FEDERICO GUTIÉRREZ B. (†), MARIO BARRANTES F. and FERNANDO M. RUDIN R. for their tireless and wide-ranging help in the field and for procuring maps and documents.

I owe a special debt of thanks to the female staff of the Geologisch-Paläontologisches Institute of Gießen University: to Frau B. LINS for carrying out the often arduous typing work, to Frau M. HÖBELER for assisting me in procuring the necessary literature, to Frau S. MORO for preparing almost all the drawings, and to M. SCHORGE for the photographic work.

Last but not least I have to thank my wife for the understanding she has shown for my work and also for her patient tolerance of my periods of absence often lasting many months.

Gießen, October 1979 RICHARD WEYL

Contents

1. Introduction

1.1 History of the geological exploration of Central America

The geological exploration of Central America has proceeded in stages. The first phase of geological research activity, which took place in the nineteenth century and carried over into the first few years of the twentieth century, can be described as **the age of the foreign explorer.** DOLLFUSS & MONT-SERRAT (1868) travelled through Guatemala and El Salvador on business connected with Napoleon III's American policy and they left behind a comprehensive report together with what is, to my knowledge, the first coloured geological map of a large area of Central America. A few years before them K. VON SEEBACH (1864/65) had devoted a detailed study to the volcanoes of Central America which was not, however, published until after his death in 1892. In 1873/74 WILLIAM GABB explored remote regions of the Talamanca of Costa Rica and recognized the post-Miocene age of the local granites (GABB 1895). Other geologists who were active in this period and who deserve mention here were MORITZ WAGNER (1870), ROBERT T. HILL (1898) and above all KARL SAPPER, whom SCHUCHERT (1935) has rightly dubbed the father of geological research in Central America. In many journeys on foot and by mule he traversed in particular the northern part of the region, accompanied only by a few Indians as bearers. He developed a conception of stratigraphy and structure, the basic principles of which are still valid even today, and he studied many of the volcanoes of the region, recording his findings in a great number of writings and maps (1890 – 1937). His pupil and successor, FRANZ TERMER, wrote a biography of SAPPER which was published first in Spanish (1956) and later also in German (1966); it contains a bibliography of SAPPER's works.

A characteristic feature of this first period of geological exploration is the universality of the research carried out by the individual explorers who, in addition to gathering geological data, made important contributions to the knowledge of the physical geography, the flora and fauna and even to the ethnography of the countries which they visited. As if that were not enough, their maps and landscape paintings have an extremely high aesthetic appeal (LÖSCHNER 1978).

The geological exploration of southern Central America was given a great impetus by the planning and construction of the **Panama Canal,** a project which is dealt with in works by C. W. HAYES (1899), BROWN & PILSBRY (1911 to 1913), D. F. MACDONALD (1913 to 1919), T. W. VAUGHAN (1918 to 1932) and in particular W. P. WOODRING. These authors above all laid the groundwork for understanding the stratigraphy of the Tertiary. A further part was not added to WOODRING's monograph on the molluscan fauna until 1973.

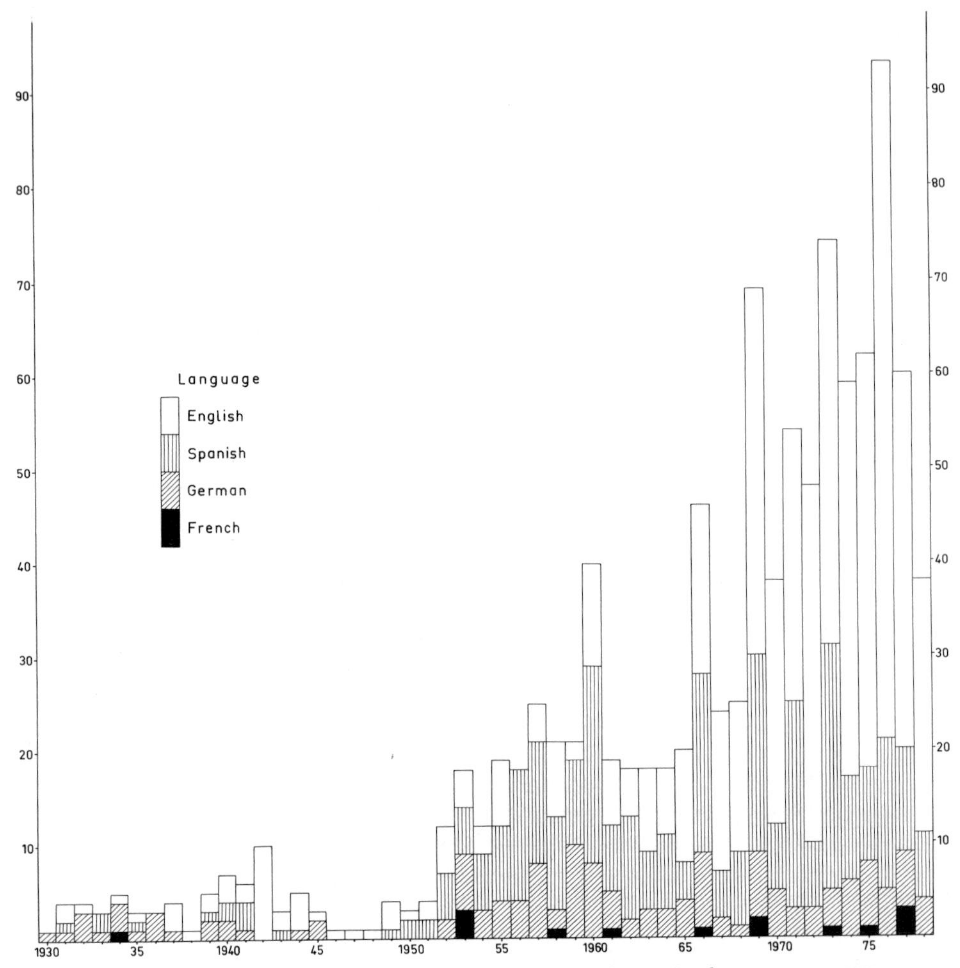

Fig 1. The number of papers written on the geology of Central America from 1930 to 1977.

The search for **petroleum** initiated a further stage in the exploration of the countries of Central America which resulted in some reports being published by A. H. REDFIELD (1923). However, with the exception of a fairly comprehensive work by R. A. TERRY (1956) on the geology of Panama and also some occasional notes, the results of the research carried out at that time have remained unpublished.

On the other hand, the search for **strategically important metals** during the Second World War led to a comprehensive inventory of the deposits of Central America being prepared by ROBERTS & IRVING (1957), whose report at the same time gives a review of the geology of Central America. The report is accompanied by an initial general map on the scale of 1 : 1,000,000 which is, however, still largely based on the maps drawn by SAPPER and TERRY.

The decades **following the Second World War** up to the present time are marked by a great variety of impetuses, some originating from foreign sources and some from researchers in Central America itself, and this period is above all notable for the development of independent geological institutions in the countries of Central America. Fig. 1 shows that as this happened so the number of publications dating from this period, and especially those in Spanish, increased thereby generating a wealth of data with which it is difficult to keep up. Only a few of the institutions can be mentioned here:

In 1951 in **El Salvador** the **Instituto Tropical de Investigaciones Científicas (ITIC)** was established at the instigation of the philosopher and biologist ADOLF MEYER-ABICH from Hamburg. This institute covered all the fields of natural science and it has initiated a series of geological studies, in particular on the Salvadorian volcanoes. At the same time the institute has given birth to a **Servicio Geológico Nacional,** one of the first in the country, which still exists today but under a different name. Four volumes of "Anales" (1955 – 1961) attest to the work performed by the staff of this office. An earthquake service was set up within the Service, and it was quickly equipped with the most up-to-date apparatus.

In 1956 another **Servicio Geológico Nacional** was set up, this time in **Managua, Nicaragua.** This Servicio was concerned mainly with exploring for ore deposits and sources of industrial raw materials and it reported on its work in 8 "Boletines" (1957 – 1964). This type of publication has been continued in the form of "Informes" which appear at irregular intervals.

The other countries in Central America followed suit with the establishment of permanent institutions to explore for deposits, to carry out geological surveying, and to provide consulting services (cf. 1.2).

The efforts of the individual countries are supported by a common Central American institution, namely the **Instituto Centroamericano de Investigación y Tecnología Industrial (ICAITI),** which is located in **Guatemala.** In 1969 its Geological Department published a 1 : 2,000,000 scale metallogenetic map of Central America; an accompanying volume contains a new inventory of the reserves of metallic raw materials. The map is at the same time a general geological map based on the facts as they are at present known. ICAITI also published the papers which are presented at the Meetings of Central American Geologists. These Meetings have taken place approximately once every three years since 1965, and they have now acquired the status of an internationally recognized Congress.

The various national endeavours are supported by bilateral research programmes and exploration projects conducted by the **United Nations.** The Federal Republic of Germany sent **Geological Missions** from the Bundesanstalt für Bodenforschung to work in Guatemala from 1967 – 1970 and in El Salvador from 1967 – 1971. Their results have been published partly in the form of maps and reports, and sometimes the data have been made available to the host countries in unpublished reports.

Fortunately the results of the **petroleum exploration work** which has been going on since the fifties in the regions of the Tertiary and Mesozoic basins have been made public and they have added considerably to our knowledge of the stratigraphy, facies, paleogeography and structure of hitherto completely unknown regions, particularly on the Caribbean side of Central America.

Co-operation with US universities, drawing on a large number of unpublished Masters' and PhD theses, has resulted in successful mapping programmes being implemented in Guatemala and Honduras. Summary reports have been published in the AAPG Bulletin and the GSA Bulletin.

It is understandable that the sharp increase in geoscientific activity should have sparked the demand for an indigenous training establishment for geologists to be set up in Central

America. After long negotiations the **Escuela Centroamericana de Geología** was founded at the Universidad de Costa Rica, and it opened its doors to students in 1970; geology had already been taught as part of the agricultural programme offered at the university. UNESCO and various individual countries helped set up the Escuela by supplying teaching staff. Since 1974 the Federal Republic of Germany has been providing two teachers, each of whom spends 2 – 3 years at the university.

Even if it is not possible in this report on the development of the last few decades to mention the names of all the geologists who are still active in this field, it is nonetheless necessary to list two of them specifically by name: HOWEL WILLIAMS and ALEXANDER MCBIRNEY. From 1952 to 1969, partly working alone and partly in co-operation with other workers, they studied the regions of Tertiary and Quaternary volcanism in Central America from a modern standpoint using modern methods; in so doing they worked a field which had lain largely fallow since SAPPER. Their studies provide an integrated review of volcanic history and its role in the structure of Central America. In the past 15 years a series of violent volcanic eruptions has triggered a large number of **volcanological studies**, and a gratifyingly large number of them have been published in the GSA Bulletin and in the Bulletin Volcanologique.

The **severe earthquakes** which have struck Central America in the last 15 years have also provided similar spurs to research. A field closely connected with earthquake research is the development of the geotectonic theory at present being discussed under the name of "plate tectonics". Central America and the adjoining Caribbean and Pacific regions play a special role in this development, and we will return to this frequently in what follows, but especially in Chapter 8.

1.2 Geological institutions in Central America

Regional institutions

Instituto Centroamericano de Investigación y Tecnología Industrial (ICAITI), Apartado 1552, Guatemala Capital, Guatemala

Geological consulting, co-ordination, publication of the "Publicaciones Geológicas del ICAITI" together with conference reports of the "Reuniones Geológicas de América Central"

Escuela Centroamericana de Geología, Universidad de Costa Rica "Rodrigo Facio", Apartado 35, San José, Costa Rica

Geological college operated jointly by the countries of Central America

Guatemala

Instituto Geográfico Nacional (Ministerio de Comunicaciones y Obras Públicas), Guatemala Capital

División de Geología: Geological consulting and mapping. Preparation and sale of topographical and geological maps of Guatemala

Dirección General de Minería e Hidrocarburos (Ministerio de Economía), Guatemala Capital

Supervision of petroleum and mining concessions. Partner of UNDP Missions carrying out exploration for deposits.

Instituto de Sismología, Vulcanología, Meterología, Hidrología (INSIDUMEH), Guatemala Capital

Instituto Nacional de Electrificación, Guatemala Capital

El Salvador

Instituto Geográfico Nacional (Ministerio de Obras Públicas), San Salvador

Preparation and sale of topographical maps

Centro de Estudios e Investigaciones Geotécnicas (Ministerio de Obras Públicas), San Salvador

Geological consulting, partner in international geoscientific missions. Has a largely independent "Servicio Sismológico, San Salvador"

Dirección de Recursos Naturales, San Salvador

Administración Nacional de Acueductos y Alcantarillados (Ministerio de Agricultura), San Salvador

Honduras

Instituto Geográfico Nacional (Ministerio de Comunicaciones y Obras Públicas), Comayaguela, D.C.

Publication and sale of topographical and geological maps

Dirección General de Minas e Hidrocarburos (Ministerio de Recursos Naturales), Tegucigalpa

Geological research, consulting and supervision of petroleum and mining concessions. Partner in international geological missions

Nicaragua

Dirección General de Cartografía (Ministerio de Obras Públicas), Managua, D.N.

Preparation and sale of topographical maps. Sale of general geological maps

Servicio Geológico Nacional (Ministerio de Economía, Industria y Comercio), Apartado Postal 1347, Managua, D.N.

Geological research and consulting with emphasis on natural raw materials

Catastro e Inventario de Recursos Naturales, Managua, D.N.

Geological, photo-geological and radar-geological mapping, publication and sale of geological maps, publication of "Informes Técnicos"

Instituto de Investigaciones Sísmicas, Managua, D.N.

Costa Rica

Instituto Geográfico Nacional, San José

Preparation and sale of topographical maps, gravimetric and magnetometric surveys, special thematic maps, publication of "Informes Semestrales" containing articles on geoscientific topics

Museo Nacional, Departamento de Paleontología, San José

Dirección de Geología, Minas y Petróleo (Ministerio de Industria y Comercio), San Francisco de Dos Ríos, San José

Geological research and consulting. Supervision of mines and petroleum concessions. Publication of "Informes Técnicos"

Instituto Costarricense de Electricidad (ICE), San José

Fundamental geological data for dam construction. Development of geothermal energy. Internal reports

Servicio Nacional de Aguas Subterraneas (SENAS), San José

Development of groundwater resources

Corporación Costarricense de Desarrollo (CODESA), Departamento de Geología, San José

Development of useful raw materials

Escuela Centroamericana de Geología, Universidad de Costa Rica "Rodrigo Facio", Apartado 35, San José

Escuela de Ciencias Geográficas, Facultad de Ciencias de la Tierra y del Mar, Universidad Nacional, Heredia

Panama

Instituto Geográfico Nacional "Tommy Guardia", Panamá 5

Preparation and sale of topographical maps

Dirección General de Recursos Minerales (Ministerio de Comercio e Industria), Panamá 5

Geological research, preparation and sale of geological maps. Partner in international geological missions

International missions

A large number of missions of limited duration from the UNDP and from various individual countries have been active in Central America since the sixties. BOHNENBERGER gave a review of the situation as of 1969 in WEYL (1971 c).

1.3 Serials, comprehensive works, bibliographies

Serials

Publicaciones Geológicas del ICAITI, Guatemala, C.A., Nr. **1** (1965) – **5** (1976).
Boletín Instituto Geográfico Nacional, Guatemala, C.A., Nr. **1** (1965) – **4** (1967).
Anales del Servicio Geológico Nacional, San Salvador, Bol. **1** (1955) – **4** (1961).
Boletín Sismológico del Servicio Geológico Nacional de El Salvador, San Salvador, **1** (1955) – **7** (1961).
Boletín del Servicio Geológico Nacional, Managua, Nicaragua, Nr. **1** (1957) – **8** (1964).
Catastro e Inventario de Recursos Naturales, División de Geología, Archivo Accesible Informe, Managua, Nicaragua, C.A.

Instituto Geográfico Nacional, San José, Costa Rica, Informe Semestral.
Dirección de Geología, Minas y Petróleo, Informes Técnicos y Notas Geológicas, San José, Costa Rica.
Brenesia, **1 – 15**, (1978), San José, Costa Rica.

Comprehensive works

BUTTERLIN, J. (1977): Géologie Structurale de la Région des Caraïbes (Méxique – Amérique Centrale – Antilles – Cordillère Caraïbe). – Paris, New York, Barcelone, Milan (Masson).
CASE, J. E. (1975): Geologic framework of the Caribbean Region. – In: WEAVER (1975).
CASE, J. E. & HOLCOMBE, T. L.: Geologic-Tectonic Map of the Caribbean Region. – (in press).
DENGO, G. (1968/1973): Estructura geológica, historia tectónica y morfología de América Central. – Instituto Centroamericano de Investigación y Tecnología Industrial (ICAITI). – México/Buenos Aires (Centro Regional de Ayuda Técnica).
HOFFSTETTER, R. (ed.) (1960): Lexique Stratigraphique International, Vol. **V**, Fasc. **2 a**, Amérique Centrale. – Paris (Centre National de la Recherche Scientifique).
NAIRN, A. E. M. & STEHLI, F. G. (eds.) (1975): The Ocean Basins and Margins, Vol. 3, The Gulf of Mexico and the Caribbean. – New York and London (Plenum Press).
ROBERTS, R. J. & IRVING, E. M. (1957): Mineral deposits of Central America. – U.S. Geol. Surv., Bull., **1034**, Washington, D.C.
SAPPER, K. (1937): Mittelamerika. – Handbuch der Regionalen Geologie, **VIII**, 4 a, Heidelberg (Winter).
SCHUCHERT, Ch. (1935): Historical Geology of the Antillean-Caribbean Region. – New York (Wiley & Sons).
STILLE, H. (1940): Einführung in den Bau Amerikas. – Berlin (Borntraeger).
WEAVER, J. (ed.) (1975): Geology, Geophysics and Resources of the Caribbean. Report of the IDOE Workshop on Geology and Marine Geophysics of the Caribbean Region and its Resources, Kingston, Jamaica (Mayaguez, P.R. 1977).
WEYL, R. (1961): Die Geologie Mittelamerikas. – Beiträge zur Regionalen Geologie der Erde, Band **1**, Berlin (Borntraeger).

Bibliographies

DENGO, G. (1962): Bibliografía de la Geología de Costa Rica. – Publ. Universidad de Costa Rica, Ser. Cienc. Naturales.
– (1968/1973): Estructura geológica, historia tectónica y morfología de América Central. – Instituto Centroamericano de Investigación y Tecnología Industrial (ICAITI). – México/Buenos Aires (Centro Regional de Ayuda Técnica).
DENGO, G. & LEVY, E. (1970): Anotaciones al mapa metalogenético de América Central. – Publ. Geol. del ICAITI, **3**, 1 – 15, Guatemala, C.A.
GÓMEZ, P., L. D. (1975): Bibliografía Geológica y Paleontológica de Centroamérica y de El Caribe. – Dept. Hist. Natural., Museo Nac. de Costa Rica, 123 p., San José.
OBIOLS, A. G. (ed.) (1966): Bibliografía Geológica de Guatemala, América Central (hasta 1965), 1a ed. – 81 p. Instituto Geográfico Nacional, Guatemala, C.A.
PROTTI M., E. (1970): Publicaciones del Instituto Geográfico Nacional, XV. Aniversario (1954 – 1969), Indice Bibliográfico, 24 p., San José.
PUJOL, G. A. & DE LEÓN, R. D. (1973): Bibliografía de la Geología de Panamá 1960 – 1970. – Dirección General de Recursos Naturales, Secc. Nac. de Geografía, 62 p., Panamá.
WEYL, R. (1971): Mittelamerika, Literaturberichte. – Zbl. Geol. Paläont. Teil I, H. 7/8, 1003 – 1051, Stuttgart.
WILLIAMS, R. L.: Bibliografía geológica, geofísica y paleontologica de Nicaragua. – Catastro e Inventario de Recursos Naturales, División de Geología, Archivo Accesible Informe, 1, 73 p., Managua, Nicaragua, C.A.

The bibliography (Chapter 11) is a continuation of that given in the first (German) edition (WEYL 1961 a). Works which were listed there are mentioned in this edition only if they are explicitly referred in the text.

2. Morphotectonic features of Central America

2.1 Definition of the region

Central America is defined here as the land and shelf area which extends from Guatemala eastwards and southwards to the Atrato Lowlands in Colombia. It thus takes in the territory of the six republics of Guatemala, Honduras, El Salvador, Nicaragua, Costa Rica and Panama as well as the politically disputed territory of Belize. The neighbouring parts of Mexico and Colombia have been considered only to the extent that they have an important bearing on the geological situation. The same applies to the adjacent marine areas whose topography and geophysical features are gaining importance in geotectonic theories.

The territory covered by the countries mentioned stretches from latitude 7° to 19° N and from longitude 77° to 92° W. Its longitudinal axis, which runs NW/SE, is 1,800 km long and the total mainland area covered is 538,000 km^2. Since the region is usually depicted on a very small scale on European wall charts and in European atlases, it is difficult to gain an accurate idea of its size. To give some idea of this, therefore, it should be noted that if Central America were superimposed on Europe it would stretch from the Oslo Fjord down to the Gulf of Genoa.

2.2 The continental regions

Central America can be divided up into two large units which differ completely from each other as regards geological history and structure. These units were recognized by SAPPER, SCHUCHERT and STILLE, and all more recent works (e.g. DENGO (1968/73) have confirmed their existence. The **northern part** containing Guatemala, Honduras, El Salvador and northern Nicaragua exhibits a continental type crust with Paleozoic or even older metamorphic rocks, anatexites and plutonites. They are overlain by Upper Paleozoic, Mesozoic and Tertiary sediments which underwent deformation following the Middle Permian and at the turn of the Cretaceous to the Tertiary. In the Tertiary northern Central America was the scene of an extremely violent continental volcanism during which large masses of ignimbrite were extruded. The **southern part,** from about southern Nicaragua to Panama, on the other hand, is formed of Cretaceous oceanic type crust on which thick marine sediments and volcanics were deposited during the Tertiary. During the Tertiary this region was converted into the present crust which occupies a position somewhere between a purely oceanic and a purely continental crust. MACDONALD (1972) introduced the very apposite term "tectonitic crust" to describe

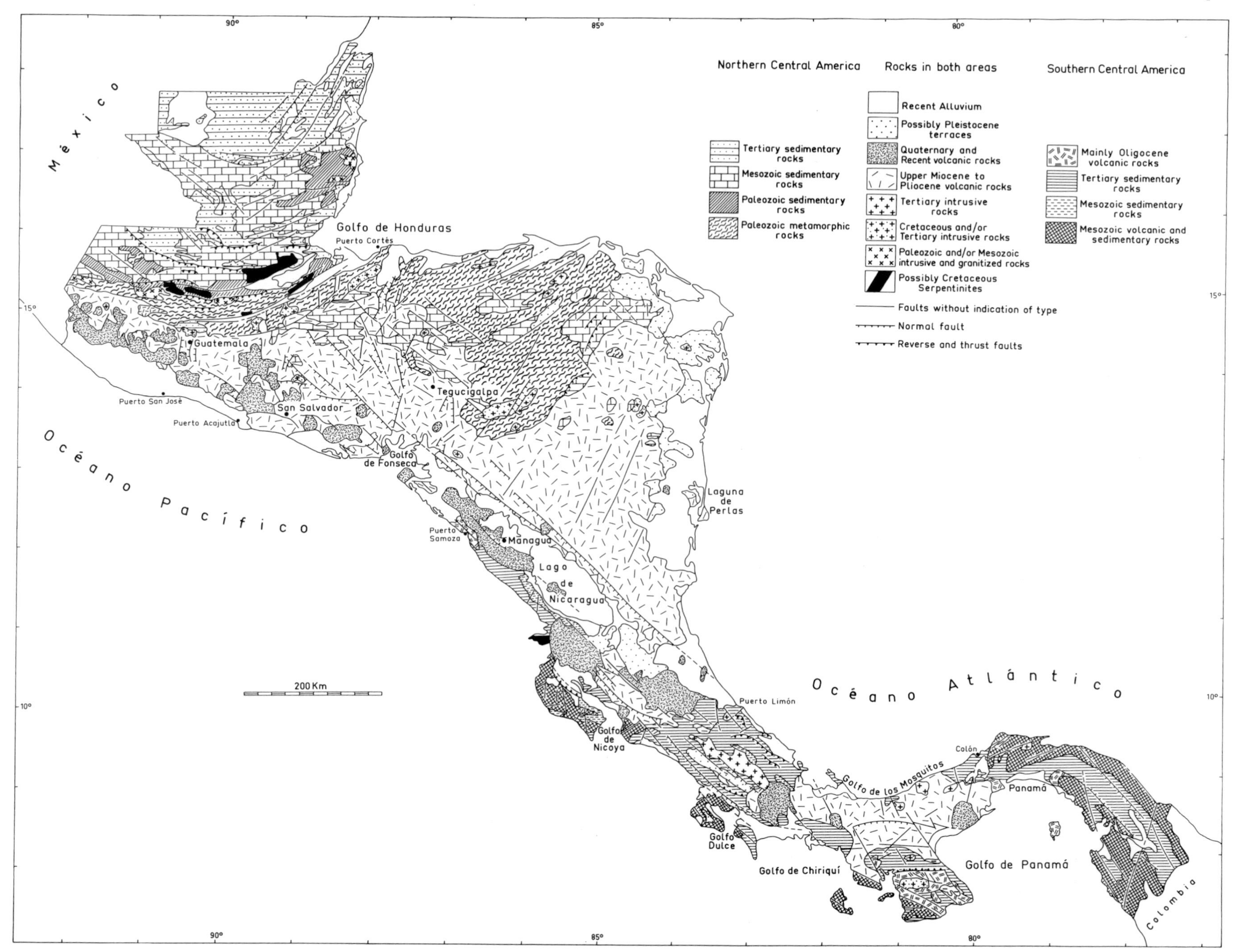

Fig. 3. Geological map of Central America. (A simplified redrawing of the "Mapa metalogenético de América Central", Guatemala 1969).

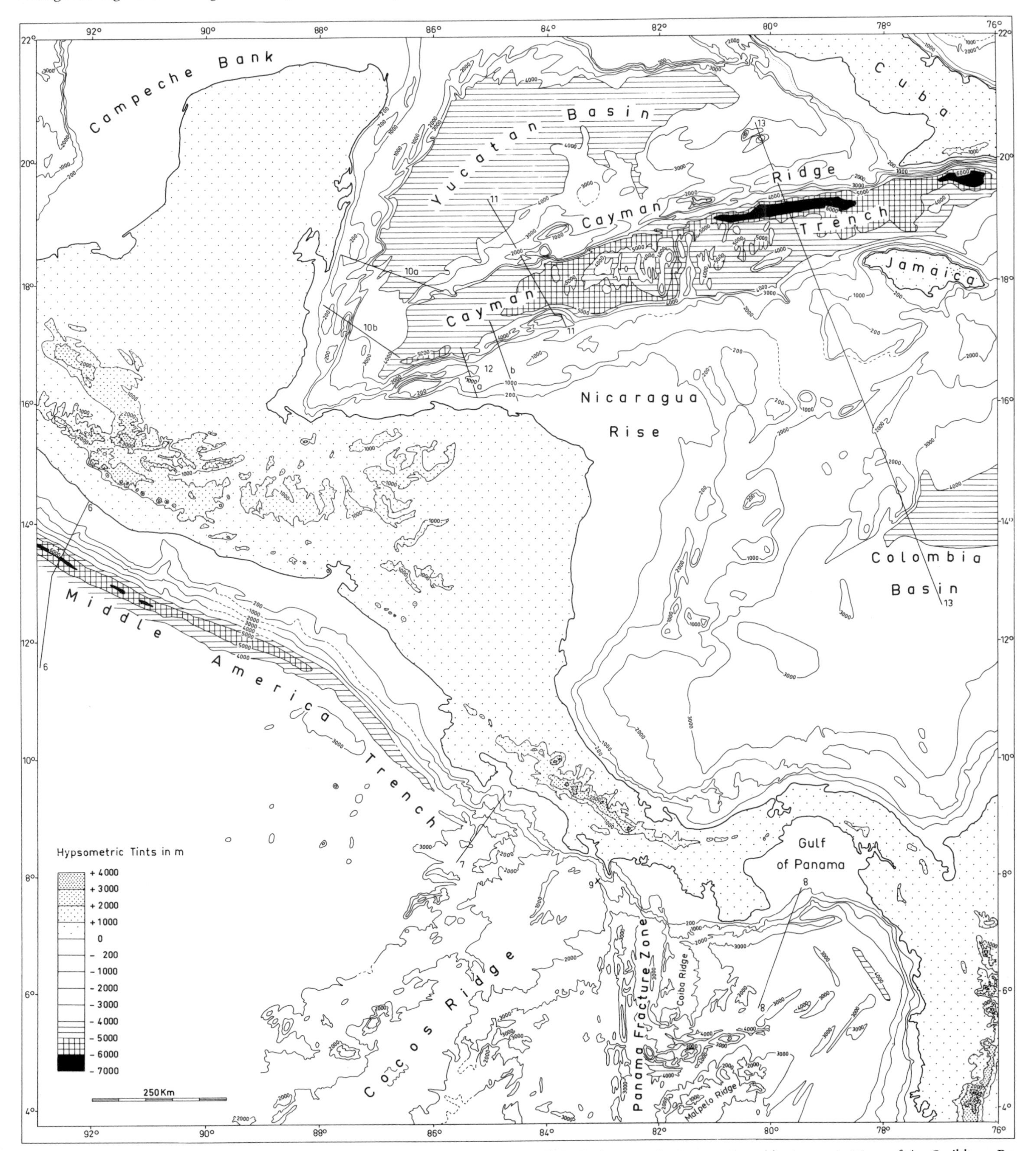

Fig. 5. Bathymetric map of the seas of Central America. (A simplified redrawing of "Preliminary geologic-tectonic and bathymetric Maps of the Caribbean Region" – U.S. Geol. Surv. open-file Map 75–146, 1975.) With positions of profile lines Fig. 6–13.

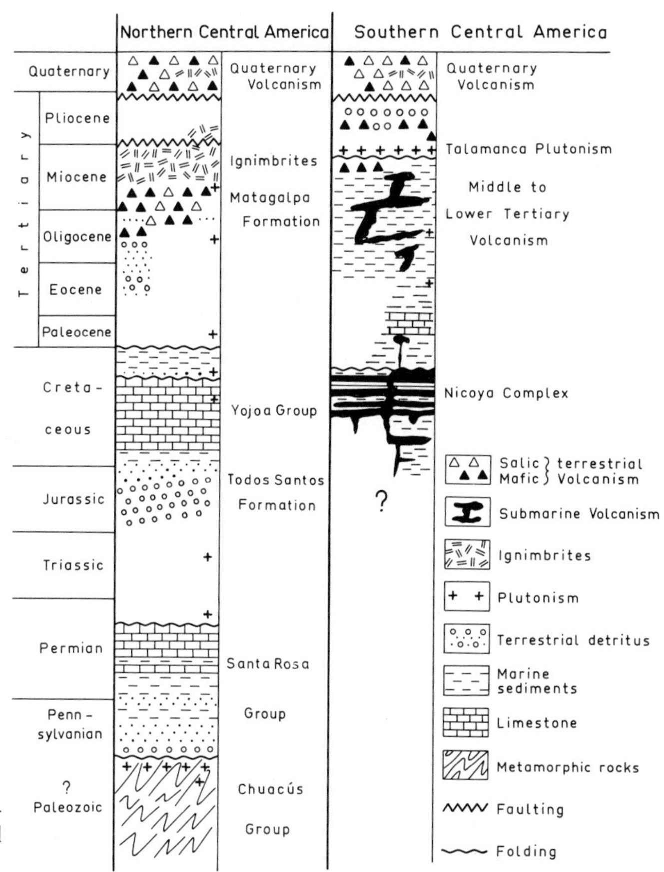

Fig. 2. Schematic crustal cross-sections through northern and southern Central America.

it. The crustal structure and the development of the two parts of Central America are contrasted with each other in highly diagrammatic form in Fig. 2. The contrast is also brought out in the general geological map shown in Fig. 3.

Central America is a tectonically active region. It is therefore appropriate to take both its inner structure and its morphology as the basis of a spatial classification. This approach was suggested by MILLS et al. (1967), DENGO (1968/73) and DENGO & BOHNENBERGER (1969) following a corresponding classification of Mexico (GUZMÁN & DE CSERNA 1963). Proceeding along essentially the same lines as these authors, the following morphotectonic units are distinguished in Central America (Fig. 4):

I. Mountain regions
   The Sierras of northern Central America
   Tertiary Volcanic Ranges and Plateaus
   The Sierras of southern Central America
   The Pacific Volcanic Chain
II. Lowlands and low hill regions
   The Lowland of Petén and the Yucatán Peninsula

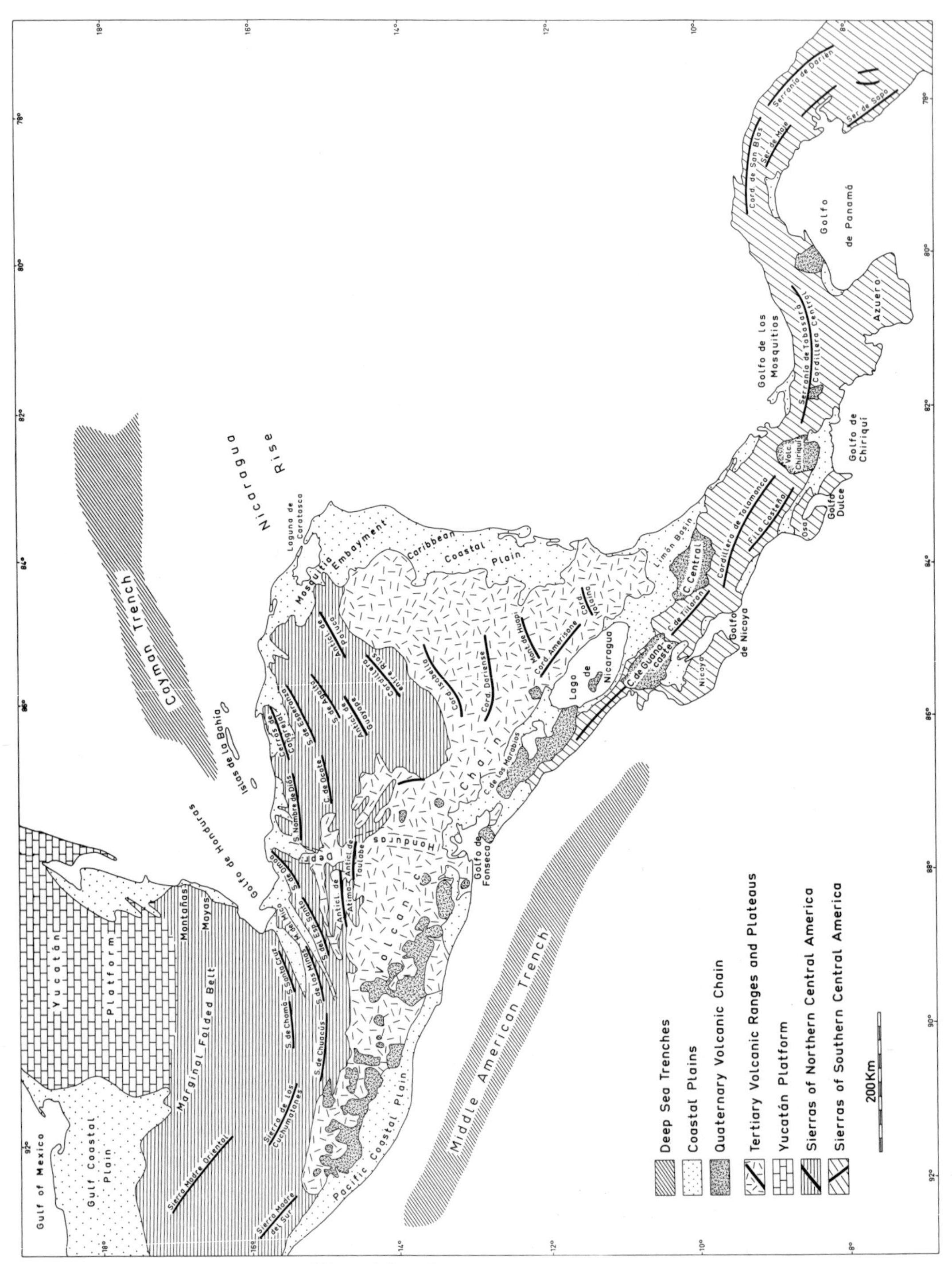

Fig. 4. Morphotectonic map of Central America.

The Gulf Coastal Plain
The Caribbean Coastal Plain
The Pacific Coastal Plain
The Nicaragua Depression

The **Sierras of northern Central America** form an arc — open to the north — of several sub-parallel high ranges extending from Mexico via Guatemala, Honduras and northern Nicaragua to the Caribbean. They can be broken down into the following separate regions: The southern chains consist mainly of Paleozoic and Mesozoic sediments of low metamorphic grade with Upper Mesozoic intrusions. A middle group is built up mainly from high rank metamorphic to anatexitic rocks, Paleozoic granitic intrusions and serpentinites with local occurrences of Paleozoic and Mesozoic sediments. The northern chains consist of folded and northwards overthrust Permian and Cretaceous limestones. To the north of these chains lie lower chains and regions of low elevation which morphologically and structurally form the transition to the Lowland of Petén. The ranges are separated from each other by faults and grabens such as the Motagua and Polochic Valleys in Guatemala, and the Chamelecón Valley in Honduras. A 1 : 250,000 scale relief map of Guatemala (Fig. 44) gives an excellent general view of the structures.

The **Volcanic Ranges and Plateaus** cover large areas of Nicaragua and Honduras and they extend to Guatemala and El Salvador. They are built up of Oligocene to Pliocene lavas, pyroclastic rocks, mostly ignimbrites, and volcanoclastic sediments which vary in substance from rhyolites to basic andesites. In Honduras and Nicaragua they form extensive plateaus and fault block mountains which continue into El Salvador.

The **Sierras of southern Central America** start out as low-altitude chains at the Pacific coast of Nicaragua and extend through Costa Rica and Panama to the Colombian border. They are built up of Tertiary but also secondarily of Cretaceous rock series. In contrast to northern Central America there are no older formations, and above all metamorphic crystalline rock is lacking. The thick Tertiary material consists of marine sediments, volcanics and volcanoclastic rocks. They contain plutonic rocks whose ages range from the Upper Cretaceous into the Upper Miocene and Pliocene. The structural style is characterized by faulting and by usually gentle folding. The Cretaceous basement crops out, mainly in the form of oceanic crust, on the Pacific peninsulas and in the mountain ranges of eastern Panama.

The **Pacific Volcanic Chain** of Quaternary volcanoes, some of which are still active, extends from the Mexican border to Costa Rica and continues — although in a much more dispersed pattern — into Panama. As a typical circumpacific structural element, the chain is closely related spatially to the zone of seismic activity and to the Middle America Trench, and it is consequently nowadays regarded as a decisive characteristic of active subduction.

The **Lowland of Petén and the Yucatán Peninsula** (=Yucatán Platform) are built up of Cretaceous limestones and evaporites and also of Tertiary limestones. This low-lying region, which in the north takes the form of a tableland and in the south merges via chains of folded mountains with the ranges of Guatemala, is dominated by karst drainage and karst topography. In geotectonic terms the region can be regarded as the foreland of the Paleozoic and Mesozoic orogens.

The **coastal plains** are in part regions of thick sedimentation of Recent age or dating back to the Mesozoic, such as the Mosquitia Basin, and in part they are low-lying plains of erosion overlying older rocks. In the regions of the river estuaries, in particular on the northern coast, the plains extend far back up the river valleys so that it is difficult to determine where the plains end and the valley floors begin.

The **Nicaragua Depression** is a graben zone running from the Caribbean coast of Costa Rica through southwest Nicaragua to the Gulf of Fonseca and continuing into El Salvador in the form of faults. It is closely related with the chain of Quaternary volcanoes.

No further subdivision or more precise characterization of the morphotectonic units of the continental regions will be given here, instead these points will be dealt with region by region in the following chapters.

2.3 The marine zones

The relief, seismic aspects, gravitational characteristics and crustal structure of the marine zones surrounding Central America are extremely important for our understanding of the continental areas. Bathymetric and geophysical studies have yielded decisively important data about the marine zones in the last 20 years. The results of these studies have been reported in a large number of specialized papers as well as in comprehensive works (BURK & DRAKE 1974, NAIRN & STEHLI 1975, HOLCOMBE 1975, CASE 1975).

The relief is depicted in the "Bathymetric Map of the Caribbean Sea", compiled by the US Naval Oceanographic Office, 1974. Fig. 5 is a simplified redrawing of this map; instead of an isobath spacing of 200 m as in the original, it was only possible to show a spacing of 1000 m in the redrawn map. A geotectonic interpretation of the relief is offered by HOLCOMBE (1975) and CASE (1975 a and b).

On the Pacific side the most important morphotectonic units are: the shelf, the Middle America Trench, the Cocos Ridge, the Panama Fracture Zone and the Panama Basin. On the Atlantic side: the Yucatán Borderland, the Yucatán Basin, the Cayman Ridge, the Cayman Trench, the Nicaragua Rise and the Colombia Basin.

The **Pacific marine area** is clearly divided into two sections of fundamentally different relief: In the northwest there is a shelf about 100 km wide, in front of which lies the Middle America Trench, followed by a relatively plain ocean floor. In the southeast, on the other hand, starting at the Cocos Ridge the relief is very pronounced with a large number of seamounts and the NS-striking Panama Fracture Zone. The boundary between the two regions is situated approximately where, on the mainland, the continental crustal structure of the northern part separates from the oceanic crust of the southern part. It is certainly significant that the seismic activity and magmatism also undergo a change in this area.

The **shelf,** which is a broad structure in the northwestern part, ends in the southeast off the Nicoya Peninsula. It is transected only once — off Guatemala — by a significant submarine canyon. Arcer profiles recorded by seismic reflection methods (ROSS & SHOR 1965, Fig. 16) revealed troughs within the shelf containing compacted sediments up to several hundred metres thick. It is not possible to decide yet whether these are individual troughs or whether they form a continuous system because the spacing between the profiles is very large. According to a crustal profile recorded by FISHER (1961) (Fig. 6) the shelf off Guatemala is underlain by 3 – 4 km of sediments; exploratory drilling has penetrated through 3,590 m of sediments. According to the model proposed by SEELY, VAIL & WALTON (1974, p. 258) the sediments should be included in the subduction zone between the Cocos Plate and the Caribbean Plate. According to GARAYAR (1977), in the Nicaragua region the Tertiary sediments

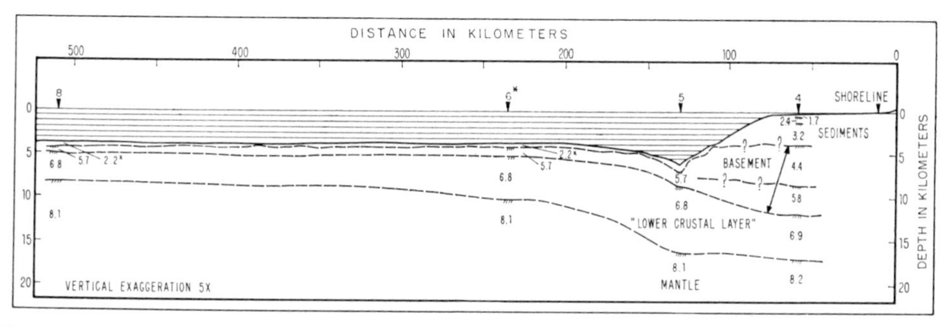

Fig. 6. Schematic crustal cross-section through the shelf, Middle America Trench and ocean floor off Guatemala. (According to FISHER 1961, Fig. 5).

of the Nicaragua Trough extend into the shelf and become incorporated into the subduction zone right at the continental slope.

The **Middle America Trench** stretches from the Tres Marias Islands of Mexico to a point off the coast of Costa Rica; it is about 2,400 km long and has an average depth of over 4,400 m with the maximum depth of 6,400 m being attained off the coast of Guatemala. Detailed studies have been carried out above all by HEACOCK & WORZEL (1955), FISHER (1961) and ROSS & SHOR (1965).

The southeastern section of the graben, which is important for Central America, begins with a sharp bend off the Gulf of Tehuantepec and ends in the region of the Cocos Ridge where the walls become less steep. The trench is asymmetrical/U-shaped in cross-section, and the floor is uneven. The sediments that it contains fluctuate in thickness between less than 200 m and 1,000 m, and their main components seem to be volcanic ashes and turbidites. The crustal profile (Fig. 6) reveals that below the shelf sediments there is an approximately 8 km thick crust exhibiting travel times of 4.4 – 5.8 km/sec, which are those of continental rock. This crust rapidly thins out as it approaches the trench. A lower crustal layer (6.8 – 6.9 km/sec) below the first attains a thickness of 7.6 km beneath the trench and slowly declines to a thickness of 5 km in the direction of the ocean. In the trench zone the boundary of the mantle zone (8.1 – 8.2 km/sec) is only slightly dented, and there is no evidence at all of the "root of a tectogene".

The ocean floor in front of the trench is also relatively featureless judging by the course of the 200 m isobaths; off Nicaragua a slight uparching can be detected and as the Cocos Ridge is approached some seamounts occur.

Geophysical and geological surveys of the landward slope of the Middle America Trench offshore Guatemala have revealed landward-dipping reflectors which are associated with high compressional wave velocities and large magnetic anomalies (LADD et al. 1978). Coring on the continental slope yielded gravels of unweathered metamorphosed basalt, serpentinite, and chert, similar to the rocks of the Nicoya Complex of Costa Rica, and unlike the andesitic volcanics of southern Guatemala and El Salvador. These data suggest that landward-dipping slices of rocks similar to those of the Nicoya Complex are embedded within the upper continental slope off Guatemala (Fig. 6 a). The continuity of a free-air gravity high traced by WOODCOCK (1975) and COUCH (1976) (vide LADD et al. 1978) from the edge of the Nicoya Peninsula all the way along the shelf edge northward to the Gulf of Tehuantepec suggests that the

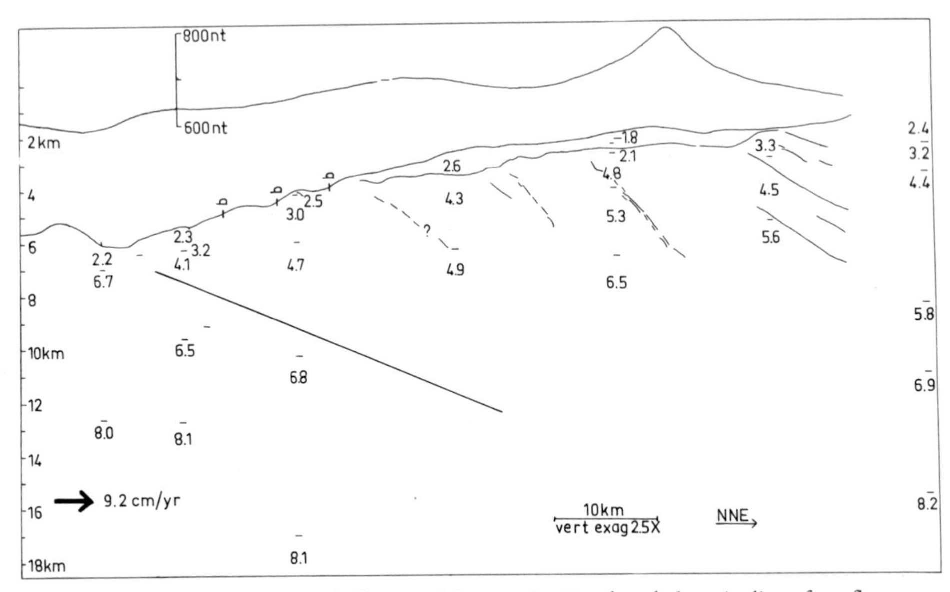

Fig. 6a. Crustal section of the shelf offshore of Guatemala. Numbers below the line of sea-floor are refraction velocities, the designations "b" above the sea-floor line are core locations. Magnetic profile is drawn above the line of sea-floor. The convergence rate of 9.2 cm/yr for the Cocos Plate with respect to the North American Plate is indicated at the left. (Simplified redrawing from LADD et al. 1978, Fig. 8).

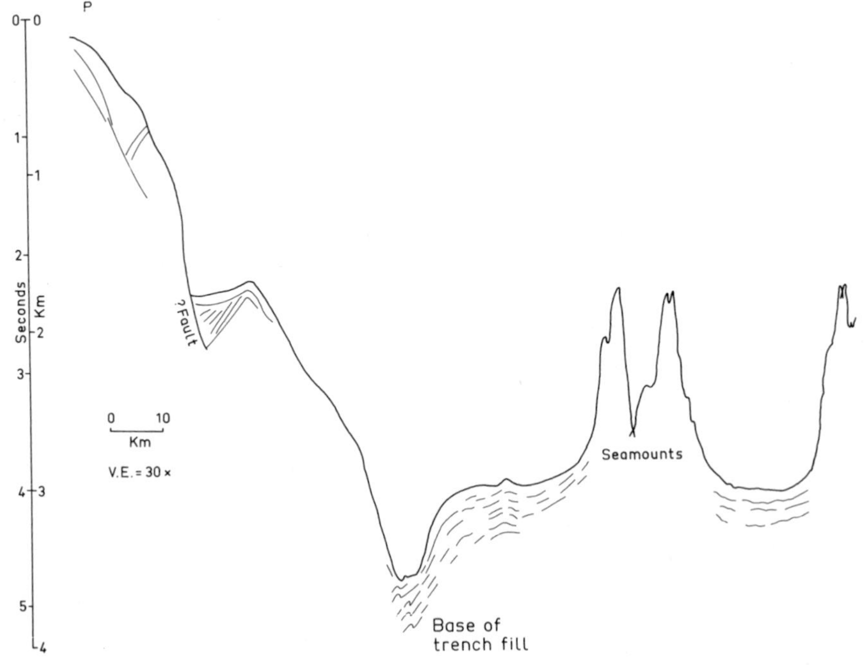

Fig. 7. Arcer profile through the continental margin southeast of the peninsula of Nicoya, Costa Rica. (Redrawn from ROSS & SHOR 1965, Figs. 19 and 20).

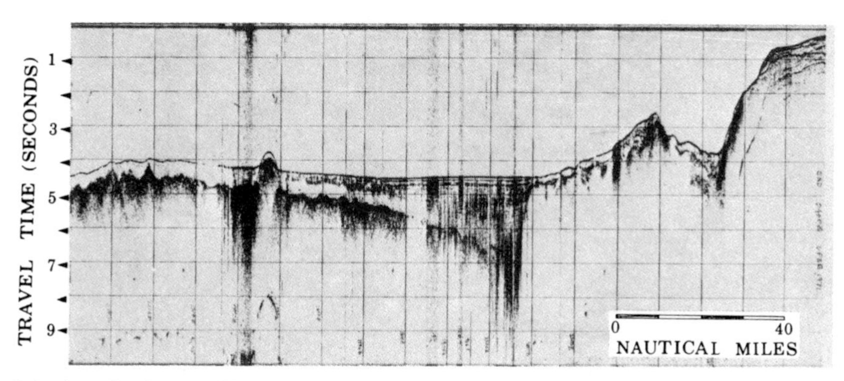

Fig. 8. Seismic reflection profile through a buried trench area south of the Gulf of Panama. (From LOWRIE 1978, Fig. 2).

same oceanic rock assemblage occurs along the continental shelf from Costa Rica to the Gulf of Tehuantepec. The lower slope is probably a tectonically deformed and consolidated sediment wedge overlying the ocean crust.

These observations confirm the former suggestions that oceanic crust of the Cocos Plate is subducted below the continental crust of the North American and Caribbean Plate (see Chapter 8).

The eastern section commences with the broad, NE/SW-striking **Cocos Ridge.** Seamounts, small elevations and narrow scarps occur on its gently undulating crest. Arcer profiles reveal narrow, sediment-filled troughs off the continental slope (Figs. 7 and 8), which can be regarded as a continuation of the Middle America Trench. Visual observation by divers revealed a series of NE-oriented scarps 10 – 20 m high (Fig. 9) and severely faulted sediment surfaces on the floor of the troughs, all of which can be interpreted as evidence of active subduction (HEEZEN & RAWSON 1977 a and b). The margins of the Cocos Ridge have a stepped structure in which the individual steps run N/S and E/W and which VAN ANDEL et al. (1971) regard as the result of two intersecting fault systems. On the eastern flank the step structure changes into a very pronounced fault system which is designated the **Panama Fracture Zone** (MOLNAR & SYKES 1969) or the Coiba Fracture Zone (VAN ANDEL et al. 1971). This fault system follows the 83rd longitude and it is characterized by a series of N/S-striking narrow troughs and ridges. As a result of vigorous seismic activity right-lateral strike-slip motion is expressed in the form of a transform fault which separates the Cocos Plate and the Nazca Plate from each other. To the east one or more troughs sometimes over 4,000 m deep follow the arc of the shelf margin in the Panama Basin. Off the Azuero Peninsula the sediment filling of these troughs attains thicknesses of around 1,200 m (HEATH & VAN ANDEL 1973, Fig. 4). According to the results of the drilling carried out in Leg 16 the sediments of the Panama Basin, which overly an acoustic basement of basalt, date from the Miocene to the Pleistocene (HEATH & VAN ANDEL 1973). In the gravity map the troughs stand out because of their strong negative free-air gravity anomaly (see Figs. 103 and 168).

On the **Atlantic side** of northern Central America the continental margin of Yucatán, the Yucatán Basin, the Cayman Ridge, the Cayman Trench and the northern flank of the Nicaragua Rise constitute a complicated puzzle which we are only just beginning to resolve with the aid of the concepts of plate tectonics.

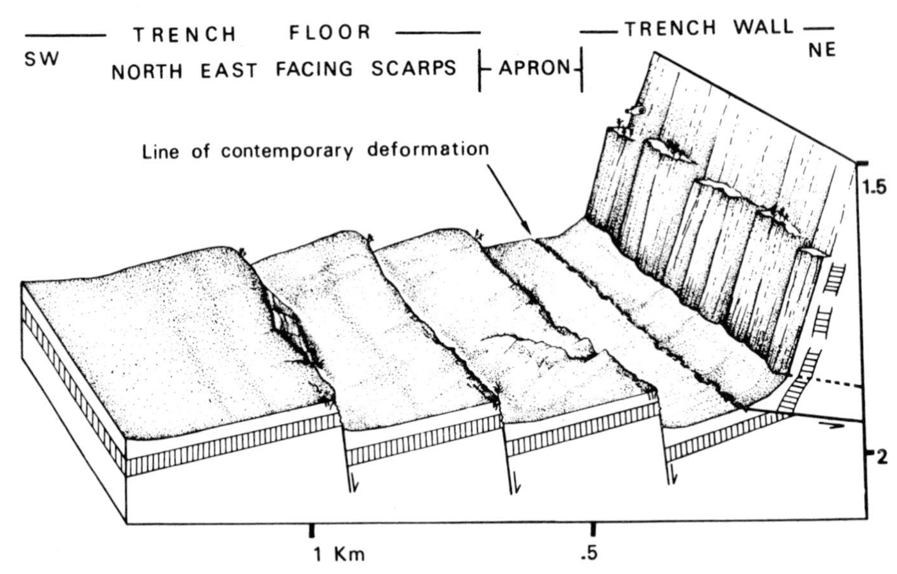

Fig. 9. Visual observations and structural inferences, southeast Middle America Trench. Northeast-facing scarps of trench floor are interpreted as normal faults which have carried oceanic crust down along trench axis. Smooth trench floor apron is bisected by zone of contemporary deformation which is interpreted as the sea-floor trace of subduction. The landward wall is to the right. (From HEEZEN & RAWSON 1977b, Fig. 2).

The continental margin to the east of the Yucatán Peninsula is divided up into five NNE-striking horsts and interlying grabens which descend in a series of steps to the Yucatán Basin. They converge in the direction of the Gulf of Honduras. In the horsts the acoustic basement crops out in the form of granitic intrusive rocks, Paleozoic metamorphic rocks, basic volcanics, and possibly also consolidated Cretaceous sediments under a thin cover of younger, unconsolidated sediments. Unconsolidated Tertiary sediments of varying thickness up to a maximum of 2 km (see profiles in Fig. 10) lie in the grabens and on the deep-sea bottom. Submarine valleys running at right angles to the structures and in particular the sedimentary fans of the rivers flowing into the Gulf of Honduras have supplied clastic deposits transported in turbidity currents, and these overlie chiefly pelagic sediments. The horsts and grabens stand out gravimetrically through positive and negative free-air gravity anomalies, and the largest graben directly at the margin of the shelf is accompanied by a strong negative Bouguer anomaly extending into western Cuba (UCHUPI 1973, DILLON & VEDDER 1973, UCHUPI 1975, CASE 1975, and cited works).

The **Yucatán Basin** is a plain attaining maximum depths up to 4,500 m, and it rises gently to the southwest towards the Gulf of Honduras. The pronounced basement relief is overlain at first by up to 1 km of pelagic sediments (presumably Cretaceous) and then by turbidites alternating with pelagic sediments. They have been laid down ever since the Central American ranges were uplifted at the end of the Cretaceous; deposition occurred in the form of sedimentary fans originating above all in the hinterland of the Gulf of Honduras. In the Yucatán Basin the crust is about 14 km thick and resembles that of the Colombia Basin; this is in keeping with a comparable Bouguer anomaly of +250 to +300 mgal (CASE 1975).

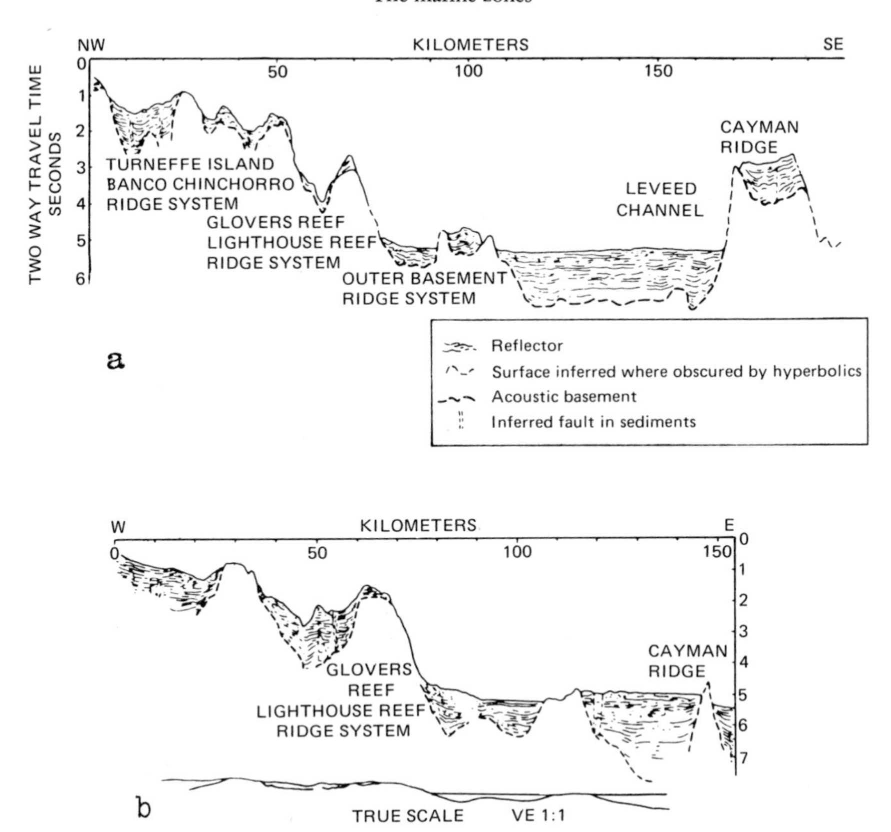

Fig. 10. Seismic reflection profiles of the continental margin of British Honduras. (From DILLON & VEDDER 1973, Figs. 5 and 6 and CASE 1975 a, Fig. 23).

The **Cayman Trench** is the largest structure in the Central American-Caribbean region. It is 1,700 km long and it stretches from the Gulf of Gonaives in Haiti to the Gulf of Honduras. The Motagua and Polochic Fault Zone in Guatemala is regarded as the continuation of this trench on the mainland. The north and south flanks of the trench are extremely steep and have a stepped structure (Fig. 11). The bottom is covered by sediments up to 1 km thick. The trench, which is regarded as the boundary zone between the North American and Caribbean plates, is characterized by oceanic crustal structure, elevated heat flow and seismic activity with left-lateral movement. In the middle the trench is broken up by steep N/S-striking ridges and troughs from whose flanks peridotites, serpentinites and basic cumulates have been dredged. An active spreading center is assumed to exist here (HOLCOMBE et al. 1973, PERFIT 1977).

The **Cayman Ridge** extends from the Sierra Maestra of Cuba to the margin of the shelf around Belize. It is asymmetrical and slopes away gradually northwards towards the Yucatán Basin and steeply to the south towards the Cayman Trench. The Cayman Islands and the Misteriosa and Rosario Banks rise from the ridge. According to seismic profiles the ridge has a continental crustal structure (Fig. 11). This has been confirmed by dredged samples (PERFIT & HEEZEN 1978) of metamorphites (amphibolites, gneisses, mica schists), plutonic rocks

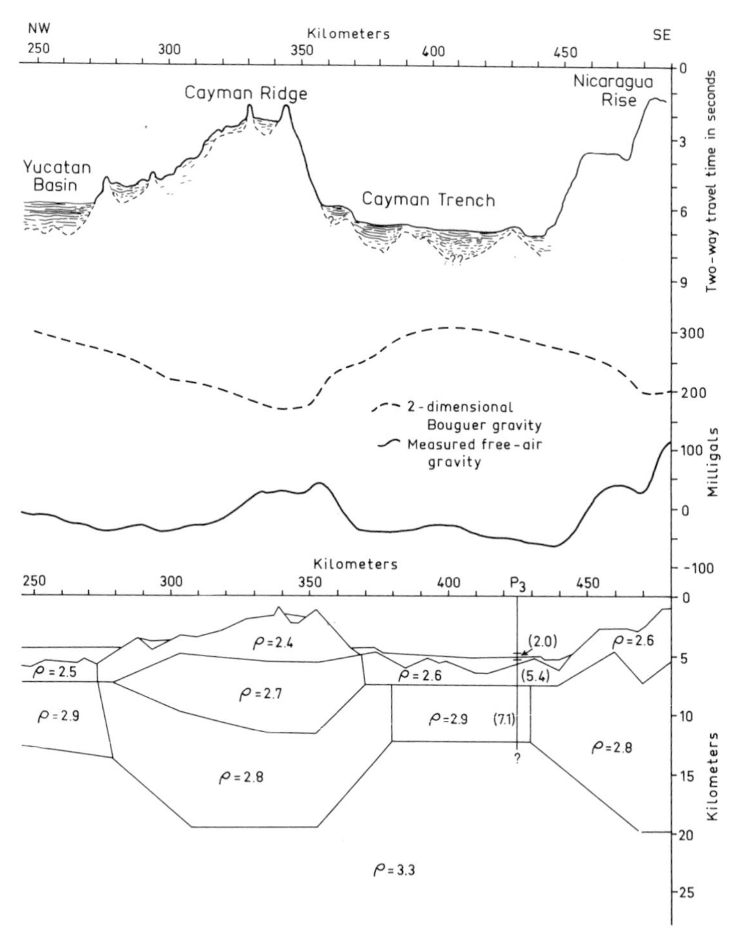

Fig. 11. Seismic reflection, gravity anomaly, and crustal profiles across the Yucatan Basin, Cayman Ridge and Cayman Trench, to the Nicaraguan Rise. (Redrawn from DILLON et al. 1972, Fig. 1).

(gabbros to adamellites, predominantly granodiorites), volcanic rocks (basalt to rhyolites, predominantly andesites), volcanoclastic rocks and Cretaceous to Pleistocene limestones. This confirms the old assumption, deduced from the relief, that the Cayman Ridge is a link between the structures of northern Central America and the structures of the Antilles. However, this theory has been modified to take account of plate tectonics, according to which the ridge is an uplifted fault block and a rotated continental fragment of Belize (DILLON & VEDDER 1973, see Fig. 173). PERFIT & HEEZEN (1978) regard the ridge as the remains of an island arc structure dating from the Laramic orogenesis (see Fig. 175).

The Cayman Trench, Cayman Ridge and Yucatán Basin together with the fault system at the eastern edge of the Yucatán Peninsula are regarded as the result of crustal stretching and rotation combined with horizontal displacements which are said to have started either in the Jurassic or in the Cretaceous. The amount of lateral displacement is, however, disputed; the assumptions range between 150 and 1,000 km. For further details see Section 3.22 and Chapter 8.

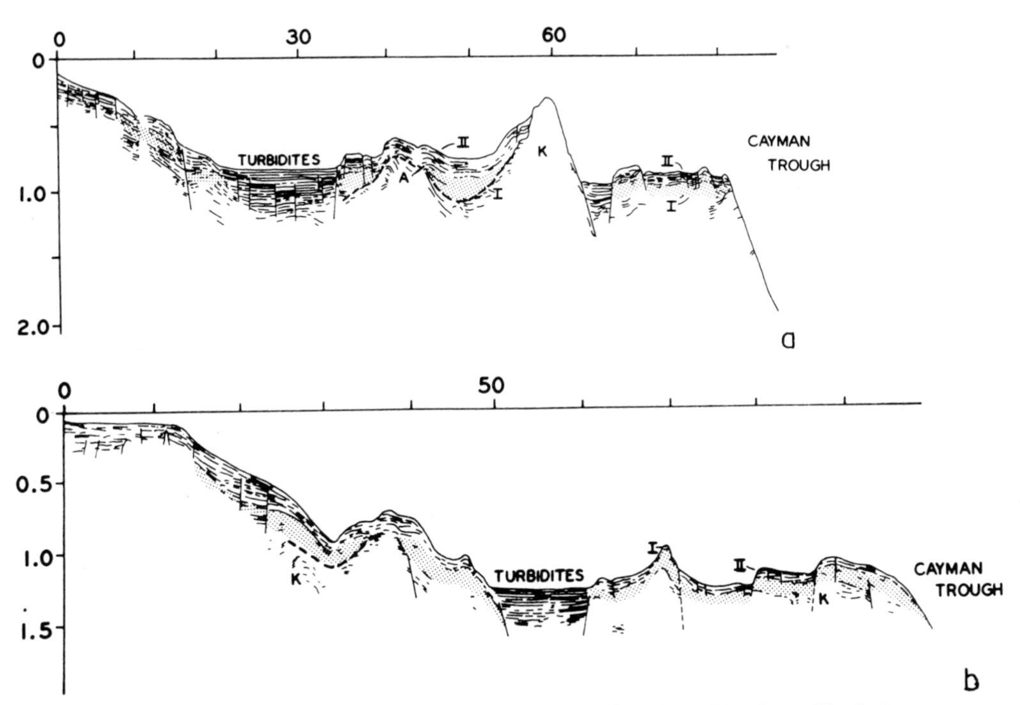

Fig. 12. Seismic reflection profiles of the continental margin of northern Honduras. Vertical exaggeration 10×. (From PINET 1975, Figs. 4 and 5).

The submarine relief off the north coast of Honduras is accented by islands, submarine ridges, steep scarps, troughs and basins. The Islas de la Bahía (Bay Islands) for example are located on the 200 km long submarine Bonacca Ridge which here forms the southern edge of the Cayman Trench. It is separated from the mainland by a basin up to > 1,400 m deep (Fig. 5). To the east of the islands a series of further basins and ridges runs parallel to the Cayman Trench and forms a block mosaic between it and the Nicaragua Rise (Fig. 12). Seismic reflection profiles (PINET 1975) permit the basement to be divided up into the following geological units: crystalline basement rocks, folded, presumably Cretaceous, strata discordantly overlain by Middle Tertiary material, which is divided into two units, and finally up to 1,000 m thick Plio-Pleistocene turbidites in the basins close to the continent. Dome-like structures in the vicinity of the continental margin are interpreted as salt plugs (PINET 1972). Strong magnetic anomalies to the south and southeast of the Islas de la Bahía (PINET 1971) are indicative of ultrabasic vein-like bodies which can be regarded as a dextrally displaced continuation of the serpentinites of Guatemala. Altogether the sea-floor to the north of Honduras exhibits a complicated block mosaic structure which was presumably folded, then covered by sediments during the Laramic orogenesis and which was ultimately broken up and sheared by lateral displacement during the Plio/Pleistocene.

The **Nicaragua Rise** links the northern Central American continent with the western end of Hispaniola and thus joins the Central American with the Antillean orogen. Its crest is formed by a series of flat banks covered with coral reefs. The Island of Jamaica rises from this rise. Towards the north the rise slopes sharply down towards the Cayman Trench while the broad

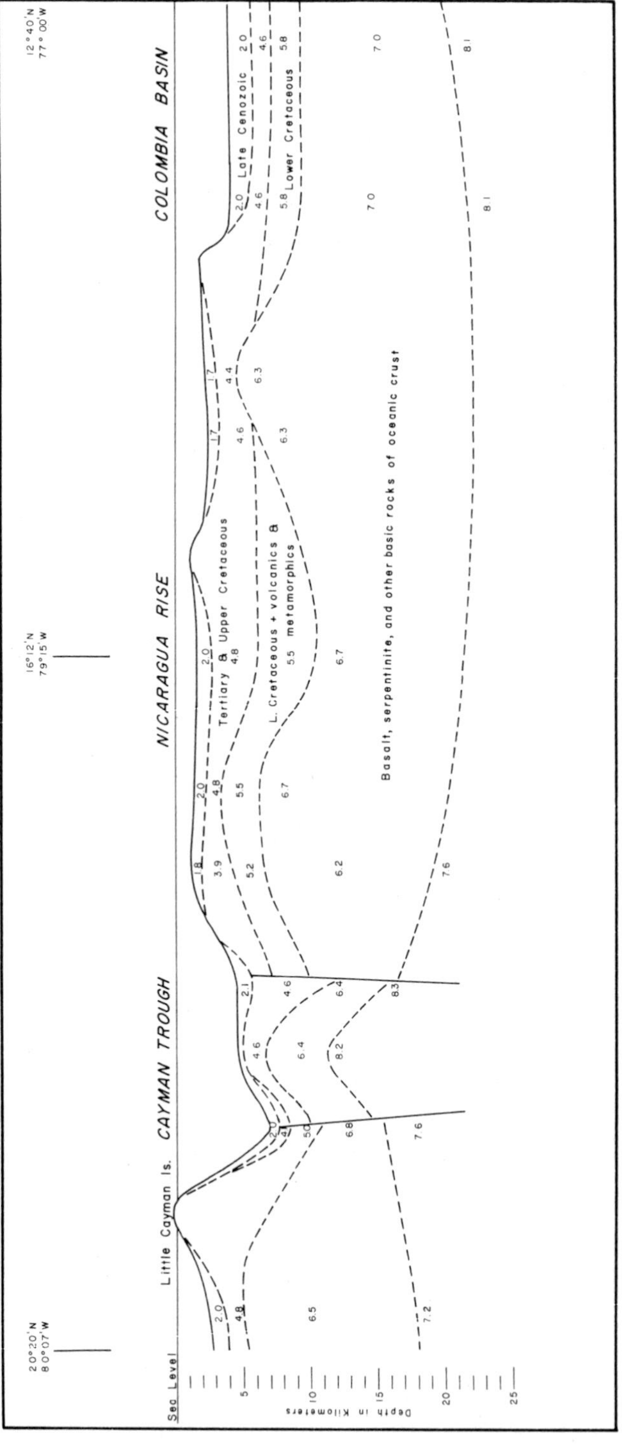

Fig. 13. Crustal section through the Nicaragua Rise. Vertical exaggeration 10×. (From ARDEN 1975, Fig. 3).

southern flank, which has a greater range of high and low regions, extends as far as the "Hess Escarpment" leading to the Colombia Basin.

Although petroleum exploration activity in the Caribbean region has yielded a wealth of seismic and geological data, most of this material has not been published. However, ARDEN (1969, 1975), EDGAR et al. (1971), CASE (1975 a and b) and UCHUPI (1975) have given summary interpretations of the material available to them. According to these authors the Nicaragua Rise, unlike the Cayman Trench and Colombia Basin, possesses a multi-layer crust up to 22 km thick, the geological interpretation of which is based on the few outcrops of Paleozoic to Cenozoic rocks in the Bay Islands at the edge of the Rise, on some drillings carried out on the Rise and on extrapolation of the geological structure of Nicaragua and Honduras in the southwest or of Jamaica in the northeast (see profile Fig. 13).

Dredge samples contain large quantities of breccias, greywackes and arenites together with detritic material derived from granites and metamorphic rocks. The carbonates date from the Tertiary, possibly also from the Cretaceous (PERFIT & HEEZEN 1978).

According to ARDEN, formation of the Nicaragua Rise commenced together with the cleavage of the Caribbean region at the end of the Jurassic and in the Lower Cretaceous, and it was associated with the extrusion of large masses of submarine lavas and basic intrusions which led to the formation of the 4.5 km thick "volcanic layer". The volcanic activity declined in the course of the Upper Cretaceous so that a mixed volcanic-sedimentary succession of beds was laid down. The Laramic orogenesis produced granodioritic intrusions and general uplift. The latter was replaced by renewed subsidence during the Middle Eocene which resulted in uniform carbonate sedimentation. The subsidence was interrupted by uplift in the Middle Oligocene and in the western part of the Rise this led to Upper Miocene following immediately upon Middle Eocene. The strata in the western part, off the Mosquitia coast, are particularly thick at around 9,000 m, and the influence of the Mosquitia Basin is discernible here. Late Tertiary movements resulted in the Rise being broken down into numerous blocks which can be detected in Jamaica and in individual areas of the western part of the Rise (see map prepared by CASE & HOLCOMBE 1975). Large fracture lines run in a northeast direction from the Central American mainland into the Rise. By these the northward-oriented upward tilting of the Rise resulted, the axis of uplift being shifted to the northwest. ARDEN (1975, p. 629) stresses that very few regions of the earth have been interpreted in such a contradictory manner as has been the Nicaragua Rise.

The Island of Providencia, which is built up of volcanic rocks (PAGNACCO & RADELLI 1962), and the Island of San Andrés, which is formed from Tertiary and Quaternary limestones (GEISTER 1975), are situated on the southeastern slope of the Nicaragua Rise facing the Colombia Basin. Only a narrow shelf covered in coral reefs and a relatively steep continental slope are found off the northern coast of Costa Rica and Panama. A powerful long drawn-out negative free-air anomaly indicates a sedimentation trough off the coast of Panama, and this is taken as a sign of subduction events (see map by CASE & HOLCOMBE 1975).

3. Northern Central America

Together with the regions of Mexico extending up to the Isthmus of Tehuantepec, northern Central America was designated by SCHUCHERT (1935) — on the basis of SAPPER's research — as Nuclear Central America and by STILLE (1940) as Sapperland. Because its structure consists of a Paleozoic crystalline basement and a superjacent continental to epicontinental series of strata, northern Central America was contrasted with the younger southern Central America which did not come into existence until the Cretaceous (see Figs. 2 – 4).

Even more modern authors such as MILLS et al. (1967, 1969), DENGO (1968/73, 1975 a) and in particular DENGO & BOHNENBERGER (1969) have considered the development of northern Central America in conjunction with that of Mexico as it is described in the works and maps of DE CSERNA (1960), GUZMÁN & DE CSERNA (1963) and LÓPEZ RAMOS (1974, 1975). According to these authors a distinction has to be made between an old Jaliscoan and a young Mexican geotectonic cycle which can be divided up into the following phases:

1. Jaliscoan geotectonic cycle
   a) Paleozoic orthogeosyncline with deposition of the Chuacús Group and its equivalents in the pre-Pennsylvanian.
   b) Anatectic phase(s) with deformation, metamorphism(s) and sialic intrusions in the pre-Pennsylvanian.
   c) Orogenic phase with the formation of a clastic (flysch) trough in the Pennsylvanian and Permian (Santa Rosa Group) and subsequent folding in the Upper Permian to Lower Triassic.
   d) Taphrogenic phase with postorogenic (molasse) sediments from the Triassic to the Jurassic (Todos Santos Formation).

2. Mexican geotectonic cycle
   a) Mesozoic geosyncline following the Mexican geosyncline from the Upper Jurassic/Lower Cretaceous to the Upper Cretaceous, containing mainly limestones and evaporites, but also oceanic volcanic rocks and eupelagic sediments.
   b) Anatectic phase with deformation and granitic intrusions.
   c) Orogenic phase with folding and deposition of clastic sediments in the uppermost Cretaceous and Tertiary (Verapaz, Petén and Valle de Angeles Group).
   d) Taphrogenic phase with uplift and block faulting in the Middle and Late Tertiary, continental clastic deposits and sialic terrestrial volcanism.

Even though it may not be possible with certainty to assign all the events in the geological development of northern Central America to one or other of these phases, thus breaking the bonds of such a rigid scheme, the latter is at least an aid to understanding and permits us to recognize the major geological relationships. Perhaps such a way of regarding geotectonic cycles also reminds us not to consider events simply from the standpoint of plate tectonics. In particular the common features of development over a large area such as northern Central America stand out. They will therefore be discussed together in the following section on stratigraphic sequence, paleogeographic development, structure and magmatism, while the more specific local details are reserved for the description of the individual areas.

3.1 Stratigraphic sequence

3.11 Crystalline basement rocks

Crystalline basement rock with metamorphic rocks and intercalated plutonic rocks is widespread in northern Central America and forms the lowermost part of the continental crust. It ranges chiefly from Chiapas in Mexico via central Guatemala to the northern and northeastern parts of Honduras and into northern Nicaragua (Fig. 14). In more limited areas it occurs in the Altos Cuchumatanes of Guatemala and in the Maya Mountains of Belize. However, according to more recent works the generally speaking only slightly metamorphosed series which crop out in the Maya Mountains do not belong to the metamorphic basement but to the folded Upper Paleozoic.

The main area of distribution of the crystalline rocks is demarcated in the north by the Chixoy Polochic Fault and in the south by the Jocotán Chamelecón Fault. According to DENGO (1969 b, p. 312) the Motagua Fault separates regions of differing rock facies from each other which in the north are referred to as the **Maya Block** and in the south as the

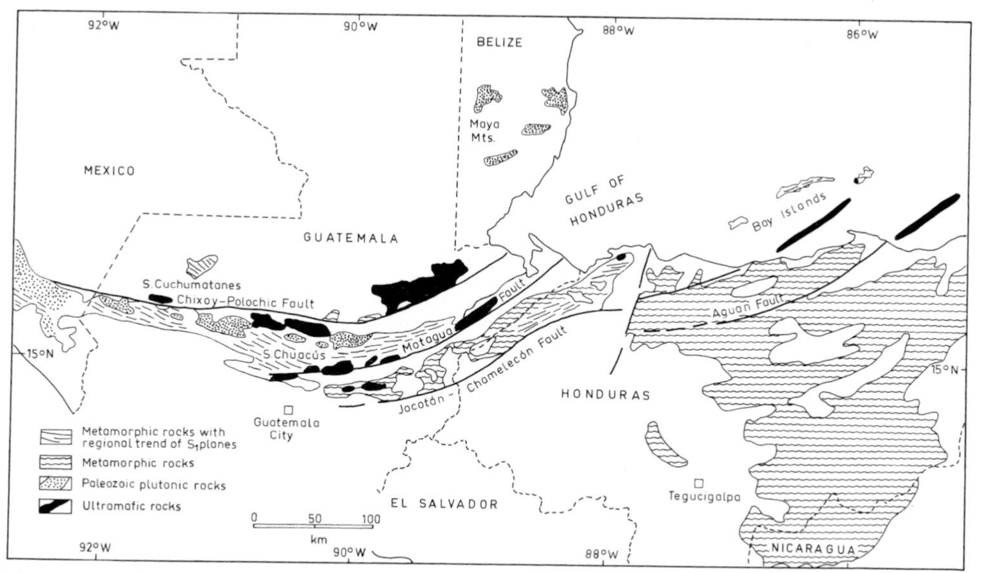

Fig. 14. Distribution of outcrop areas of pre-Mesozoic metamorphic and plutonic rocks in northern Central America.

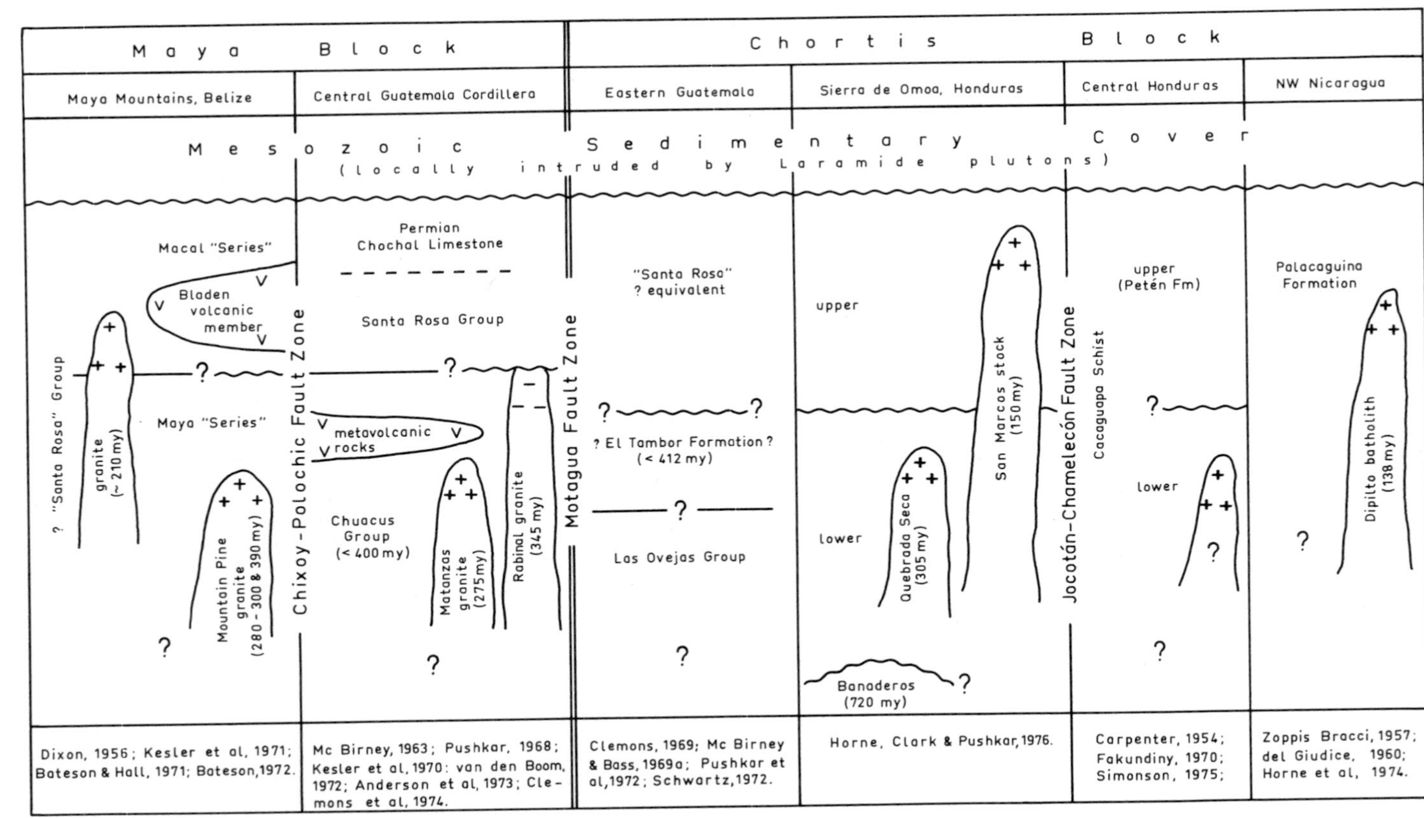

Fig. 15. Generalized schematic stratigraphic chart of basement rocks in northern Central America. (Redrawn from HORNE et al. 1976, Fig. 2).

Chortis Block. According to more recent works (HORNE, CLARK & PUSHKAR 1976) the differences in the rock facies of these blocks are, however, smaller than was previously assumed.

Up until the start of the sixties the knowledge available on the crystalline basement rock did not go beyond the works of SAPPER (1899, 1937) and POWERS (1918), but then such decisive progress was made that HORNE, CLARK & PUSHKAR (1976) in the course of a study of the Sierra de Omoa in Honduras were able to give a general description from which the comparative overview in Fig. 15 is taken.

The rock material of the crystalline basement comprises para-rocks derived from a wide range of initial materials and exhibiting various degrees of metamorphism on the one hand, and metavolcanic rocks and metaplutonic rocks on the other. The most thorough studies have been carried out in the central area of Guatemala, in the **Sierra de Chuacús,** from which the name the **Chuacús Group** has been derived.

The first new studies and genetic theories were initiated by LJUNGGREN (1959) and the present author, and they were reported on in the first (German) edition of this book (WEYL 1961 a, pp. 15 – 19). A little later in the Sierra de Chuacús MCBIRNEY (1963) distinguished between amphibolites, granite amphibolites, mica schists, quartz-albite-epidote-muscovite-chlorite-biotite schists and gneisses of various composition, which he combined together in the Chuacús Group. His example was followed by VAN DEN BOOM (1972) and the "General Geological Map of Baja Verapaz and the southern part of Alta Verapaz 1 : 25,000" Hannover 1971. In addition, VAN DEN BOOM classified the material into zones according to the BARROW type and was able to provide more detailed data on the original material and degree of metamorphism.

He distinguished between the following facies zones:

Chlorite-Sericite Zone with
- sericite schists and sericite-chlorite schists
- metagreywackes and metaarkoses
- quartzites and quartzite schists
- epidote-actinolite schists
- muscovite-stilpnomelane schists
- crystalline limestones
- para-amphibolites
- granitoids

Biotite Zone with
- biotite-muscovite schists and muscovite schists
- muscovite-epidote schists
- para-amphibolites
- marbles
- quartzites

Garnet Zone with
- garnet-muscovite schists ± chloritoid
- garnet-disthene schists
- garnet-biotite schists
- garnet-amphibolites
- disthene-quartzites

Fig. 16. Metatexitic rock of the Chuacús Group. El Chol, Sierra de Chuacús (Guatemala). Photo: R. WEYL.

It can be assumed that the parent rocks of the typical para-rocks were chiefly greywacke-like, argillaceous but also sandy, marly and calcareous sediments.

The most recent regionally active metamorphic event is granitization. This commenced in the chlorite zone with a feldspar metablastesis, with no mobile components being separated out. In the biotite zone partial mobilization occurred and this resulted in segregation of the blastic feldspars together with quartz in bands, strips and schlieren running mainly parallel to the main cleavage plane (Fig. 16). Pegmatoid intercalations occur in conjunction with this process (e.g. near El Chol). The final product of metamorphism is the formation of the "Rabinal granite" which has a radiometric age of 345 ± 20 million years. MCBIRNEY (1963) and VAN DEN BOOM (1972) regard this not as a magmatic intrusion but as a metagreywacke modified by granitization.

In a strip around Huehuetenango, which is located further to the west, KESLER, JOSEY & COLLINS (1970) described para-rocks together with muscovite gneisses, banded schists, marbles, graphite schists, chlorite gneisses and intercalated orthogneisses, which they regarded as the "**western Chuacús Group**" parallelling MCBIRNEY's Chuacús series, which they elevated to the status of a group. Their observations are complemented by mapping carried out in the western Altos Cuchumatanes by ANDERSON et al. (1973); apart from describing the various metamorphic rocks, these authors in particular draw attention to cataclasis and blastomylonite formation along the Chixoy Polochic Fault Zone. BLOUNT (1967) had also found strong cataclasis in the metamorphic rocks in the region around Chiantlá and had concluded that the Polochíc Fault Zone originated during the Paleozoic.

In the Río Hondo area of the **Sierra de las Minas** NEWCOMB (1975, 1977, 1978) has mapped an areally extensive quartz-monzonitic gneiss, characterized by a variety of cataclastic textures, ranging from ultramylonite to blastomylonitic gneiss. This gneiss has been considered as the structurally lowest member of the Chuacús Group ("San Agustín Formation"). This apparent orthogneiss is separated from overlying phyllites of the "Jones Formation" and marbles of the "San Lorenzo Formation" by a south-dipping low-angle thrust.

NEWCOMB points out that "retrograde mineral assemblages characteristic of the upper-greenschist facies developed in response to cataclastic deformation adjacent to the Motagua Fault Zone, and do not represent regional metamorphism in the classic sense. The areal extent of the gneiss suggests that it may be a fundamental basement unit north of the Motagua Fault Zone in Guatemala's Cordillera Central. It has not been recognized south of the fault zone" (1978, p. 271).

ROPER (1976) in the western Sierra de las Minas distinguished at first between four types of metasediment of increasing metamorphic rank: 1. muscovite schist and micaceous quartzite, 2. micaceous schist and gneisses with carbonate horizons, 3. migmatized micaceous schist and gneisses, 4. hornblende gneiss. The rocks in this sequence range from the greenschist facies at the top to the lower amphibolite facies at the bottom. Presumably these para-rocks are a stratigraphic succession to the Chuacús Group, but no new information could be obtained on the latter's age. On the other hand, the sequence of metamorphic stages in the field would seem to indicate a possible refolded nappe.

Later (1978), ROPER by and large adopted NEWCOMB's classification and divided the metamorphic series of the Sierra de las Minas in accordance with the map (Fig. 17) which he adopted from the latter author.

To the north of the Chixoy Polochic Fault metamorphic rocks are widespread in the centre of the Altos Cuchumatanes and in the Maya Mountains. In the Altos Cuchumatanes these rocks are low-grade metamorphosed schists, chloritic schists and metaquartzites which ANDERSON et al. (1973) classify as part of the lower Santa Rosa Group. On the other hand, they compare light feldspar-biotite-gneisses, micaceous schists and amphibolites with the crystalline rocks of the western Chuacús Group which crop out to the south of the Chixoy Polochic Fault. Since, however, there are no radiometric datings available for the crystalline rocks of western Guatemala, these classifications are still somewhat uncertain, and BOHNENBERGER (see ANDERSON et al. 1973, p. 823) thinks it more likely that an older series of higher metamorphic rank and a younger series of lower metamorphic rank can be detected in the northeast part of the area studied.

In the **Maya Mountains of Belize** DIXON (1955) distinguished between an older highly folded and moderately metamorphosed Maya series and a younger and discordantly superjacent, relatively undeformed Permo-Carboniferous Macal series. More recent studies (KESLER et al. 1971, BATESON & HALL 1971, HALL & BATESON 1972) have shown, however, that there is no discordance present, that the pattern and degree of deformation is the same in both series and that furthermore it is not possible to detect any great differences in the degree of metamorphosis, which is in any case only slight. On the other hand, there are a great many common points between these series and the Permo-Carboniferous Santa Rosa Group of Guatemala; consequently the Maya and Macal series must be assigned to this latter group and not to a pre-Upper Carboniferous basement.

South of the Motagua Fault LAWRENCE (1975), within the Paleozoic (?) **Las Ovejas Complex,** detected gneisses of amphibolite facies, mica schists, marbles and foliated diorites and other calc-alkaline intrusions. In addition the complex contains metabasalts which are lithologically sharply distinguished from the acid metasediments of the Chuacús Group. The boundary between the two complexes is formed by the Motagua Fault Zone. SCHWARTZ (1976), in the Zacapa region, also describes the Las Ovejas Complex with schists, marbles, gneisses and intruding diorites, tonalites and granodiorites and he stresses that there are no equivalents to be found north of the Motagua Fault Zone.

South of the Motagua Fault MCBIRNEY & BASS (1969 b) recognized a series of low-grade metamorphosed greywackes, cherts and basic effusive rocks more than 5,000 m thick which they interpreted as a more volcanically influenced equivalent of the Chuacús Group and, like this, as the original geosynclinal deposit. They gave this series the name **El Tambor Formation.** However, following WILSON's (1974, pp. 1363 – 1365) discovery in eastern Guatemala of a very similar series which he was able to classify as Cretaceous on the basis of miliolid-bearing calcareous pebbles, the stratigraphic position of the El Tambor Formation needs to be revised. LAWRENCE (1975), in his study of the Sanarate-El Progreso region, already takes this into account and places the El Tambor Formation in the Mesozoic, presumably in the Cretaceous. I will return to this in more detail later in Section 3.14.

The **Sierra de Omoa,** which is situated in the northern part of Honduras, is built up of very different metamorphic rocks (Anonymous 1972 e, HORNE, CLARK & PUSHKAR (1976). At the northernmost end of the range low-grade metamorphosed metavolcanics and metagreywackes are found. In the centre of the range the predominant rocks are those of the almandine-amphibolite facies while in the south chiefly metasediments in greenschist facies occur. In the southwestern part of the Sierra the researchers distinguish between an "old sequence" containing epidote-rich quartz-feldspar-mica schists and hornblende-andesine-tremolite-amphibolite and a "young sequence" containing low-grade metamorphosed and argillaceous-calcareous schists, calcareous phyllites, calcitic and sericitic schists and fine-grained marble.

Far fewer data are available from the southeastern basement region of Honduras and Nicaragua than from Guatemala. In **central Honduras** CARPENTER (1954) described under the name of **Petén Formation** a series of phyllites, sericite schists and graphite schists which occur in the San Juancito Mountains. They are discordantly overlain by Mesozoic, and radiometric dating (PUSHKAR et al. 1972) indicates a maximum age of 412 million years. From the Comayagua Mountains FAKUNDINY (1970, see HORNE et al. 1976, p. 572) described as **Cacuapa Schist** para-rocks which can be divided into an upper (?) series of metapelites in greenschist facies and a lower (?) series of mylonitic augen gneisses. The other series presumably correlates with the Petén Formation.

Similar series of para-rocks have been described by ZOPPIS BRACCHI (1957) and DEL GIUDICE (1960) as the **Palacaguina Formation** (Esquistos de Nueva Segovia/Palacaguina) in northwestern Nicaragua. Taking critical account of the age, ENGELS (1964 b, p. 765) assigns the name "epizonale Metamorphite" to phyllites in the narrow sense, to conglomerates with elongate pebbles, to quartzitic greywackes, to sometimes schistose or caulescent quartzites, to siliceous schists, and to marbles and he gave the name metavolcanics to tuffites, diabase-greenschists and other intermediate volcanics as well as to contact-metamorphic schists and hornfelses — all from the region between the Río Coco and the Río Bocay.

Finally, the crystalline basement also includes a number of **plutons** on some of which recrystallization has been superimposed as a result of regional metamorphism, while others have only been cataclastically deformed and a third group has undergone no deformation at all. Their age needs to be mentioned here in the context of the metamorphic crystalline rocks. The following dates have been determined:

Rabinal-granite, Sierra de Chuacús (according to VAN DEN BOOM 1972, in situ granitized metagreywackes) (GOMBERG, BANKS & MCBIRNEY 1968)	345 ± 20 m.y.
Granite gneiss, Sierra de Chuacús (PUSHKAR 1968, MCBIRNEY & BASS 1969 b)	368 m.y.
Schistose plutonite, Quebrada Seca, Sierra de Omoa (HORNE et al. 1976, p. 578)	305 ± 12 m.y.
Granodioritic gneiss, Banaderos, Sierra de Omoa (HORNE et al. 1976, p. 578)	460 – 980 m.y.
Mountain Pine granite, Maya Mountains (BATESON 1972, p. 962)	280 – 300 m.y. 390 m.y.

Younger Mesozoic and Tertiary plutons will not be discussed here but will be examined in detail in Chapter 5.3.

The **age of the crystalline basement** was long disputed and even today, despite an increasing number of radiometric datings, it still has not been settled. HORNE et al. (1976) have incorporated the existing dates into their survey (Fig. 15). Datings of max. 400 m.y. for rocks from the Chuacús Group and of max. 412 m.y. for the Petén Formation indicate Middle Paleozoic, while three datings of metamorphic magmatites (Banaderos gneisses) in the Sierra de Omoa have yielded ages of between 460 and 980 m.y., with a very poor accuracy of ± 260 m.y., and these rocks thus extend back to the Upper Precambrian. The very high age of a zircon from a gneiss in the Chuacús Group, namely 1,075 m.y., is interpreted as indicating that this mineral was re-deposited from an older complex (Grenville) (GOMBERG et al. 1968). The Yucatán No. 1 borehole struck metamorphic crystalline rock with a radiometric age of 290 ± 30 m.y. (Pennsylvanian) (BASS & ZARTMANN 1969).

As far as the age of the metamorphism is concerned VAN DEN BOOM (1972, pp. 42 – 43) came to the following conclusion in the Sierra de Chuacús: The continuity of the facies series is evidence for only one metamorphism. The "Rabinal granite" which is intercalated in the Chuacús Group is regarded not as a magmatic intrusive rock but as the end product of a meta-arkose modified by granitization, so that the granitization was the last event in the metamorphic process. Its age can be put at 345 ± 20 m.y. and it thus belongs to the Lower Carboniferous.

Indications of the minimum age of the metamorphic series are also provided by datings of the plutonic rocks which are intruded in them (cf. Fig. 15).

HORNE et al. (1976, p. 580) summarize the present knowledge of the age as follows, taking the Sierra de Omoa as an example: The oldest rocks are metagranites of the Banaderos series, possibly from the Precambrian, but at least from the Ordovician and they thus permit the continental crust to be dated back to the Early Paleozoic. Thick sedimentary and volcanic series from the Early to Middle Paleozoic underwent deformation and metamorphism prior to the Middle Carboniferous and were intruded in the Middle Carboniferous by plutonic rocks of intermediate composition. The thus characterized basement rock complex is discordantly overlain by sediments of the Upper Carboniferous to the Permian, and these sediments initiated the more recent development of northern Central America.

This consequently confirms the conception derived by earlier authors from in-the-field observations; the following sentence is quoted from SAPPER (1937, p. 100) as a typical representative of this group: „We can definitely assume that some of the crystalline schists are converted Paleozoic stratified rocks; but probably a considerable portion of them are very old, perhaps Archean . . ."

With regard to the **structure of the metamorphic rocks** it was long assumed, mainly on the basis of the morphological pattern of the mountain chains, that the internal structures follow the same course and form an arc open towards the north so that Paleozoic, Mesozoic and Cenozoic structures run more or less parallel to one another. On this point SAPPER (1905, p. 63 et seqq.) stated: "Despite many differences, both the crystalline schists and the strata of the younger overlying rock for the most part strike predominantly in an E/W or ENE direction and thus show that they belong to the mountain system of northern Central America."

This conception was to a large extent confirmed by systematic structural measurements carried out above all in the central crystalline rock zone of Guatemala between the Sierra Madre del Sur in Mexico and the Sierra de las Minas (MCBIRNEY 1963, KESLER & HEATH 1970, KESLER 1971). According to these authors the axial planes and associated linear elements of an older and much more violent folding and foliation process do in fact form the assumed arc open to the north (Fig. 14). The folds are inclined to the north as KESLER established in the eastern part of the chain and VAN DEN BOOM (1972) discovered in the centre. Folds with approximately N/S- to NNW/SSE-striking axial planes constitute a younger structural element. They occur only in the vicinity of large fault zones and they appear to indicate that movement was initiated along these fault zones thus resulting in transverse folding.

NEWCOMB (1975) discovered at least two deformation phases within the Chuacús Group of the Río Hondo region. The older phase includes large recumbent isoclinal folds which may possibly be evidence of decken structure. The younger deformation phase led to asymmetrical slip folds and to cataclasis of the San Augustín Formation.

ANDERSON et al. (1973) surveyed foliation planes and linear elements in the metamorphic rocks of the Altos Cuchumatanes and showed that the foliation together with a concordantly intercalated plutonic rock strike N 40° W while the linear elements dip to the SW at an angle of about 30°. Fold axes in the Mesozoic of the Altos Cuchumatanes and the upfaulting accompanying the southern margin generally strike in the same direction. A comparison of foliation planes and bedding of the Upper Paleozoic and Mesozoic strata reveal a high degree of parallelism (Fig. 48). ANDERSON therefore correctly points out (p. 824) the astonishing constancy of the stress pattern. Near the Chixoy Polochic Fault the grain fabric is oriented approximately parallel to the fault.

In the southwestern section of the Sierra de las Minas ROPER (1976) assumes **decken structure** in the crystalline schists of the Chuacús Group. He deduces this from the stratification of four elements of different metamorphic degree in which he also sees a stratigraphic sequence of unknown age.

According to HORNE et al. (1976, p. 574) it is possible to detect three deformation phases in the structure of the Sierra de Omoa of northern Honduras whose elements generally strike ENE. The Chamelecón Fault, which is closely accompanied by severe cataclasis, strikes in the same direction. Together with different rock facies on both sides of the fault, these phases are regarded as signs of significant strike-slip faults.

ENGELS (1962, 1964 a, b) recorded microtectonic structural elements in **northern Nicaragua** but these do not give a clear picture of the rock structure. He summarized his results as follows: "The inner tectonic structure is extremely complicated and in many respects is reminiscent of regions containing crystalline schists (polymetamorphic rocks). In the west the folds and overthrusts . . . verge mainly towards the north; in the central part of the country . . . the orientation is towards the east; and finally, in the eastern part the orientation is once more predominantly towards the north and northwest . . . Overall the vergence seems to be mainly to the north . . . It is quite certain that several deformations or stresses occurred at different times and in different directions" (1964 b, p. 791).

The present state of our knowledge of the crystalline basement may be summarized as follows: The sedimentary initial material presumably dates from the Paleozoic up to and including the Devonian. The area in which it was deposited was described very cautiously by DENGO & BOHNENBERGER (1969, pp. 209 – 210) as an E/W-striking orthogeosyncline which transected Central America during the Lower Paleozoic. The age of the metamorphism is delimited in an upward direction by the overlying unmetamorphosed Santa Rosa Group which belongs to the Pennsylvanian. Radiometric datings indicate Upper Devonian (MCBIRNEY & BASS 1969, p. 273) or Lower Carboniferous (VAN DEN BOOM 1972, pp. 42/43), which also ties in with the conception advanced by HORNE et al. (1976, Fig. 2) (cf. Fig. 15). The degree of metamorphism extends from greenschist facies to anatexis and granitization. Only traces of a cohesive pattern can be discerned in the structures, which probably belong to several deformation phases. In Guatemala and northern Honduras the structural elements run almost parallel to the present course of the mountain chains. Severe cataclasis in the vicinity of the Motagua and Polochic Fault Zones is evidence of their origin in the Paleozoic.

3.12 The Upper Paleozoic

The Upper Carboniferous and Permian are the oldest unmetamorphosed formations in Central America. The Permian is biostratigraphically well documented by fusulines, brachiopods and ammonites, while the Upper Carboniferous can be identified only on the basis of stratigraphic sequence. The strata which merge with one another without any sharp boundaries are grouped in an E/W-striking belt on the northern flank of the metamorphic rocks of Guatemala and they extend as far as Chiapas. Further extensive occurrences are found in the Maya Mountains (cf. the map in Fig. 18).

The Upper Paleozoic is divided into a lower, clastic sequence, a middle shaly sequence and an upper carbonate sequence, the naming and classification of which have been confused and not always consistent. BOHNENBERGER (1966) together with ANDERSON et al. (1973) and CLEMONS et al. (1974) have tried to correct this situation and to re-classify these strata. The classification scheme proposed by them is probably closest to our present knowledge, and it is therefore placed ahead of the previous classifications in Table 1. By taking the Upper Paleozoic together as the **Santa Rosa Group** they in addition follow the lead in naming this formation which was given by DOLLFUS & MONT-SERRAT (1868), although the latter's type locality is nowadays assigned to the Mesozoic Todos Santos Formation.

In **Guatemala** the subjacent **Chícol Formation** consists of a distinctive sequence of interbedded greenish-grey and light bluish-grey conglomerates and sandstone with greyish-green, grey and maroon tuff and volcanoclastic beds and less common andesite breccia about 1,000 m thick. The contact between this and the older crystalline rock is usually obscured by

Table 1. Correlation chart for the Upper Paleozoic and the Todos Santos Formation in Guatemala. (From ANDERSON et al. 1973, Fig. 4).

		Anderson et al. (1973)	Dollfus & Mont - Serrat (1868)	Sapper (1897 – 1899)	Roberts & Irving (1957)	Walper (1960)	Bohnenberger (1966)	Bateson & Hall (1971)	van den Boom Müller Nicolaus Paulsen (1971)
Lower Cretaceous Upper Triassic (?)		Todos Santos Formation	? — ?	Todos Santos Formation	Todos Santos Formation	Todos Santos Formation			Todos Santos Formation
Permian	Leonardian	Tuilán mbr. / Chóchal Limestone (Santa Rosa Group)	Santa	? / Karbonkalke	Chóchal Limestone	f / Chóchal Formation	?	?	Chóchal Formation
Permian	Wolfcampian	Esperanza Formation (Santa Rosa Group)	Rosa	? / Santa Rosa Formation	Santa Rosa Formation		Tactic Formation (Santa Rosa Group)	Santa	Upper Tactic Formation
Permian or older		Tactic Formation (Santa Rosa Group) / f	Group		- - - - - - -	Tactic Formation		Bladen Volc. mbr. / Rosa	Lower Tactic Formation
Pre - Permian		Chicol Formation / f / Western Chuacús Group f Crystalline Rocks	(oldest unit)	Crystalline Basement	Lower Santa Rosa Formation / Crystalline Basement	? / Base not exposed	Sacapulas Formation (Santa Rosa Group) / Chuacús Group	Group	Sacapulas Formation / Chuacús Group

fractures or contact-metamorphism, but near Sacapulas it is seen to be discordant. In addition the crystalline basement is contained in the form of pebbles in the conglomerates.

The following **Tactic Formation** consists of shales over 800 m thick which are sometimes schistose, sometimes phyllitic and occasionally contain beds of limestone and dolomite. Following the large-scale process of deforestation which has taken place, these beds are particularly subceptible to erosion in the deeply incised river valleys and they are therefore dissected by deep ravines or exposed by slides.

The **Esperanza Formation** is lithologically similar and contains fusulines; as a result it is classified in the Wolfcampian. Its thickness is given as 470 m. The overlaying **Chóchal Limestone,** which is at least 200 m thick, stands out in the landscape as a result of the steep slopes and kegelkarst that it forms. On the basis of its rich fauna, and in particular the fusulinids (KLING 1960), it can be placed in the Leonardian. In the upper section the fossil-rich silty/sandy **Tuilán member** attains thicknesses of 180 m and can in certain places be treated as a separate entity. Its brachiopod fauna, which has been dealt with by STEHLI & GRANT (1970), like an ammonite fauna described by ANDERSON et al. (1973, p. 813) also belongs to the Upper Leonardian.

In the **Maya Mountains** DIXON (1955) distinguished an older Maya series which he assigned to the metamorphic rocks. Above this, in his view, comes the discordant Macal series consisting of conglomerates, sandstones, shales and crinoidal limestones. On the basis of a mollusc and brachiopod fauna this series has been classified in the Upper Pennsylvanian to Permian [1]. According to a new survey of the Maya Mountains, however, it is not possible to distinguish the Maya and Macal series from each other and in addition they exhibit the same type of deformation. They are therefore listed in the works by BATESON & HALL (1971), KESLER et al. (1971), HALL & BATESON (1972) and BATESON (1972) under the designation Santa Rosa Group. They are interbedded with the rhyolitic-felsitic volcanics of the **Bladen Volcanic Member** which according to radiometric dating are 300 m.y. old.

[1] For a detailed description and list of fauna see the 1st (German) edition (WEYL 1961 a, pp. 21 – 22), because DIXON's original is difficult to obtain.

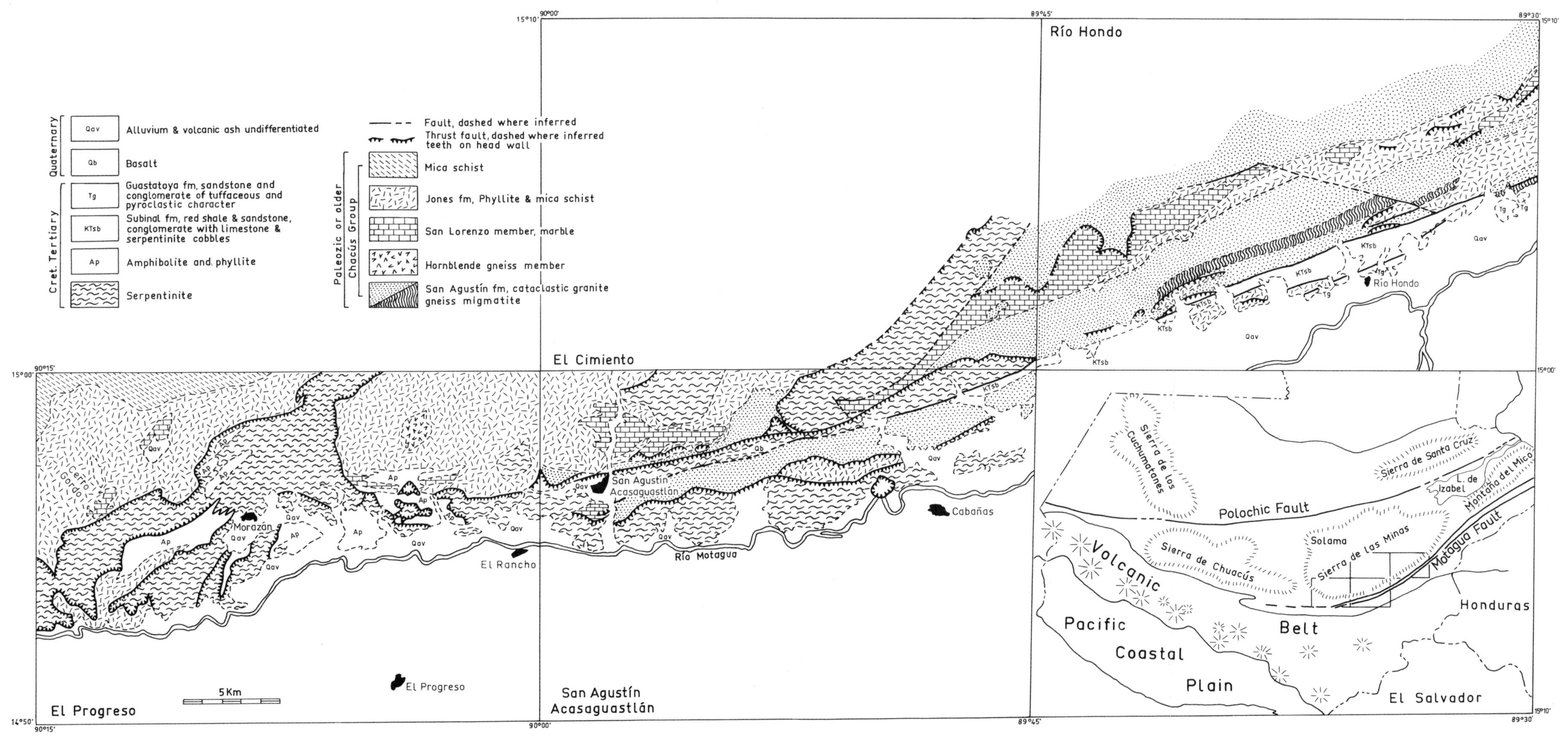

Fig. 17. Geological map of the south side of the Sierra de las Minas Range. (Simplified redrawing from ROPER 1978, Fig. 2).

As regards the **paleogeographic position** of the Upper Paleozoic SAPPER's (1937, p. 26) theory is still valid, namely that it is the "erosion product of an Upper Carboniferous folded mountain chain over which the Middle Permian sea transgressed following peneplanation". It is not made clear where these mountain chains were located in relation to the present system of co-ordinates and SAPPER certainly did not have in mind the shifting of continental blocks by hundreds or thousands of kilometres. But even more recent attempts to reconstruct the paleogeographic situation proceed essentially from the assumption that at least Mexico and northern Central America were at some time connected with each other (e.g. DENGO 1975 a, Fig. 3). The paleogeographic sketch (Fig. 18) is based on this belief. On the other hand, if one accepts the theory that extensive lateral shifts were still taking place in the Tertiary (see Chapter 8), then the southern boundary of the Upper Paleozoic sedimentation basin should coincide with the present Motagua Fault Zone. I am not prepared to hazard a guess on how far all this coincides with the following observations and considerations.

DENGO & BOHNENBERGER (1969, p. 211) described the Santa Rosa Group as the "clastic wedge" which belongs to the Paleozoic Jaliscoan cycle. HALL & BATESON (1972, p. 954) referred to a "molasse type" in the region of the Maya Mountains, while they compared the occurrences in Guatemala with flysch. Between the two regions lay a barrier built up of acid volcanics of the Bladen Volcanic Member. In keeping with this classification they postulated a miogeosyncline in the northern part and a eugeosyncline in the southern part (Fig. 19).

In the broader context (see map Fig. 18) it is probably correct, along with LÓPEZ RAMOS (1969), to assume that the Paleozoic sedimentation basins of Chiapas and Guatemala were connected with the Paleozoic geosynclines of Central Mexico. DENGO 1975 a, Fig. 4) also adopts LÓPEZ RAMOS's view. According to the description given by VINIEGRA (1971) a Pa-

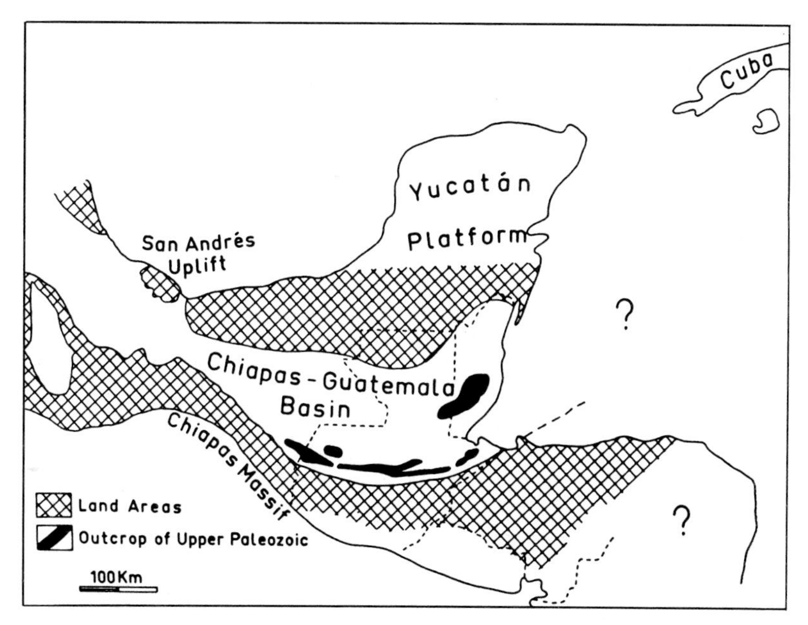

Fig. 18. Paleogeographic sketch of Upper Paleozoic in northern Central America. (Essentially from LÓPEZ RAMOS 1969). (From WEYL 1974 a, Fig. 3).

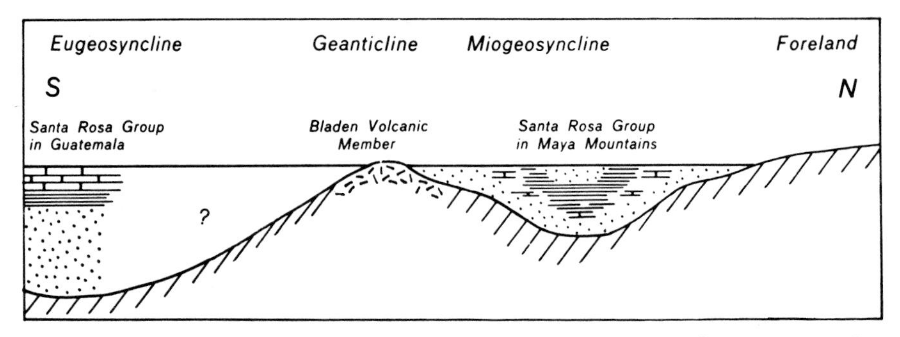

Fig. 19. Diagrammatic section through the Upper Paleozoic geosyncline of Guatemala and Belize. (From HALL & BATESON 1972, Fig. 3 C).

leozoic Guatemalan geosyncline bent around from Chiapas via Tabasco towards the north in the direction of the Gulf of Mexico and was bounded in the west by the Oaxaca craton. It is generally agreed that the southern boundary was formed by the Chiapas Massif and further east by a land region in the area of southern Guatemala and Honduras, whence the clastic material can probably also be derived except in those cases where it is made up of local debris of erosion. In the north the Yucatán Platform is regarded as a continental area following the discovery here, by drilling, of Lower Cretaceous and red beds of the Todos Santos Formation directly superjacent upon metamorphic rocks and rhyolites with a radiometric age of 410 m.y. (cf. LÓPEZ RAMOS 1975, p. 267).

The eastern continuation of the geosyncline has not been positively established but it certainly did not end abruptly with the entry of the present Caribbean Sea into the region. A. A. MEYERHOFF (1966 and in KHUDOLEY & MEYERHOFF 1971) has listed countless reasons why he believes that the Paleozoic orogen of Central America, and thus also the Paleozoic geosyncline, extended to the east in the Nicaragua Rise and the Cayman Ridge and that its structures thus must have reached as far as Jamaica or eastern Cuba. ARDEN (1975, p. 631) doubts this: "I believe much of this orientation is more anticipated than real because of map trends associated with young topographic and structural features." According to his interpretation of the Nicaragua Rise, Lower Cretaceous volcanics and metamorphic rocks are immediately superjacent here on oceanic crust (Fig. 13) and also, despite all the assumptions, the presence of Paleozoic rocks in the Antilles has not been proven either.

The Upper Paleozoic of northern Central America has undergone varying degrees of deformation. It is not possible to classify the deformation phase exactly because the very uppermost strata of the Permian and Triassic are missing and only the Upper Jurassic with the clastic deposits of the Todos Santos Formation has been proven. Therefore it is assumed that the uppermost Permian or Triassic was the period when deformation took place, and it was followed in the Maya Mountains by a granite intrusion aged 206 – 213 m.y. (BATESON 1972, p. 962). In any event, the result of the Upper Paleozoic crustal movements was the uplifting of the entire northern part of Central America and the lack of marine deposits in the Triassic. It is not known how far this Early Mesozoic continental block extended into what are today marine regions; geophysics and drillings carried out in the Caribbean do not yet provide any clues.

3.13 Jurassic to Lower Cretaceous

The series of Mesozoic strata in northern Central America commences in Honduras with the **El Plan Formation** (CARPENTER 1954, pp. 25 – 27). The occurrences of the formation extend in a relatively narrow belt from San Juancito, approximately 30 km NE of Tecucigalpa, towards the northeast into the Mosquitia. MILLS et al. (1967, p. 1769 and Fig. 30) therefore believe that a taphrogenic block fault occurred here in which deposits over 900 m thick of chiefly limnal, but also to some extent of marine sandstones, pelites and small lentils of conglomerates containing streaks of coal were deposited.

The classification into the Upper Triassic to Lower Jurassic still rests on the old flora determinations of NEWBERRY (1888, p. 342) and KNOLWTON (1918, pp. 607 – 614). In addition, unconfirmed ammonite finds (SCHUCHERT 1935, p. 355 and IMLAY 1952, pp. 969 – 970) were supplemented by the occurrence of *Trigonia* cf. *quadrangularis, Gervillia* spec. and ? *Meretrix* spec. in fine-grained sandstones and siltstones found in the vicinity of San Juancito, which together point to the Jurassic age of part of the strata. Since, according to ELVIER (1974, p. 23), plant finds are common in the El Plan Formation, it would seem that much could be gained from making a new collection of flora or from carrying out palynological dating.

The strata in the formation are sharply folded (Fig. 20) — according to MILLS et al. (1967, p. 1769) this occurred during a Middle Jurassic deformation phase accompanied in the area of the Mosquitia Mountains by volcanic activity which is not described in any further detail. ELVIER (1974, p. 24) stresses that the next youngest formation, namely the (Upper Jurassic to Lower Cretaceous) Todos Santos Formation, discordantly overlies the El Plan Formation while WILSON (1974, p. 1352) puts the folding later, namely in the Middle Cretaceous. Furthermore he stresses that data regarding the original distribution of the formation, which he regards as exclusively limnal, are pure assumption. FINCH & CURRAN (1977), in their revision of the Mesozoic in central Honduras, also stress that both the stratigraphic position as well as the deposition region and deposition conditions still present unsolved problems and that the El Plan Formation presumably constitutes the oldest sedimentary unit.

Fig. 20. Folded sandstones and siltstones of the El Plan Formation. San Juancito, Honduras. Photo: R. WEYL.

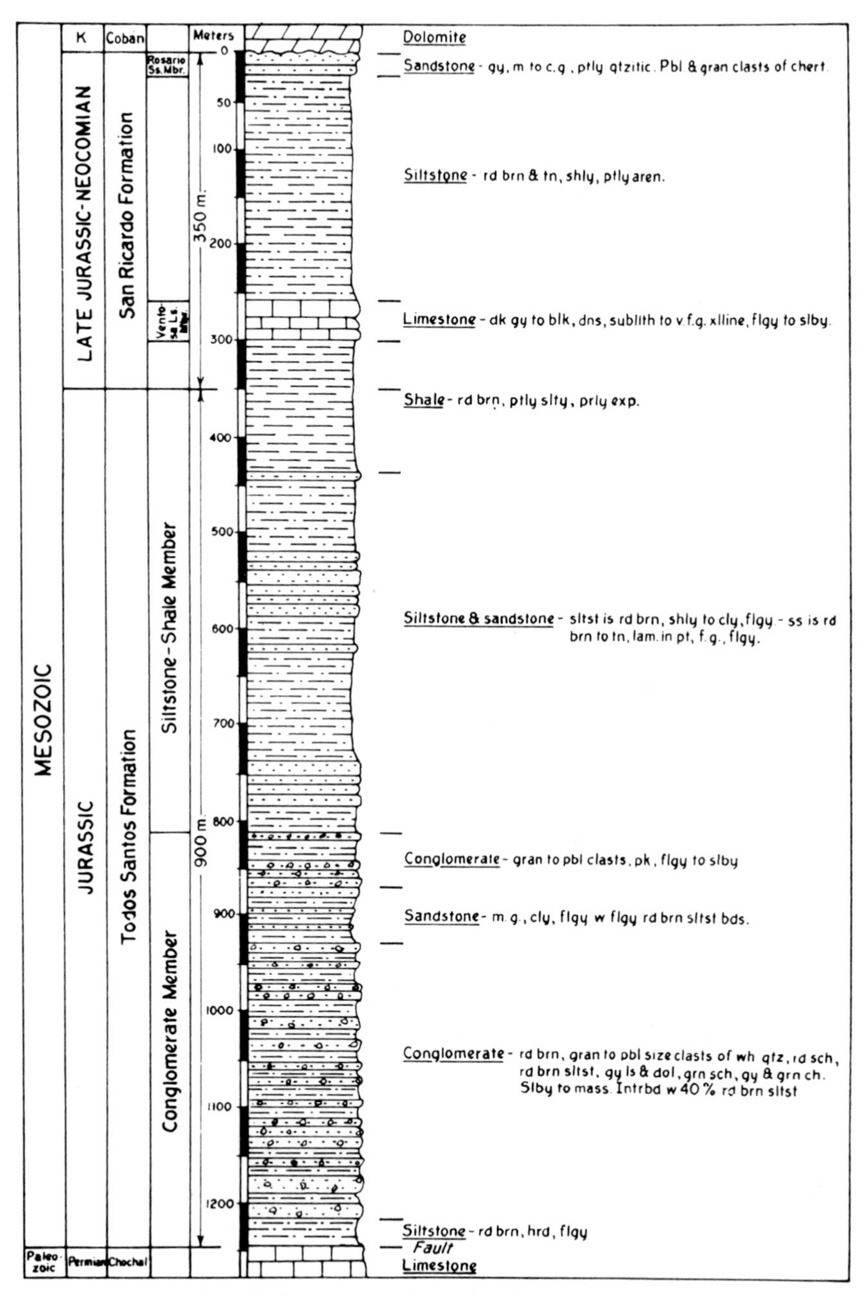

Fig. 21. Ventosa section of the Todos Santos and San Ricardo Formations. Huehuetenango (Guatemala). (From RICHARDS 1963, Fig. 3).

Red continental conglomerates, sandstones and argillaceous shales are widespread over large areas of northern Central America and of the adjoining Mexican state of Chiapas. They are discordantly superjacent on metamorphic crystalline rock and on Upper Paleozoic strata. SAPPER (1894, pp. 8 – 9; 1937, pp. 26 – 28) described these strata and gave them the name **Todos Santos Formation.** A great many more recent papers have provided further data on the structure, classification (Table 2) and paleogeographical situation of this deposition area (RICHARDS 1963, MILLS et al. 1967, VINIEGRA 1971, ANDERSON et al. 1973, CLEMONS et al. 1974, WILSON 1974, and others).

The most important occurrences are situated in western Guatemala to the north of the Chixoy Polochic Fault at the base of the Altos Cuchumatanes (the type locality in that region), in the Baja Verapaz, in central Honduras and in the Mosquitia. RICHARDS (Fig. 21) described a characteristic profile which by and large agrees with that given by ANDERSON et al. (1973, Fig. 6). RICHARDS treated the mainly silty upper strata, with their intercalations of limestone, separately as the **San Ricardo Formation,** but CLEMONS et al. (1974, p. 318) do not go along with this and again fall back on SAPPER's definition.

The lithology and thickness of the Todos Santos Formation vary greatly. In the Altos Cuchumatanes two basins with deposits up to 1,200 m thick are separated by an uplift with deposits less than 10 m thick. In northern Honduras, in the region of Atima and Taulabé, two basins contain over 800 m of clastic sediments, and in the Mosquitia region the thickness of these sediments fluctuates between 180 and 650 m. The pebble content of the conglomerates depends on the respective environment and it consists of metamorphic material, granites and Paleozoic limestones. It is particularly worth noting the occurrence of andesitic lava sheets and re-worked andesitic volcanics in the Mosquitia region.

The age is certainly not uniform and it is therefore still the subject of discussion in the above-mentioned works. On the basis of fossil finds in the marine limestones of the upper strata (San Ricardo Formation), these strata can probably fairly certainly be assigned to the Jurassic or Lower Cretaceous, and there is therefore general agreement that the underlying beds can be assigned to the Jurassic.

In Honduras the name Todos Santos-Formation was given to red beds underneath the Cretaceous limestones of the Yojoa Group (see below). Their exact age can only be given on the basis of the overlying strata, and these would indicate that the sediments date from the Jurassic to Lower Cretaceous. However, FINCH & CURRAN (1977) point out that many of the red beds which are shown as belonging to the Todos Santos Formation in reality form part of the Lower Valle de Angeles Group. They furthermore point out that on the basis of the paleomagnetic measurements carried out by GOSE & SWARTZ (1977 a and b) the deposition region was located on a different lithosphere plate from the region of Guatemala. This makes it difficult, if not to say impossible, to reconstruct the paleogeographic situation.

Along with DENGO & BOHNENBERGER (1969, p. 212), we can interpret the Todos Santos Formation as post-orogenic molasse of the Upper Paleozoic (Jaliscoan) tectonic cycle which was deposited in the form of detrital fans in intramontane basins and grabens. The volcanic activity in the Mosquitia region completes the picture of a post-orogenic phase. Towards the end of this phase the beginnings of a transgression are evident in the limestone beds of the San Ricardo Formation; this reached its climax in the Middle Cretaceous and initiated the Mexican geotectonic cycle. According to VINIEGRA (1971, Fig. 10) it originates in neighbouring Mexico and this puts western Guatemala in the boundary zone of salt basins in Chiapas; the oldest of these basins belongs to the Oxfordian and one of the younger basins can be clas-

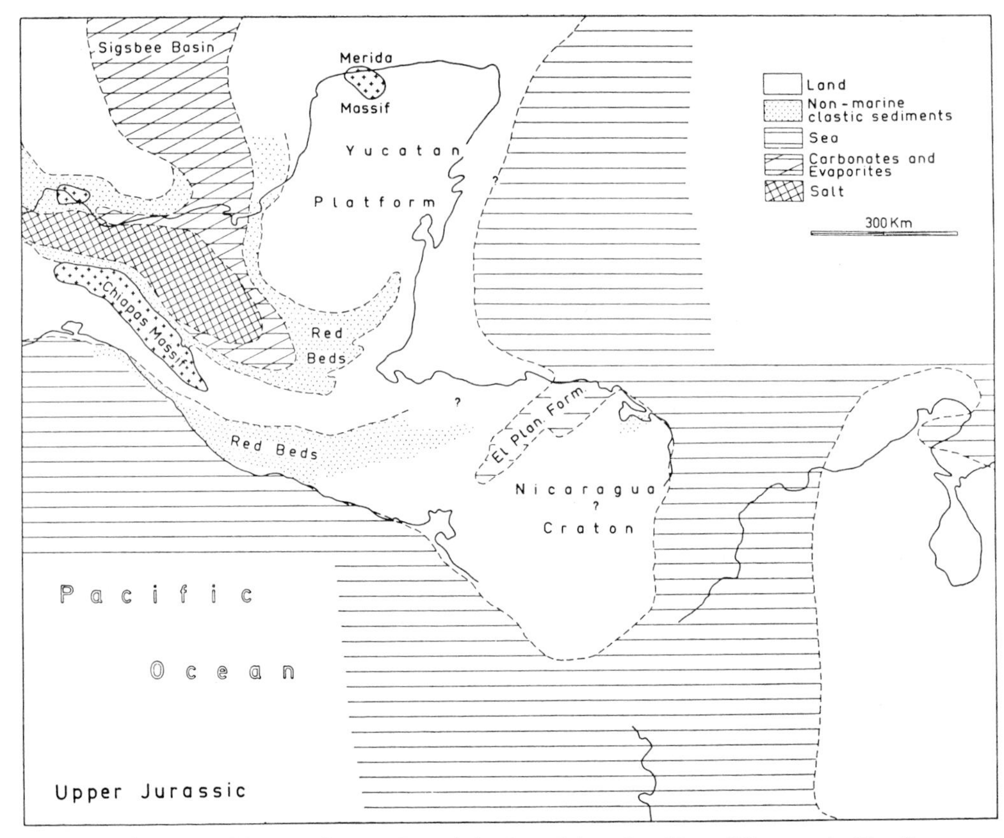

Fig. 22. Paleogeographic map of Upper Jurassic in Central America. (From WEYL 1974a, Fig. 4).

sified in the uppermost Jurassic and the lowermost Cretaceous. An attempt has been made in the map in Fig. 22 to show the paleogeography of the Jurassic, although it was not possible to divide up Central America in the way that MACDONALD (1976 a and b) did, nor could the plate movements described by GOSE & SWARTZ (1977) be taken into account.

3.14 The Cretaceous

The carbonate components in the upper sections of the terrestrial deposits of the Todos Santos Formation increase and as a result the deposits merge discordantly in places with the limestone formations of the Cretaceous. They attain thicknesses in excess of 3,000 m, are widespread and form the landscape in the Altos Cuchumatanes, the Verapaz and the Petén of Guatemala. In addition they form the material of the limestone mountains of northern Honduras (cf. general maps of Guatemala and Honduras, Figs. 43 and 58) and they crop out further south through the Cenozoic volcanic cover in the form of inliers. Very probably they covered the whole of northern Central America in an uninterrupted sheet before the more recent uplift and dissection of the mountain ranges occurred.

Table 2. Correlation chart of the Mesozoic and Cenozoic in northern Central America.

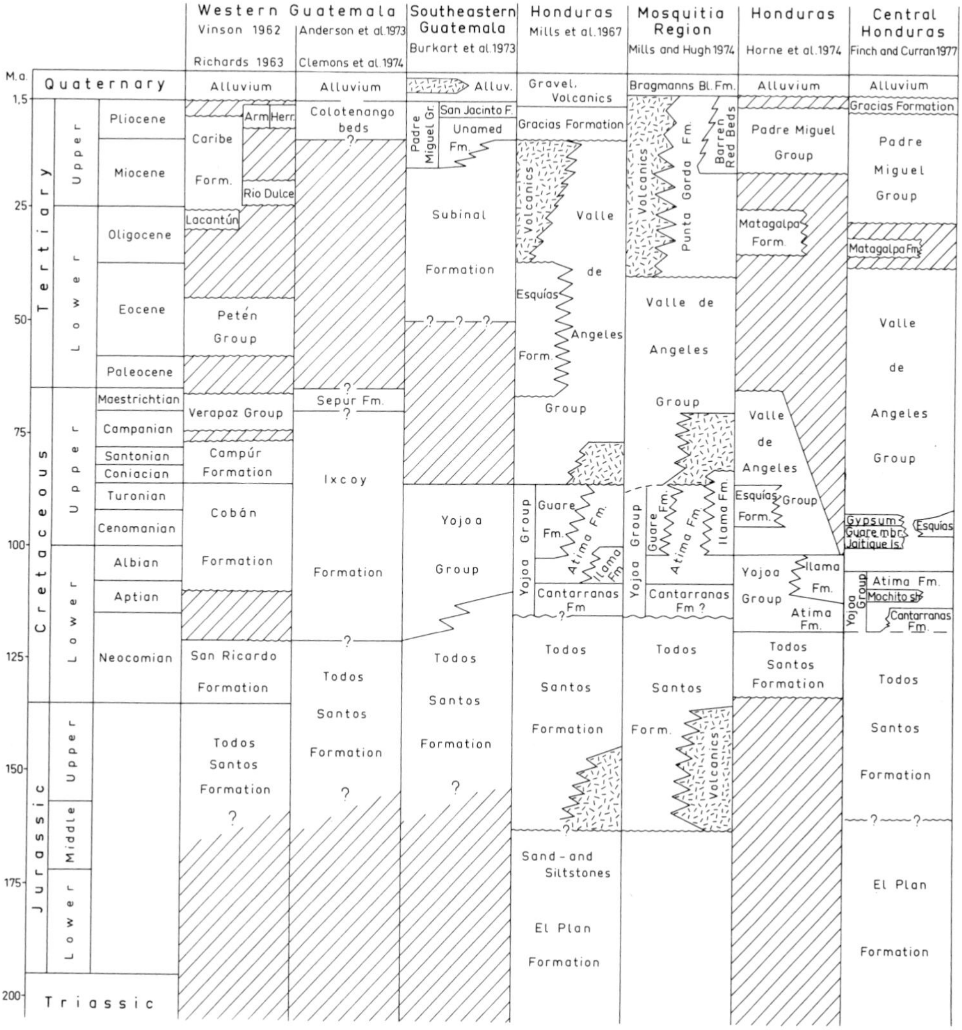

M.a.	Period	Epoch	Stage	Western Guatemala Vinson 1962, Richards 1963	Western Guatemala Anderson et al. 1973, Clemons et al. 1974	Southeastern Guatemala Burkart et al. 1973	Honduras Mills et al. 1967	Mosquitia Region Mills and Hugh 1974	Honduras Horne et al. 1974	Central Honduras Finch and Curran 1977
1.5	Quaternary			Alluvium	Alluvium	Alluv.	Gravel, Volcanics	Bragmanns Bl. Fm.	Alluvium	Alluvium
	Tertiary	Upper	Pliocene	Arm, Herr, Caribe	Colotenango beds ?	Padre Miguel Gr.: San Jacinto F., Unamed Fm.	Gracias Formation	Volcanics, Punta Gorda Fm., Barren Red Beds	Padre Miguel Group	Gracias Formation, Padre Miguel Group
			Miocene	Caribe Form., Rio Dulce			Volcanics	Volcanics, Punta Gorda Fm.		Padre Miguel Group
25		Lower	Oligocene	Lacantún		Subinal	Valle de Angeles		Matagalpa Form.	Matagalpa Fm.
			Eocene	Petén Group		Formation ?—?—?	Esquías Form., Valle de Angeles	Valle de Angeles		Valle
50			Paleocene				Group	Angeles		de
	Cretaceous	Upper	Maestrichtian		Sepur Fm. ?			Group	Valle de Angeles	Angeles
75			Campanian	Verapaz Group	?				Group	Group
			Santonian	Campúr Formation	Ixcoy					
			Coniacian	Campúr Formation						
			Turonian	Cobán		Yojoa	Yojoa Group: Guare Fm., Atima Fm.	Yojoa Group: Guare Fm., Atima Fm., Ilama Fm.	Esquías Form., Group	
			Cenomanian				Ilama Fm.			Gypsum, Guare mbr., Jaitique ls., Esquías
100		Lower	Albian	Formation	Formation	Group			Yojoa Group: Ilama Fm., Atima Fm.	Yojoa Group: Atima Fm., Mochito sh., Cantarranas Fm.
			Aptian				Cantarranas Fm ?	Cantarranas Fm ?		
125			Neocomian	San Ricardo Formation	? Todos Santos Formation	Todos Santos Formation	Todos Santos Formation	Todos Santos Form.	Todos Santos Formation	Todos Santos Formation
150	Jurassic	Upper		Todos Santos Formation ?	Todos Santos Formation ?	?	Formation ?	Volcanics		Formation ?—?
		Middle					Sand- and Siltstones			
175		Lower					El Plan Formation			El Plan Formation
200	Triassic									

Designation and classification of the Cretaceous deposits have undergone many changes in the course of research and from region to region, as is reflected in Table 2. This table shows that even today views still differ on these deposits. SAPPER (1899, p. 65; 1937, pp. 28/29) referred to the Cretaceous limestones in Guatemala as **Cobán-Kalke,** from which he eliminated rudistan limestones; he separated these out in eastern Chiapas as belonging to the Upper Cretaceous. TERMER (1932, p. 247) introduced the name Ixcoy Formation. WALPER (1960, p. 1298) called the lower limestone strata the Ixcoy Formation and the upper strata the Cobán Formation. ANDERSON et al. (1973) and CLEMONS et al. (1974, pp. 318 – 319) attempted a revision of the stratigraphic nomenclature and in the process fell back upon TERMER's desig-

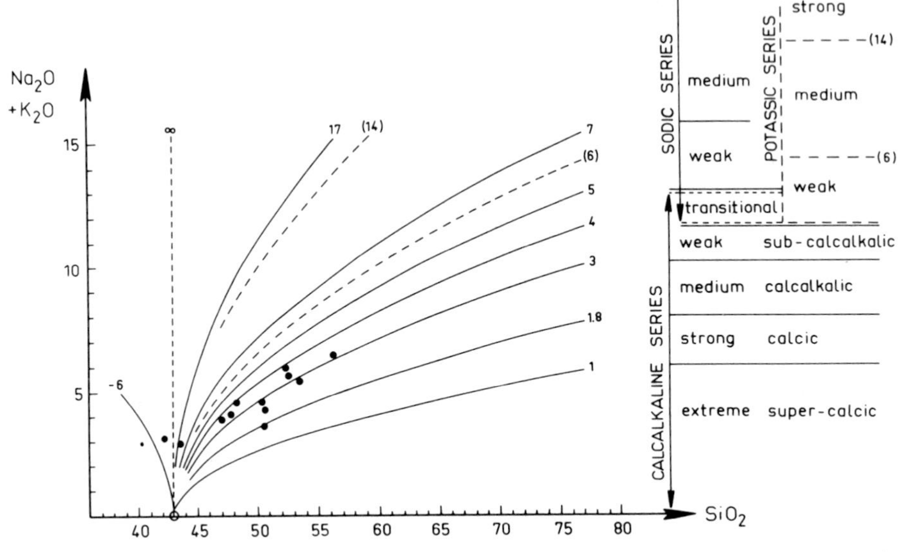

Fig. 23. Serial index plot (alkali versus SiO_2) of pillow lavas of the El Tambor Formation. Analytical data from BERTRAND & VUAGNAT (1975).

Fig. 24. Massive bedded limestone of the Atima Formation. Quarry of the Mochito mine, Honduras. Photo: R. WEYL.

nation **Ixcoy Formation,** because according to their in-the-field surveys in the Altos Cuchumatanes there were no detectable mappable boundaries between the limestones.

Prior to this, however, VINSON (1962) had separated the upper portion of the limestones as the **Campúr Formation** from the Ixcoy or Cobán Formations, because it was lithologically easy to distinguish. He was followed in this by BONIS (1969, p. 79) in a review of the geology of Guatemala.

In Guatemala the uppermost sections of the Cretaceous (the Campanian and Maestrichtian), following SAPPER's (1899) lead, are classified separately as the **Sepur Formation.** This formation contains marls, calcarenites and clastic red beds. Coccoliths and foraminifera of the type locality belong to the Maestrichtian (CĚPEK 1975). VINSON (1962) subdivided the Sepur Formation and separated out the contemporaneous but regionally and facially distinct **Chemal** and **Lacandón Formations** and put them together as the **Verapaz Group.**

Thanks to the recent studies and mappings carried out in the course of petroleum exploration and by US universities, not only has the stratigraphy of the Cretaceous been studied on the basis of the microfauna, it has also been possible to distinguish between very different types of facies. These have been described taking typical profiles not only from Guatemala (VINSON 1962, BLOUNT & MOORE 1969, ANDERSON et al. 1973, BURKART et al. 1973) but also from Honduras (MILLS et al. 1967, 1969, HORNE, ATWOOD & KING 1974, FINCH & CURRAN 1977) and from both countries by WILSON (1974), and they form the basis of still highly divergent conceptions regarding the depositional conditions, paleogeography and tectonics. Only a few examples can be chosen here from the wealth of data.

In the western **Altos Cuchumatanes** of Guatemala the limestones of the **Ixcoy Formation,** according to ANDERSON et al. (1973), attain thicknesses of around 2,500 m. The lower 1,000 m consist chiefly of fossil-free, dark grey, semi-crystalline dolomites and limestones with intercalated calcareous breccias. The breccias contain fragments of limestones, siliceous limestones, dolomites and occasionally of shales and siltstones. According to a study made by BLOUNT & MOORE (1969) these are lithoclastic breccias, the material for which comes from eroded fault scarps in the vicinity of the deposition region. In addition, evaporite-solution breccias and tectonic breccias are widespread. Fine-grained clastic strata occur in the central part of the Ixcoy Formation. The upper 1,000 m consist of fine-grained, greyish-brown stratified limestone with thick intra-formational breccias and conglomerates. The limestones contain layers which are rich in microfossils, in particular miliolids and locally large numbers of rudistans. The overlying **Sepur Formation** consists of red sandstones and argillaceous shales with fossil-rich limestones. It is interpreted as an in-shore shallow-water deposit.

In **Petén** the limestones of the Ixcoy Formation grade into **evaporites** which were found to be up to 3,000 m thick in the course of drilling carried out in the search for petroleum. Above this formation, in the south, comes the **Campúr Formation** with rudistan-rich limestones about 1,000 m thick. According to VINSON (1962, pp. 437 – 442), the final **Verapaz Group** of the Campanian to Maestrichtian, which is up to 1,500 m thick, can be divided up from south to north into the following facially different formations: The **Chemal Formation,** which consists of red shales, calcarenites and conglomeratic limestones, is situated at the northern edge of an uplifted land mass in central Guatemala. To the north follows the **Sepur Formation** which consists of fine clastic and calcareous-marlaceous rocks which in turn give way towards the north to the carbonate shelf deposits of the **Lacandón Formation.** In Yucatán thick evaporites are intercalated in the carbonates of the Upper Cretaceous.

A heterogeneous chaotic series of coarse breccias and conglomerates, greywackes, radiolarites, serpentinites, spilitic lavas and volcanogenic detritus, which WILSON (1974, pp. 1363 – 1365) described as "**Jalapa Basinal Melange**" from Jalapa in eastern Guatemala, is of particular significance for the pattern of tectonic development. The Cretaceous age of this series has been verified on the basis of limestone detritus with miliolids and also on the basis of the genus *Cuneolina.* WILSON described the series as typically eugeosynclinal and he regarded the association of radiolarites with basic lavas as a deep-sea formation. He interpreted the occurrence of massive conglomerates as the expression of severe tectonic dislocation which resulted in the limestones of the Albian/Cenomanian being eroded away and embedded as slide masses in the melange. When this Cretaceous orogenic series was discovered, WILSON thought it necessary to revise the age of the neighbouring and lithologically similar El Tambor Formation which MCBIRNEY & BASS (1969 b) had placed in the Paleozoic. In a discussion with HORNE et al., WILSON (1977) pointed out that similar series with microfaunas of the Campanian to Maestrichtian were also to be found under the metasediments in the Sierra de Omoa, and this was in keeping with his conception of a violent Late Cretaceous orogenesis.

LAWRENCE (1975, 1976) described a similar rock series from the region of Sanarate in central Guatemala south of the Motagua Fault Zone. It contains siliceous phyllites, thin layers of chert and metavolcanic rocks with chemical characteristics similar to those of oceanic tholeiites (normative hypersthene, low alkali content and a TiO_2 content of on average 1.54%). LAWRENCE, with some reservation, places the series together with the El Tambor Formation in the Upper Cretaceous and comes to the conclusion that it forms a residue of an oceanic crust which became deformed and metamorphosed in a Mesozoic subduction zone. According to LAWRENCE the location of the subduction zone is indicated by the present Motagua Fault Zone.

BERTRAND & VUAGNAT (1975) came to a similar conclusion, on the basis of pillow lavas which they discovered at the Carretera al Atlántico (km 85) within highly sheared serpentinites. They are metamorphosed into pumpellyite-prehnite facies. According to the authors they are slightly spilitic in character. The majority of the analyses, however, yield serial indices between 2 and 4 (Fig. 23) which are indicative of the medium to weak calc-alkaline character. Two analyses which yielded strikingly high alkali contents were based on highly metamorphosed rocks and are therefore not representative. BERTRAND & VUAGNAT include these lavas in the melange of the Motagua Fault Zone, and together with the El Tambor Formation they still place them in the Paleozoic, but they see in them also the relic of an earlier subduction zone.

The observations made by LAWRENCE and BERTRAND & VUAGNAT very much support WILSON's (1974) view that during the Upper Cretaceous a eugeosynclinal trough ran through northern Central America and underwent a violent orogenesis in the Upper Cretaceous (Fig. 27). According to the plate tectonics model the most likely factors in the formation of such a trough would be rifting of an oceanic fissure, thus elongating the present Cayman Trench, and a collision of the North American and Caribbean plates in the region of the present Motagua Fault Zone. In the process the El Tambor Formation and its analogues were squeezed out and metamorphosed. It also seems reasonable to correlate the emplacement of peridotite-serpentinite massifs in Guatemala with these events.

In **Honduras** petroleum exploration activity has yielded decisive new knowledge about the Mesozoic and in particular the Cretaceous stratigraphic sequence, which was published for the first time by MILLS et al. (1967/69). These authors distinguished within the Cretaceous between a **Cantarranas Formation** as a backreef facies, an **Atima Formation** as a reef facies, a **Guare Formation** as a forereef facies and an **Ilama Formation** as a conglomerate facies. They grouped these formations together as the **Yojoa Group** and classified them in the Albian to Turonian (cf. Table 2). Above the Yojoa Group comes the mainly terrestrial **Valle de Angeles Group** with the intercalated limestones of the **Esquias Formation.**

This classification is reflected in, among other places, the general geological map of Honduras and its accompanying text which were prepared by ELVIER (1974). However, in the course of the following years this map has had to be revised several times, and WILSON (1974), HORNE et al. (1974) and most recently FINCH & CURRAN (1977) have given summary reports on this process. But even their classifications do not coincide completely, as the list in Table 2 shows.

According to FINCH & CURRAN the strata above the Todos Santos Formation should be classified and characterized as follows:

The **Yojoa Group** commences with the **Cantarranas Formation** which consists of calcareous schists with intercalated massive limestones. Its thickness varies between 0 and 190 m. It can be placed in the Neocomian to Lower Aptian and is to be regarded as a backreef facies.

The **Atima Formation** is the most widespread element in the Yojoa Group. It consists of thick-bedded micritic to biomicritic limestones (Fig. 24) with interlayers of silt and bands of flint. The thickness varies between 90 and 1,430 m. According to MILLS et al. (1967) the environment was a "patch reef facies". In contrast to these authors it is to be assumed that the Atima Formation extends only into the Upper Albian, at least in central Honduras.

In the region around Santa Bárbara and Lake Yojoa the 115 m thick layer of shales is intercalated in the limestones; the best exposures of these shales occur at the Mochito Mine and hence they are called "**Mochito-Shale**".

The Ilama Formation proposed by MILLS et al. (1967) must be dropped because the conglomerates and red beds described by those authors are not situated below the Atima Formation but within the Valle de Angeles Group.

On the basis of its fossil content the **Guare Formation** described by MILLS et al. (1967) is classified in the Cenomanian and thus belongs already to the Valle de Angeles Group. It consists of thin-bedded bituminous limestones and is interpreted as pelagic facies.

The **Valle de Angeles Group** is a thick sequence of clastic red beds which is exposed between Tegucigalpa and the mining district of Valle de Angeles and which stands out clearly in the landscape due to its colouration. The intercalated limestones permit the lower part of the group to be classified in the topmost strata of the Upper Cretaceous. In contrast, the upper limit is not biostratigraphically attested and it is indicated only by the discordant overlying layer of volcanics of the (Oligocene?) Matagalpa Formation. The intercalated limestones are the already mentioned **Guare Formation** and a **Jaitique Limestone** which has been newly described by FINCH & CURRAN (1977). The latter is built up of shallow marine deposits about 100 m thick, the lower part of which consists of thick-bedded and the upper part of thin-bedded laminated limestones. These are interpreted as hypersaline lagoonal deposits. Within the red beds the lower strata are coarse-grained and contain quartz conglomerates while the upper sections are more fine-grained and locally gypsiferous. Altogether, the Valle de Angeles Group can be regarded as a type of deposit occurring at a continental margin and containing intercalations of marine shallow-water and evaporitic formations. Facially it is thus located between the marine deposits of the Yojoa Group and the continental Matagalpa Formation.

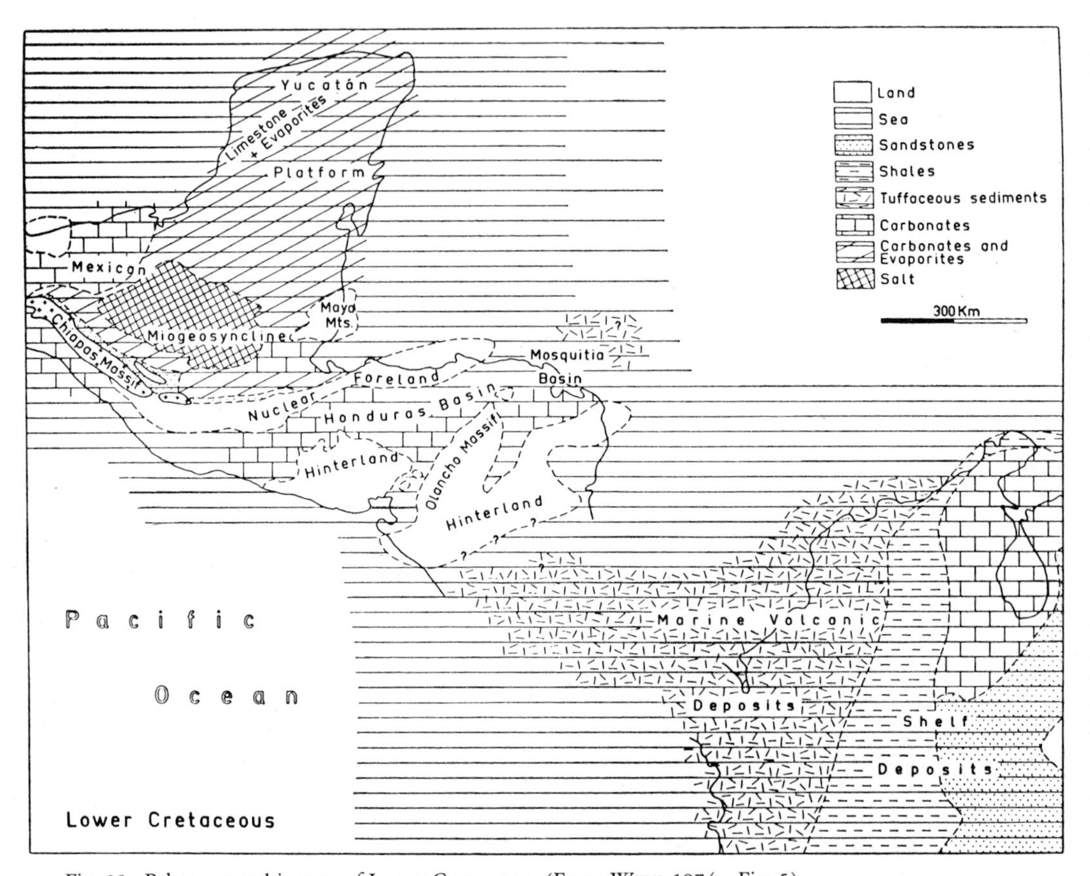

Fig. 25. Paleogeographic map of Lower Cretaceous. (From WEYL 1974a, Fig. 5).

The discontinuous and thin occurrences of Cretaceous limestones and conglomerates in **El Salvador** and **central Nicaragua** still bear the name **Metapán Formation** or Metapán strata which was given them by SAPPER (1899) (see Section 5.51 and Table 12).

There are two very different approaches to assessing and describing the **paleogeography of the Cretaceous:** The approach represented in particular by MILLS et al. (1967), 1969) derives from the present distribution of the Cretaceous a very distinct subdivision of northern Central America into basins and interspersed areas of uplift (Fig. 25), while in contrast to this WILSON (1974) is of the opinion that a more or less complete cover of marine sediments existed which would indicate a uniform deposition area even if the facies varied in time and space (Fig. 27). Since the present state of our knowledge probably does not permit us to decide in favour of either of these approaches, both conceptions will be compared with each other.

The Mesozoic transgression over northern Central America attained its climax in the Middle Cretaceous. According to the works published by VINSON (1962), MILLS et al. (1967, 1969), DENGO & BOHNENBERGER (1969) and VINIEGRA (1971), the following structuring occurred in the process: In the north lay the **Yucatán Platform,** a gradually sinking shelf on which drilling has revealed Cretaceous strata up to almost 3,000 m thick and the Lower Cretaceous is represented by evaporites and tuffitic limestones, and the Middle and Upper Creta-

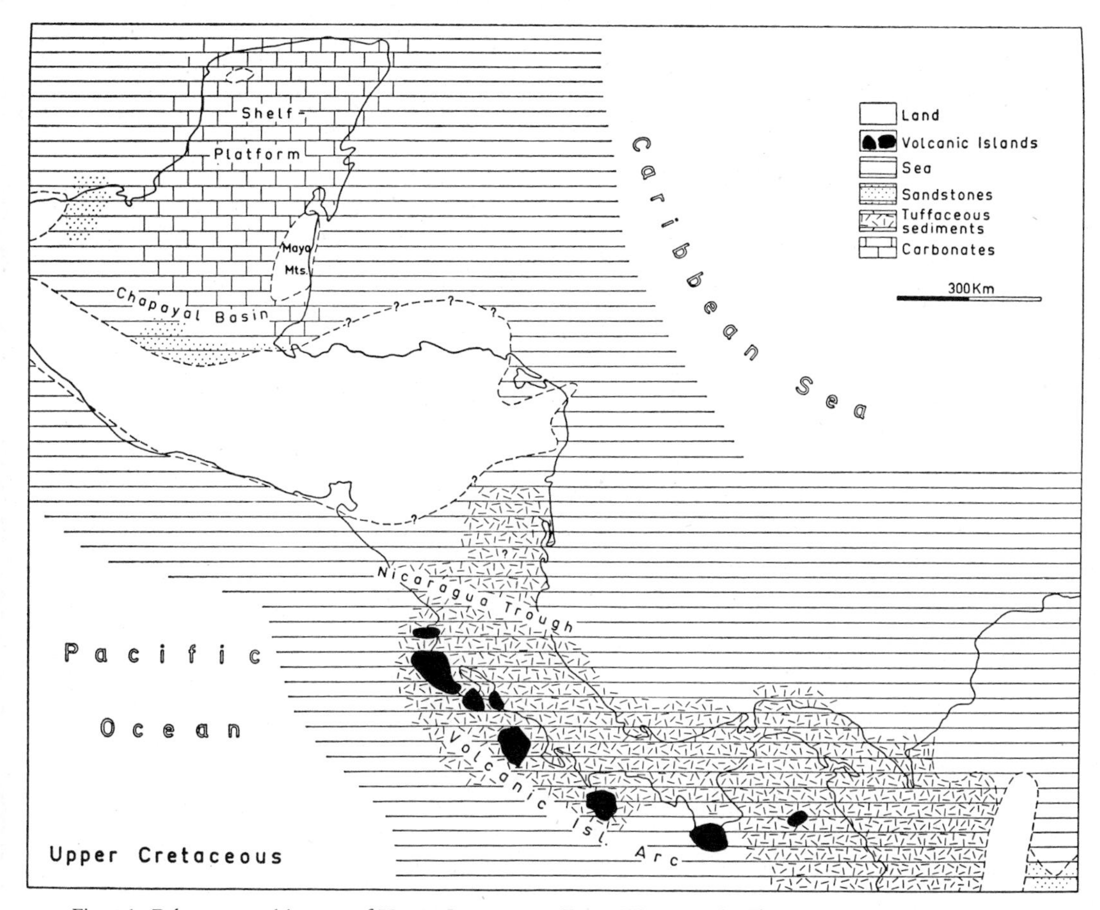

Fig. 26. Paleogeographic map of Upper Cretaceous. (From WEYL 1974a, Fig. 6).

ceous are represented by alternating layers of limestones, dolomites and anhydrites. To the south of this extended the eastern projection of the **Mexican miogeosyncline** with its centre in northern Chiapas and Petén. VINSON (1962) spoke of a Mesozoic-Cenozoic basin, the **Chapayal Basin,** which he subdivided into the Yucatán Platform, the La Libertad Arch, Chapayal Trough and the Amatique Embayment. This basin was formed from the Upper Neocomian to the Turonian in the form of a saline basin extending from Chiapas to Guatemala and Belize and containing over 3,000 m of evaporites (see map in Fig. 25). At the same time, to the south of this basin the low-fossil carbonate sedimentation of the Cobán or Ixcoy Limestones occurred; these limestones extend over large areas of northern Central America and indicate the climax of the transgression. The upper limestone strata of the Campúr Formation, which are rich in foraminifera and rudistans, are, however, limited to the region north of the Polochíc Valley and probably indicate the Upper Cretaceous regression. According to VINIEGRA a barrier extending from the Chiapas Massif towards the east via the present crystalline mountain chains marked the boundary of a southern marine region in which deep water predominated, and limestones, turbidites and volcanics were deposited.

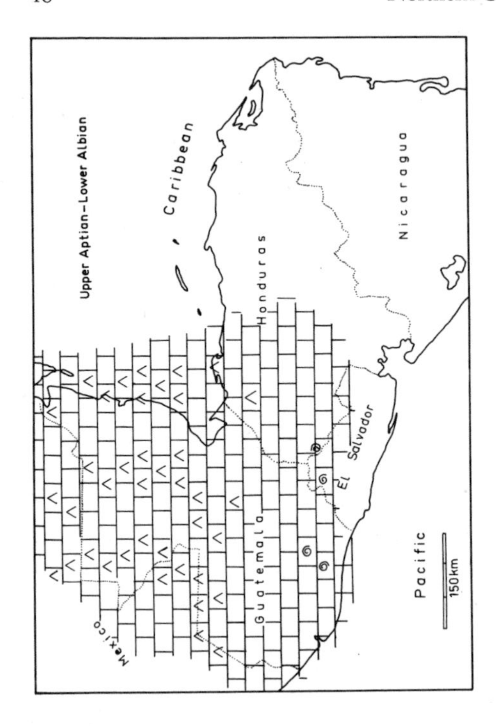
Upper Aptian–Lower Albian
Caribbean
Honduras
Nicaragua
Mexico
Guatemala
El Salvador
Pacific
150km

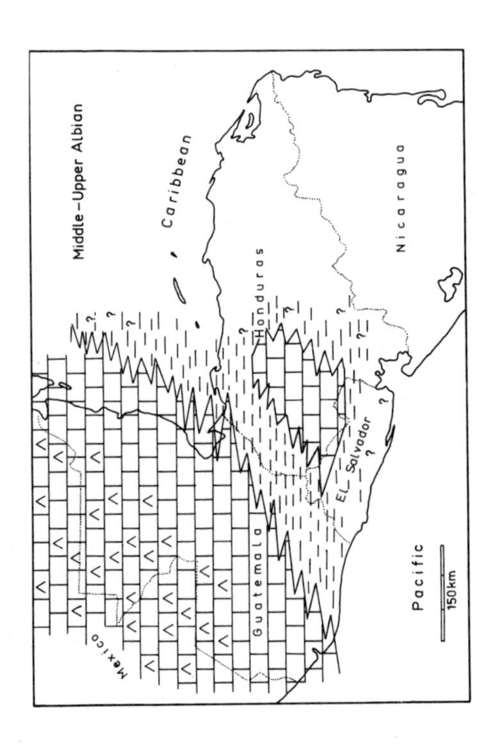
Middle–Upper Albian
Caribbean
Honduras
Nicaragua
Mexico
Guatemala
EL Salvador
Pacific
150km

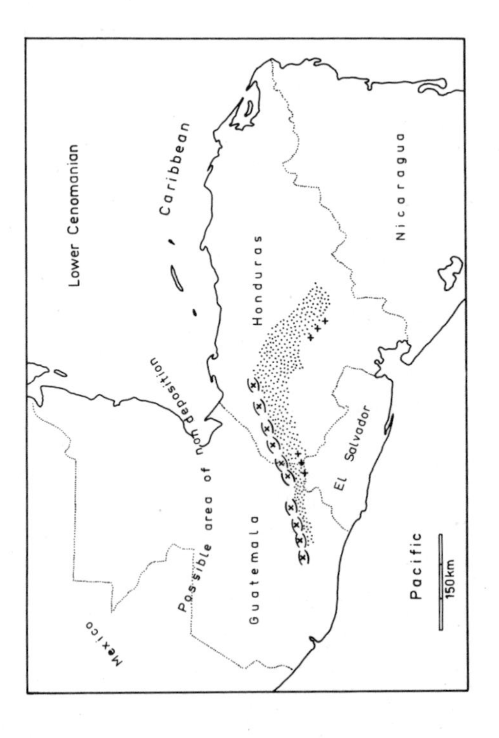
Lower Cenomanian
Caribbean
Honduras
Nicaragua
Possible area of non deposition
Mexico
Guatemala
El Salvador
Pacific
150km

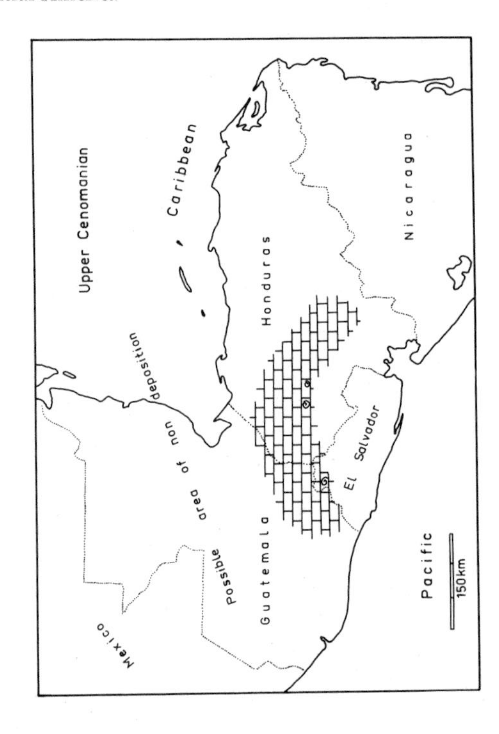
Upper Cenomanian
Caribbean
Honduras
Nicaragua
possible area of non deposition
Mexico
Guatemala
El Salvador
Pacific
150km

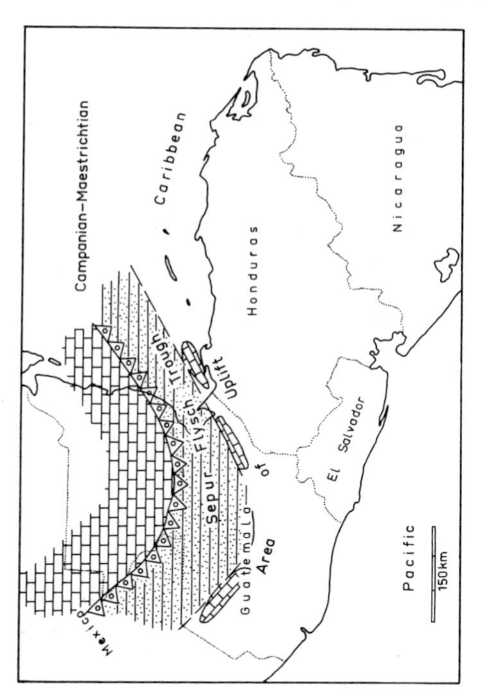

Fig. 27. Facies maps and Upper Mesozoic orogenesis in northern Central America. (Redrawn from WILSON 1974).

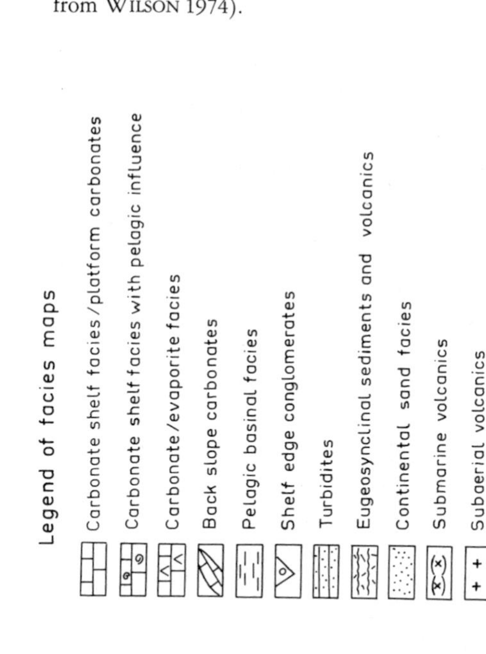

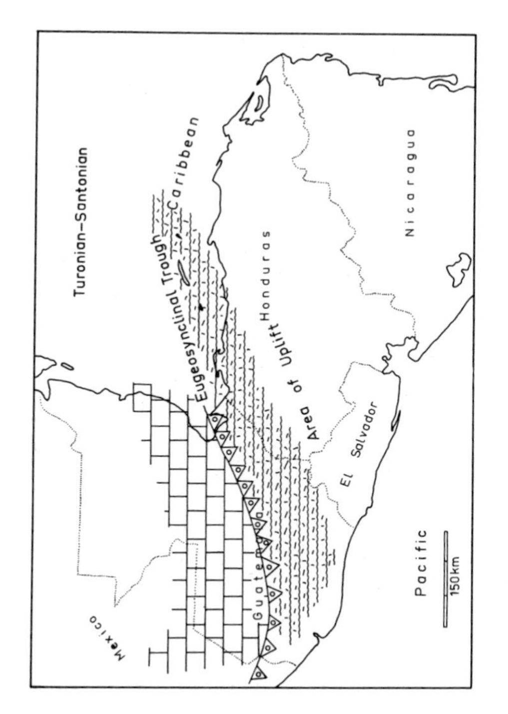

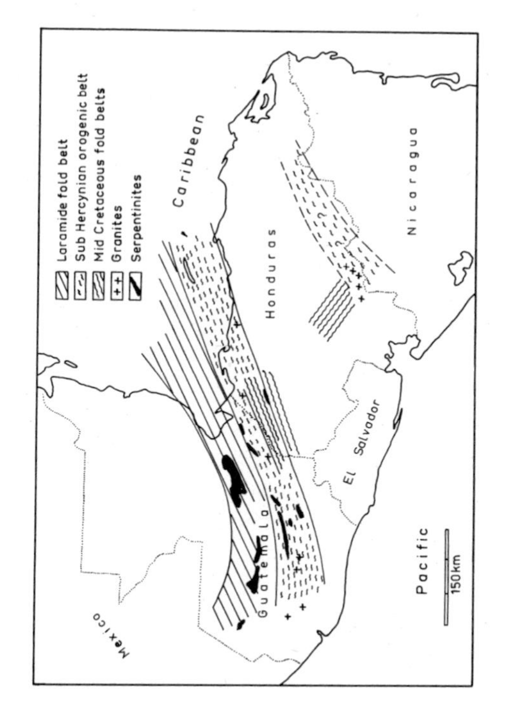

Crustal movements commenced in the uppermost Upper Cretaceous and continued on into the Lower Tertiary, thus resulting in a fundamental change in the distribution of land and sea. In Guatemala and Chiapas a foredeep formed at the northern edge of the emerging central landmass. In this **foredeep** were deposited the conglomerate limestones, calcarenites and shales of the Chemal and Sepur Formations which, according to BONIS (1969), together with their Lower Tertiary component attain thicknesses of over 3,000 m. They are often discordantly superjacent on the Lower Cretaceous material, but in some cases they are also said to emerge continuously from the latter. In northern Petén and on the Yucatán Peninsula the carbonate-evaporitic sequence of sediments extends into the Upper Cretaceous and the Lower Tertiary without any hiatus. This indicates a triple structure of shelf platform, foredeep and uplifted orogen (Fig. 26).

The formation of the **Honduras Basin** began in the Aptian and reached its final form in the Albian with the deposition of the facially highly differentiated Yojoa Group. The Honduras Basin lay to the south of a central highland region which was referred to by VINSON (1962) as the **Nuclear Central American Geoanticline,** and by MILLS et al. (1967) as the **Nuclear Foreland.** The highly subdivided Honduras Basin extended to the south of this ridge as an intra-continental basin in which the separate basins of Jocotán, Ulua and Mosquitia can be distinguished. To the east the Honduras Basin was connected with the Caribbean and at times it was also linked with the Pacific via the region of the Gulf of Fonseca and El Salvador with its Metapán Limestones. In the south the Honduras Basin was bounded by a landmass known as a **hinterland** in the region of Nicaragua; this landmass in turn bordered on the intra-oceanic link in the region of southern Central America (Fig. 25). Intensive crustal movements began in the Upper Cretaceous, the Turonian, and this resulted in folding of the basin deposits into E/W-striking anticlinoria running parallel to the foreland. Together with this went uplift and erosion which resulted in the continental deposition of red beds of the Valle de Angeles Group which extended from the Cretaceous to the Tertiary period. At the same time **volcanic activity** started in the Juancito Mountains of central Honduras and in the northwestern part of the present Mosquitia shelf, and the products of this activity are found in the conglomerates in the Valle de Angeles Group.

According to WILSON (1974) the transgression commenced in the **Upper Aptian** and resulted in deposition of shelf-carbonates from Petén to El Salvador and Honduras (Fig. 27). In contrast to the conception outlined so far, WILSON assumes that there was a uniform marine region without any subdividing land barriers. Towards the north evaporites became intercalated in the limestones. With the start of the **upper Middle Albian** there was major subsidence of a SW/NE-striking trough extending from the region of Guatemala City into the region of Jocotán and the coastal area of Honduras, and a further area of subsidence was formed in the region of Tegucigalpa. Here, ammonite-bearing radiolarian-rich limestones and cherts were deposited up to the Lower Cenomanian, while in the north the carbonate-evaporitic mixed facies persisted.

In the **Lower Cenomanian** lava flows and coarse conglomerates are indications of widespread crustal movements in southern Guatemala and Honduras, which led to interruption of sedimentation or to the deposition of continental sands. There followed in the **Upper Cenomanian** a period of tectonic quiescence and renewed extensive deposition of carbonates. The **Turonian** and **Santonian** are characterized by the formation of a eugeosynclinal trough which extended from southern Guatemala via northern Honduras to the Gulf of Honduras. The heterogeneous rocks of the Japala Melange were deposited in this trough while the volcanics of the **Los Planes Formation** erupted on its southern flank. Folding, granite intrusions and metamorphism are expressions of violent orogenesis in the Lower Campanian which resulted in the trough becoming landlocked, and in the Campanian/Maestrichtian the depression shifted towards the north together with the Sepur flysch trough in which flysch-like sediments with turbidites were deposited, while in northern Guatemala the carbonate shelf remained. In contrast, in the region of Honduras the purely marine sequence was concluded at the end of the Turonian and it was succeeded by the mainly continental red series of the Valle de Angeles Group. According to HORNE, ATWOOD & KING (1974) this sequence was deposited in the form of talus fans on a mature relief.

Summing up, it can be said that from the Aptian to the Campanian a marine region of subsidence and sedimentation extended from Mexico to the eastern part of northern Central America; depending on the interpretation of the facies and the distribution of its deposits, this region is regarded as uniform or as subdivided into separate basins. Its basement was formed by the crystalline rocks of the Lower Paleozoic and by the folded sedimentary series of the Upper Paleozoic, i.e. by rocks of a typically **continental crust.** However, in the region of the Motagua Fault Zone an oceanic basin temporarily opened up, and the El Tambor Formation or Jalapa Melange were formed.

3.15 The Tertiary

According to DENGO & BOHNENBERGER (1969, p. 213) the crustal movements which started in the Upper Cretaceous lasted until the Lower Tertiary and attained their climax in the Early Eocene. The structures which were formed in the process will be discussed in connection with the overall structure of northern Central America. As a result of these crustal movements and the associated uplift the sea was pushed back towards the north. During the Tertiary the sea penetrated only occasionally, and then only in limited areas, into northern Central America which already to a large extent existed in outline, while the Yucatán Peninsula and the states of Tabasco and Chiapas adjoining to the west still remained largely a marine region.

The following discussion of the marine or estuarine Tertiary is largely based on the works of VINSON (1962), BUTTERLIN & BONET (1966), LÓPEZ RAMOS (1969, 1973, 1975), MILLS & HUGH (1974) and FINCH & CURRAN (1977). The continental Tertiary, which was for the most part formed by volcanic activity, will be discussed later in Chapter 5 in connection with the magmatic history of Central America.

During the **Paleocene** Guatemala was a continental region of erosion. MILLS et al. (1967, Fig. 32) had expected to find marine sedimentation of the Esquias Formation ony in the Ulua Basin in northern Honduras. But since this formation has been dated back to the Cenomanian to Turonian by HORNE, ATWOOD & KING (1974) it is probable that this was not a marine deposition area either.

On the other hand, the continental deposition of the Valle de Angeles Group in central Honduras extended far into the Tertiary, and in the Eocene alone reached a thickness of 3,000 m in the Mosquitia region. It was at this time that the formation of the **Mosquitia Basin,** which extended into the Upper Tertiary, commenced. This is one of the largest Tertiary basins in Central America extending as it does over an area of 62,500 square miles and with Tertiary deposits revealed by drilling to be 5,000 m thick in the offshore region; our knowl-

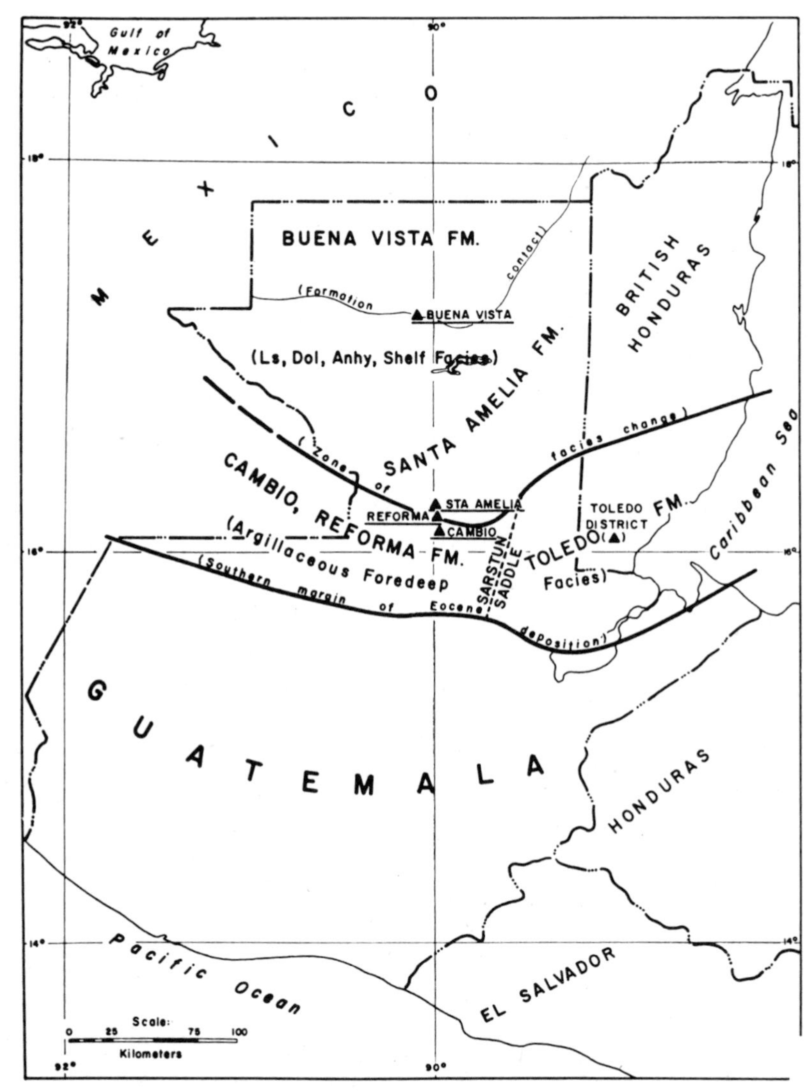

Fig. 28. Areal distribution of formations of the Petén Group, showing dominant facies characteristics, northern Guatemala. (From VINSON 1962, Fig. 11).

edge of this basin is derived mainly from some oil-well drillings which have been evaluated by MILLS & HUGH (1974) (see Fig. 30).

During the Lower **Eocene** the foredeep in the northern part of Guatemala, which had already been the sedimentary basin for the Verapaz Group during the uppermost stages of the Cretaceous, had subsided again so much that it was flooded and became the deposition area for the **Petén Group,** which is several thousand metres thick. This group was subdivided by VINSON (1962, pp. 442 – 448) into a series of facially different formations (Fig. 28). The **Cambio, Reforma** and **Toledo Formations,** which are built up of clays and shales, are depos-

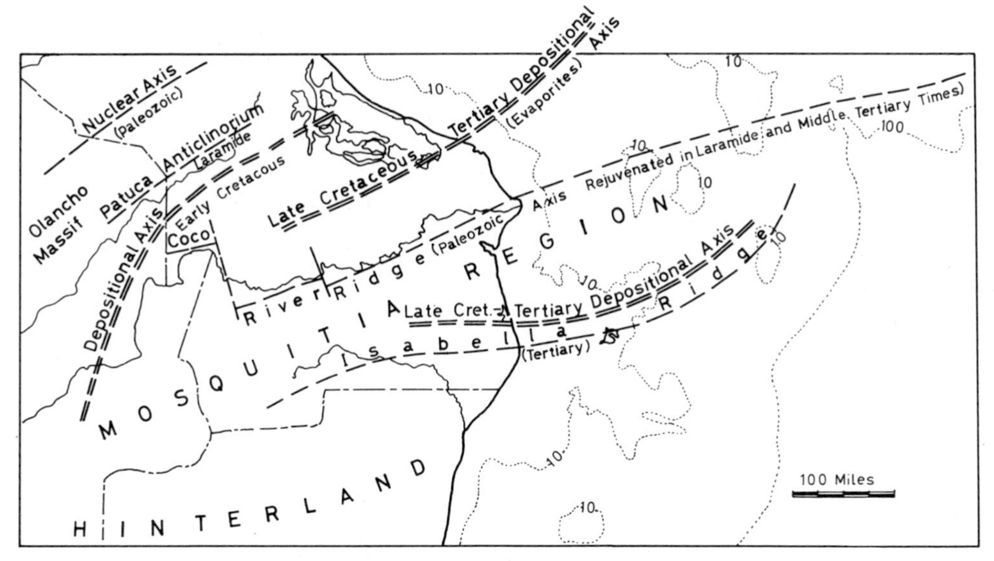

Fig. 29. Structural sketch of the Mosquitia region. (From MILLS & HUGH 1974, Fig. 6).

its of a deeper basin within the foredeep, while the carbonate **Santa Amelia Formation** and the carbonate-sulphate Toledo Formation are shelf sediments on the Yucatán Platform. Uplift and regression occurred from the **Middle Eocene** onwards, affecting also the Yucatán Platform and giving rise here to a stratigraphic gap extending into the Upper Oligocene. Only the area of the Gulf Coastal Plain remained a marine deposition area. During the Upper Oligocene local deposition of several hundred metres of terrestrial conglomerates occurred in western Petén and eastern Chiapas.

In the Mosquitia Basin the Coco River Ridge was uplifted as a central ridge which was probably laid down in the Paleozoic and it divided the basin into a northern Honduran and a southern Nicaraguan Basin which continued to exist up to the Pleistocene (cf. the map in Fig. 29 and the stratigraphic profile in Fig. 30). The **Punta Gorda Formation,** which is built up of approximately 1,800 m of fine-grained sandstones and limestones, was deposited and overlays the 1,500 m thick Valle de Angeles Group on the Mosquitia Banks. During the **Oligocene** the sedimentation and subsidence in the Mosquitia Basin continued in similar facies and together with the Upper Tertiary a thickness in excess of 1,500 m was attained.

During the **Lower Miocene** advances of the sea in the western continuation of the Cayman Trench led to an ingression in the region of the Amatique Bay into which the present Río Dulce flows and where reef limestones up to 1,000 m thick were deposited. Also on the Yucatán Peninsula the sea transgressed for the last time in the Upper Miocene so that large areas of the peninsula were covered by the **Carillo Puerto Formation** which extends into the Pliocene (LÓPEZ RAMOS 1975, Fig. 12 F).

During the **Pliocene** the large grabens which intersect the mountain region of Guatemala filled up with the clastic deposits of the over 2,000 m thick **Armas** and **Herreria Formations.** These deposits are to a large extent continental but they also still contain some marine inclusions.

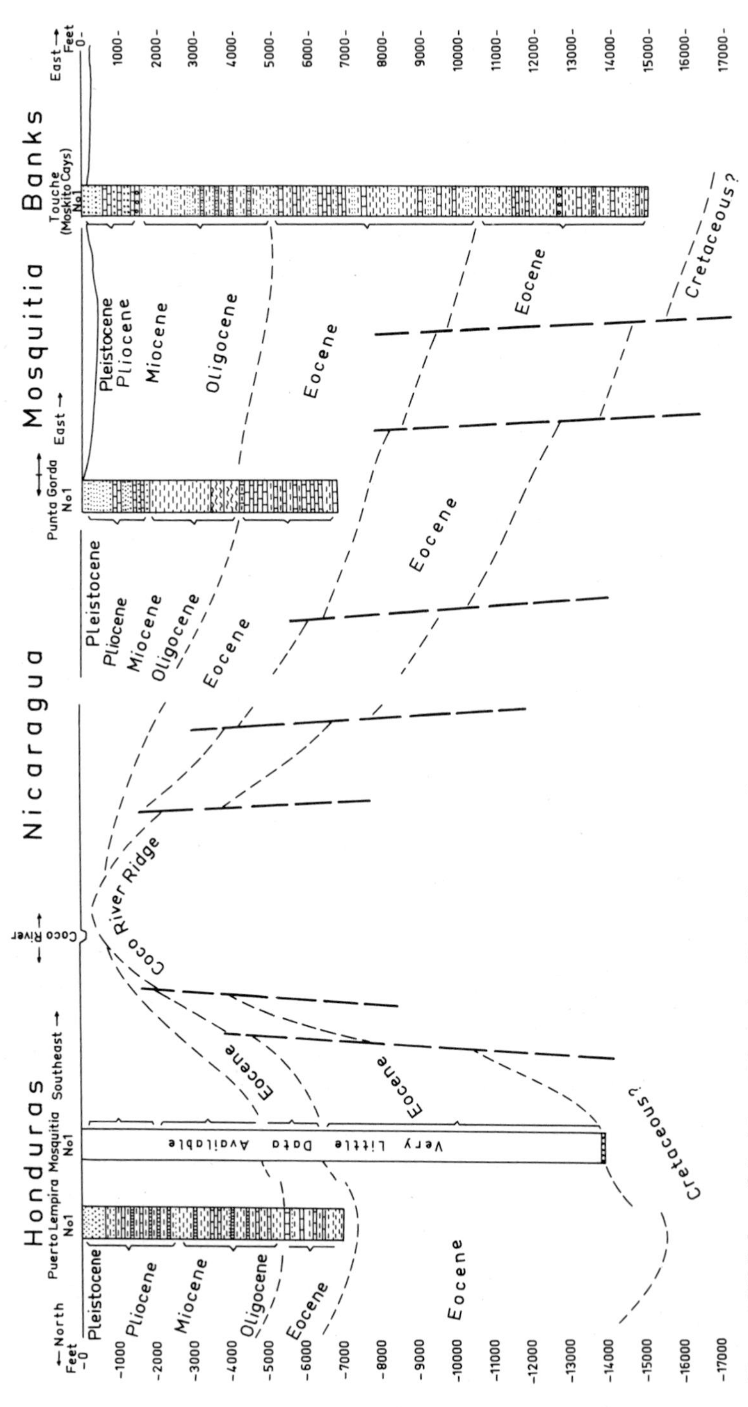

Fig. 30. Well data of the Mosquitia region. Vertical exaggeration 25×. (From MILLS & HUGH 1974, Fig. 7).

The deposition of **clastic red beds** which had commenced in the Upper Cretaceous continued in the continental region of northern Central America. Deposition in separate basins resulting in separate occurrences, lithological similarity of strata of varying age, and the lack of continuous biostratographically identified horizons make it difficult to classify these red beds with any certainty and also result in local variations in their designation (cf. Table 2). There is agreement that they should be placed between the marine Cretaceous and the Tertiary volcanic covers and that they thus represent the erosional debris of the Cretaceous orogen.

In central Guatemala HIRSCHMAN (1963) singled out the **Subinal Formation** which was later classified, with reservations, between the Maestrichtian and the Miocene (BURKHART et al. 1973). In central Honduras CARPENTER (1954) described the red beds as the **Valle de Angeles Formation** and placed them in the Upper Cretaceous to Lower Tertiary. MILLS et al. (1967) raised them to the status of a group, the lower sections of which contained interbedded Upper Cretaceous marine sediments. Later revisions (HORNE, ATWOOD & KING 1974, FINCH & CURRAN 1977) led to the marine intercalations referred to as the Esquias Formation or Jaitique Limestone being placed in the Cenomanian. These authors furthermore divide the Valle de Angeles group into a lower coarse clastic and an upper fine-grained sequence which extends from the upper strata of the Lower Cretaceous to the Middle Tertiary. In northwestern Nicaragua red conglomeratic series were described by DEL GIUDICE (1960, p. 26) as the **Totogalpa Formation.** According to later surveys (Anonymous 1972 b, pp. 56 – 57) this formation interlocks with the upper part of the volcanic Matagalpa Formation and therefore can be placed, with reservations, in the Oligocene.

3.2 The geological structure of northern Central America

The geological features of northern Central America include structures from the Paleozoic crystalline basement, from the sedimentary Upper Paleozoic, from the Mesozoic and finally from the Cenozoic. In broad terms it is possible to assign these structures to certain deformation phases, and this has resulted in the subdivision into two large geotectonic cycles with a series of developmental phases as discussed at the start of this chapter. However, it is still not possible to date many of the structures with certainty, particularly since it must be expected that each region has undergone several deformations.

The structural features of the crystalline basement were described in Chapter 3.1, as far as this is possible on the basis of present knowledge. Similarly, it was possible to give some data on the structure of the sedimentary Upper Paleozoic in connection with the discussion of the stratigraphic sequence during this era. But it is still in many cases uncertain, precisely here, to what extent these structures are of Late Paleozoic or Upper Mesozoic age.

The block folding which was formed in the Upper Cretaceous and Lower Tertiary and the fault structures, which are probably to a large extent of Cenozoic age, are much more important for the structure of northern Central America. They will be discussed in the following sections, and this will sometimes mean that volcanological, seismological and plate tectonics problems will have to be touched uppon, although these will be dealt with in detail in Chapter 8.

3.21 The structure of the Mesozoic orogen

The Upper Mesozoic orogen of northern Central America is the eastern continuation of the Sierra Madre Oriental and of the Cretaceous folded chains of Chiapas. This holds true not only for the history of the deposits, which has been discussed above, but also for the coherency of the structure which exhibits great similarity in particular with Chiapas. In the first German edition of this book (1961) it was to a large extent necessary to extrapolate from this point and to refer to the older observations of SAPPER, but nowadays a wealth of observation data has become available in particular from Guatemala, so that an attempt can be made to elaborate the basic features. The works of DENGO have been of particular assistance in this respect.

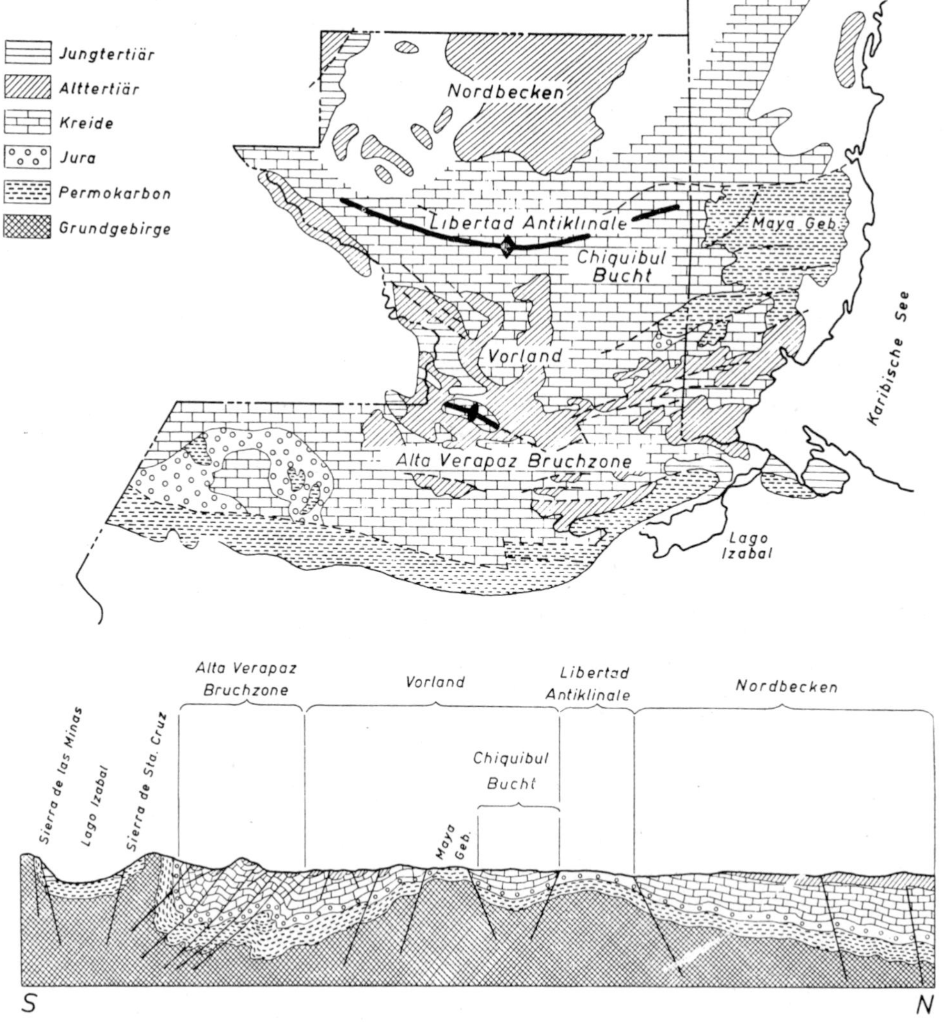

Fig. 31. Geological map and section across the Petén and Verapaz, Guatemala. (Redrawn from LLOYD & DENGO 1960).

In the course of temporal development the climaxes of deformation were the Paleozoic in general, the end of the Upper Paleozoic, the Upper Cretaceous and the Early Tertiary. The more important phases are indicated in the stratigraphic tables. Many, however, probably possess only local significance so that the Middle and Upper Cretaceous and Early Tertiary deformations are the more important with regard to recent developmental history. They were followed in the Cenozoic by the faulting which played such a decisive role in shaping the landscape, in conjunction with the Cenozoic magmatism. WILSON (1974) distinguishes fold systems which are assigned to these deformation phases in the geological structure of northern Central America (Fig. 27), but it would appear that, compared with regions such as Central Europe which have long been the subject of study, positive delineation of regions associated with different deformation periods is still in the early stages.

SAPPER (1937, pp. 102 – 103), had already described the structure of the orogen in Guatemala as a fold system of widely varying folding intensity in whose core metamorphic crystalline rock and Paleozoic folds occur. The longitudinal faults which run throughout the entire length of the range are particularly important in this connection. A large number of more recent works provide a more accurate picture of individual regions which are discussed in Chapter 3.3.

According to LLOYD & DENGO (1960) and DENGO (1968/73 and 1975 a) the major structural elements in a cross-section from the lowland of Petén to the Pacific coast of El Salvador can be identified as follows (Figs. 31 and 32):

1. **In the northern foreland** which is called the "Nordbecken" in the map in Fig. 31 and which extends from southern Petén to the Yucatán Peninsula, the Mesozoic-Upper Tertiary strata are flat-bedded and intersected only by faults. Towards the south gentle folding commences with salt and sulphates thickened into pillows in the cores of the folds (MURRAY 1966).

2. **The Libertad Arch** (Libertad Antiklinale in Fig. 31) is an E/W-striking anticlinal structure extending through the Mesozoic basin in an arc open to the north. It is claimed that this Arch was formed as early as in the Paleozoic and that in the Mesozoic it was a region of reduced sedimentation. Along the geographical extension of the Arch the Maya Mountains rise up where faults occur. The Libertad Arch separates the foreland from the

3. **Fold belt of the Alta Verapaz** (Alta Verapaz Bruchzone in Fig. 31). Here the fold structure is tight, the folds are intersected by a large number of southward-dipping imbricate structures, and the rigid limestones and dolomites have undergone decollement on the more plastic evaporites of the Lower Cretaceous. In detail, a large number of special maps and profiles from this region exhibit a widely varying fracture-fault structure. The style of the folding depends to a large extent on the reactive capacity of the rocks. The massive limestones of the Cobán Formation exhibit a predominantly fault structure with weak folding; the incompetent claystones and sandstones of the Todos Santos Formation on the other hand are severely folded. The most highly stressed rocks are the evaporites which in many cases have been forced out of their original position (PAULSEN 1969) (Fig. 33). It seems logical to make a comparison with similarly structured fold systems. TRUSHEIM (1976) has done this for the Sierra Madre Oriental of Mexico. In my opinion this comparison could also be extended to take in Chiapas and northern Central America.

Systems of various orientation and various age which intersect with each other to form a lattice pattern, can be detected in the faults. The **Chixoy Polochic Fault Zone** and the **fault zones of the Río Motagua** further to the south as well as the **Chamelecón Jocotán Fault**

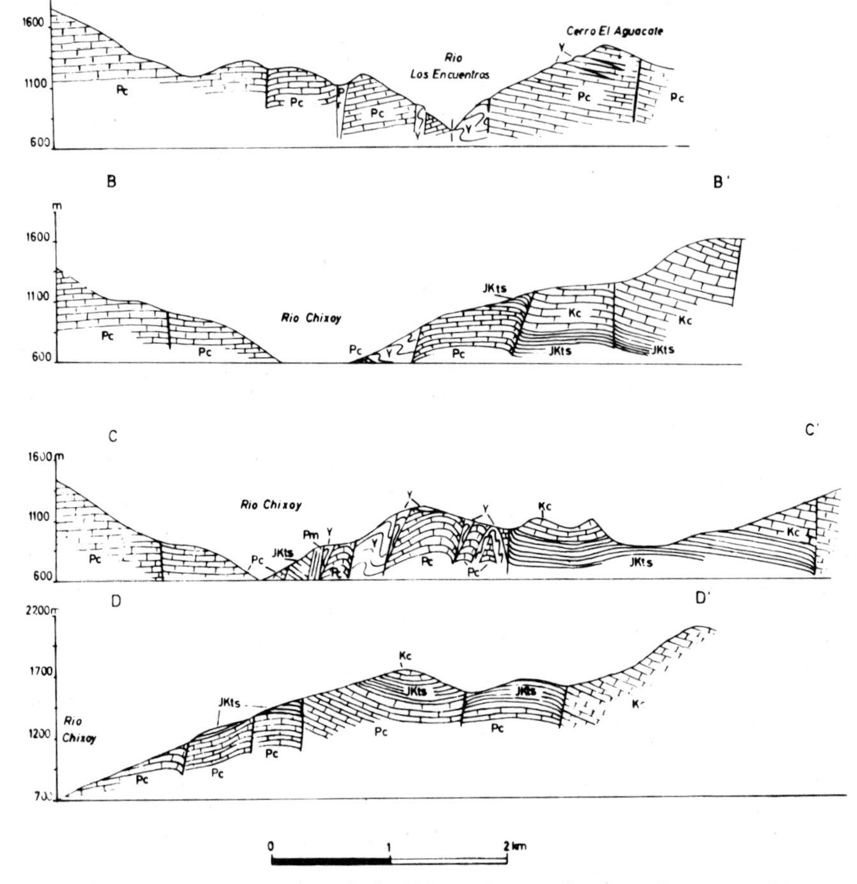

Fig. 33. Cross-sections through the area of Río Chixoy, Guatemala. (From PAULSEN 1969).
Kc Cretaceous limestone and dolomite
JKts Jurassic-Cretaceous red beds
Pc Permian limestone
Pm Metamorphic rocks
Y Gypsum

Zone are of regional importance. Not only do they separate different stockworks of the orogen from each other, but, as extremely extensive lineaments and horizontal shifts of great age, they form the boundary between the North American and Caribbean plates.

The internal structures are reflected in the morphological pattern of the mountain chains of southern Petén and Verapaz, which describe an arc open to the north. These structures stand out extremely well in aerial photographs and in a large-scale relief map of Guatemala. The course of the large fracture zones can also be clearly detected (Fig. 44).

4. DENGO (1973, p. 17) treats the mountain chain of the Sierra de Chuacús, Sierra de las Minas and Sierra Espiritu Santo, which is built up of crystalline rock and sedimentary Paleozoic, separately as an **intermediary massif** or rigid central zone ("zona central rigida"). The

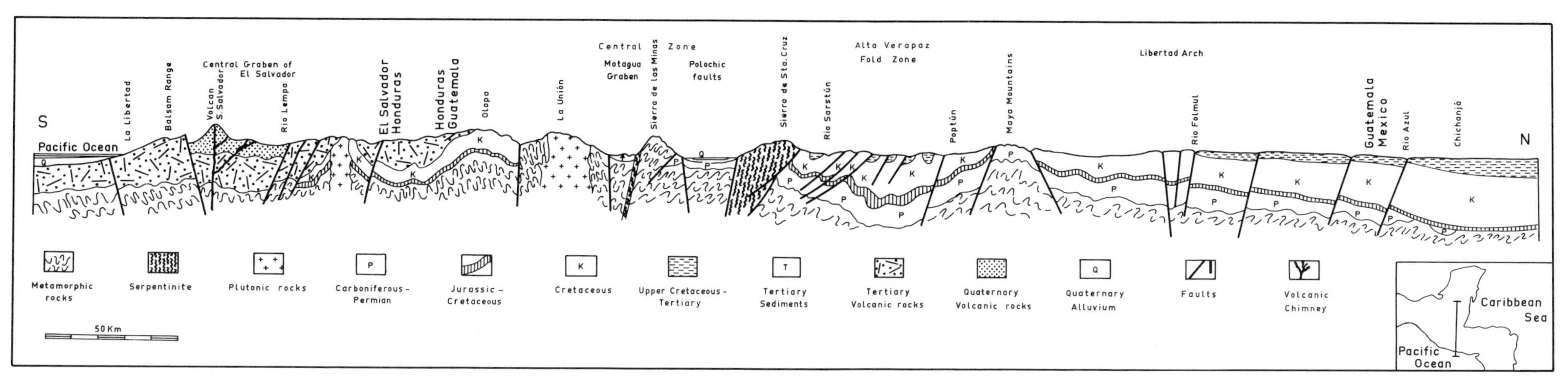

Fig. 32. Idealized cross-section through northern Central America. (Redrawn from DENGO 1968/73, Fig. 7).

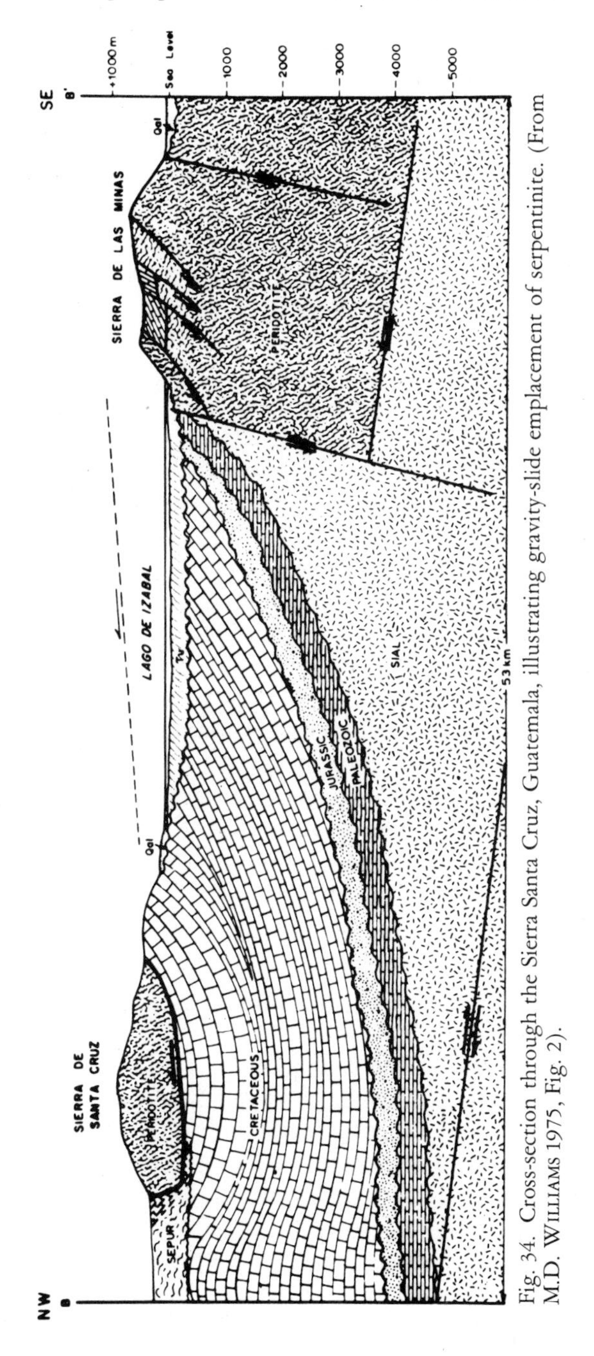

Fig. 34. Cross-section through the Sierra Santa Cruz, Guatemala, illustrating gravity-slide emplacement of serpentinite. (From M.D. WILLIAMS 1975, Fig. 2).

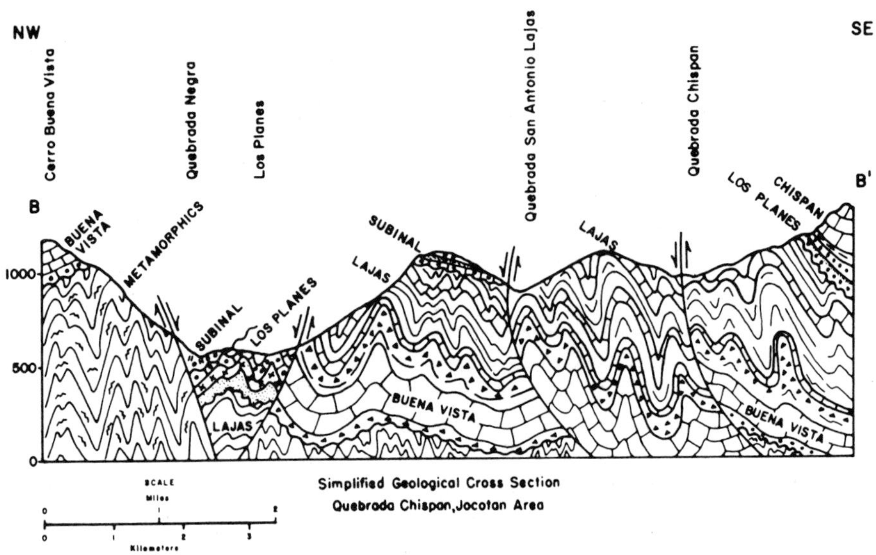

Fig. 35. Geological cross-section Cerro Buena Vista – Quebrada Chispan, Jocotán area, Guatemala. (From WILSON 1974, Fig. 14).

structure of the metamorphic crystalline rock was discussed in connection with the basement rock. The Upper Paleozoic strata are vigorously folded and intersected by a large number of faults, above all along the Chixoy Polochic Fault Zone (cf. the profiles in Figs. 32 and 33). A special role is played in this zone by the large masses of serpentinite running with the strike of the range which were forced up during the Cretaceous orogenesis (BONIS 1968) and which according to MCBIRNEY (1963) represent mantle material that was mobilized along the major faults — i.e. along what is now regarded as the boundary of the plate. According to the conception developed by WILSON (1974) and M. D. WILLIAMS (1975) the ultrabasites of Sierra Santa Cruz are supposed to have slid down towards the north from the more strongly uplifted Sierra de las Minas and are now lying as an allochthonous cover on the Upper Cretaceous of Sierra Santa Cruz (see profile in Fig. 34).

5. The **hinterland,** which is largely covered by Tertiary and Quaternary volcanics, extends to the south of the Chamelecón Jocotán Fault. According to DENGO (1973, p. 17) the Cretaceous sediments are less intensively folded here than in the fold zone in the north, but a profile recorded by WILSON (1974, Fig. 14) in the southeastern part of Guatemala reveals very tight folds and an imbricate structure (Fig. 35). The hinterland is the region where most Cretaceous-Tertiary granitic plutons are found.

In the eastern part of northern Central America, i.e. **Honduras, Nicaragua** and **El Salvador,** the Mesozoic is not so consistently distributed as in Guatemala. The occurrences take the form either of large or small erosional outliers between regions of methamorphic cystalline rock, or they exist as inliers in the cover of Cenozoic volcanics (cf. maps in Figs. 43 and 58). Where the thick Atima Limestones crop out, the mountain relief is steep and exhibits the forms of tropical kegelkarst, as for example in the region of the Mochito mine to the west of Lago de Yoja.

According to the works of MILLS et al. (1967, 1969) the Mesozoic strata in Honduras form a series of E/W- or NE/SW-striking anticlines which are subdivided by folds, shingle blocks and faults. The profiles published by them are compiled in the table shown in Fig. 36, and they reveal the pattern of block faulting. In the Mosquitia region the structure is dominated by broad anticlinoria and synclinoria together with a large number of steep wrench faults. The temporal development of the structure according to MILLS & HUGH (1974) is given in Fig. 37.

It is not yet possible to subdivide this part of the orogen in the same way as in Guatemala. It extends to the northeast in the Nicaragua Rise, thus forming a link with Jamaica. According to ARDEN (1975) its formation commenced in the Jurassic. However, Jamaica, as the next element in the orogen of the Greater Antilles, has undergone a much different development from that of the Central American continent.

3.22 Fault structures in northern Central America

Our knowledge of the fault structures of Central America is at present undergoing a change brought about by the work which has now commenced on special large-scale mapping and by the evaluation of aerial photographs and radar imagery, and it is a process which will continue for some time to come. This change is reflected in the general geological maps of the in-

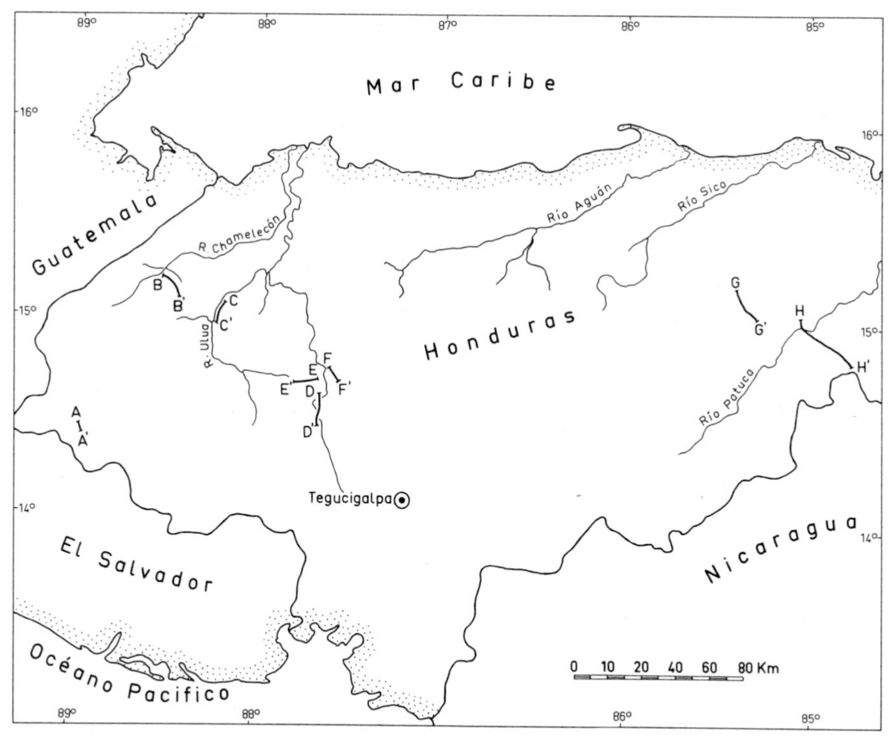

Fig. 36. Structural cross-sections through the Mesozoic of Honduras. (Redrawn and arranged from MILLS et al. 1967). (See following pages).

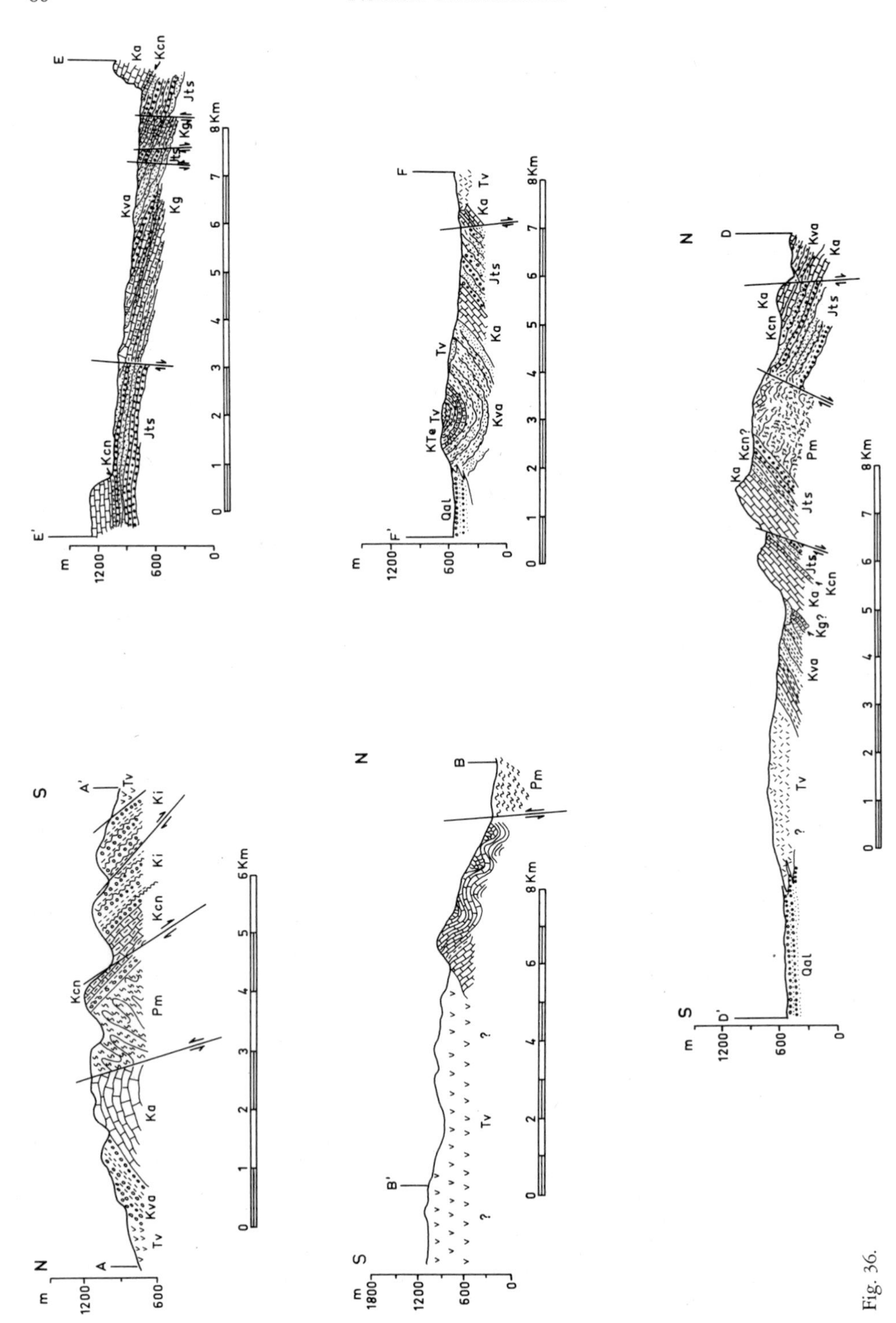
E
E'
Ka
Kcn
Jts
Kg
Kva
8 Km
m
1200
600
0
F
F'
Tv
KTe
Qal
N
S
A
A'
Ki
Pm
6 Km
B
B'
D
D'
Kcn?
Kg?

Fig. 36.

Fig. 36.

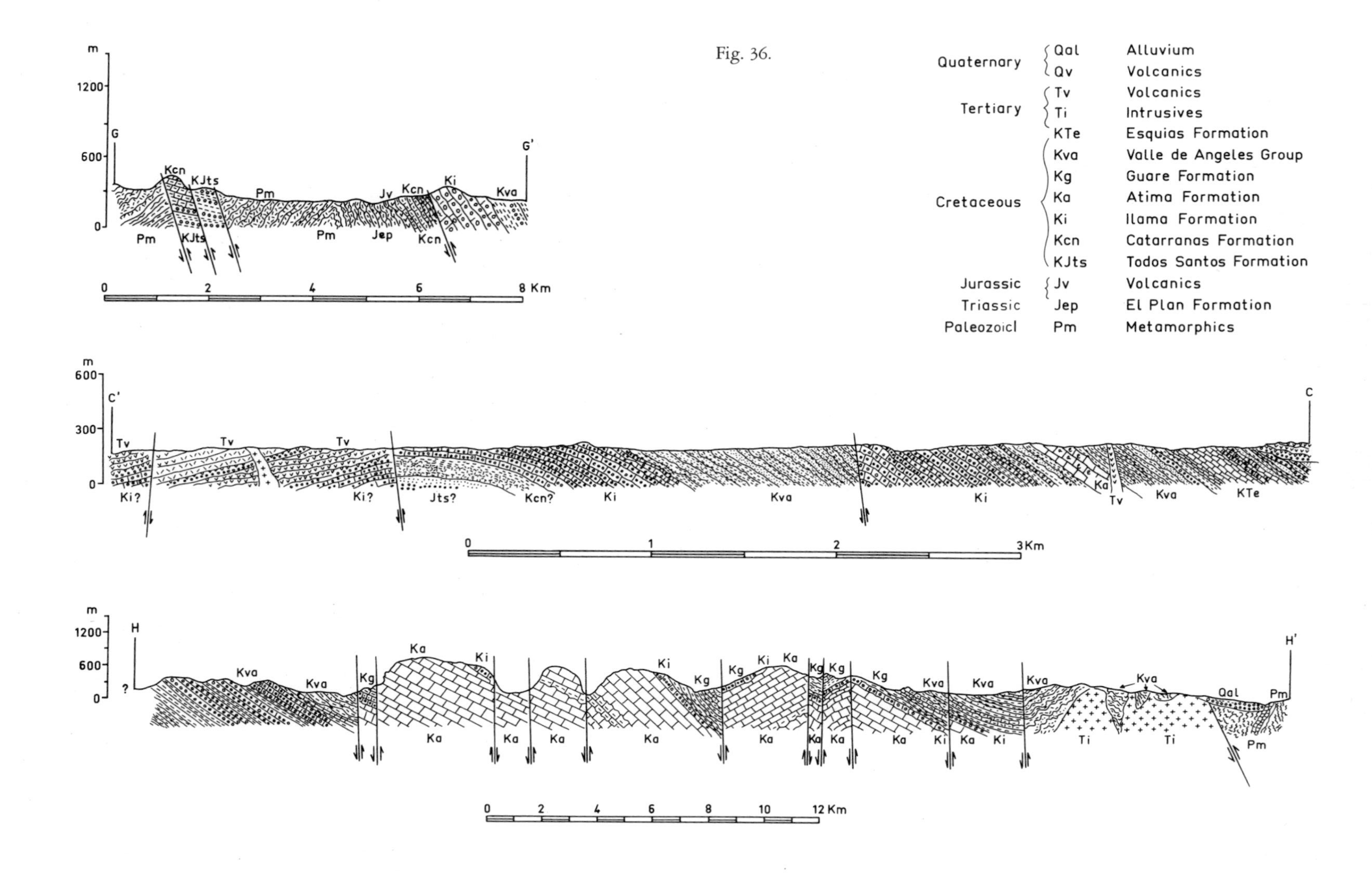
Quaternary
Qal Alluvium
Qv Volcanics
Tertiary
Tv Volcanics
Ti Intrusives
KTe Esquias Formation
Kva Valle de Angeles Group
Kg Guare Formation
Cretaceous
Ka Atima Formation
Ki Ilama Formation
Kcn Catarranas Formation
KJts Todos Santos Formation
Jurassic
Jv Volcanics
Triassic
Jep El Plan Formation
Paleozoicl
Pm Metamorphics
G
G'
C'
C
H
H'
m
1200
600
300
0
0 2 4 6 8 Km
0 1 2 3 Km
0 2 4 6 8 10 12 Km

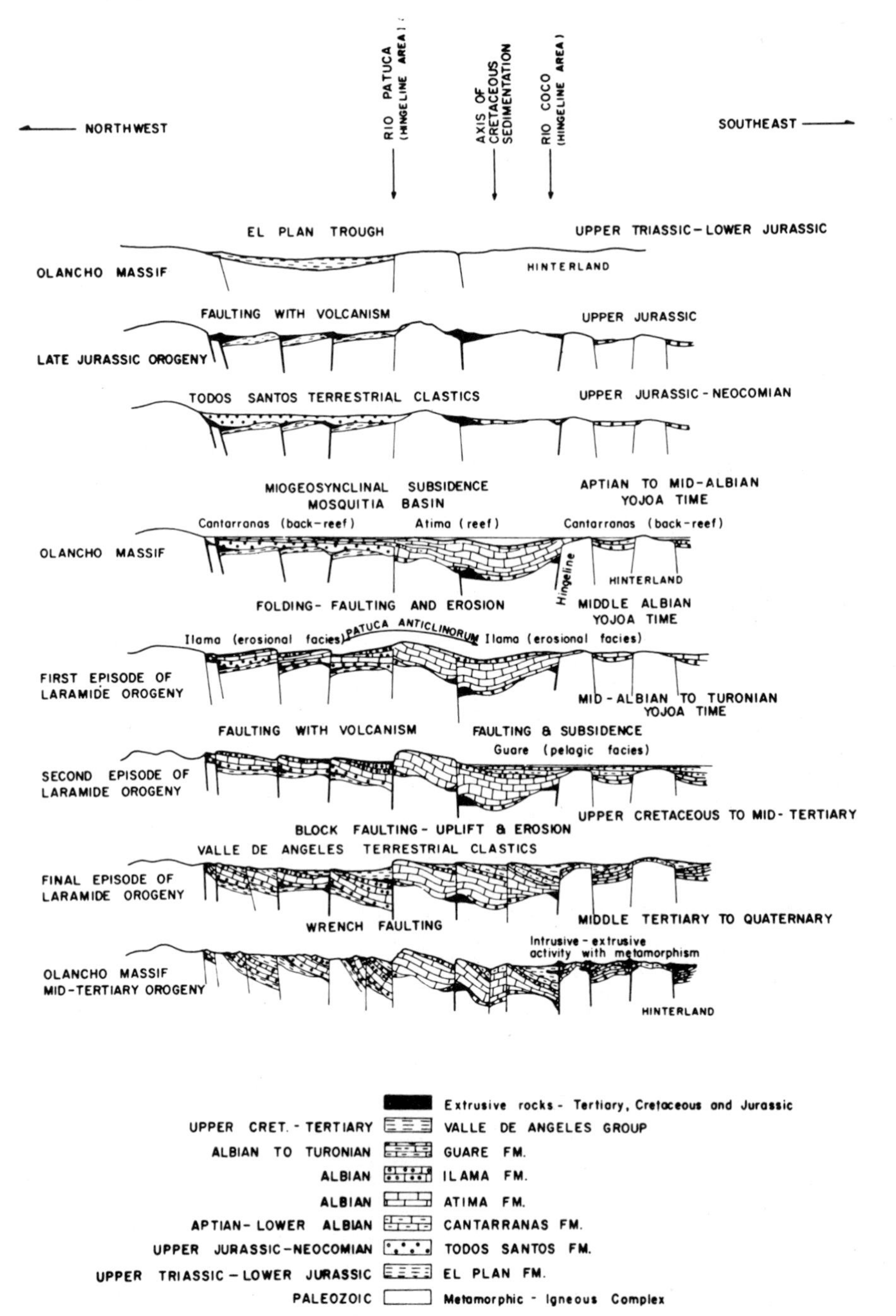

Fig. 37. Cross-sections and evolutionary sketch of the west end of the Mosquitia Basin. (From MILLS & HUGH 1974, Fig. 8).

dividual countries on which our maps of the fault structures (Figs. 38 and 39) are based. The areas in which special studies have been conducted stand out clearly in these maps from those areas in which only very large-scale surveys have been made. We only have to compare Nicaragua or El Salvador with broad regions of Honduras to see that. The differences reflect the present state of our knowledge and not, say, a different structure.

The older maps, some of which still go back to SAPPER (1937) and TERRY (1956) and traces of which can still be found, for example, in the metallogenetic map of Central America (DENGO et al. 1969), show long drawn-out fault lines, which in many cases only very approximately match up with the morphological pattern that has been identified. Let me cite as examples the Nicaragua Graben, the escarpments of the Cordillera Costeña of Costa Rica, and faults in the northern part of Guatemala and Honduras. This conception of the structure was and still is to some extent even today valuable in working out regional analyses of the kind published by DENGO (1968/1973) or DENGO, BOHNENBERGER & BONIS (1970). The change in our knowledge is evident particularly also in this last work where specialized maps of Guatemala are contrasted with the general maps; the special maps, which were compiled on the basis of detailed studies, show instead of long, continuous faults a large number of smaller fault lines which run in different directions.

Any interpretation of the fault structures will have to distinguish between macro-structures or first-order structures and smaller, second-order structures; this is the principle which FERRARI & VIRAMONTE (1973) used in an analysis of Nicaragua and on which WIESEMANN (1975) also based a structural analysis of El Salvador.

I should like here to list the following **first-order structures** in **northern Central America:**

1. The longitudinal fault lines running parallel to the margin of the Pacific, with which the volcanic chain and the zone of frequent and violent shallow-focus earthquakes of Central America are connected.
2. NNE- to NE-striking transverse faults which intersect the circumpacific system and which run parallel to the segment boundaries of the Cocos Plate which were discovered by STOIBER & CARR (1974).
3. N/S-striking fault and graben structures.
4. The large faults in northern Honduras and central Guatemala which are grouped together as the Motagua Fault Zone and which are nowadays regarded as the boundary zone between the North American and Caribbean plates.

The **Pacific Marginal Fault Zone** is most impressively developed in the **Nicaragua Depression,** which runs in an almost straight line from NW to SE from the Gulf of Fonseca in the northwest to the Atlantic Coastal Plain of Costa Rica. However, once this region had been properly charted, the fault lines shown as extending over hundreds of kilometres in the older maps had to be replaced by a large number of individual faults which mark the boundaries of the depression. Details on this subject are given in the review of the geological structure of Nicaragua in Chapter 3.34.

To the southeast the depression continues on to Costa Rica, and its boundaries are only very approximately known. The volcanoes of the Cordillera de Guanacaste are located, offset by about 40 km, on the southwestern margin, while smaller volcanoes are to be found inside the depression and close to the northern boundary.

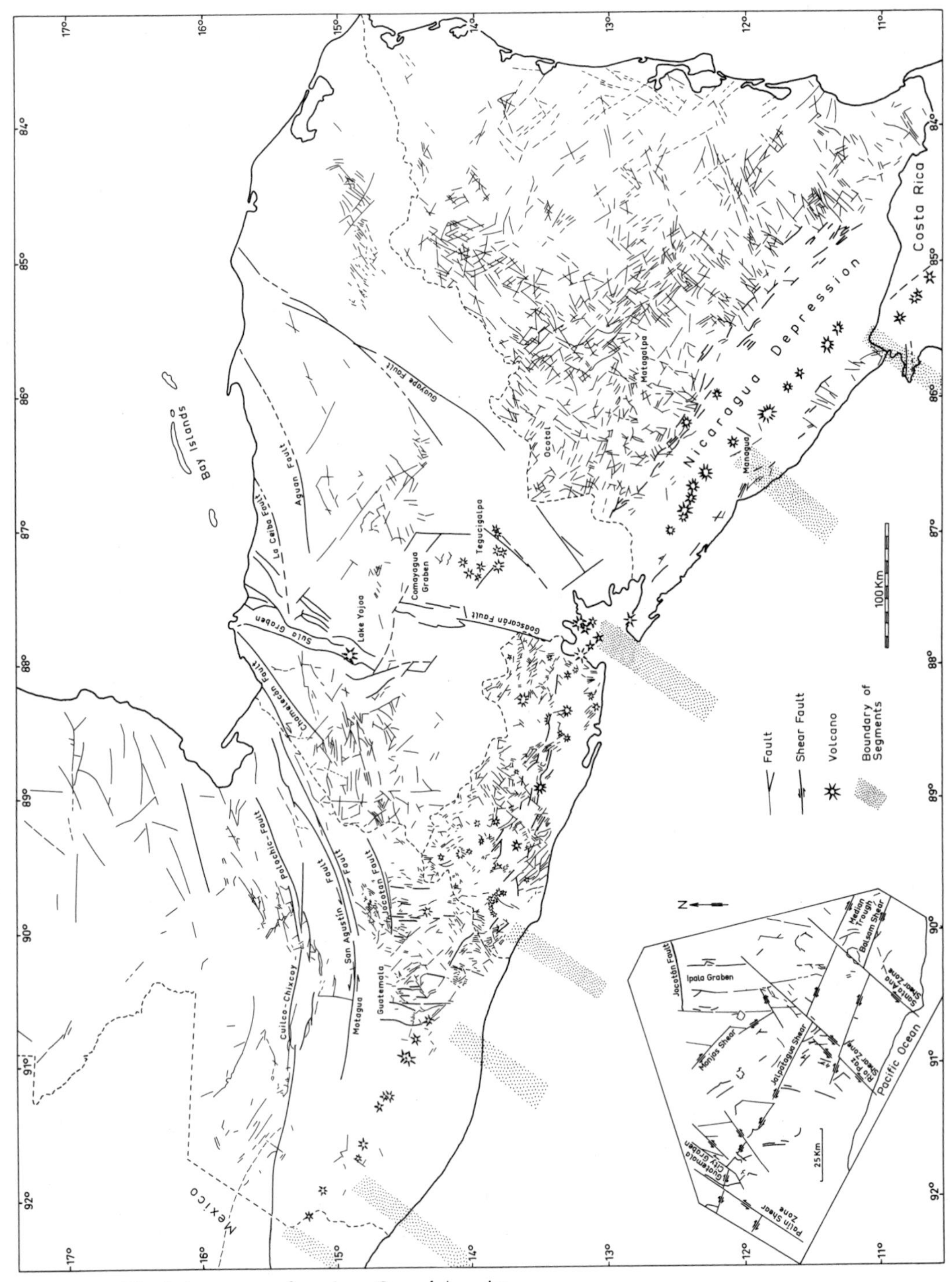

Fig. 38. The fault patterns of northern Central America.

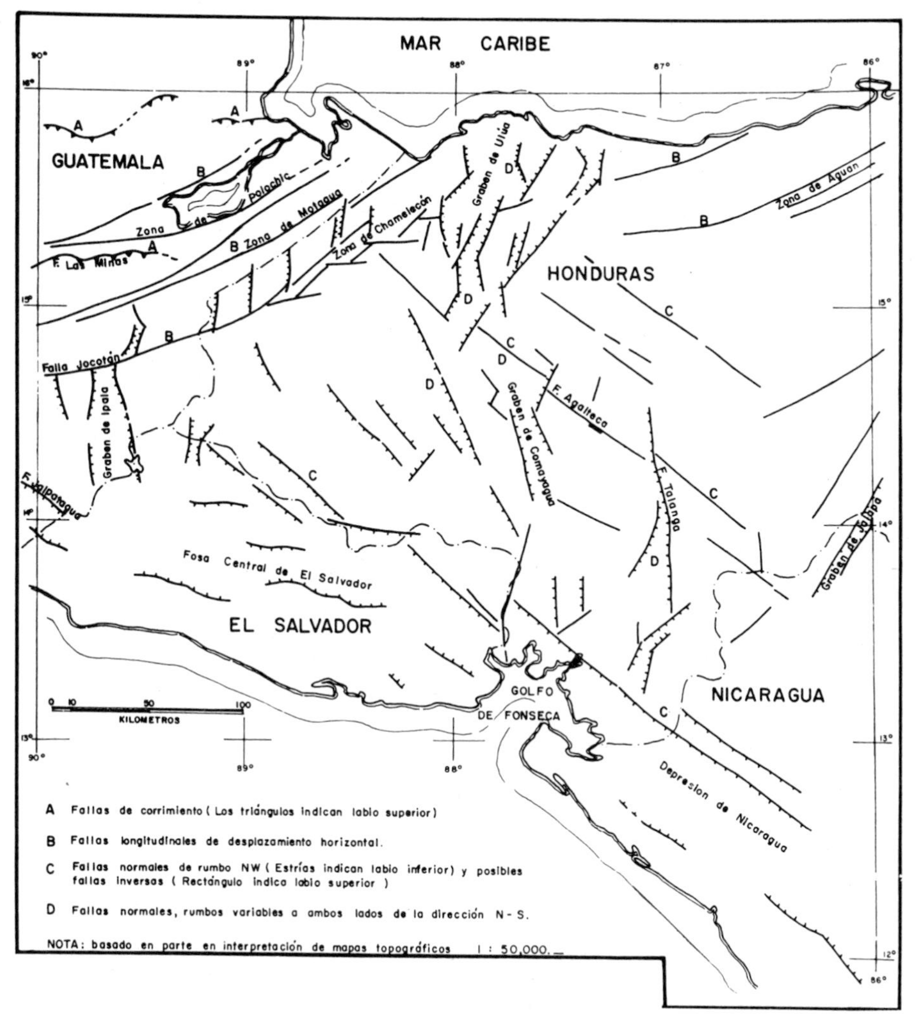

Fig. 39. The fault patterns of northern Central America. (From DENGO 1968/73, Fig. 9).

The northwestern continuation of the Nicaragua Depression into **El Salvador** is interpreted in different ways. From the time of WILLIAMS & MEYER-ABICH (1955, p. 5) up to DENGO, BOHNENBERGER & BONIS (1970) it was regarded as a WNW/ENE-striking graben structure which had formed as a result of the collapse of the crest of a gentle uparching (see profile sequence in MEYER 1967 b, Fig. 12). The large active volcanoes are located on the southern flank of the graben and they frequently mask the structure of the graben beneath their eruptive material. Older, extinct volcanoes are located either within the graben or on the northern flank. In addition, fault systems have been detected running transverse and diagonal to the graben, and some of them also carry secondary series of volcanoes.

This conception had to be revised following the geological survey of El Salvador, which was carried out by the Bundesanstalt für Bodenforschung. Although WIESEMANN (1975) also regards a WNW-striking fault system as one of the main structural elements in El Salvador, he regards the echeloned, SSE-NNW-oriented Plio-Pleistocene basins of Olomega, Titihuapa, Río Lempa and Metapán as the continuation of the Nicaragua Depression. The Lempa Basin was described by K. RODE (1975) who recognized that its boundaries were marked by NW- and NE-striking faults (see map Fig. 152).

A major fault system, in fact the "Salvador Graben" referred to by the authors mentioned so far, which runs WNW at an angle to the strike of the Nicaragua Depression, can be traced from the northern edge of the Olomega Basin to the Guatemalan border, even though it does not always remain a proper graben. To the east of the Río Lempa the downfaulted block is located on the southern side, while to the west as far as the Balsam Range it is located on the northern side. Further to the west a true graben, which WIESEMANN designated the "Salvador-Graben sensu stricto", runs as far as the Guatemalan border. Leaving aside the large volcanoes, this fault zone is the one in which the most frequent and most violent shallow-focus earthquakes of El-Salvador occur (SCHULZ 1965). Because of the large number of transverse and diagonal faults this zone does not stand out as clearly in the structural map (Fig. 151) as it does in the landscape, which exhibits clear and sharply outlined fault scarps.

In **Guatemala** the Pacific Marginal Fault Zone disappears under the deposits of the large volcanoes. However, the very striking linear configuration of the volcanic chain and the sharp drop in the highland region of Guatemala of sometimes more than 2,000 m down to the Pacific Coastal Plain must be regarded as evidence of the continuation of the fault system.

The **NNE- to NE-trending transverse fault system** clearly shows up in the map of Nicaragua and El Salvador together with many fracture lines. The earthquakes that struck Managua in 1931 and 1972 occurred along faults having this orientation (see Chapter 6). In El Salvador WIESEMANN (1975) regards this system as merely of secondary importance while CARR (1974, 1976, Figs. 4 and 5) considers it is an important zone of left-lateral strike-slip faulting, such as for example the „Santa Ana Shear Zone" and the "Río Paz Shear Zone" in western El Salvador to which DÜRR (1960, p. 19) and MEYER (1967 b, p. 209) had already drawn attention. According to STOIBER & CARR (1974) and CARR (1976) they run in the direction of the break in the volcanic chain and of the segments in the subduction zone (Fig. 38, inset map).

According to CARR the **N/S-striking fault and graben systems** are normal faults which in many cases determine the position of the volcanic eruption centres running at right angles to the main direction of the chain. Examples of this are the volcanoes Atitlán-Tolimán and Fuego-Acatenango. The **Guatemala City Graben,** whose peripheral faults were partly activated during the Guatemalan earthquake of 1976 (cf. Chapter 6, Figs. 162 and 163), strikes north-south and is filled with thick layers of pumice. The basin of Lago Amatitlán ingresses and the Pacaya group of volcanoes rises up at the southern end of the graben. In eastern Guatemala the large **Ipala Graben** also trends north-south accompanied by countless parallel faults and chains of volcanoes which were recorded by H. WILLIAMS et al. (1964, Fig. 2) and also discussed by DENGO et al. (1970, Fig. 3) (Fig. 40).

The largest of the transverse N/S structures is the **Honduras Depression.** MILLS et al. (1967, p. 1718) regarded this as a long series of north-south trending horsts and grabens extending from the Maya Mountains to the Gulf of Fonseca. DENGO (1968/73, Fig. 9) and WILLIAMS & MCBIRNEY (1969) stressed that this depression consists of a series of echeloned depressions which require special study before any detailed remarks can be made about the structure. Such studies were undertaken by MUEHLBERGER and his pupils, and their work has

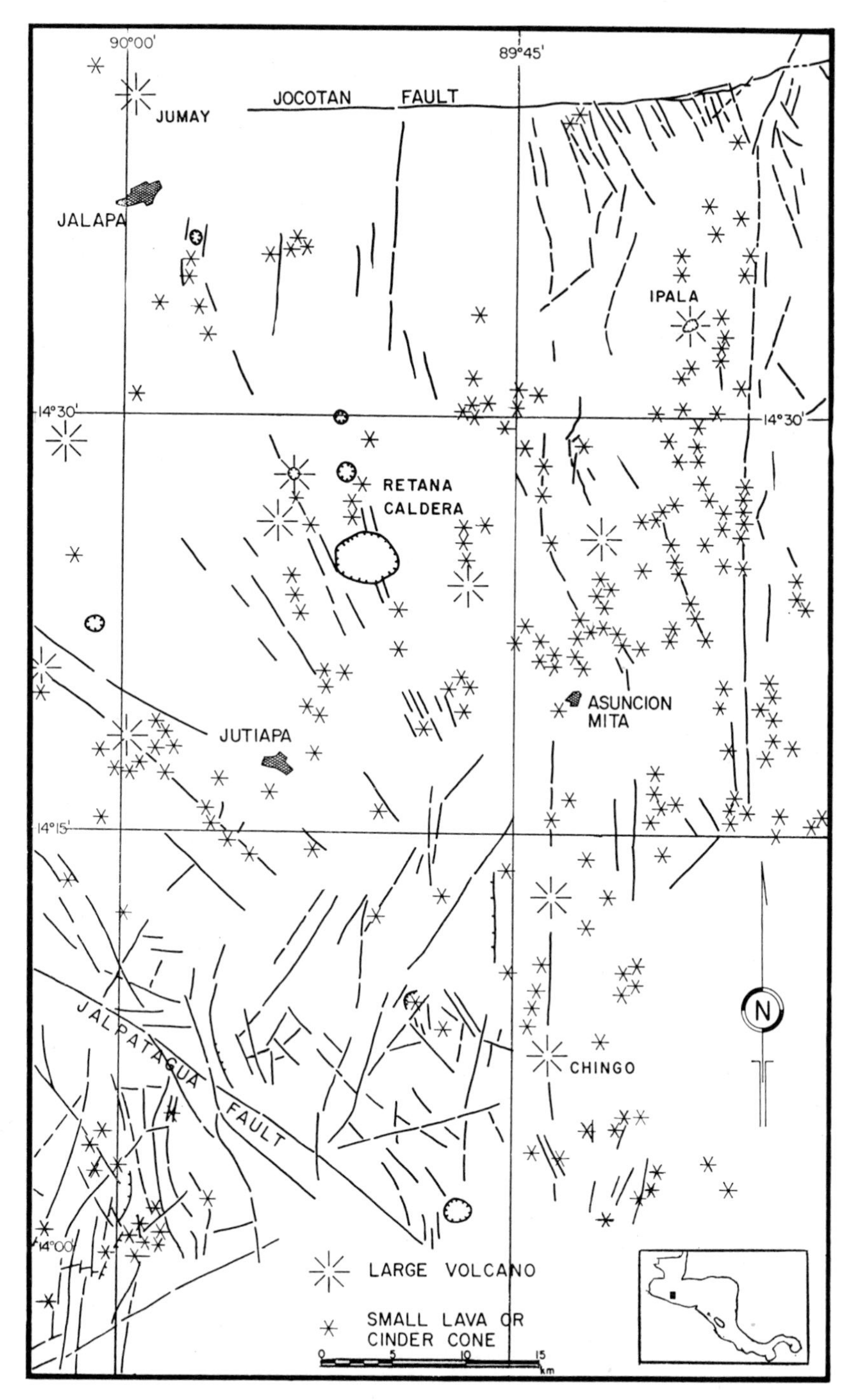

Fig. 40. Structural trends and related volcanic vents in the Ipala Graben area, southeastern Guatemala. (From DENGO, BOHNENBERGER & BONIS 1970, Fig. 3).

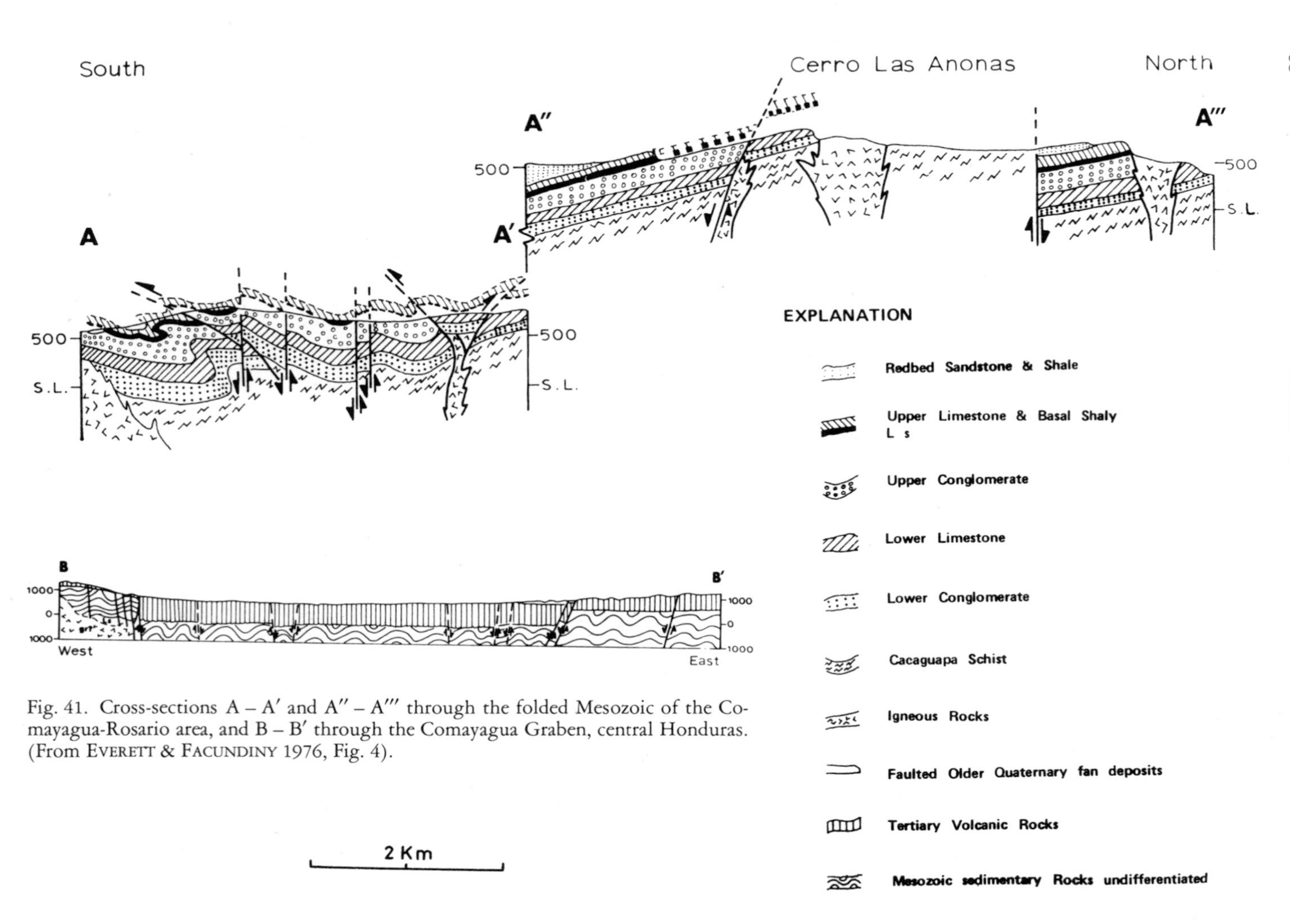

Fig. 41. Cross-sections A – A′ and A″ – A‴ through the folded Mesozoic of the Comayagua-Rosario area, and B – B′ through the Comayagua Graben, central Honduras. (From EVERETT & FACUNDINY 1976, Fig. 4).

resulted in the compilation of the geological map sheets for El Rosario and Comayagua and the publication of summary reports by EVERETT & FACUNDINY (1976) and MUEHLBERGER (1976).

According to these researchers the depression consists of a large number of individual, in general N/S-striking grabens and faults which run from the northern coast to the Gulf of Fonseca, although they are widely offset in relation to each other (Fig. 38). The deepest case of subsidence discovered so far is in the Comayagua Graben where a vertical fault of up to 2,000 m exists (Fig. 41). The character of the depression as a tension zone is underscored by Pleistocene volcanoes in the Gulf of Fonseca, in the vicinity of Tegucigalpa and above all at Lago de Yojoa. MUEHLBERGER interprets the depression as a tension zone between differently moving portions of the Caribbean Plate; he believes the western part may have rotated slightly in a counter-clockwise direction and/or the eastern part may have moved both northwards and eastwards. However, it is scarcely possible to accept his view that the depression is the reflection of the different structure of the plate — i.e. continental in the west and oceanic in the east — because both eastern Honduras and central Nicaragua exhibit a clearly continental-type crust.

The largest structure within the depression is the Sula or Ulua Valley in the northern section which is assumed to be a long N/S-striking graben. HORNE et al. (1976 b, p. 86) have vigorously contested this view and stress that their mapping work in this region has not revealed any indication of a tectonic subsidence and that basement rock extends throughout the valley at the same height as at the flanks. While these authors accept that the eastern side of the valley is a fault zone, they were unable to find any fault structures on the western side, and therefore they regard the depression as a pure product of erosion.

The fault tectonics, earthquakes and volcanism of the Pacific Marginal Zone are closely related and must be considered nowadays, with the theory of **plate tectonics** in mind, as the outcome of plate tectonic events. STOIBER & CARR in particular have in recent years combined these factors into an illuminating conception, especially since they take account of the entire Pacific margin of Central America from Guatemala to Costa Rica as well as of the structures of western El Salvador and eastern Guatemala which serve as a model (CARR 1974, CARR & STOIBER 1974, 1977, STOIBER & CARR 1974).

Their views are based on the discovery that the Cocos Plate is underthrusting Central America, not along a uniform front but in segments which are characterized by different strikes and dips in the earthquake zone (Fig. 166). The result of this is that both the underthrust plate as well as the overlying lithosphere are broken up by **transverse faults** which separate the individual segments from each other. It is assumed that eight such segments exist in Central America, and their boundary zones are shown in our maps in Figs. 38 and 120. The following surface phenomena are regarded as the chief characteristics of the boundaries of the segments:

Left-lateral strike-slip fault zones (perpendicular to the line of volcanoes) (Palín Shear Zone, Río Paz Shear Zone, active fault lines in Managua).

Break of the volcanic chains or abrupt change in their directional strike (cf. map in Fig. 120).

Differences in the type of volcanic activity, volumes of erupted material, form, elevation and structure of the volcanic centres (cf. Fig. 121).

Differences in the prevailing chemical properties of the erupted material (cf. Fig. 143).

The position of particularly large volcanic centres in the boundary areas (Santa Maria, Pacaya) and catastrophic eruptions of volcanoes that were for a long time quiescent (Santa Maria, Coseguina, Arenal).

Concentration of shallow to intermediate-depth earthquakes (< 70 km) and of moderate to medium intensity earthquakes in the segment boundaries, indicated by maxima in the energy release curve along the Pacific Marginal Zone (STOIBER & CARR 1974, Fig. 5).

A **longitudinal system** of right-lateral strike-slip faults, i.e. running parallel to the volcanic chain, is assigned to the transverse system. The strike slip is confirmed at the Monjas and Jalpatagua Fault Zone in Guatemala (Fig. 38). In El Salvador the 1965 earthquake had a right-lateral horizontal component, and WIESEMANN (1975), too, does not exclude the possibility that corresponding movements took place. However, a much more striking feature is the vertical component which is documented in the subsidences of the Nicaragua Depression, the grabens or half-grabens of El Salvador and volcano-tectonic depressions in Guatemala. They are partly regarded as the result of the eruption of large masses of volcanic products. However, in order to explain the **uplift** of the volcanic highlands of Guatemala or of the southward tilted fault block of the Balsam Range and of the Jucuarán Mountains of El Salvador it is also necessary to consider the possibility of isostatic adjustment of a continental-type crust with a diffused negative gravity anomaly.

The **north-south-striking transverse faults** and the accompanying volcanic chains (Guatemala City Graben, Ipala Graben, Honduras Depression, Nejapa Pits Chain in Nicaragua) are regarded as tension structures arising from the longitudinal and transverse horizontal movements.

The amount of slip generated by the horizontal movements is said to be approximately 10 km in the course of several million years. This rate of movement is one power of ten less than that of plate convergence. The structures in the near-surface crust are therefore regarded as **second-order structures** in the Pacific Marginal Zone of Central America. It is still uncertain to what extent they are important for the structural pattern beyond the Pacific Marginal Zone towards the northeast. At least the fault systems deep in the hinterland of Nicaragua run in similar directions to those in the Pacific Marginal Zone. Faults noted by CASE & HOLCOMBE (1975) striking in a northeasterly direction in the hinterland of Honduras and Nicaragua as a continuation of the segment boundaries of the Pacific Marginal Zone, and extending as far as the Nicaragua Rise, still require an explanation. The fault tectonics in central Guatemala and adjacent areas of Honduras are completely different.

In this region a major fault system runs through northern Central America in an arc open to the north. This system stands out just as clearly in the landscape, and thus also in the relief map of Guatemala (Fig. 44), as it does in the satellite images (MUEHLBERGER & RITCHIE (1975, Fig.2)). The three main fault lines, or probably more correctly fault systems, are in the north the **Cuilco Chixoy Polochic Fault,** in the centre the **Motagua** with the parallel **San Agustín Fault** and in the south the **Jocotán Chamelecón Fault** (Fig. 38). They are grouped together as the Motagua Fault System.

The faults have predetermined the course of the large rivers, and between the mountain ranges of the Sierra de Chuacús and Sierra de las Minas, the Sierra Espiritu Santo and the Sierra Santa Cruz they form deep and in the east graben-like broad furrows (cf. profile in Fig. 32 after DENGO 1968/1973). In the west the Cuilco Chixoy Polochic Fault intersects the southern edge of the Altos Cuchumatanes and separates Paleozoic from the Mesozoic strata of the range. On the other side of the Mexican boundary it forks and peters out in a system of echelon faults. The Motagua Fault disappears in the west under the Tertiary and Quaternary volcanic cover and in the east under the alluvial deposits from the Río Motagua.

Ever since TABER as early as 1922 pointed out that the faults merge into the large fault system of the Bartlett/Cayman Graben, their interpretation has been closely linked with the latter. A. A. MEYERHOFF (1966) reviewed the many interpretations which were discussed right up to and into the sixties — interpretations ranging from a compression form with buckling of the earth's crust, via tension forms and strike-slip faults extending as far as the boundary between the North American and Caribbean plates. In the process he himself came to the conclusion that the Cayman Fault System is a zone of weakness which formed in the Paleozoic geosyncline and which extends from Central America to Jamaica. No significant fault formation occurred before the Middle Cretaceous orogenesis and the present graben form was not attained until the Miocene and Pleiocene. He also accepts that both horizontal and vertical movements took place and that the horizontal dislocation was presumably less than 50 km and in no case more than 200 km.

In recent years countless facts and hypotheses relating to the age and genesis of the fault system as well as to the direction of motion have been advanced as further contributions to the discussion which was initiated mainly by H. HESS (1938) and HESS & MAXWELL (1953); these contributions are particularly important for a study of the aspect of plate tectonics.

KESLER (1971) was able to show that in the pattern of structures of the metamorphic crystalline rocks the arc-shaped course of the younger structures of northern Central America, such as Mesozoic folds and the faults under discussion, is predetermined. Violent cross-folding, particularly in the vicinity of the Cuilco Chixoy Polochic Fault, can best be interpreted as the earliest expression of a tension field from which later the large faults developed. Left-lateral strike-slip faulting cannot, however, have amounted to more than 150 km because intensive regressive dislocation in their structures would bring diverging sections of the metamorphic rock together.

In the metamorphic rocks north of the Motagua Fault NEWCOMB (1977) found widespread cataclastic structuring in ultramylonites to blastomylonites. They point to highly intensive deformation in the faults and also to their having been formed in the Paleozoic.

DONNELLY et al. (1968) showed that the Jocotán Fault in the Albian delimits various facies regions and that coarse detrital sediments gathered at its foot, a process which would indicate considerable relief energy and erosion. No horizontal dislocation was detected. WALPER (1960) was also unable to find any evidence of horizontal movement within the Mesozoic series in the region of Alta Verapaz. On the other hand, transverse folding in the Miocene Río Dulce Formation and of young sediments between Lago Izabal and the Gulf of Honduras and the dragging of folds at partial faults in the Chixoy Polochic Fault Zone are interpreted as the results of left-lateral movement (DENGO & BOHNENBERGER 1969, p. 215).

Mapping carried out in Verapaz (general geological map of Alta Verapaz) and in the Altos Cuchumatanes (ANDERSON et al. 1973) recorded details of the fault zones (cf. profile in Fig. 46). It revealed the pattern of vertical dislocation involving a vertical component of 2.5 km and a horizontal component of max. 1 km.

On the other hand, evaluation of earthquakes in the region of the Cayman Graben itself revealed left-lateral movement (MOLNAR & SYKES 1969, p. 1654).

In plate tectonics MALFAIT & DINKELMAN (1972) regard the Motagua Fault as the plate boundary between the northeastwards-moving Caribbean Plate and the North American Plate, and again left-lateral motion corresponds to the motion of the plates.

PINET (1972) suspected the existence of particularly large strike-slip motions of 1,000 km after he had interpreted diapir-like structures in the northern shelf region of Honduras as salt structures and tried to relate these spatially with the saliniferous rocks of the Chiapas Basin.

From an examination of morphological characteristics KUPFER & GODOY (1967) and SCHWARTZ (1976, 1977) had detected considerable left-lateral motion along the Motagua Fault during the Lower Quaternary. They found sag ponds, shutter ridges, offset of streams and river terraces, scarps and springs which indicated left-slip of on average between 30 and 50 m. Terraces reveal left-lateral dislocation between 24 m in the case of the youngest and up to 59 m in the case of the oldest mappable terraces, while the vertical throw fluctuates between 60 cm and 2.5 m. Similar observations were made along the eastern Chixoy Polochic Fault. The slip rate during approximately the last 40,000 years is given as 0.5 cm/year.

BURKART (1978) researched the offset across the Polochic Fault of Guatemala and Chiapas and summarizes the result: "The Polochic Fault is seen on LANDSAT imagery to continue its westward path from northwestern Guatemala across the Chiapas Massif to the Pacific coastal plain. The fault has had 132 ± 5 km of left-lateral displacement that is recorded in the offset of Cenozoic fold and thrust belt structure and stratigraphy. The trace of the Polochic Fault has been folded into what approximates a sinusoidal curve of about 130-km wave length and 7-km amplitude by an essentially east-west compressive stress. The curious similarity in displacement and fold wavelength results in a premovement reconstruction that reveals not only a match across the fault in geology, but an almost perfect fit of one block against the others. Some segments of the fault across which recorded slip took place are probably locked and not active. Strain has shifted to other shears in western Guatemala and Chiapas and to the Motagua Fault. Parts of the Polochic and the newer shears may be alternating with the Motagua as the Caribbean-North American plate boundary. With the 132 km of slip removed, segments of the now fragmented Cenozoic fold belt can be brought into coincidence in a clear-cut arcuate trend that is convex southwest. The major fault displacement is believed to have occurred within the interval from middle Miocene to middle Pliocene time. Eastern Guatemala has undergone a counter-clockwise rotation of about 25° that reoriented the fault trace and all other structural fabric from due east to the present east-northeast azimuth."

A widely distributed or smeared-out plate margin between the North American and the Caribbean Plate is suggested.

The southernmost situated Chamelecón Fault Zone runs in an arc approximately parallel to the Motagua Fault and extends as the **Jocotán Fault** as far as Guatemala. According to RITCHIE (1976) Paleozoic phyllites in the north are dislocated in relation to Cretaceous limestones in the south. Sometimes covered by Tertiary and Quaternary volcanics, the fault appears to run as far as the region north of Guatemala City where it is dislocated by the N/S-striking faults of the Guatemala Graben. The fault no longer appears in the landscape and thus not in satellite images either. Therefore MUEHLBERGER & RITCHIE (1975) think that it is at present inactive.

The faults are revealed to be deep-reaching fracture zones because of the accompanying or at least parallel-running **serpentinite bodies** which were detected already by SAPPER. According to the new maps and to ANDERSON et al. (1973, p. 809) the axis of the serpentinite belt of Chiantlá is located a few kilometres north of the Chixoy Polochic Fault. Inclusions of unaltered country rock and the limestones at the discordant contact indicate that the serpentinite ascended as a cold intrusion along shear planes which do not coincide with the present course of the fault. DENGO (1968, 1973) regarded them and the fractures as a single expres-

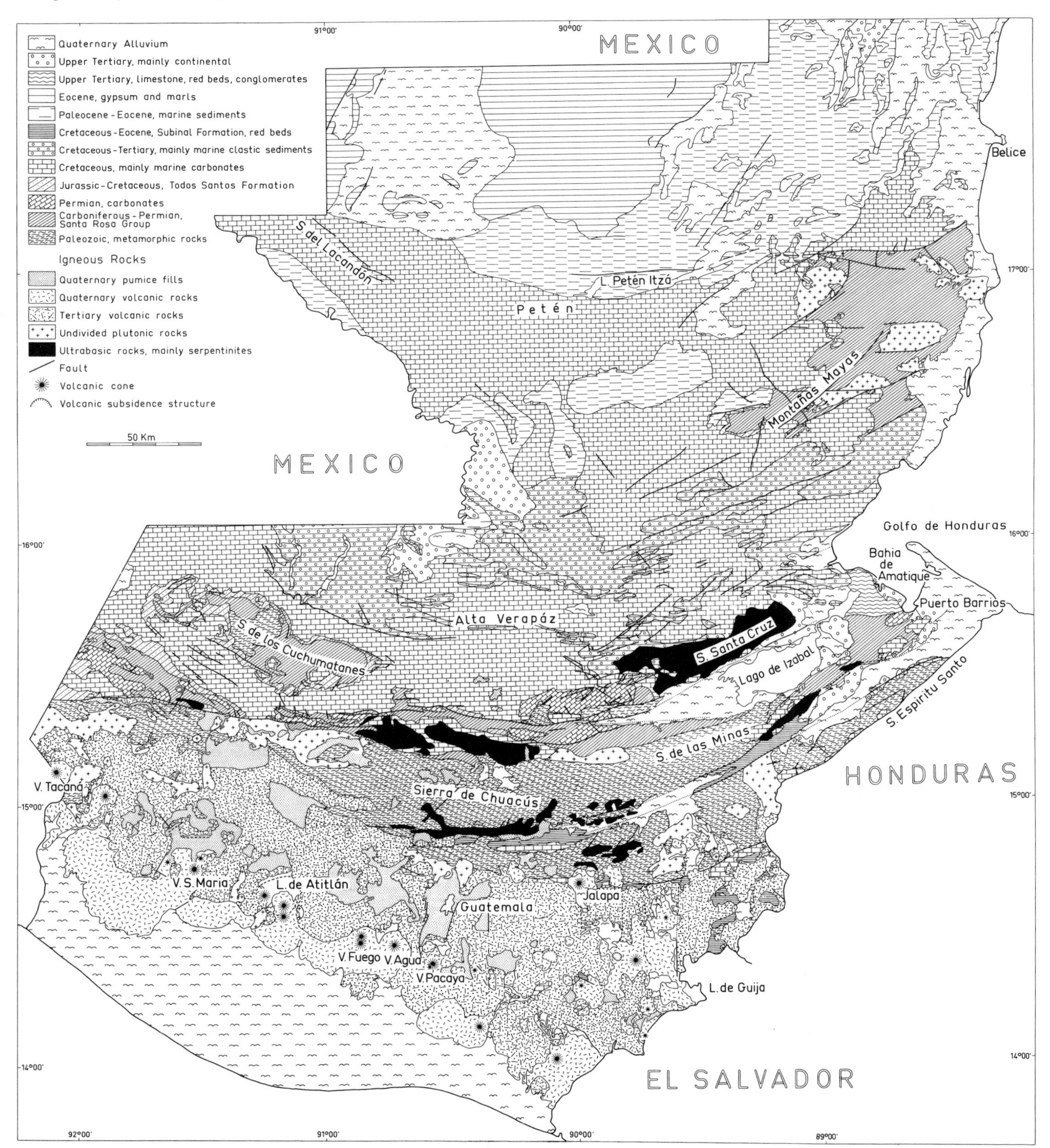

Fig. 43. Geological map of Guatemala. (Slightly modified redrawing from "Mapa Geológico de la República de Guatemala 1 : 500,000", Primera Edición, Guatemala 1970).

sion of the very complicated development from the Paleozoic to the present. MCBIRNEY (1963) had interpreted them as representing uplifted mantle rock. They were probably emplaced during the Cretaceous because no serpentinite pebbles are found in the conglomerates of the Todos Santos Formation, while according to an observation made by BONIS (1968) they appear for the first time in the Upper Cretaceous Sepur Formation.

The observations listed here prove beyond doubt that we are dealing with very large and repeatedly active fault zones in the crust. The most sensible interpretation to place on the contradictions in the apparent or assumed movements — vertical or lateral — would be that the same fault or the same fault zone moved in different directions at different times. KUPFER & GODOY (1967) sought conclusive proof for the most recent movements in seismology and their statement "seismological evidence for the sense of movement on Guatemalan faults, particularly those associated with the Polochic and Motagua zones, should be sought" was certainly borne out, in a way which the authors undoubtedly would not have wished to see, by the Guatemala earthquake of February 4, 1976. At that time horizontal displacements of on average 1 m (max. 3.4 m) were observed in the area of the Motagua Fault Zone (PLAFKER 1976). What happened is that the northern block moved to the west and the southern block to the east. This should be sufficient proof, at least for the present tectonic situation, that the Motagua Fault is a horizontal displacement of extremely large extent which fits into the pattern of recent plate tectonics (Fig. 170). See Chapters 6 and 8 on the earthquake and its geotectonic evaluation.

The **uplift of large parts of Central America** to their present height is closely related with the Upper Tertiary and Quaternary fault tectonics. The associated morphological phenomena such as plateaus, deeply incised river courses, and the attendant systems of terraces on the one hand, and the formation and filling of intramontane basins on the other were noticed very early on but, with few exceptions, they have still not been explored.

SAPPER (1937, pp. 59 – 65) described a number of large, uplifted blocks such as the Sierra Madre of Chiapas, the Altos Cuchumatanes, the Alta Verapaz and the Maya Mountains or the Sierra de las Minas. Crossing these mountains on foot he could scarcely fail to notice the contrast between the steeply dropping edges and the uplifted peneplain. In similar fashion, TERMER (1936) had observed in the Altos Cuchumatanes broad dry valleys and alluvial plains through which nowadays only tiny streams flow (Fig. 49). In Honduras BENGSTON (1926) had detected planations at elevations of 1,500 and 1,000 m and distinguished between an upper and a lower peneplain. HELBIG (1959, pp. 156 et seqq. and Plate 11) conducted confirmatory observations. The more recent geological literature contains evidence, e.g. in MCBIRNEY (1963, Plate 20) from the Sierra de Chuacús and in WILLIAMS & MCBIRNEY (1969, pp. 73 – 74), of regional uplifting of Honduras. According to these authors the Miocene ignimbrite sheets extruded over a low-lying flat landscape which was not uplifted and intersected, together with other ranges such as the Balsam Range of El Salvador or the highlands of Guatemala and northern Nicaragua, until the Pliocene and early Pleistocene. At the same time the foreland subsided so that only the Bonacca Ridge surmounted by the Bay Islands remained.

To sum up, it must be stated that we are still far from having achieved a geologically based understanding of the landforms and their morphogenesis.

3.3 Individual areas

In the preceding sections northern Central America, that is to say SCHUCHERT's Nuclear Central America or STILLE's Sapperland, was regarded and treated as a single unit. It is, however, understandable that individual landscapes should stand out in such a large and multiform region. Some of these areas, on which adequate material is available and which the author himself has, where possible, visited, will be discussed in the following.

Political boundaries do not delineate natural regions, let alone geological structures. However, they do to a large extent represent the boundaries of our knowledge of such features and in particular of their cartographic representation, as is apparent from the general geological maps which are reproduced here in simplified form. Nevertheless, it seemed sensible to give a brief review of the individual countries and thereby offer the reader a more rapid means of orientation than would have been possible by reviewing the entire region. As far as southern Central America was concerned, it in any case proved necessary to deal with Costa Rica and Panama separately.

3.31 Guatemala [2]

Guatemala can be divided into four main structurally and physiographically different provinces (Fig. 42). They manifest themselves clearly in the distribution of the geological formations and thus in the general geological map (Fig. 43) on the one hand, and in the landscape and thus in a relief map of the country (Fig. 44) on the other. They are from south to north: The **Pacific Coastal Plain,** the **Volcanic Belt,** the **Cordillera Central** and the **Petén Lowland** with the **Chapayal Basin.**

The Volcanic Belt with the chain of Quaternary and active volcanoes located on its southern edge strikes WSW/ESE. In the west it ends at the Mexican border, but in the east it extends with the same strike direction into El Salvador. The coastal plain bordering the Volcanic Belt to the south is part of the alluvial coastal plain extending over a length of about 700 km from the Isthmus of Tehuantepec to Acajutla in El Salvador. Ahead of this coastal plain towards the ocean comes a broad shelf and the Middle America Trench.

The axis of the country is formed by the Cordillera Central which is built up of many individual mountain ranges and which describes an arc open towards the north from the Gulf of Honduras to the Mexican province of Chiapas. The Petén Lowland, which is located to the north, possesses structures running parallel to the Cordillera Central, but on the basis of its development during the Mesozoic and Tertiary it is also regarded as a boundary region of the

---

[2] NAGLE, ROSENFELD & STIPP (1977), in an excursion report written for the University of Miami, give a brief review of the geology of Guatemala and describe the route followed by the excursion. This report can be recommended as an introduction to the subject, and it may be obtained from the "Miami Geological Society". The authors divide the country up into tectonic units corresponding to the modern concept of plate tectonics. From north to south they distinguish between:

The North American Plate corresponding to the area of Mesozoic folds and the Lowland of Petén.

The Caribbean Plate which with its northwestern end composed of metamorphic rocks, igneous rocks and sediments, makes up the Central Cordilleras.

The Cocos Plate which is subducted into the mantle below the Middle America Trench and is thus responsible for the belt of Tertiary and Quaternary volcanoes and the young sediments derived from them.

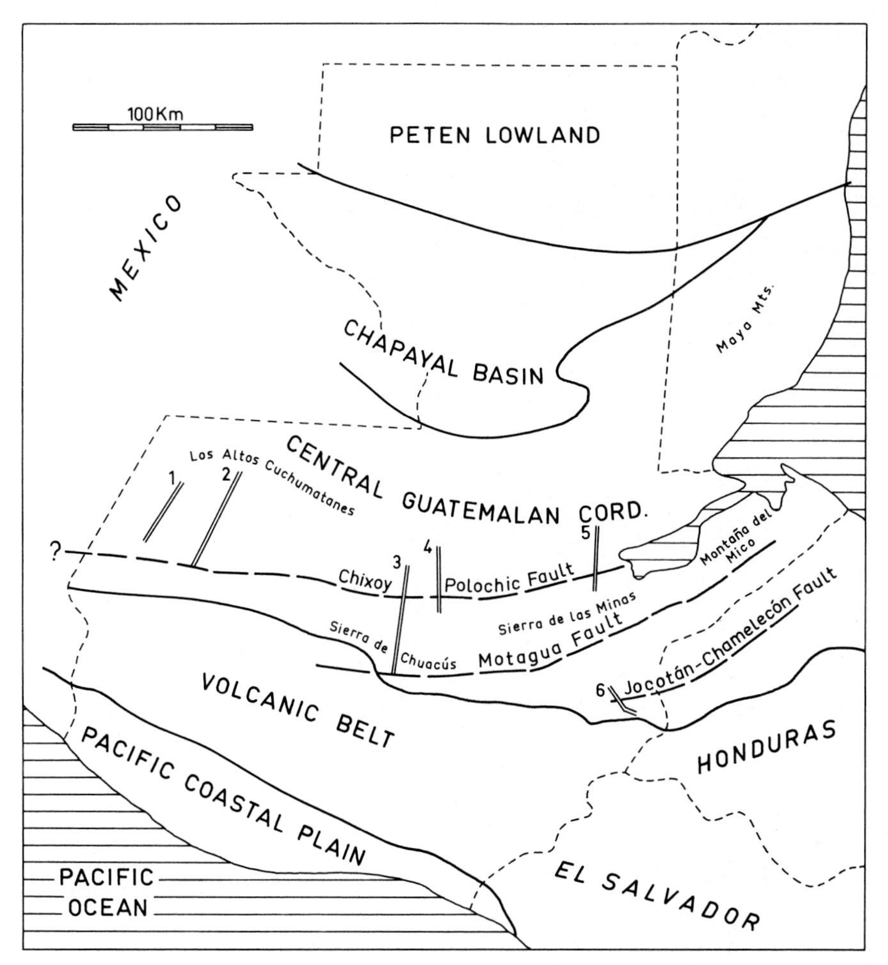

Fig. 42. The structural provinces of Guatemala, with an index of the cross-sections. 1 and 2 – Fig. 51; 3 – Fig. 46; 4 – Fig. 33; 5 – Fig. 34; 6 – Fig. 35.

Gulf of Mexico. To the east the Maya Mountains of Belize rise up out of the Lowland, and these must be regarded as a structural province in their own right.

Large fault zones trending parallel to the chains of the Cordillera Central and containing the Cuilco Chixoy Polochic Fault, the Motagua and San Agustín Fault and the Jocotán Chamelecón Fault stand out as very striking structures. They possess supra-regional importance as the boundary zone between the North American and Caribbean plates, and they are dealt with in detail in Chapters 3.22 and 6.

The **Coastal Plain** is up to 50 km wide. It rises from the coast at a shallow angle to the foot of the volcanoes and the upper limit is marked by a clear nickpoint at the 500-m line. Along the coast the plain is broken up by deltas, sand bars and behind them esteros (tidal flat areas). Their sediments are for the most part fluviatile, but lahars occur in the upper portion while delta, sand bar and estero sediments are located off the coast. The material comes al-

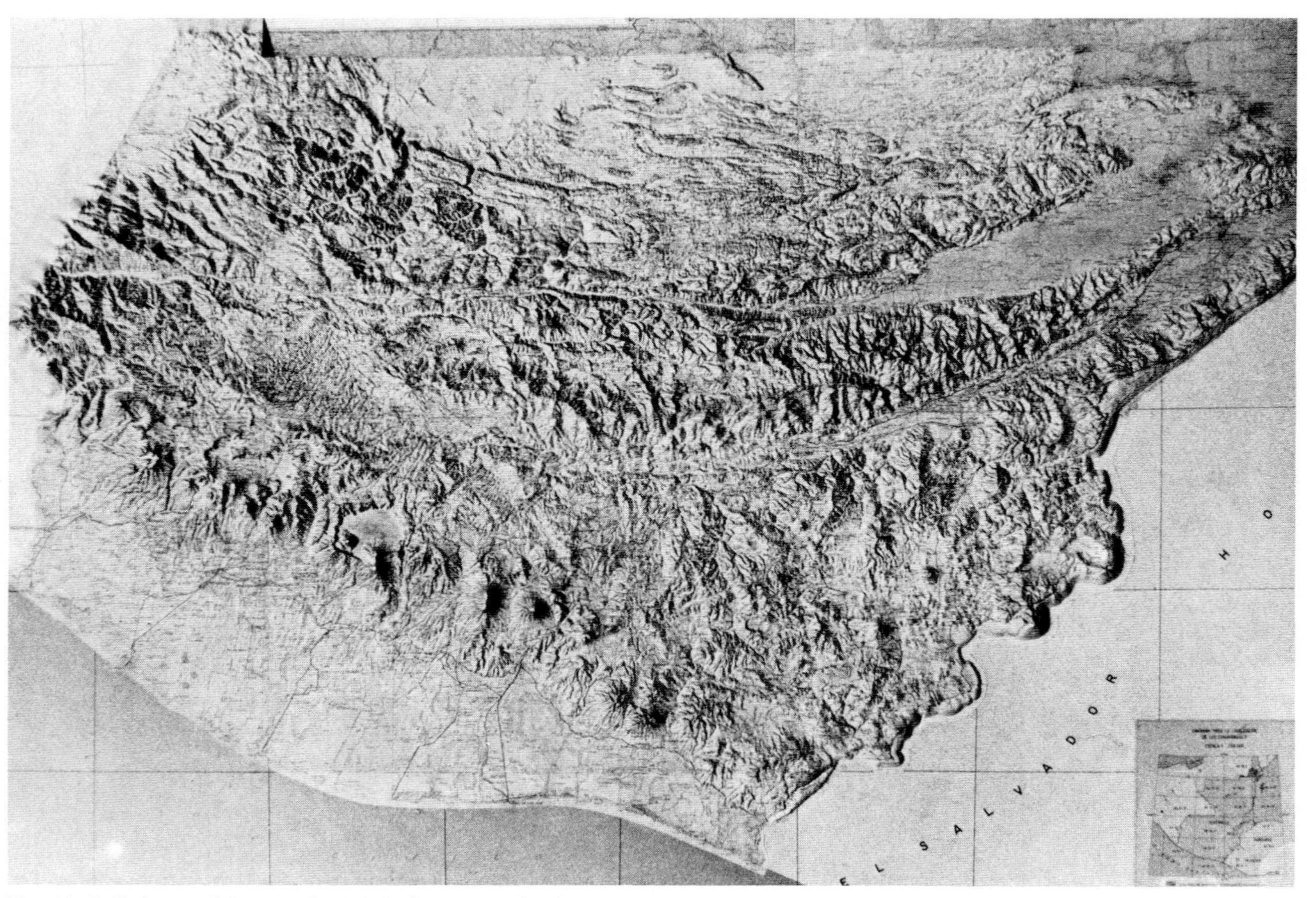

Fig. 44. Relief map of Guatemala. Original on the scale of 1 : 250,000. Photo: R. WEYL.

Fig. 45. The volcanoes of Agua (right), Fuego and Acatenango seen from the western slope of Pacaya, another volcano. Photo: R. WEYL.

most exclusively from the Volcanic Belt and therefore consists of pebbles of andesites, basalts, dacites, etc. and minerals such as plagioclase, basic silicates and magnetite. The average grain size decreases towards the coast, and the grains are less rounded and scarcely weathered. HORST, KUENZI & MCGHEE (1979) provided an analysis of the sediments which are similar to those of the Salvadorian coastal sediments. Further details are given in Chapter 5.52.

Since geophysical surveys had revealed sediments of considerable thickness in the Coastal Plain and Shelf, exploratory drilling was carried out at the end of the sixties by Texaco and Esso; on the mainland the drillings reached a depth of approximately 3,820 m and in the shelf area 3,590 m. Although no information was given on the sequence of strata through which the drills passed, the data indicate that one or more sedimentation basins of great depth extend below the Coastal Plain and Shelf.

The **Volcanic Belt,** emphasized by the 4,000 m high Quaternary volcanoes, rises steeply from the Coastal Plain (Fig. 45). Below the volcanic edifices and ahead of them to the north extends the area to Tertiary volcanism containing the pumice-filled basins and grabens. With its towering volcanoes and the large volcano-tectonic depressions filled with lakes, the Volcanic Belt of Guatemala forms one of the most impressive volcanic landscapes on earth. Details will be discussed in Chapter 5 in connection with the Tertiary and Quaternary volcanism.

The **Cordillera Central** of Guatemala is the core of the mountain chains in northern Central America. It consists of a series of individual mountain ranges which are separated from each other by broad valleys or grabens in the fault zones. Large intramontane basins are also present. The southern ranges are built up of crystalline schists and plutonites. To the north come belts of Upper Paleozoic sediments ahead of which further to the north comes the large region of Cretaceous limestones which are underlain by clastic deposits of the Todos Santos Formation. Block folding and imbricate structures predominate, and the chains of folds stand out as subparallel mountain chains, in particular on the northern slope of the Alta Verapaz, and their morphological pattern and structure are reminiscent of the chains of the Swiss Jura (Fig. 44). Viewed overall, the formations plunge in stages towards the north, and in the Petén Lowland and on the Yucatán Peninsula they are located at depths up to 3,000 m. The intensity of deformation also decreases towards the north.

Characteristic individual regions of the Cordillera Central were surveyed by MCBIRNEY (1963) and a little later by the Geological Mission from the Bundesanstalt für Bodenforschung. Their 1 : 125,000 scale map covers a region from the Motagua Fault Zone in the south via the central **Sierra de Chuacús** to the southern edge of the Alta Verapaz near Cobán. Unless other works are referred to, it is on this map and the unpublished reports which were generously made available to me that the following review is based. As far as was permitted by the necessary reduction in size, the map was incorporated into the general map of Guatemala (Fig. 43), and the accompanying cross-section is reproduced in Fig. 46.

The metamorphic and plutonic rocks that crop out in the Sierra de Chuacús were discussed in Chapter 3.11. The subsequent Upper Paleozoic and Mesozoic strata were divided up a little differently from the division made by the American authors; the stratigraphic sequence is shown in Fig. 47 together with data on the lithology and thicknesses.

According to MCBIRNEY (1963, pp. 182 – 187) the **ultrabasic rocks** are 75% composed of more or less serpentinized olivine together with diallage and bastite pseudomorphs after bronzite/enstatite. They are completely serpentinized in the shear zones, in particular along the Motagua Fault Zone, and in the interior of the large peridotite masses they are coarse-grained and massive. Chromite lenses have given rise to sporadic mining activity. The ultrabasic rocks form elongate bodies trending parallel to the axis of the range and are in fault contact with the country rock. They are regarded as multiply mobilized mantle rocks.

The structure of the region is explained by the cross-section shown in Fig. 46. In the south, this cross-section reveals the graben-like Motagua Fracture Zone which is filled with sediments from the Tertiary Subinal Formation and by a cover of pumice. According to MCBIRNEY (1963, pp. 206 – 210) the pumices were produced by the volcanic zone to the south.

The series of metamorphic rocks with intrusions of gabbros and serpentinites follows at steep faults. These metamorphic rocks are discordantly overlain in the north by the Upper Carboniferous Sacapulas Formation which is directly succeeded by folded limestones of the Middle to Upper Cretaceous Cobán Formation. They are penetrated and overlain by a broad serpentinite body. Severely faulted Permian strata, which are permeated by gypsum diapirs, crop out in the fault zone of the Río Chixoy which follows to the north (Fig. 33). In the north they are faulted, sometimes in relation to metamorphic rocks and sometimes in relation to the Jurassic/Cretaceous Todos Santos Formation and the Cretaceous carbonate series. WALPER (1960) has described them as having a typical fault-fold structure.

The forms of the Sierra de Chuacús are characterized by the contrast between a medium-relief at altitudes between 1,700 and 2,200 m and the steep, deeply incised flanks – this contrast was first mentioned by TERMER (1936, p. 109) and after him by MCBIRNEY (1963, p. 181 and pl. 20). There are several intramontane basins embedded in the Sierra, and their young, unconsolidated sediments and filling of pumice were described by K. RODE (1965), although he was unable to explain fully the circumstances under which they were formed.

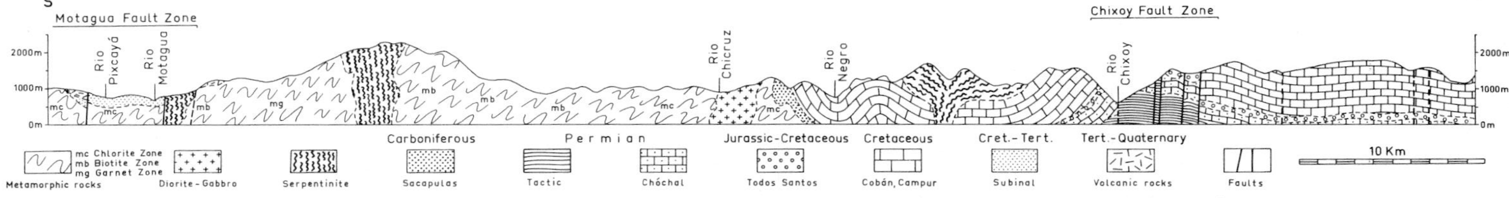

Fig. 46. Cross-section through the Baja Verapaz (Sierra de Chuacús) and the southern part of the Alta Verapaz. (Redrawn from the "Geologische Übersichtskarte 1 : 125 000 der Baja Verapaz und Südteil der Alta Verapaz, Guatemala; Hannover 1971).

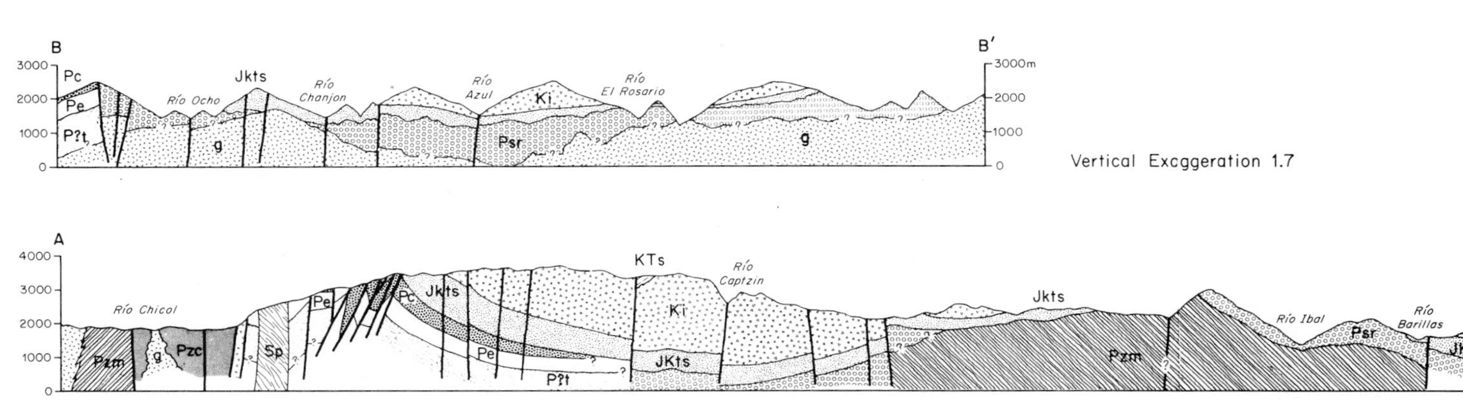

Fig. 51. Structural cross-sections through the western Altos Cuchumatanes, Guatemala. (From ANDERSON et al. 1973, Fig. 2). KTS = Sepur Formation, Ki = Ixcoy Formation, Jkts = Todos Santos Formation, Psr = Santa Rosa Group, Pc = Chóchal Limestone, Pe = Esperanza Formation, P?t = Tactic Formation, Pzc = Chicol Formation, Pzm = Undifferentiated metamorphic rocks, g = Undifferentiated intrusive rocks, Sp = Serpentinite.

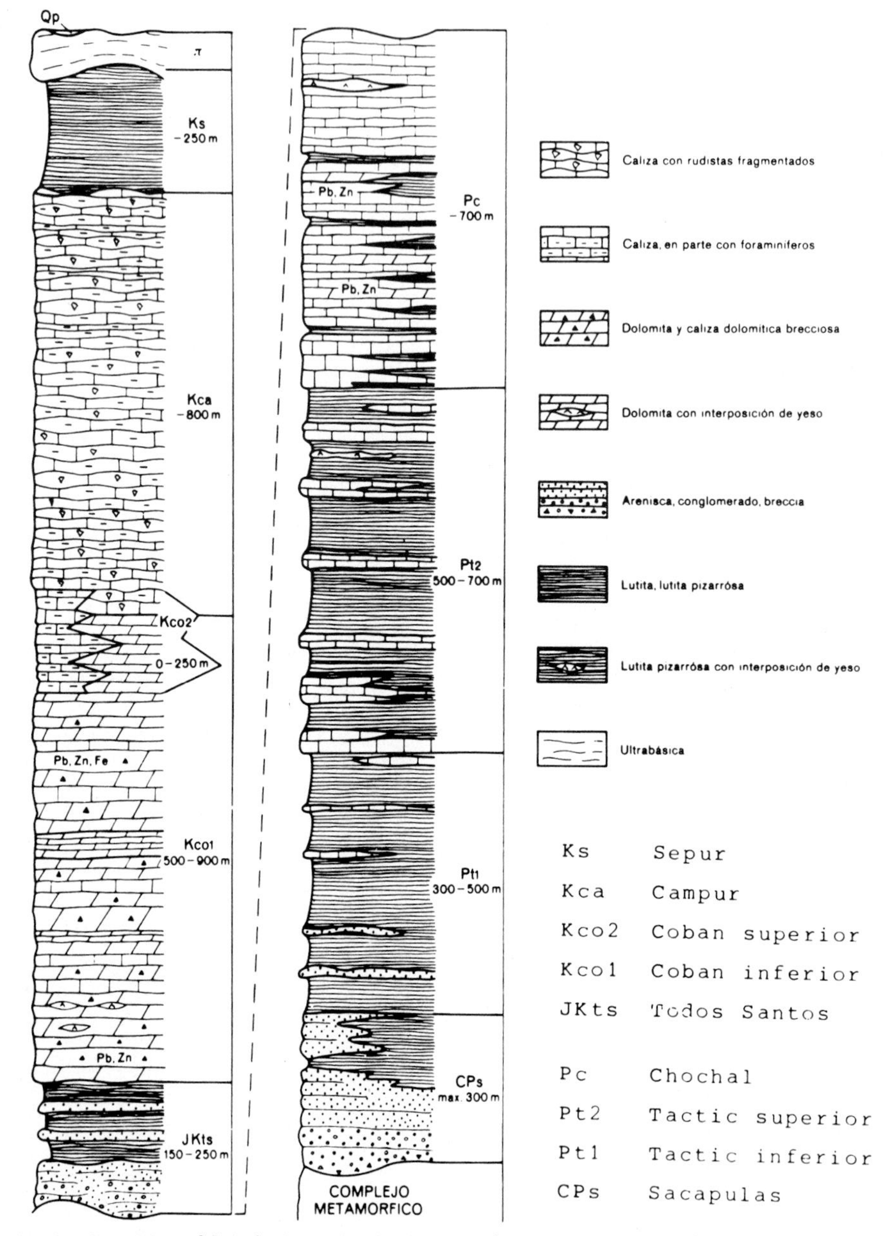

Fig. 47. Stratigraphic and lithologic section in the area of the Baja Verapaz and the southern part of the Alta Verapaz, Guatemala. (From NICOLAUS 1971).

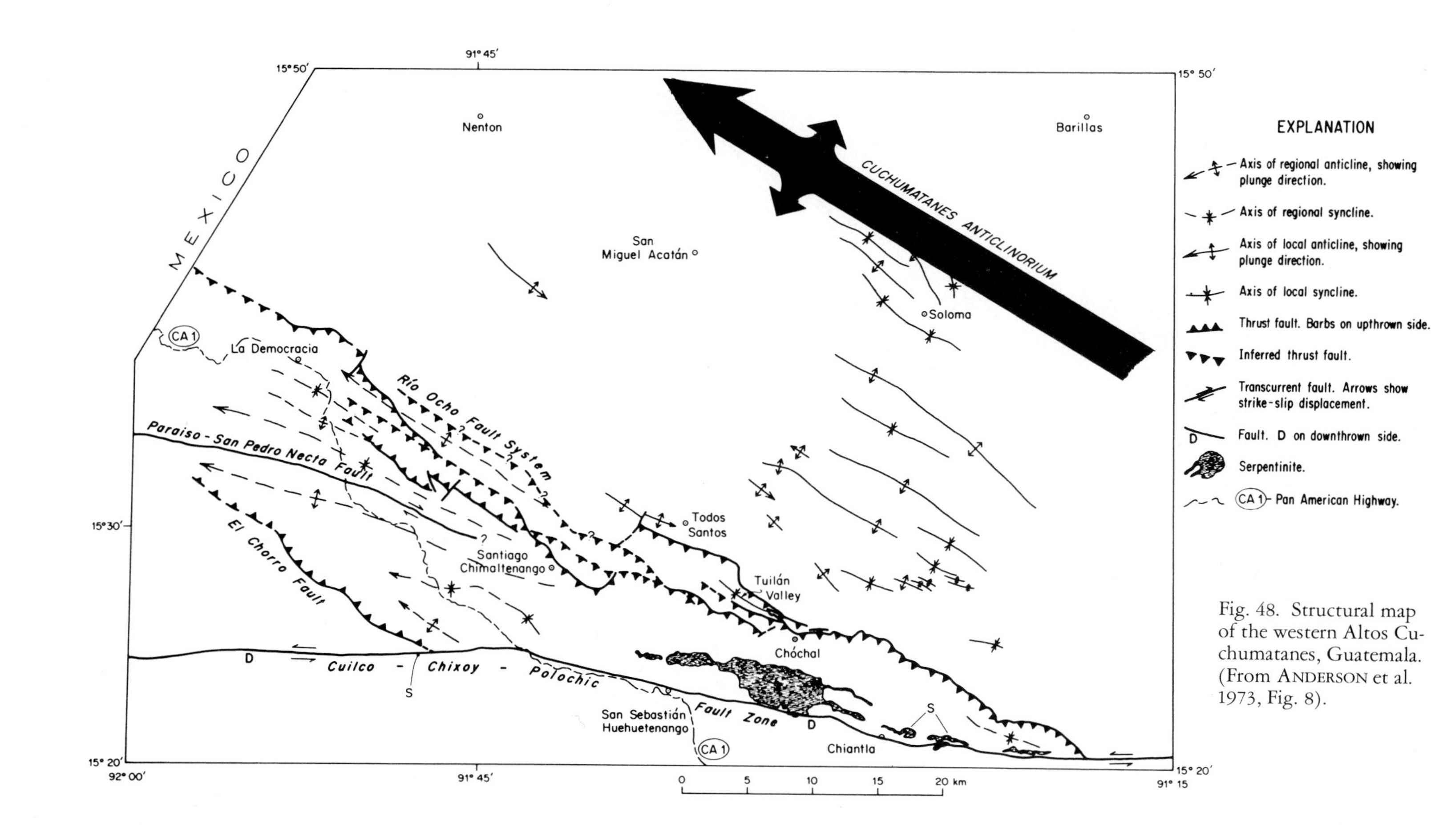

Fig. 48. Structural map of the western Altos Cuchumatanes, Guatemala. (From ANDERSON et al. 1973, Fig. 8).

The serpentinite regions stand out clearly from their surroundings in the landscape. Their low-nutrient soils can support only scanty vegetation, mostly sparse pine forest which has been severely damaged by clearing operations.

Karst relief prevails in the **Alta Verapaz** which follows to the north: "An irregular accumulation of cone-shaped hills with steep slopes all around alternates with funnel-shaped or cylindrical hollows, true dolines of significant depth, dry elongated basins, oval poljes, valley-like depressions. Water drains away mainly underground in the cavities and joints in the rock. Rivers running on the surface are rare; when they are present they are important arteries carrying large quantities of water to the Polochic or northwards to the Usumacinta system." (TERMER 1936, p. 105.) Deep-red loams, which are overlain by yellow and brown soil horizons and are very probably pre-Recent formations, are found in the depressions of the karst landscape. It would be worthwhile examining them to see if they contain any deposits of mineable bauxite.

The karst relief, cave systems and karst hydrography in the region of the Chixoy Loop to the west of Cobán were studied in detail by a Franco-Guatemalan mission in connection with the hydro-electric project on the Río Chixoy (DEUX & COURBON 1976, TRIPET 1977). They found poljes, dolines, dry valleys and limestone domes all mixed up together but without any of the steep forms of the typical tropical cockpit karst.

The **Altos Cuchumatanes** [3] are located in the western part of the Cordillera Central. They rise steeply above their southern foreland to heights of more than 3,500 m and are thus the highest non-volcanic mountains in northern Central America. Near Chiantlá and Huehuetenango the southern boundary is formed by the Chixoy Polochic Fault Zone (Fig. 48); to the west follows the NW-striking Río Ocho Fault System, and the northern and northeastern margins of the range are also accompanied by faults.

The landscape and history of the forms of the Altos were described by TERMER in a highly impressive way. In their present form "we must imagine the Cuchumatanes as a gigantic uplifted block which was raised higher in the W and S than in the E and NE so that a slope was formed to the E and NE. These events were associated with fractures at the margins in the S and W; but in the N and E they were probably associated with bending in the form of flexures. In the interior, the range is broken up by longitudinal and transverse faults which divide it up into individual sub-blocks. As the young river terraces prove, the processes of uplift continued into the recent geological past. Usually two such terraces can be found in most valleys and in many cases they are joined by a third. The youthful nature of these uplift events is demonstrated to us by the unbalanced longitudinal profile of the valleys, most impressively perhaps in the westward-opening valleys on the western slope." (TERMER 1936, p. 238.)

Broad **dry valleys** and alluvial plains, through which nowadays only feeble creeks flow, are found in the interior of the Altos Cuchumatanes. This type of landscape is particularly impressive along the road over the Altos to San Juan Ixcoy (Fig. 49). Although the karst character of the flat upland area, which is composed of limestones, may play an important role in the formation of the dry valleys, the flattish relief at high altitude is nevertheless sharply contrasted with the deeply incised valleys, in particular of the rivers draining to the north and northeast.

---

[3] Also called the Sierra de Cuchumatanes and recently the Cordillera de Cuchumatanes.

Fig. 49. Medium relief and dry valley in the Altos Cuchumatanes near Paquix. Photo: R. WEYL.

Fig. 50. Moraines in the Altos Cuchumatanes. SW side of the valley "Llanos de San Miguel – Llano Ventura". (From HASTENRATH 1974, Fig. 4). Photo: HASTENRATH.

"The present macroforms of the range were created by erosion. This process dissolved the originally cohesive limestone cover of the high plateau to such an extent that this cover now remains intact in fragmentary form only in the W and S. Further east a few fragments have been preserved as residual blocks or peaks at just two points ... In the other regions, however, in the area of Ixcán and Xacbal, the process of dissection has been so severe that it has already reached the crystalline basement." (TERMER 1936, pp. 239, 240.)

The great height of the Altos Cuchumatanes was indicative of **Quaternary glaciation** similar to that in the Cordillera de Talamanca of Costa Rica. However, this glaciation was not discovered until ANDERSON (1968) and was described in detail for the first time by HASTENRATH (1974). I am grateful to the latter for the photograph of a moraine landscape in the high-altitude valley of Llanos de San Miguel — Llano Ventura (Fig. 50).

Our geological knowledge of the Altos Cuchumatanes has been greatly enlarged in the last decade by a series of mapping projects, the results of which have been summarized in a paper by ANDERSON et al. (1973). The following description is based on this work.

The **basement of the range** is formed by feldspar-biotite gneisses, mica schists, amphibolites, metaquartzites and metaconglomerates, which are exposed in a large area to the east of Soloma. They are presumably the equivalents of the metamorphic series that are widely distributed in the foreland to the south of the Chixoy Polochic Fault. The presence of chiastolite indicates contact metamorphism brought about by an underlying pluton. Granitic plutonites

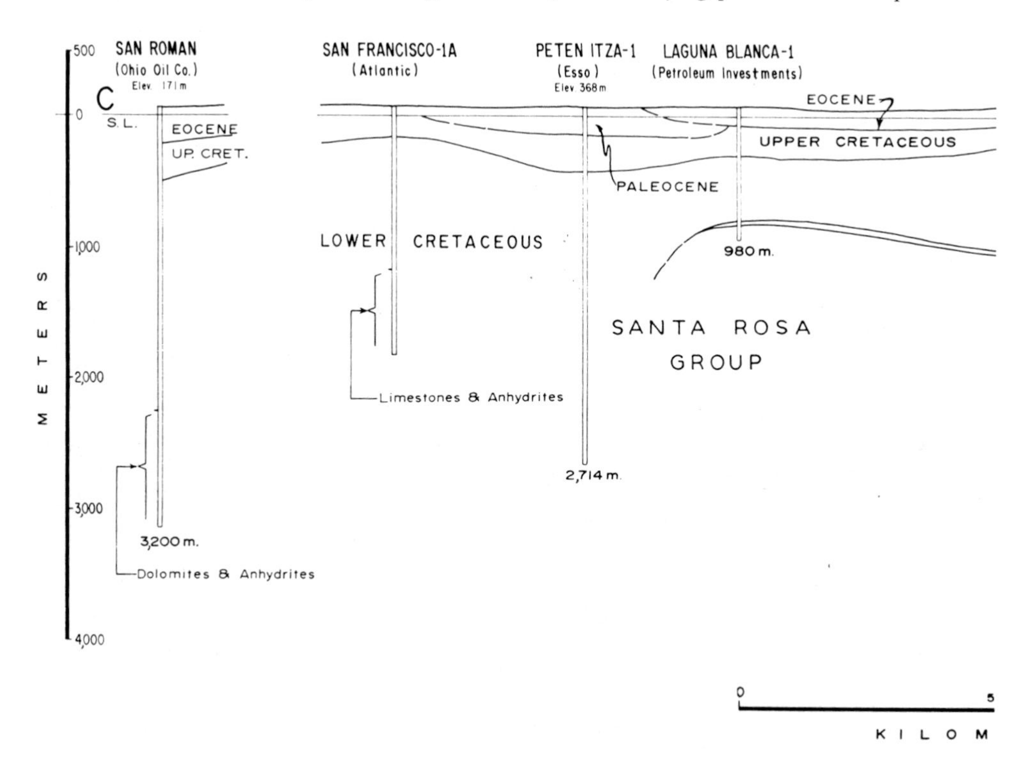

Fig. 52. SW-NE cross-section showing the subsurface stratigraphic sequence of northeastern Guatemala and Belize. (From VINIEGRA 1971, Fig. 6).

are exposed over a distance of 14 km in the valley of the Río Catarina. They intrude into the strata of the Upper Paleozoic Santa Rosa Group and are discordantly overlain by the Todos Santos Formation whose basal conglomerate contains granite pebbles. Their age can therefore be put at post-Lower Permian to pre-Upper Jurassic. The mineralization in the limestones of the Lower Cretaceous Ixcoy Formation (see p. 301) would also, however, seem to indicate post-Lower Cretaceous intrusion activity.

Peridotites and **serpentinites** crop out along the Chixoy Polochic Fault Zone at the foot of the Altos. Internally they consist of slightly modified lherzolite, while at the edges they are serpentinized and they have undergone a great deal of shearing. Their age is unknown, but chromite grains occur in the Upper Cretaceous Sepur Formation, which is indicative of the erosion of an exposed ultrabasite. It is likely that multiple mobilization occurred. The lack of re-crystallization of embedded or directly adjacent limestone points to cold intrusion.

The **sedimentary cover** extends from the Permo-Carboniferous Santa Rosa Group to the Upper Cretaceous Sepur Formation and to the Plio-Pleistocene Colotenangeo strata. The stratigraphic sequence was described in Chapter 3.1 (cf. Tables 1 and 2). The unusually large thickness of more than 3,000 m in the Santa Rosa Group and of 2,500 to 3,000 m in the Cretaceous carbonates is typical of the Altos Cuchumatanes. There are considerable differences in thickness ranging between 10 and 1,200 m within the Todos Santos Formation, which indicates an eminence in the sedimentation region, namely the "Poxlax Uplift". The thick pack-

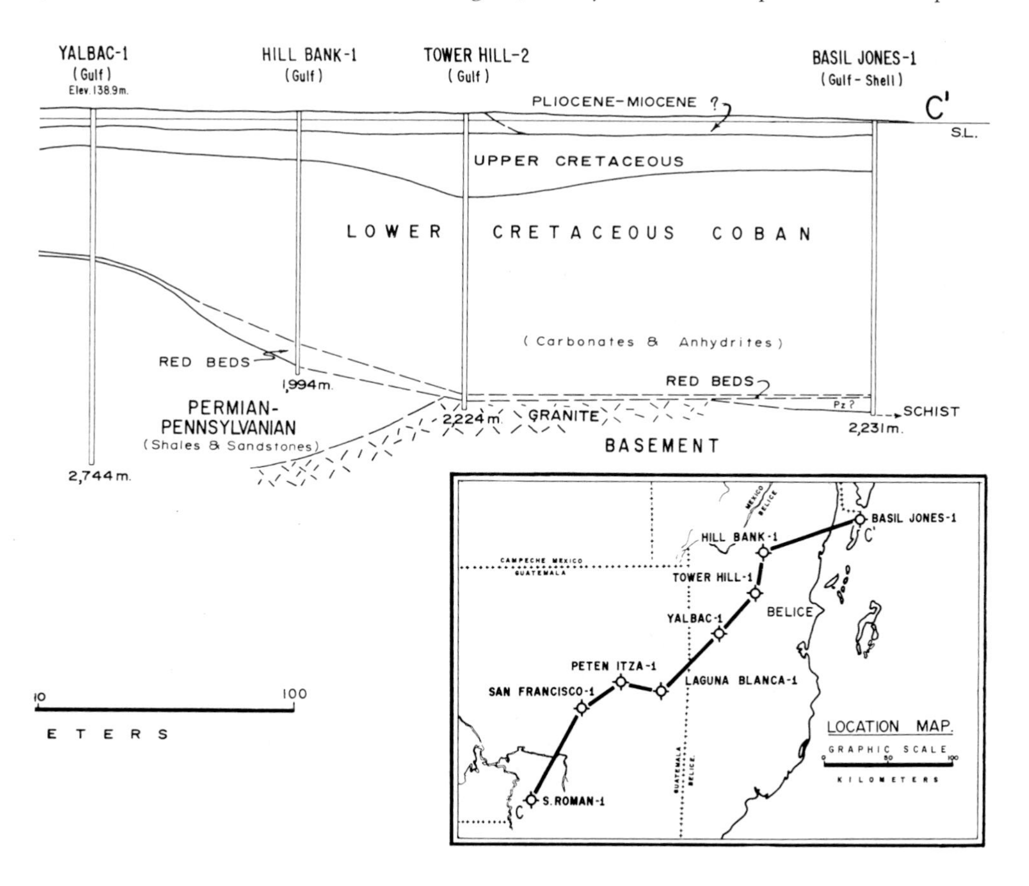

age of Upper Paleozoic and Mesozoic sediments, in association with the pluton which is assumed to exist in the basement, is probably the cause of a strong negative gravity anomaly in the region of the Altos Cuchumatanes and no doubt also plays a role in the recent strong uplift of the range (Fig. 169).

The **structure of the Altos Cuchumatanes** is characterized by a 40° W striking foliation of the metamorphic rocks, paralleled by slight folding in the sediments and major fracture tectonics with predominantly N 55° W and N 30 – 40° E trending faults. The Chixoy Polochic and Río Ocho Fault Systems dominate the southern and southwestern margin at which both the northern and southern oriented upthrow and downthrow faulting must have taken place. WALPER (1960) & BLOUNT (1967) had postulated that the northern blocks had subsided by amounts of between 1,500 and 2,500 m, while later movements raised the northern blocks. The tectonic sketch given in Fig. 48 and two cross-sections in Fig. 51 illustrate the structure.

It is possible to distinguish between three main deformation phases; the first, presumably a two-phase event, involved deformation of the Chuacús Group and therefore occurred in the Paleozoic, the second occurred between deposits of the Permo-Carboniferous sediments and of the Todos Santos Formation, i.e. between the Permian and the Jurassic, and the third took place towards or after the close of the Cretaceous.

The **Petén Lowland** is a tropical plain with an average altitude of around 100 m. It is built up of horizontally bedded or gently dipping Mesozoic and Tertiary sediments. Karst has developed on the Cretaceous carbonates.

The structure of this region has already been discussed in Chapter 3.21 in the review of the Mesozoic structure of northern Central America based on the work by LLOYD & DENGO (1960) and DENGO (1968/73). A cross-section of the structure in northern Guatemala and Belize (VINIEGRA 1971), which is based on information yielded by drilling, is given in Fig. 52. LÓPEZ RAMOS (1975) has provided a summary description of the geology of the region of the Yucatán Peninsula, which adjoins to the north.

3.32 The Maya Mountains

The highest peak in the Maya Mountains of Belize is Victoria Peak at 1,023 m. The region is covered by a peneplain above which rise isolated monadnocks such as the Cockscomb Range. The peneplain is transected by short but deeply incised river valleys.

The mountains are formed by an uplifted block which is bounded in the north and south by large faults striking approximately E/W. It slopes away gently towards the west following the dip of the geological structures. This block contains sedimentary folded Upper Paleozoic, volcanics and plutonic rocks which are surrounded by Cretaceous limestones, and, in the west, by Quaternary coastal sediments (cf. map Fig. 53).

The description of the Maya Mountains given in the first German edition of this book was based above all on the fundamental studies and maps by OWER (1928) and DIXON (1955). Recent studies by BATESON & HALL (1971, 1977), KESLER et al. (1971), HALL & BATESON (1972) and BATESON (1972) necessitate significant modification of this description. Consequently, the general map given by HALL & BATESON (1972, Fig. 1) also deviates greatly from that given by DIXON (1955). It has been incorporated in the general geological map of Guatemala (Fig. 43).

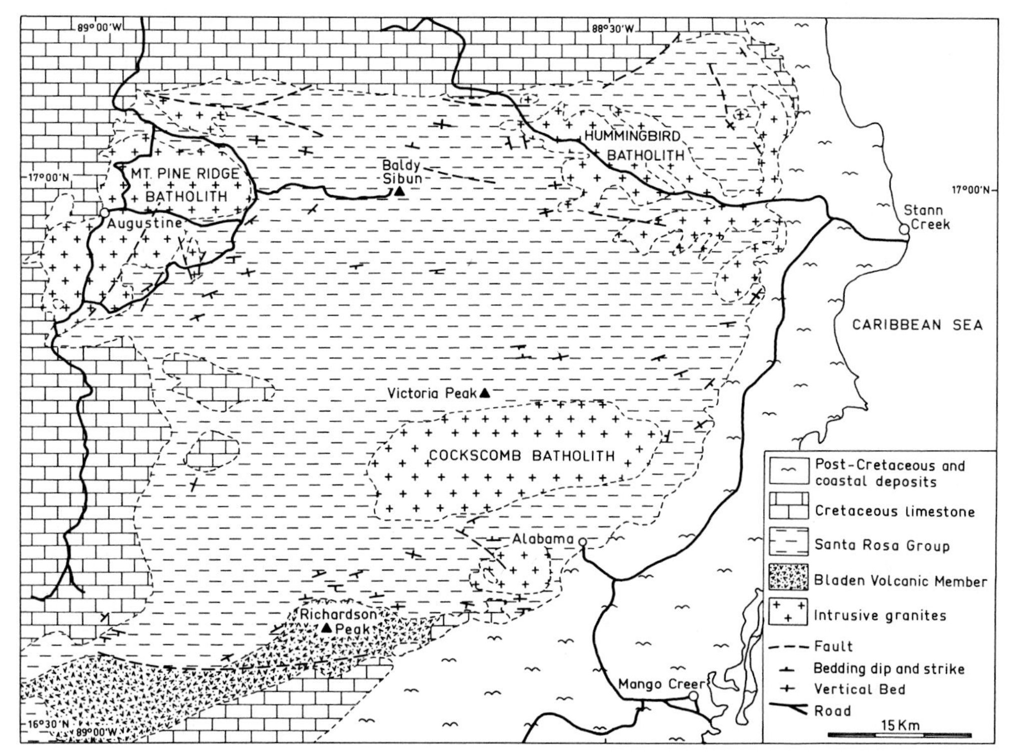

Fig. 53. Geological map of the Maya Mountains, Belize. (Redrawn from HALL & BATESON 1972 and KESLER et al. 1974).

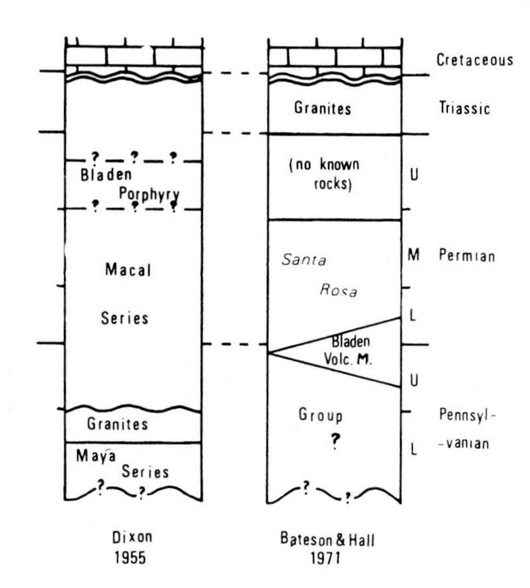

Fig. 54. Stratigraphical correlation of the Maya Mountains after DIXON 1955 and BATESON & HALL 1971. (From BATESON & HALL 1971).

The decisive point on which the views expressed by the new authors differ from the opinions of the earlier researchers concerns the grouping together of DIXON's Maya and Macal series, which is discussed in the section on stratigraphy (p. 32), and their allocation to the Santa Rosa Group of Guatemala. This had to be done because the discordances between the two series were unconfirmed, their structures coincide and the fossil content of both series is evidence of an Upper Carboniferous to Middle Permian age. DIXON's old classification and the new one by BATESON and HALL are compared in Fig. 54.

The most spectacular rocks of the **Santa Rosa Group** are coarse conglomerates with pebbles of hard sandstone and quartzite, some phyllite and contact-metamorphosed sediments. No evidence was seen by BATESON and HALL to indicate the existence of a major regional unconformity between the conglomerate units and the earlier sediments. The conglomerates are associated with sandstones, most of which are subgreywackes. Only some of the fine, well-sorted sandstones can be classified as orthoquartzites. Fine-grained argillaceous rocks occupy the largest part of the Santa Rosa Group. They show effects of regional and thermal metamorphism. A common feature of these rocks is a well-developed close jointing pattern with two or three directions. Dark limestone bands are associated with the argillaceous rocks. Mostly these bands are composed of crinoid fragments, some bivalves and, in a few instances, both solitary and colonial corals. DIXON regarded the limestones as being confined to and characteristic of the "Macal series". Later work, however, has proved the existence of many more crinoidal limestone bands at various stratigraphic levels. BATESON & HALL (1977) summarized the lithofacies distribution within the Santa Rosa Group in a map and cross-section reproduced in our Fig. 55. Paleontological data from DIXON (1955), ROSS (1962), and BATESON & HALL (1977) indicate a general age for the Santa Rosa Group of Pennsylvanian to Middle Permian.

The paleogeographic framework controlling the sedimentation during the Upper Paleozoic has been discussed in Chapter 3.12 and it is illustrated by Figs. 18 and 19. According to BATESON & HALL (1977, p. 28) it "appears to have been an elongate basin trending roughly east-west, with coastlines to the north and east, partially cut off from contemporary sedimentation to the south by the Bladen Volcanic island, but freely connected to the west with the main area of sedimentation".

The volcanics referred to as the **Bladen Volcanic Member** were regarded by DIXON (1955) as a porphyry intrusion while the studies carried out by BATESON & HALL (1971, 1977) revealed a variegated sequence of acid rhyolitic lavas, pyroclasts and volcanic sediments. The interlocking with the sediments of the Santa Rosa Group and a radiometrically determined date of 300 m.y. place the formation of these rocks around the turn of the Upper Carboniferous to the Permian (Fig. 54). Corresponding to our Fig. 19, the paleogeographic situation of these rocks was described by HALL & BATESON (1972) as a volcanic island in the Upper Paleozoic geosyncline.

Six chemical analyses of rocks from these volcanic series reveal the following range of scatter in the main constituents (BATESON & HALL 1977, Table 1):

SiO_2	71.39 – 79.32 %	CaO	tr – 0.92 %
TiO_2	0.07 – 0.53 %	Na_2O	1.13 – 4.90 %
Al_2O_3	10.75 – 13.71 %	K_2O	1.44 – 6.52 %
Fe_2O_3	0.41 – 3.14 %	H_2O^+	1.07 – 4.52 %
FeO	0.26 – 2.26 %	H_2O^-	0.14 – 1.05 %
MnO	tr – 0.37 %	P_2O_5	0.03 – 0.19 %
MgO	0.08 – 1.64 %		

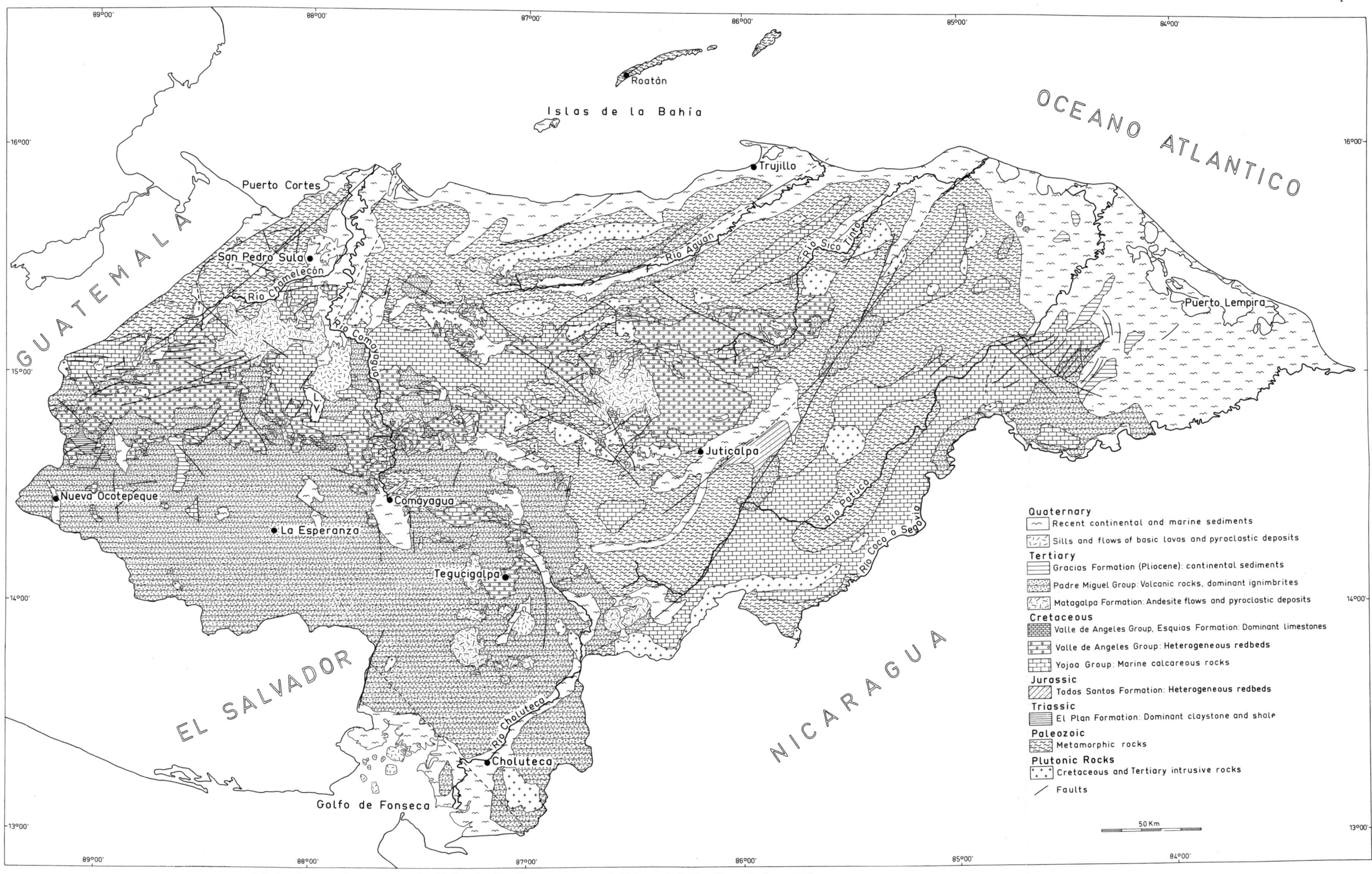

Fig. 58. Geological map of Honduras. (Simplified redrawing from "Mapa Geológico de Honduras 1 : 500,000", Primera Edición, Tegucigalpa 1974).

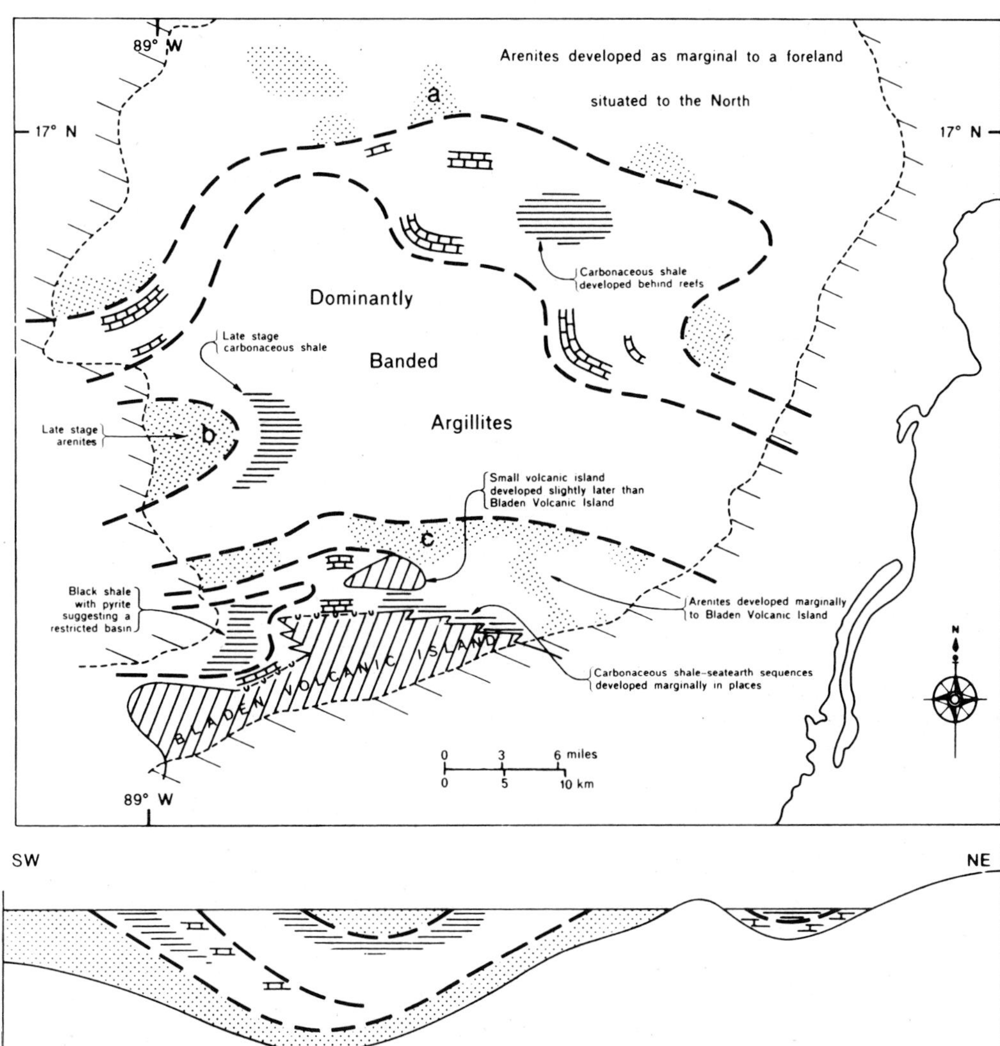

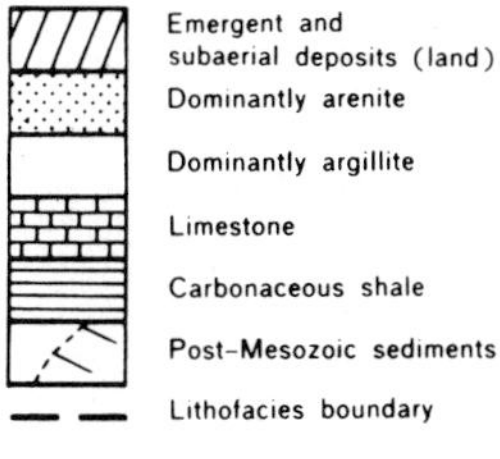

Fig. 55. Lithofacies map and section of the Santa Rosa Group in the Maya Mountains. (From BATESON & HALL 1977, Fig. 3).

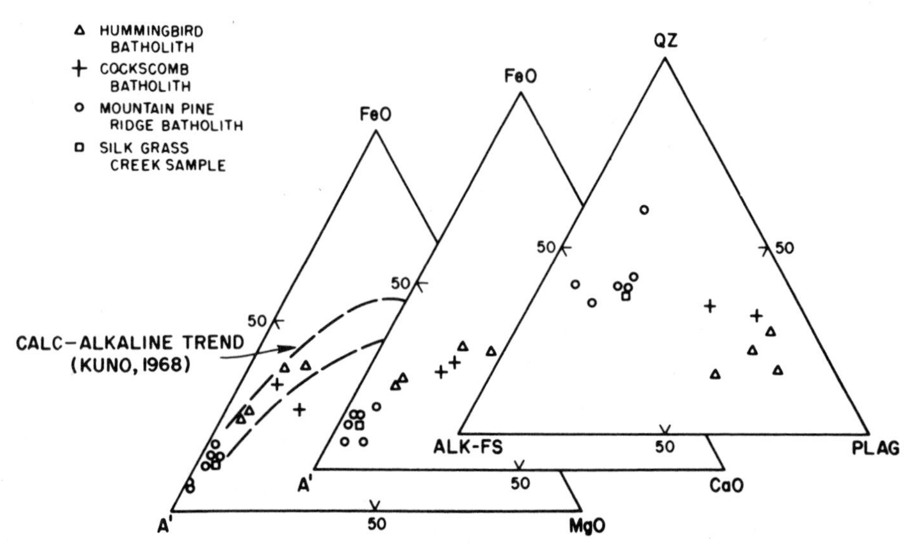

Fig. 56. Modal and chemical variations of plutonic rocks in the Maya Mountains. (From KESLER, KIENLE & BATESON 1974, Fig. 3).

On the basis of these analyses the rocks were classified as alkali rhyolites which, however, underwent secondary chemical conversion.

The **intrusive rocks** of the Maya Mountains occur as three separate batholiths: the Mountain Pine granite, the Cockscomb-Sapote granite and the Hummingbird-Mullins River granite (cf. Fig. 53). Radiometric dating yielded an age of 320 ± 10 m.y. for the Mountain Pine batholith (Rb : Sr whole-rock analyses) and 225 m.y. for the Cockscomb and Hummingbird batholith (argon ages on biotite).

Petrological studies conducted by KESLER et al. (1974) showed that the Mountain Pine batholith is built up of highly differentiated granite, and the other two batholiths are made up of muscovite-containing granodiorite. Judging by their chemical composition they all follow the calc-alkaline trend, but they exhibit finer differences in the modal mineral composition and chemical make-up (Fig. 56). The muscovite content and the high SiO_2 and K_2O values are characteristic and clearly distinguish these rocks from those of the Caribbean islands.

The **structure of the Maya Mountains** is an uplifted fault block consisting of a synclinorium trending east-northeast and plunging to the west at about 10°. According to BATESON & HALL (1977, p. 20) "the area is clearly bordered on the north and south by major faults (Fig. 57). The eastern boundary is probably also formed by a major fault striking north-northeast, parallel to the present coastline. The structure of the western boundary of the fault block is completely obscured by gently folded Cretaceous-Eocene limestones which lie unconformably on the Santa Rosa Group. This simple regional pattern is complicated by the intrusion of several granitic masses arranged in a semicircular arc and by the presence of several faults, some of which have a considerable throw . . . All the lava flows and the bulk of the associated sediments of the Bladen Volcanic Member dip northward, but a few southerly dips were recorded in the metasediments along the southern margin of the area. These may be the southern limb of a small anticline within the main synclinorium, but are probably more likely to be related to drag folding associated with the Southern Boundary Fault."

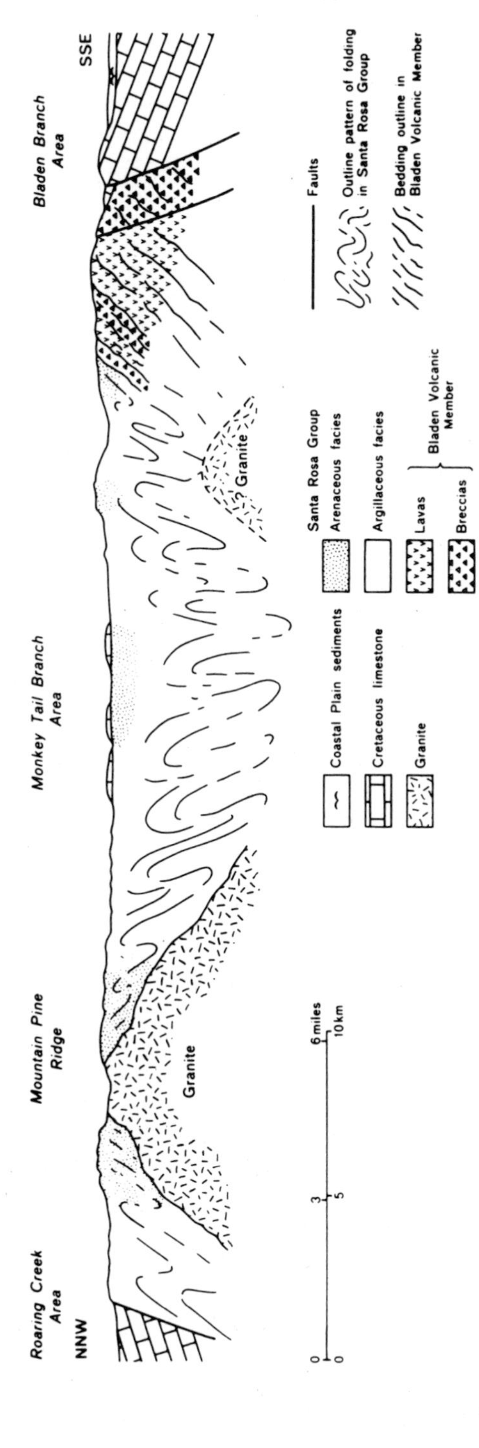

Fig. 57. Generalized geological section across the western flank of the Maya Mountains. (From BATESON & HALL 1977, Fig. 6).

"The main orogeny which deformed the Maya Mountain Block took place in late Permian and early Triassic times. The present horst structure is believed to have been initiated in Middle Cretaceous times, although the major regional lineaments were inherited from features of at least Paleozoic age. Since the Middle Cretaceous the block has been subjected to periodic relative uplift until the mid-Tertiary" (BATESON & HALL 1977, p. 22 and 25).

3.33 Honduras

Up until the sixties Honduras was one of the geologically least known areas of Central America. Since that time, however, a great change has taken place. The petroleum industry has published some of the results of its explorations, particularly of those carried out in the northern regions (MILLS et al. 1967, 1969, MILLS & HUGH 1974). WILLIAMS & MCBIRNEY (1969) reviewed the volcanic history of the country. But the main achievement has been that of the Dirección General de Minas e Hidrocarburos which, in co-operation with the Instituto Geográfico Nacional, ICAITI, the United Development Program, the University of Texas, the Wesleyan University and the Peace Corps, was able to prepare geological maps of large areas (for index map see Fig. 198) and to interpret the stratigraphic sequence, structure and magmatism of these regions. As a result, by 1974 a completely new picture of the geology of Honduras was presented to the IVth Reunión de Geólogos de América Central, and this new knowledge is reflected in a general geological map (scale 1 : 500,000), with explanations by ELVIER (1974), and a comprehensive volume of congress papers (Publicaciones Geológicos del ICAITI, **5**, 1976) (Fig. 58).

The stratigraphic, lithological and paleogeographic knowledge thus obtained about the metamorphic basement rock, the sedimentary Mesozoic and Tertiary have been incorporated into the review of northern Central America in Chapter 3.1. The lineaments of the Upper Tertiary and Quaternary fracture tectonics are also discussed in the review in Chapter 3.2. Honduras proved to be a key region with regard to the volcanic history of the Tertiary, and it is dealt with in Chapter 5.2.

Apart from the coastal plains on the Caribbean and the Gulf of Fonseca, the landscape of Honduras is very broken and mountainous. Its northern part is occupied by mountain chains striking generally WSW/ENE and made up of Paleozoic metamorphics and Mesozoic strata. To the southwest follow mountains and plateaus made up of Tertiary volcanics, and only the extreme southern part of the country forms part of the Pacific Volcanic Belt. The following units are identified in a detailed classification which was carried out by MILLS et al. (1967 and 1969) (Fig. 59):

Caribbean Coastal Plain (Planicie Costera del Caribe): A narrow belt of Tertiary and Quaternary alluvia between the Río Motagua in the west and the Río Negro in the east.

Mosquitia Embayment: Morphologically this is an almost plane hilly landscape transected by gentle ridges; from the geographical point of view it has been described as the "Mosquitia Savanna", in particular by HELBIG (1959). In geotectonic terms this is a zone of subsidence that has existed since the Lower Cretaceous; since that time at least 7,000 m of sediments have accumulated in this area, and it has been greatly fragmented by countless deformation phases (cf. Fig. 37). To the southwest the southern Cordilleras rise up, and to the northeast the Mosquitia continues in the Nicaragua Rise.

Northern Cordillera (Cordillera del Norte): This is part of the Paleozoic crystalline core of Central America to which belong the Sierras de Espiritu Santo, Omoa, Nombre de Diós

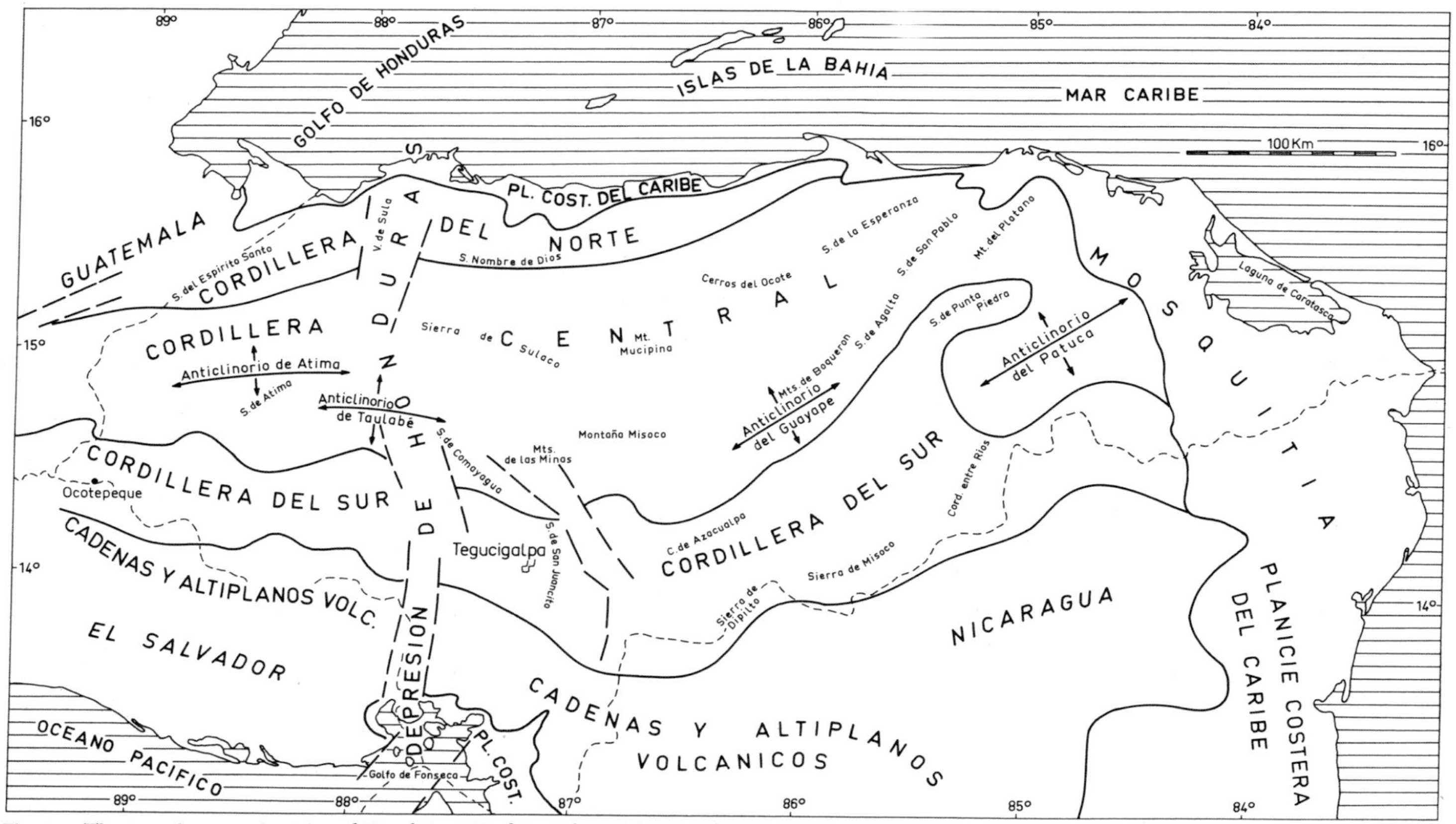

Fig. 59. The morphotectonic units of Honduras. (Redrawn from MILLS et al. 1969, Fig. 2).

and Cangrejal which are made up of gneisses, schists, phyllites and granitic intrusions. The Sierra de Omoa which runs along the border with Guatemala has been studied in detail (Anonymous 1972 e, HORNE et al. 1976 a–c). The individual mountain ranges are delineated by large faults following the general trend and are the precursors of the broad valleys of the Río Chamelecón and the Río Aguán. Wherever Cretaceous limestones were deposited, they have already been eroded again, and the relief of the mountain chains is therefore relatively gentle.

Central Cordillera (Cordillera Central): This range constitutes the core of the intracontinental basin of Honduras, in which 3,000 to 6,000 m of Mesozoic sediments, in particular the limestones of the Yojoa Group, were deposited. These sediments were intensively folded and fractured during the Laramide orogenesis, and the cross-sections in Fig. 36 illustrate the structural style which can probably best be characterized as fault-folding. Individual anticlines, namely those of Atima, Taulabé, Guayapa and Patuca, strike E/W or ENE/WSW. The limestones, in particular the massive deposits of Atima Limestone, give rise to a pronounced relief, often combined with tropical kegelkarst.

The SE/NW-trending structure of the Sierra de Comayagua occupies a key position in the geology of Honduras; it comprises the entire stratigraphic sequence from Paleozoic crystalline rock through the Mesozoic to the sedimentary and volcanic Tertiary. It was therefore examined in more detail in connection with the mapping studies (FACUNDINY & EVERETT 1976, EVERETT & FACUNDINY 1976). Single or multiple deformations occurring during the Paleozoic gave rise to a southwards-dipping isoclinal folding and slight metamorphism. In the course of the Late Cretaceous-Early Tertiary orogenesis the Mesozoic sequence was broken down into E/W-trending folds and upthrusts, and the Comayagua Structure was formed at this time. Following the Tertiary volcanism the N/S-striking Comayagua Graben formed through subsidence. The structural style is illustrated in the cross-sections in Fig. 41 which are based on EVERETT & FACUNDINY (1974, Fig. 4).

Southern Cordillera (Cordillera del Sur): This range is built up of Paleozoic metamorphic rocks, widely scattered erosional remains of Mesozoic and Tertiary sediments and of Tertiary volcanics, and it passes through Honduras from Ocotepeque in the west to Gracias a Diós in the east where it submerges below the Mosquitia Basin. The mountain chains trend NE/SW in the east, while in the western part of the country on the other hand they run parallel to the Honduras Depression from NNW to SSE. The boundary between the Central Cordillera and the volcanic Range and Plateau Province to the south appears somewhat arbitrary because the regions which are referred to as the Southern Cordillera are in fact more a transitional region.

The Volcanic Ranges and Plateaus (Cadenas y Altiplanos Volcánicos) cover the southwestern part of the country and form high-lying plateau landscapes and mountain ranges. Where they are built up of ignimbrites, plateaus bounded by scarps are a characteristic feature. Countless faults determine the course of the deeply incised valleys. The Sierra de Opalaca, Sierra de Montecillos, Montaña de Azacualpa, Montaña de Chile, de Misoco and Curarén are individual mountain ranges composed primarily of Tertiary volcanics. ELVIER (1974, p. 17) also includes the Montañas de Comayagua, although this depends on how the different geological units are demarcated.

The deeply incised volcanic plateau dips towards the Gulf of Fonseca, and the landscape is marked by plateaus gently inclined to the south. WILLIAMS & MCBIRNEY (1969) carried out a detailed study of the volcanic province.

The Pacific Volcanic Belt contributes only a few extinct volcanoes in and around the Gulf of Fonseca to the structure of Honduras.

The **Honduras Depression** (Depresión de Honduras) is a zone of more or less NS-trending faults and grabens extending through the country from the mouth of the Río Ulua to the Gulf of Fonseca, and it was dealt with in detail in Chapter 3.22. The occurrence of basic, in part alkali-rich Pleistocene to Holocene volcanics at Tegucigalpa, on Lago de Yojoa and on the western flank of the Sula Valley emphasizes that this is a tension zone (WILLIAMS & MCBIRNEY 1969, MUEHLBERGER 1976).

3.34 Nicaragua

The northern part of Nicaragua belongs to northern Central America and the Pacific coastal region to southern Central America. Between or over both parts lie the thick cover of Tertiary volcanics, the depression of the Nicaragua Graben and the Quaternary Volcanic Chain, and they thus mask the boundary or the transition between the two main units of Central America.

MCBIRNEY & WILLIAMS (1965, Fig. 1) had identified the following as the most important physiographical units: the Atlantic Coastal Plain, the Interior Highland, the Nicaragua Depression and the Pacific Coastal Plain. They explained this structuring in a geological cross-section running NE/SE, and they joined an Ignimbrite Plateau to the Interior Highland (Fig. 60).

GARAYAR (1977) proposed that the country be divided into structural provinces corresponding to the distribution of the various differently aged formations, and this scheme is reflected in the general geological map of the country (Figs. 61 and 62). By and large in concurrence with GARAYAR we can distinguish between the following features:

The region of crystalline schists of Nueva Segovia
The Bocay Basin
The Mosquitia Basin
The Volcanic Province of Matagalpa
The Volcanic Province of Coyol
The Depression of the Nicaragua Graben
The Quaternary Volcanic Chains
The Cretaceous-Tertiary sedimentation Basin of the Pacific Coastal Region (Nicaragua Trough)
The Caribbean Coastal Plains.

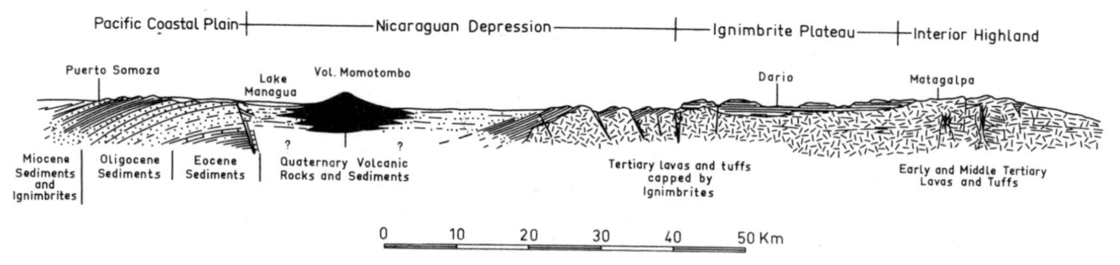

Fig. 60. Generalized geological section through southwestern Nicaragua. (Redrawn from MCBIRNEY & WILLIAMS 1965, Fig. 10).

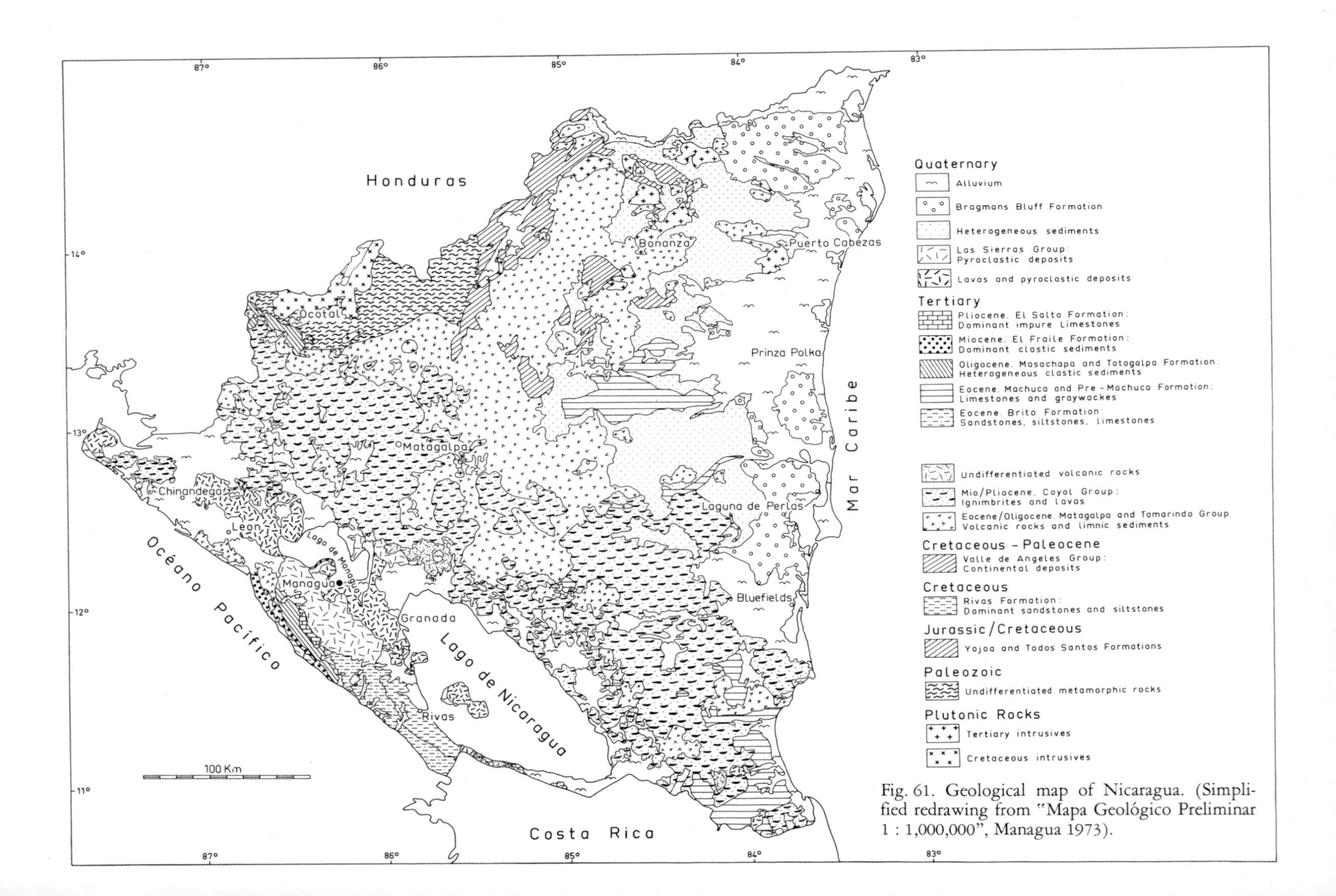

Fig. 61. Geological map of Nicaragua. (Simplified redrawing from "Mapa Geológico Preliminar 1 : 1,000,000", Managua 1973).

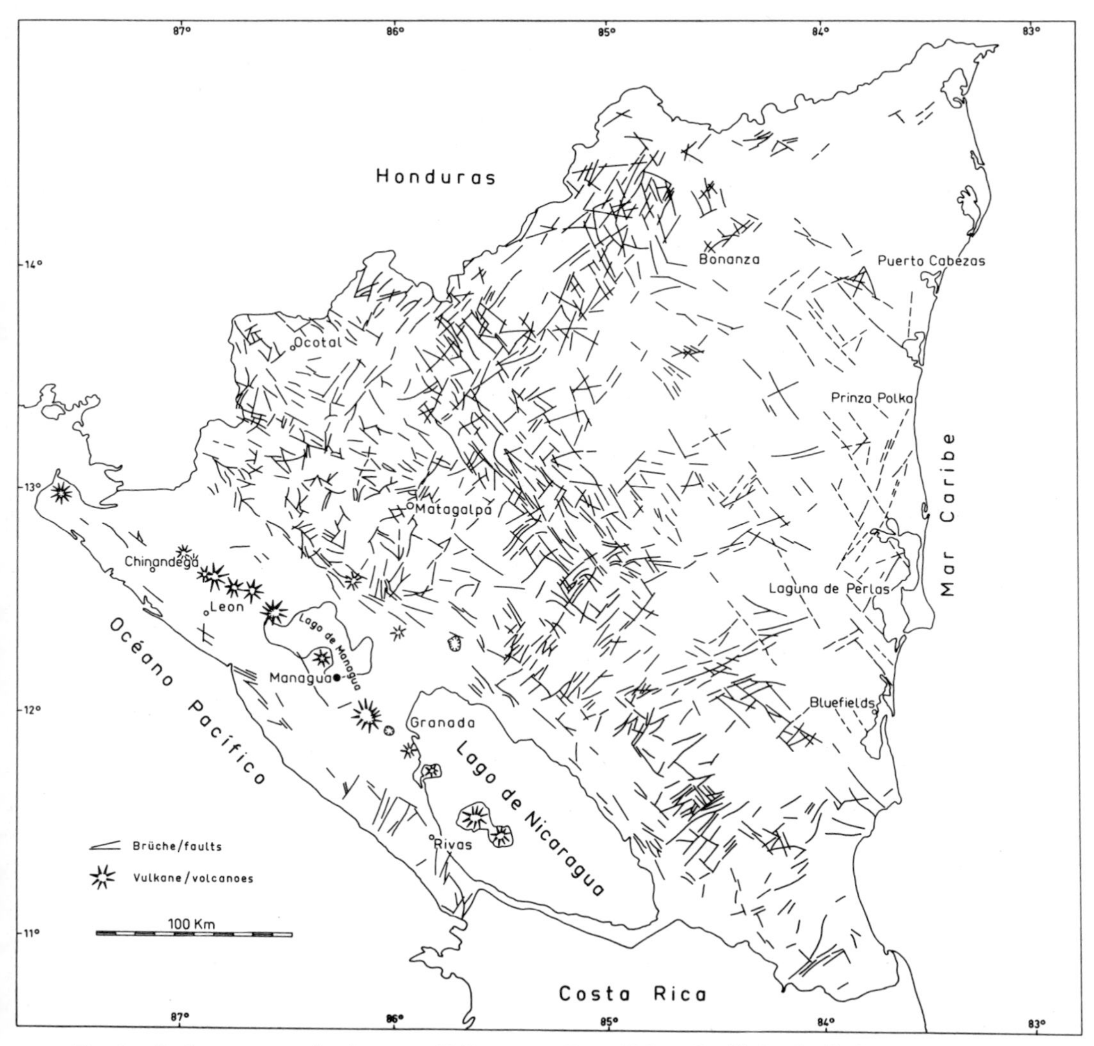

Fig. 62. Fault patterns and volcanoes of Nicaragua. (From "Mapa Geológico Preliminar 1 : 1,000,000", Managua 1973).

The region of crystalline schists of **Nueva Segovia** is situated in the area of the border with Honduras. The crystalline schists consist for the most part of phyllites in which are embedded conglomerates with elongated pebbles, siliceous shales, quartzites, marble, tuffites and diabase-like greenstones. According to ENGELS (1964 a and b) the rocks underwent not only epizonal regional metamorphism but also intensive deformation and — in the vicinity of the contact with the pluton of Dipilto — contact-metamorphism as well. The structural elements recorded by ENGELS strike mainly NE/SW, but many of them may also deviate from this direction.

The age of the metamorphic rocks is the subject of much debate. Some people assign them to the Paleozoic crystalline rock of Honduras and Guatemala (PAZ RIVERA 1962, GARAYAR

1971), and some people regard them as slightly metamorphic Mesozoic. FIGGE (1966) in particular drew attention to the similarity between the rocks which he distinguishes as the **Ocotal Series** and the Cretaceous of Metapán in El Salvador, and he regarded the latter as a post-orogenic deposit of the Lower Cretaceous which was intensively folded and metamorphosed during the Upper Cretaceous and may even have been heated up by the pluton of Nueva Segovia. GARAYAR (1971) points out the similarity between these rocks and those of the El Tambor Formation of Guatemala, although he does not state that they are of the same age.

Since the work carried out in recent years in Guatemala and Honduras has shown that the Cretaceous orogenesis was an extremely intensive process, it does not seem all that wrong to assign a Mesozoic age to these rocks, and this ties in with an earlier observation by ZOPPIS BRACCHI & DEL GIUDICE (1958 b) on the Río Coco; these two authors noticed a gradual transition of Lower Cretaceous sediments into "esquistos". However, the question of age can probably only be settled after detailed studies have been carried out on the metamorphic rocks and their milieu — something which has not yet been done.

The **Bocay Basin** extends in an NE/SW direction along the valleys of the Río Bocay and Río Coco in the northwestern frontier area of Nicaragua. Two expeditions into this rather inaccessible territory (ZOPPIS BRACCHI & DEL GIUDICE 1958 b and PAZ RIVERA 1962) found terrestrial and marine sediments in the eastern part of the metamorphic zone and, following SAPPER's example, they compared these sediments with the Metapán strata of El Salvador. The rocks in question comprise, in the lower strata, terrestrial conglomerates, sandstones and siltstones and thick limestones containing Albian foraminifera and, in the upper strata, a conglomerate series. According to PAZ RIVERA basic and intermediate volcanics occur in the lower strata of these series of sediments and are thus to be taken as proof of a Mesozoic volcanism. The latter could correspond to the volcanics in the Todos Santos Formation of the Mosquitia Basin.

According to GARAYAR (1977) modern stratigraphy distinguishes between the following: Metapán Formation, Todos Santos Group, Yojoa Group and deep parts of the Valle de Angeles Group. Folds follow an NE/SW trend, and their age is given as Laramide.

There are many large and small **intrusive bodies** oriented E/W to ENE/WSW in the region of the Bocay Basin and of the metamorphic rocks. They resemble each other in composition, being made up of granites, quartz monzonites, tonalites, and diorites to peridotites and, according to recent radiometric dating (GARAYAR 1971, Anonymous 1972 b, Table IV), they are between 112.4 ± 1.5 m.y. and 107.5 ± 2.2 m.y. old. Earlier datings by ZOPPIS BRACCHI (1961) had yielded ages of 83 ± 3.0 m.y., and biotite datings ages between 91.6 to 93.3 m.y., but they are regarded as less reliable.

The largest massif is the **Dipilto pluton** in the northwestern part of the country near the border with Honduras; it covers an area of about 400 km². In his mapping of one particular area ZOPPIS BRACCHI (1961) had distinguished an approximately vertical sequence of different rock types, while PIÑEIRO & ROMERO (1962) did not attempt any differentiation in their map. The pluton is characterized by a rich accompaniment of dike rocks such as aplites, lamprophyres, pegmatites and pegmatitic-pneumatolytic lodes which have triggered intensive but so far unsuccessful exploration for useful ore deposits.

GARAYAR & VIRAMONTE (1971, 1973) found an unusually wide range of rocks in a small intrusive stock 13 km long near San Juan del Río Coco containing hornblende-peridotites, hornblende-gabbrodiorites, hornblende-diorites, biotite-tonalites and granodiorites. They constitute a sequence of basic to acid rocks (SiO_2 55.0 – 74.1%) which, on the basis of in-the-field findings, thin-section-examination and chemical analysis, may be interpreted as the result of gravitational differentiation in the originally vertically aligned stock. The present horizontal configuration is taken to be the result of tectonic deformation.

The plutonic rocks which are dated as Cenomanian are seen as being connected with Middle Cretaceous deformations in the northern part of Nicaragua, and they may have heated up the crust to such an extent that the right conditions were created for slight metamorphism of at least some, and possibly all, of the metamorphic rocks of the age in question. They are spatially and temporally separated by Tertiary intrusive stocks in the region of the Matagalpa and Coyol Formations adjoining to the south and southeast so that, together with WILLIAMS & MCBIRNEY (1969, p. 6), we have to assume that the plutonism (and volcanism) shifted southwards in the course of time. This may provide useful information for reconstructing older subductions in the light of plate theories, and such information has already been used by MALFAIT & DINKELMAN (1972, Figs. 1 and 2) to determine the position of earlier plate boundaries.

In the northeastern continuation of the Bocay Basin, the **Mosquitia Basin** extends on both sides of the Río Coco over the plains of the Honduran and Nicaraguan coastal regions, the Mosquitia, and continues in the Nicaragua Rise. The continental and shelf regions account for approximately equal proportions of the basin which covers about 62,500 square miles and which was recognized by the petroleum exploration expeditions of the sixties as one of the largest Mesozoic-Tertiary sedimentation basins in Central America (KARIM et al. 1966, MILLS et al. 1967, MILLS & HUGH 1974, ARDEN 1969 and 1975).

From the turn of the Triassic/Jurassic onwards the Mosquitia Basin developed between the "Nuclear Foreland" of Honduras, a "Hinterland Region" along the border between Honduras and Nicaragua, and a "Nicaragua Foreland", which may have stretched from eastern Nicaragua into the region of the islands of San Andrés and Providencia. In Honduras the axis of Mesozoic sedimentation was situated between Río Patuca and Río Coco, and crustal movement at the end of the Cretaceous divided the basin into two separate basins on both sides of a Coco River Ridge (see map in Fig. 29 and the cross-sections in Figs. 30 and 37). The Río Coco Ridge continues in the Nicaragua Rise and thus represents a link with Jamaica as an advance guard of Central American structural elements.

Drilling in the shelf and coastal regions as well as mapping activity has shown that the Mesozoic stratigraphic sequence of the Mosquitia Basin is more than 2,000 m thick and the Tertiary sequence of the shelf region is more than 5,000 m thick. These sequences are summarized in the correlation chart in Table 2. The oldest post-Paleozoic sediments are fine-grained dark sandstones and pelites of the Triassic/Jurassic El Plan Formation, which is about 300 m thick. They are followed by 300 – 700 m thick red beds and andesitic volcanics of the Todos Santos Formation (Jurassic to Neocomian). The latter are overlain by the limestones of the Yojoa Group which extends from the Aptian to the Turonian and attains maximum thicknesses of up to several thousand metres.

Due to intensive fracture tectonics and uplift the Upper Cretaceous and Early Tertiary take the form of renewed terrestrial red beds, the Valle de Angeles Group with andesitic volcanics. The basin is now divided by the Río Coco Ridge, and Middle and Late Tertiary shallow-sea and continental deposits accumulated on both sides of the ridge; according to drilling profiles (see Fig. 30) these deposits together with the Valle de Angeles Group attain thicknesses up to 5,000 m.

Drilling and seismic data indicate the presence of up to 6,000 m thick Tertiary and Upper Cretaceous, and below that a further 5,000 m of Lower Cretaceous, volcanics and metamorphics in the Nicaragua Rise, i.e. a thick continental series of rocks between the semi-oceanic crust in the Cayman Trench and the Colombia Basin (Fig. 13).

The tectonic development of the Mosquitia Basin has been described by MILLS & HUGH (1974) in a serial section extending from the El Plan Trench through a Late Jurassic orogenesis and a geosynclinal period in the Cretaceous to the Laramide crustal movements and Tertiary fracture tectonics (Fig. 37). Together with the profiles from Honduras published in particular by MILLS et al. (1967), WILSON (1974) and EVERETT & FAKUNDINY (1976), they provide evidence of very much stronger Laramide tectonic activity than was previously realized. The associated volcanism, which was also only recently recognized, and in particular vigorous plutonic activity during the Cretaceous, complete the picture of violent orogenesis in northern Central America.

The **Volcanic Provinces** of Matagalpa and Coyol are discussed in Chapter 5 in connection with the Tertiary volcanism of northern Central America. However, the reader's attention is drawn here to the correlation chart (Table 3).

The **Depression of the Nicaragua Graben** forms part of the Pacific Marginal Zone of Central America, which is characterized by deep-sea trenches, shelf regions, chains of active volcanoes, earthquakes and still active fracture tectonics. This elongate depression filled with lakes extends between the Tertiary volcanic plateaus in the northeast and the Pacific mountain and hill country in the southwest, and its central section presents the appearance of a graben which is sharply delineated on both sides. Correspondingly the edges were described in the literature, in particular in regional surveys, as faults running NW/SE in an absolutely straight line (e.g. SAPPER 1937, Table VIII, DENGO 1968, Fig. 9, DENGO, BOHNENBERGER & BONIS 1970, Figs. 1 and 2). The geological mapping of western Nicaragua (Anonymous 1972 b) revealed, however, that the edges of the graben are formed by a highly fragmented block mosaic and that transverse structures within the graben play an important role. The presentation in Figs. 62 and 64 is based on this knowledge. A corresponding cross-section was drafted by MCBIRNEY & WILLIAMS back in 1965; these authors regarded the northeastern margin as an antithetic fault and the southwestern margin in the region of Managua as the fault scarp of an uparching inclined towards the Pacific (Fig. 60). The term Nicaragua Depression, which is commonly used in the literature, is perhaps more appropriate to the structure of this macro-feature than the more precisely defined concept of a "graben".

As a morphological-tectonic structure the Nicaragua Depression extends from the Gulf of Fonseca into the Atlantic Coastal Plains of Costa Rica, i.e. it is over 500 km long and about 50 km wide. The floor of the Depression is about 35 to 50 m above sea level, and it is approximately half covered by Lake Managua and Lake Nicaragua, both of which drain via the Río San Juan into the Caribbean.

The Depression appears most clearly as a morphological graben in the region of the City of Managua. In the southwest the **Coastal Range,** with the 250 m high, sharp cuesta of the Mateare Fault, rises to a height of over 900 m and then slopes gently away from its flat crest down to the Pacific. It does not appear to be referred to by any one particular name; instead, terms such as "Sierras", "Sierras de Carazo", "Cuestas de Diriamba" and "Cordillera del Pacífico" are used to describe it. The Mateare Fault can be traced for 70 km, then it flattens out at both ends and submerges beneath the alluvial deposits which fill the Depression. The Coastal Range is interpreted as a tilted fault block dipping towards the southwest. Its uplifting and thus the subsidence of the Depression took place after the deposition of the Plio/Pleistocene Sierras Group, and these processes are still going on even today. Opposite the Coastal Range the terrain on the northeast side of the graben in the Cuesta Coyol rises to altitudes of about 500 m; and above this in the distance the mountain ranges of the Highland of Nicaragua rise to an altitude of over 1,500 m.

To the northwest the Coastal Range drops away to the **Mesas de Tamarindo** (see map in Fig. 64), and these disappear below the alluvial plain between the coast and the chain of Marabios volcanoes. To the southeast as well the sharp Mateare Escarpment is replaced by the hilly landscape of the Serranías de Brito in front of which lies the Rivas Plain. Although it is more dissected, the northeastern boundary of the Depression can be traced into the hinterland of the Gulf of Fonseca; to the southeast it fades away as it approaches the Caribbean coast so that the Depression comes to an end in the lowland plains of northern Costa Rica. It is possible that the Depression abuts against the Cordillera de Guanacaste and the Cordillera Central with extremely large-throw faults, but little is known about the latter.

A series of extinct Pleistocene volcanoes is situated on the northeastern flank of the Depression (McBIRNEY 1964); the inside is followed by the volcanic Marabios Range and the range of volcanoes southeast of Managua. These will be discussed in connection with the Quaternary volcanism of Central America.

Faults running transversely and diagonally to the longitudinal orientation of the Depression — McBIRNEY & WILLIAMS (1965, p. 43) draw special attention to these — reveal themselves in scarplets, chains of small volcanoes, offset of Holocene and Pleistocene deposits in the subsoil of Managua and in a particularly dramatic way in the faults which were activated by the Managua earthquakes (see map, Fig. 159). CARR & STOIBER (1977) regard the structures striking N 30° E diagonally to the Depression as the expression of deep-reaching segment boundaries within the Pacific Marginal Zone which result in variations in the dip of the Benioff zone, offset of the chains of volcanoes and varying volcanic activity. Such segment boundaries run through the Gulf of Fonseca, through the region of Lake Managua and between the volcanoes of Lake Nicaragua and the Cordillera de Guanacaste of Costa Rica.

Interpretation of new information (SCHWARTZ et al. 1975, CLUFF et al. 1976) suggests that Managua is located in a major north-south structural depression which is referred to as the **Managua Graben.** The western boundary is defined by a zone of intensive volcanic collapse striking N/S to N 15° W and containing the Nejapa Pits (Fig. 131); the eastern boundary is an alignment of normal faults, striking N/S to N 15° E, extending from the Masaya Caldera through Tipitapa. Within the graben are dominantly tensional N/S faults and left-slip faults striking N 35 – 40° E (Fig. 159) which were partly activated during the 1972 Managua earthquake.

The volcanogenic deposits of the Sierras Group belong to the sedimentary filling of the Depression. They are overlain by similar but less consolidated ashes, pumices, lapilli, lahars and their derivatives which have been re-deposited by fluviatile action. A series of fossil soils marks interruptions in the process of sedimentation. One of these soils dated at 5,945 ± 145 years has acquired fame because of the human footprints found at Acahualina. In order to plan the re-building of Managua in a reasonably earthquake-proof manner, it is important to distinguish the boundary between the Pleistocene and Holocene deposits and in particular to establish to what extent tectonic faults run through either just the Pleistocene or also the Holocene thus indicating their possible seismic activity. A diagram of the recent de-

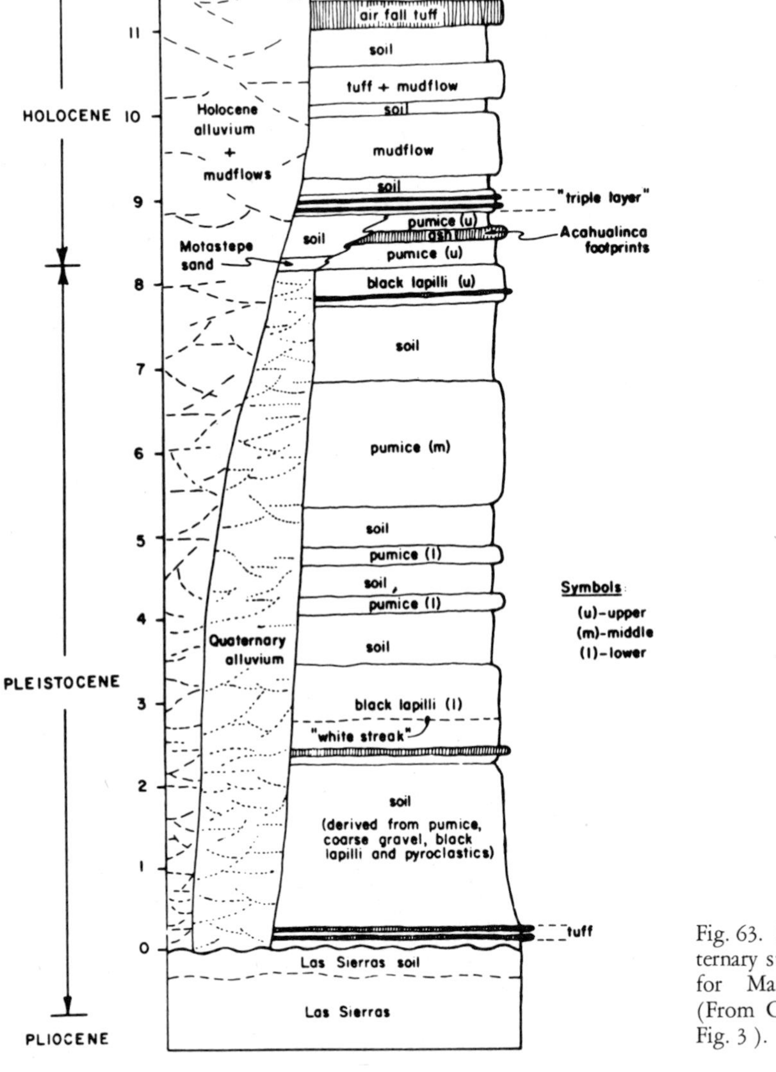

Fig. 63. Preliminary Quaternary stratigraphic column for Managua, Nicaragua. (From COLLINS et al. 1976, Fig. 3).

posits (Fig. 63) is taken from a provisional report by COLLINS, NICCUM & BICE (1976). SWAIN (1961, 1966) has published studies on the recent lake sediments.

Altogether the Nicaragua Depression is one of the most striking elements in the structure of Central America, and it marks the boundary between the northern and southern part of the land bridge. The Tertiary volcanic plateaus of the northern part end at this Depression, and to the south follows the Cretaceous-Tertiary Nicaragua Trough which can already be regarded as an element of southern Central America. The question whether the Nicaragua Depression extended northwestwards in the region of El Salvador and Guatemala was dealt with in Chapter 3.22 in connection with the fault structures of northern Central America.

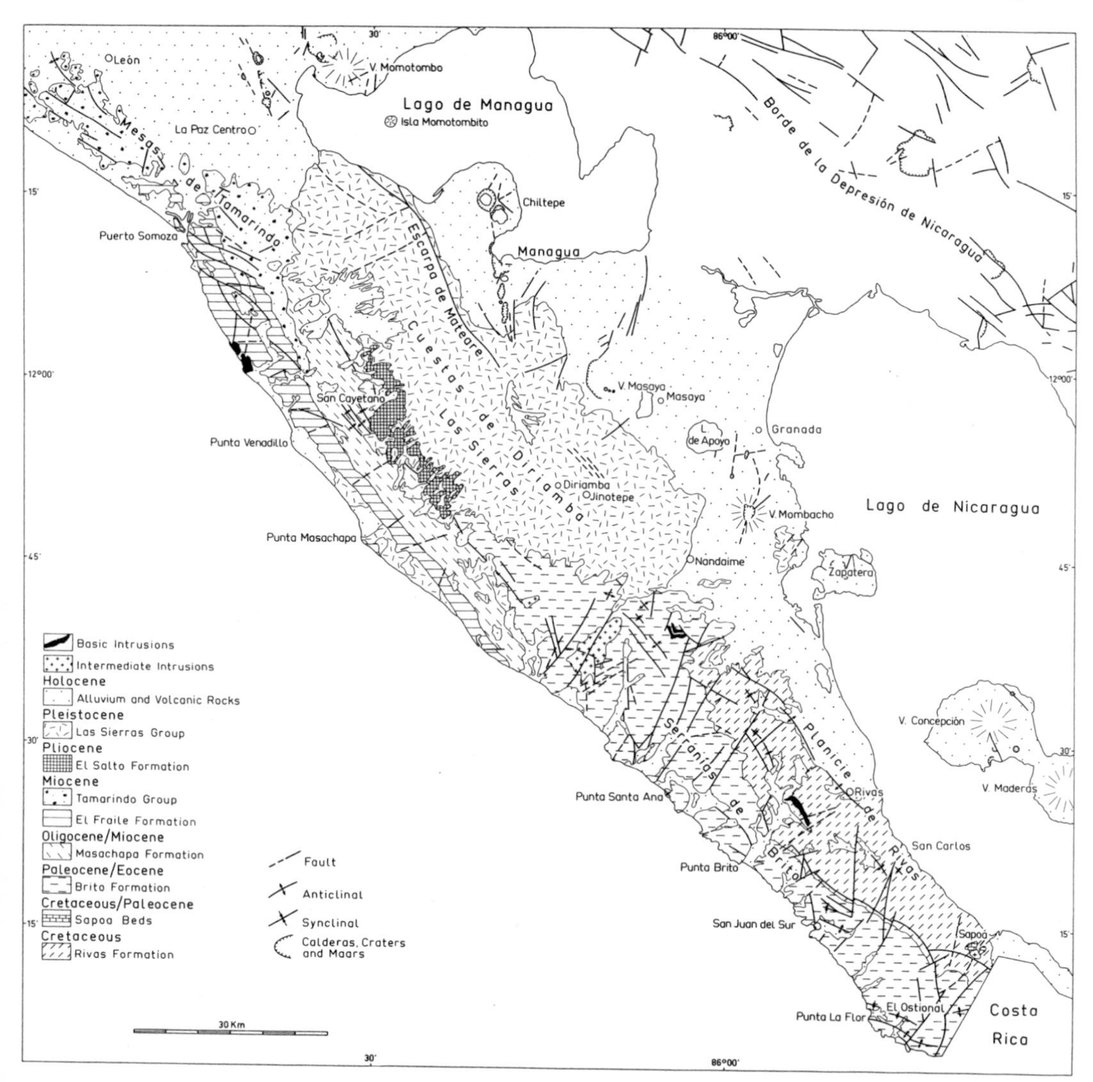

Fig. 64. Geological map of southwestern Nicaragua. (Redrawn from "Mapa Geológico de Nicaragua Occidental 1 : 250,000", Managua 1972).

The Cretaceous-Tertiary sedimentation basin of the Pacific Coastal Zone of Nicaragua, which GARAYAR (1977) briefly refers to as the **Nicaragua Trough ("Cuenca de Nicaragua")**, extends between the Nicaragua Depression and the Pacific shelf and continues to the southeast in Guanacaste province and in the Limón Trough of Costa Rica. It is filled with a series of strata almost 10,000 m thick ranging from the Cenomanian to the Upper Miocene. This stratigraphic sequence crops out in the Serranías de Brito, the Planicie de Rivas, the Cordillera del Pacífico and the Mesas del Tamarindo, and it was described for the first time by HAYES (1899) in studies made for a canal route. More detailed investigations were carried out in the course of petroleum exploration, and the results of this have been summa-

Table 3. Stratigraphic correlation chart of southwest Nicaragua. (From "Mapa Geológico de Nicaragua Occidental 1 : 250,000", Managua 1972).

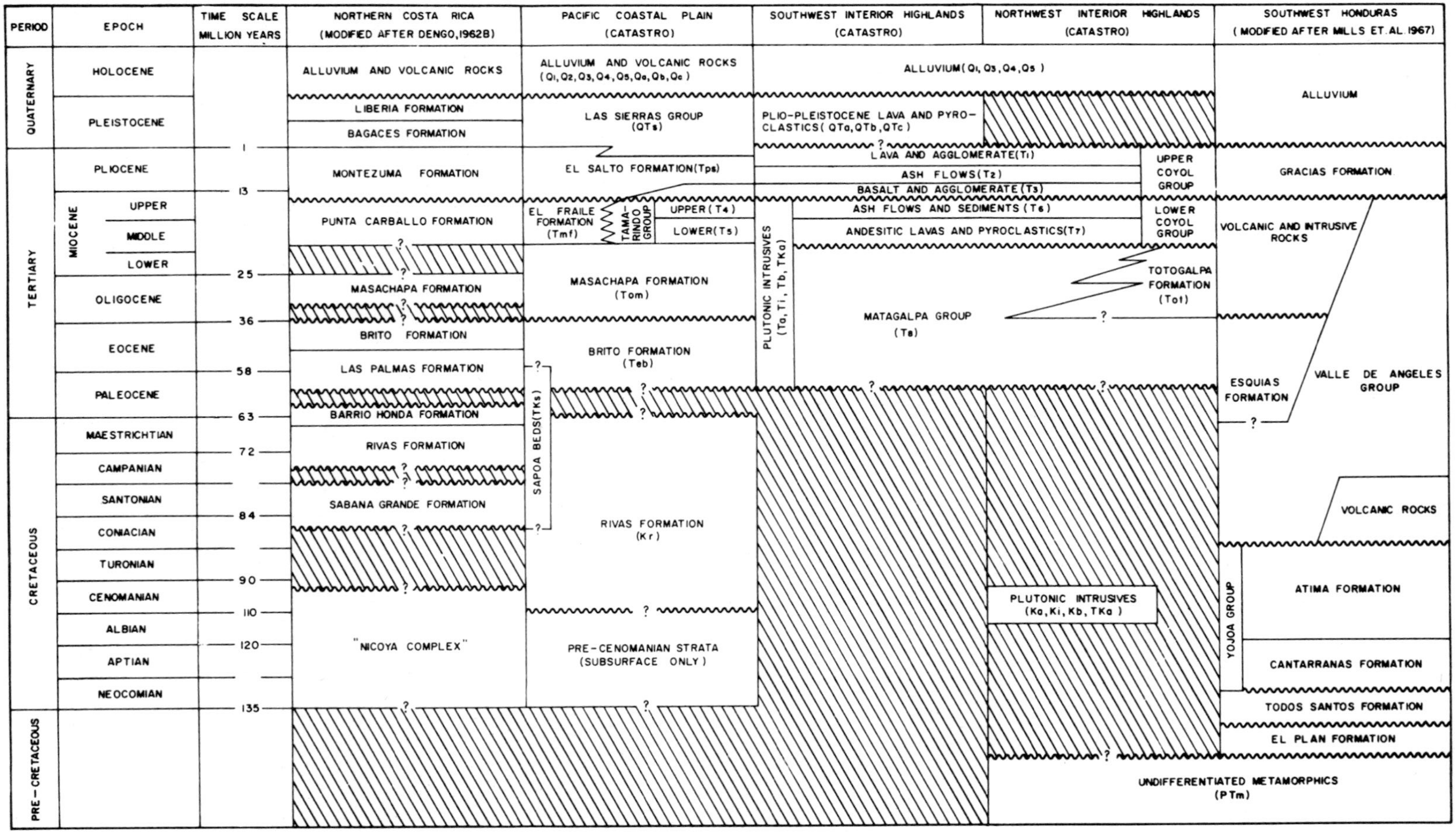

PERIOD	EPOCH	TIME SCALE MILLION YEARS	NORTHERN COSTA RICA (MODIFIED AFTER DENGO, 1962B)	PACIFIC COASTAL PLAIN (CATASTRO)	SOUTHWEST INTERIOR HIGHLANDS (CATASTRO)	NORTHWEST INTERIOR HIGHLANDS (CATASTRO)	SOUTHWEST HONDURAS (MODIFIED AFTER MILLS ET. AL. 1967)
QUATERNARY	HOLOCENE		ALLUVIUM AND VOLCANIC ROCKS	ALLUVIUM AND VOLCANIC ROCKS (Q1, Q2, Q3, Q4, Q5, Qa, Qb, Qc)	ALLUVIUM (Q1, Q3, Q4, Q5)		ALLUVIUM
	PLEISTOCENE	1	LIBERIA FORMATION BAGACES FORMATION	LAS SIERRAS GROUP (QTs)	PLIO-PLEISTOCENE LAVA AND PYROCLASTICS (QTa, QTb, QTc) ?		
TERTIARY	PLIOCENE	13	MONTEZUMA FORMATION	EL SALTO FORMATION (Tps)	LAVA AND AGGLOMERATE (T1) ASH FLOWS (T2) BASALT AND AGGLOMERATE (T3)	UPPER COYOL GROUP	GRACIAS FORMATION
	MIOCENE UPPER		PUNTA CARBALLO FORMATION	EL FRAILE FORMATION (Tmf); TAMARINDO GROUP: UPPER (T4)	PLUTONIC INTRUSIVES (Ta, Ti, Tb, TKa); ASH FLOWS AND SEDIMENTS (T6)	LOWER COYOL GROUP	VOLCANIC AND INTRUSIVE ROCKS
	MIOCENE MIDDLE		?	LOWER (T5)	ANDESITIC LAVAS AND PYROCLASTICS (T7)		
	MIOCENE LOWER	25	?	MASACHAPA FORMATION (Tom)	MATAGALPA GROUP (Tm)	TOTOGALPA FORMATION (Tot)	
	OLIGOCENE	36	MASACHAPA FORMATION ?			?	
	EOCENE	58	BRITO FORMATION	BRITO FORMATION (Teb)			VALLE DE ANGELES GROUP
	PALEOCENE	63	LAS PALMAS FORMATION BARRIO HONDA FORMATION	SAPOA BEDS (TKs) ?	?	?	ESQUIAS FORMATION ?
CRETACEOUS	MAESTRICHTIAN	72	RIVAS FORMATION	RIVAS FORMATION (Kr)			
	CAMPANIAN		?				
	SANTONIAN	84	SABANA GRANDE FORMATION				VOLCANIC ROCKS
	CONIACIAN		?	?			
	TURONIAN						YOJOA GROUP: ATIMA FORMATION
	CENOMANIAN	90	?			PLUTONIC INTRUSIVES (Ka, Ki, Kb, TKa)	
	ALBIAN	110	"NICOYA COMPLEX"	? PRE-CENOMANIAN STRATA (SUBSURFACE ONLY)			
	APTIAN	120					CANTARRANAS FORMATION
	NEOCOMIAN						TODOS SANTOS FORMATION
PRE-CRETACEOUS		135	?	?		? UNDIFFERENTIATED METAMORPHICS (PTm)	EL PLAN FORMATION UNDIFFERENTIATED METAMORPHICS (PTm)

rized by ZOPPIS & DEL GIUDICE (1958 b). More recent publications are those by PAZ RIVERA (1962), KUANG (1971) and an anonymous report (Anonymous 1972 b) which was prepared together with the general 1 : 250,000 scale map of western Nicaragua (Fig. 64).

Table 3 provides an overview of the stratigraphic sequence which commences with the **Rivas Formation.** It consists of tuffitic shales and siltstones, tuffitic sandstones and conglomerates, marly shales and individual beds of limestone. Interbedded with these are sills of basic volcanics and an andesitic agglomerate which is taken to be an intrusive product. The amount of volcanic material increases with depth. The thickness fluctuates – in a type section at the surface it is given as 2,660 m, and in a borehole (Rivas 1) as 3,435 m. The lowermost sections of the Rivas Formation, which crop out in the core of the anticline of the same name, consist for the most part of laumontite, low-temperature albite, quartz and green amphibole. BASS (in MCBIRNEY & WILLIAMS 1965, p. 9) therefore includes them in the zeolith facies of the metamorphism. No higher metamorphic stages were encountered.

The age is attested by foraminifera as Cenomanian to Maestrichtian, and according to BANDY (vide Anonymous 1972 b, p. 43) *Rotaliopora appeninica* is an important key fossil for Cenomanian. SCHMIDT-EFFING (oral communication) states that some of the strata of the Rivas Formation date from the Tertiary; given the lithological similarity of the entire stratigraphic sequence, only micropaleontological analysis can confirm this fact. The relationships of the facies to the stratigraphic sequence of northwestern Costa Rica also still remain to be explained, particularly since the sequence probably encompasses the entire Cretaceous and also parts of the Early Tertiary.

At Peñas Blancas, near the Costa Rican frontier, massive reef limestones up to a maximum of 30 m in thickness crop out and were given the name **Sapoá strata.** They consist chiefly of algal limestones but so far no identifiable fossils have been found. Their stratigraphic position is therefore still unclear. Either they are a neritic facies of the Rivas Formation or of the following Brito Formation. According to observations made by SCHMIDT-EFFING (oral communication) these are slide masses in which the limestone blocks are embedded in an argillaceous matrix.

The **Brito Formation** has been the subject of several studies since its first description by HAYES (1899), and these have been summarized by Anonymous (1972 b, pp. 49/50) to show that the Formation contains the entire Eocene and parts of the Upper Paleocene. It is separated from the Rivas Formation by a conglomerate which can be interpreted as evidence that sedimentation was interrupted. There does not seem to be any discordance in Nicaragua, while DENGO (1962 b, p. 20) observed a discordance in Guanacaste.

The thickness of the Brito Formation is given as 2,400 to 3,400 m; in a profile recorded by KUANG (1971) the thickness was determined to be 2,570 m, and this is taken as representative. The profile provides an impression of the lithological sequence (from the upper to the lower strata):

420 m	Calcareous sandstones, thin limestone lenses, marls, tuffitic sandstones, thick tuffs and agglomerates
220 m	tuffitic sandstones (greywackes) with beds of carbonate pelites
550 m	greywackes with lenses of calcarenites, tuffites, limestone lenses, marls
300 m	argillaceous-calcareous sandstones, greywackes, conglomeratic sandstone
600 m	greywackes interbedded with pelites, lenses of conglomeratic sandstones, pelites
120 m	pelites with thin beds of greywacke
360 m	fossiliferous tuffites, lenticular limestone beds, marls, greywackes and calcareous sandstones
2,570 m	

The **Masachapa Formation,** which follows, overlies the slightly folded Brito Formation with an angular unconformity, while it itself dips at a uniformly shallow angle to the southwest. It consists of a monotonous sequence of tuffitic and carbonate shales, siltstones and sandstones, with the latter increasing towards the upper strata in the sequence. The presence of fossilized wood indicates proximity to land. Intercalations of volcanic rocks are less frequent than in the Brito Formation. The thickness, which is given by KUANG (1971) as 1,683 m, deviates from earlier data (2,600 m), but it is taken to be accurate. Because key fossils are lacking and because of the lithological similarity with the overlying El Fraile Formation, the stratigraphic position of the Masachapa Formation is not quite certain; however, it is placed in the Oligocene to Lower Middle Miocene.

The **El Fraile Formation,** which according to KUANG (1971) commences with a volcanogenic conglomerate and an almost contemporaneous agglomerate, follows concordantly and without any distinct lithological boundary. In the central section of the Formation the grain size declines and the upper strata are composed of tuffitic sandstones and siltstones. A great deal of fossil wood is present, and this was earlier interpreted as the criterion for identifying the boundary with the Masachapa Formation. The total thickness is given as more than 2,700 m. The middle and lower elements of the El Fraile Formation interlock with the volcanic Tamarindo Group which is identical with the lower Coyol Group of central Nicaragua. Accordingly it is possible to date both groups in the Middle Miocene. Marine fauna determined by FAGGINGERAUER (1942) (vide Anonymous 1972 b, p. 60) are also supposed to be indicative of Middle Miocene while prior to that date WOODRING (in SCHUCHERT 1935, p. 60) put the Formation in the Lower Miocene.

The El Fraile Formation is regarded as an inshore neritic and delta deposit with a rich accumulation of woods and pyroclastic material. In a northeastern direction, i.e. as it approaches the Miocene continent, the Formation merges into the **Tamarindo Group,** which is built up of a basal andesite with agglomerates and overlying dacitic to rhyodacitic tuffs and ignimbrites. Their thickness is given as 586 m. From the interlocking with the El Fraile Formation on the one hand, and according to radiometric dating of 13.7 ± 0.5 m.y. on the other, the age is Middle Miocene. With its volcanics the Tamarindo Group is fully equivalent to the Coyol Group of the central highland of Nicaragua. A detailed description of the rocks can be found in MCBIRNEY & WILLIAMS (1965, pp. 10/11). The ignimbrites stand out clearly in the landscape because of their characteristic scarps and plateaus.

The residues of eroded marine neritic sediments in the form of shell limestones, marls and impure limestones, which are overlain by tuffitic sandstones and siltstones, are found on the slope of the Sierras or Cuestas de Diriamba. Near San Rafael del Sur this material is exploited for the production of cement and it has yielded a rich molluscan fauna, which according to WOODRING indicates Lower Pliocene and a Pacific character with a Peruvian-Ecuadorian mixture of species (WOODRING 1973, pp. 181/182, 1976). The strata, which are altogether a maximum of 100 m thick, were described by ZOPPIS & DEL GIUDICE (1958 a, p. 43) as the **El Salto Formation.**

With this Formation are interlocked the basal sections of a thick series of basic pyroclastic deposits which, on the basis of their wide distribution in the Sierras, were designated the **Sierras Group.** According to KUANG (1971) they attain thicknesses of 680 m. Deposition occurred mainly on the continent and only locally in shallow water where the tuffitic sediments are cemented with carbonate and contain a marine molluscan fauna, which according to WOODRING (vide Anonymous 1972 b, p. 66) has the same age as the fauna of the El Salto Formation. However, the upper sections of the Sierras Group are placed in the Pleistocene. On the basis of its origin and composition the Sierras Group belongs to the Quaternary volcanism of the Nicaragua Depression.

Nearly all the formations of the Nicaragua Trough contain dikes, sills and stocks of basic to intermediate **igneous rocks** which were referred to as diabases, porphyric basalts and andesites, phenoandesites, dioritic porphyrites, diorites, gabbros, orthoclase gabbros (ZOPPIS & DEL GIUDICE 1958 b, pp. 51 – 54, WEYL 1961 a, p. 55, MCBIRNEY & WILLIAMS 1965, pp. 55 – 56, KUANG 1971, pp. 71 – 76). One of the sills, the "Loma las Mesas Sill", was radiometrically dated as 49.3 ± 4.0 m.y. old (BARR & ESCALANTE 1969); a chemical analysis is available of the gabbrodiorite of the Cerro Abejonal (WEYL 1961 a, p. 146).

Although detailed petrographical and chemical analyses of these sub-volcanic to plutonic rocks have still to be performed, this material, just like the volcanics in the filling of the Trough, indicates basic to intermediary chemical make-up of the magmas. The sialic magmatism of the central highland of Nicaragua extends to the margin of the Trough only in the tuffs and ignimbrites of the Tamarindo Group, and it is then in turn replaced by the basic flows of the Sierras Group. This may be a reflection of the end of the subsidence of the Trough during the Late Tertiary and of the beginning of tension tectonics during the Quaternary.

The **structure of the Nicaragua Trough** in its present form is determined by slight folding faults and some intrusive bodies (Fig. 64). The general strike of the folds is NW/SE. The predominant structures are the fragmentary anticline of Rivas, the group of folds in San Cayetano north of Masachapa, and the El Ostional anticline. The deepest, low-grade metamorphic Rivas Formation crops out in the Rivas anticline, the San Cayetano folds are contained in the Masachapa Formation, and those of El Ostional in the Brito Formation. The folding took place at the end of the Miocene, and it no longer affected the Pliocene El Salto Formation.

Some of the fracture structures run parallel to the fold axes in a NW/SE direction, but many also run at right angles to them. The structures extend into the Pleistocene deposits of the Sierras Group and form in places very clear fault scarps. The concentration of the centres of shallow earthquakes within the Nicaragua Graben on the one hand and the NE/SW-striking faults which were opened up during the Managua earthquake of 1973 on the other are evidence of ongoing tectonic activity. The, on the whole, gentle slope of the Cuesta de Diriamba down to the Pacific — the road to Masachapa runs down the slope — indicates that the Cuesta is a tilted block, the face of which is directed towards the Nicaragua Graben. The altitude of the marine El Salto Formation near San Rafael, namely about 100 m, is the result of post-Pliocene uplift.

Looking back, there are several features about the Nicaragua Trough which are of more general significance:

From the Cenomanian to the Miocene, subsidence and sedimentation of around 10,000 m prevailed almost continuously. The sediments are for the most part clastic to volcanoclastic. The coarse sandstones which are referred to as "greywackes" are probably for the most part tuffites, and closer examination of them would probably reveal some important information about the character of the volcanism. It is not known from what eruption centres the tuffitic material emerged, but the presence of agglomerates and effusives, sub-effusives and plutonic rocks make it seem likely that these centres were located within the Trough or in its immediate neighbourhood.

The northeastern boundary of the Trough is not known, but it is unlikely to have extended very far beyond the present Nicaragua Depression. To the southwest the Trough extended far into the present Pacific shelf region. To the southeast the Trough extends into the Basin of Limón in Costa Rica and can therefore be included in the region of southern Central America.

Stratigraphic gaps, conglomerates and the occasional inclusion of reef and shallow-water limestones indicate a neritic to in-shore deposition milieu, and towards the end the large quantities of fossil woods and the interlocking with terrestrial volcanics are all evidence of alluviation.

The stratigraphic gaps between the Rivas and Brito Formations and the slight discordance between the Brito and Masachapa Formations as well as the inclusion of conglomerates can be taken as evidence of the probably temporary emergence of islands. This seems to be important with regard to the migration of terrestrial fauna down into South America at the end of the Cretaceous and for the appearance of North American ungulates in the Lower Miocene of the Canal Zone in Panama.

It seems logical to regard the Nicaragua Trough and its eastern continuation as a precursor of the present Middle America Trench whose subsidence was, however, compensated by a filling of sediments. Such a Cretaceous-Tertiary trough can in turn be taken as evidence of an early subduction zone on the southern and southeastern edge of northern Central America. The reconstructions of such old subduction zones (MALFAIT & DINKELMAN 1972) follow this theory through to the Paleocene.

4. Southern Central America

Southern Central America takes in the Republics of Costa Rica and Panama. It is fundamentally different in its geological development and thus in its structure from northern Central America (cf. Figs. 2 – 4). There is no Paleozoic crystalline basement nor any cover of Upper Paleozoic and Mesozoic continental and epicontinental deposits. Instead, the deepest stockwork is formed by oceanic sediments and volcanic rocks dating chiefly from the Cretaceous. They were formed as a result of the rifting of the North American and South American plates.

In various troughs this basement is overlain by deposits of chiefly marine Tertiary sediments. They are only slightly deformed but are permeated by large quantities of volcanic and plutonic rocks which exhibit signs of a gradual transition from a basic oceanic volcanism to an intermediate to sialic island-arc magmatism.

Violent uplifting during the Late Tertiary and Quaternary caused the isthmus to rise up to form the present mountainous countries, and the arcuate course was determined by various plate movements and transform faults.

Each of the two countries forming southern Central America has been described separately, because it would be too confusing to deal with them both together. The chain of Quaternary volcanoes which, with a few exceptions, ends in Costa Rica, has been included in Chapter 5.4 (Quaternary volcanism).

4.1 Costa Rica

4.11 General morphology

The morphological classification of Costa Rica is based, in the northwestern part of the country, on DENGO (1962 b) and in the rest of the country on maps prepared by the Instituto Geográfico Nacional and on my own knowledge of the country. A detailed description is given in WEYL (1971 a und b). A classification into "morphological regions", which is proposed by BERGOEING & BRENES (1978), will be examined at the end of this chapter.

At first sight Costa Rica can be divided into three main regions:

a central mountain range striking from northwest to southeast;
a northeastern Atlantic-Caribbean Lowland;
a Pacific coastal region broken up by peninsulas and bays.

A comparison of the general geological and morphological maps (Figs. 65 and 66) shows that the main regions consist of very differently constructed parts on which the more detailed classification into morphological-tectonic units (inset map in Fig. 65) is based.

For example, the **central mountain range** in the northwest is composed of Quaternary and also in part still active volcanoes or of Tertiary magmatic rocks while the southeastern part comprising the Cordillera de Talamanca and the Fila Costeña consists of Tertiary sediments, volcanic and plutonic rocks. The individual regions are, from northwest to southeast:

The Cordillera de Guanacaste

Cordillera de Guanacaste is the name given to a chain of volcanoes trending NW/SE which separates the Atlantic and Pacific lowlands from each other. The most important of the volcanoes are Orosi (1,487 m), Rincón de la Vieja (1,825 m), which consists of an aggregate of several volcanoes, Miravalles, Tenorio and the cone of Arenal off to the southeast, which in 1968 surprisingly produced massive eruptions of nuées ardentes, while the activity of Rincón de la Vieja is limited to occasional eruptions of steam and ash. Aerial photographs and the new topographic maps provide information on the shapes of the volcanoes which otherwise have been little studied (cf. WEYL 1970). On the basis of these photographs and maps HEALY (1969) assumes that the structure of these volcanoes consists of the remains of large calderas or cauldron subsidences. A number of studies have been carried out (cf. Chapter 5) on the eruptions of Arenal.

In front of the Cordillera de Guanacaste to the southwest is a broad plateau of Pleistocene ignimbrites, pumice tuffs and re-worked products of these materials which DENGO (1962 b) has described in detail. He gave the name **Meseta Volcánica de Santa Rosa** to this plateau and, because of its position in the lowland of Guanacaste, he assigned it to the Tempisque Depression. On the basis of the rocks (DENGO 1962 b, WEYL 1969 c) and its association with recent volcanism, I would like to group it with the Cordillera de Guanacaste. From the foot of the volcanoes at an altitude of 400 to 500 m the plateau gradually slopes down to around 100 m at the edge of the Tempisque Valley.

The Cordillera de Tilarán

To the southeast, with an interruption caused by a depression near Arenal, follows a mountain range made up of Late Tertiary volcanic and plutonic rocks; despite a certain lithological and morphological uniformity, this range still does not have a name of its own. Parts of it are called the Sierra de Tilarán, the Cerros Cedral de Miramar, the Cerros de San Antonio, the Cerros de Abangares and the Fila de Aguacate. It has been proposed that they all be grouped together under the name „Cordillera de Tilarán" (WEYL 1971 a and b).

In the southwest the Cordillera de Tilarán is bounded by a very striking fault scarp, the "Falla de Las Juntas", and in the north by a similarly NW/SE-trending fault, the "Falla de Arenal".

The Cordillera Central

The next portion that stands out is the Cordillera Central, which is built up of volcanoes and whose main edifices in the west, namely Viejo (2,050 m), Poás (2,722 m) and Barba, are arranged in a WNW/ESE direction, while the two large eastern volcanoes Irazú (3,423 m) and

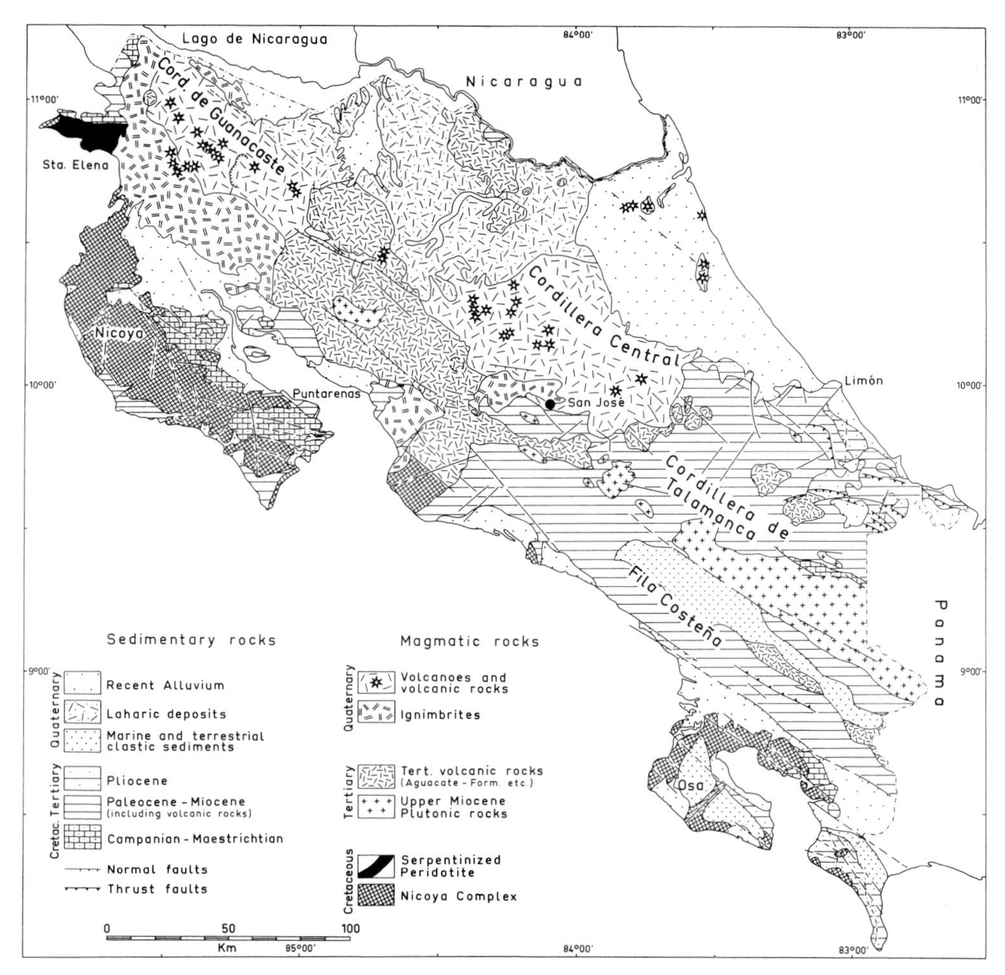

Fig. 66. Geological map of Costa Rica. (Simplified redrawing from "Mapa Geológico de Costa Rica 1 : 700,000", 1968).

Turrialba (3,328 m) exhibit a fissure running diagonal, i.e. ENE, to the latter configuration. With the development of the Atlantic Lowland, traffic is increasingly being routed through the depressions between the volcanoes. In particular the volcanoes on the rainy northern and eastern flanks are dissected by deep gorges, and large quantities of water are discharged by their rivers either into the Río San Juan or directly into the Atlantic.

Of the volcanoes, Poás in 1954 and Irazú in 1963 had up to 64 severe ash eruptions. Otherwise the craters of both volcanoes exhibit varying degrees of fumarole activity (cf. Chapter 5).

The Valle Central

South of the Cordillera Central runs the Valle Central — the centre of settlement in Costa Rica — in an E/W direction and thus diagonal to the axis of the mountain range. The valley is divided into an eastern and western basin by the oceanic watershed at an altitude of 1,500 m and the slope down to both oceans. Both basins are transected by deeply incised river systems with broad terraces which unite in the east in the Río Reventazón, and in the west in the Río Grande de Tárcoles. The basement of the Valle Central is built up of slightly folded marine Oligocene and Miocene over which lavas, tuffs and ignimbrite sheets from the volcanic Cordillera Central have been laid down. In the south the Valle Central is bounded by outliers of the Cordillera de Talamanca rising steeply at faults; of these outliers the Cerros de Escazú towers 2,300 m high above the capital of San José.

The Cordillera de Talamanca

The Cordillera de Talamanca is the largest and highest mountain range in Costa Rica and it stretches from the Valle Central as far as and into western Panama. The construction of the Interamerican Highway has opened up the western part of the range to traffic and thus to settlement, but large parts in particular of the eastern section and of the northeastern slope are still today unexplored territory.

The Cordillera de Talamanca is built up of Tertiary marine sediments with intercalated volcanic and Upper Miocene plutonic rocks. It slopes away sharply at faults to the SW down to the Valle del General, while the slopes to the northeast are very much flatter. The watershed is therefore clearly shifted to the southwest, and the valleys draining to the Río General are correspondingly much shorter than those draining into the Atlantic.

Fig. 67. Glacial relief in the Chirripó Massif (3,820 m), Cordillera de Talamanca, Costa Rica. Photo: R. WEYL.

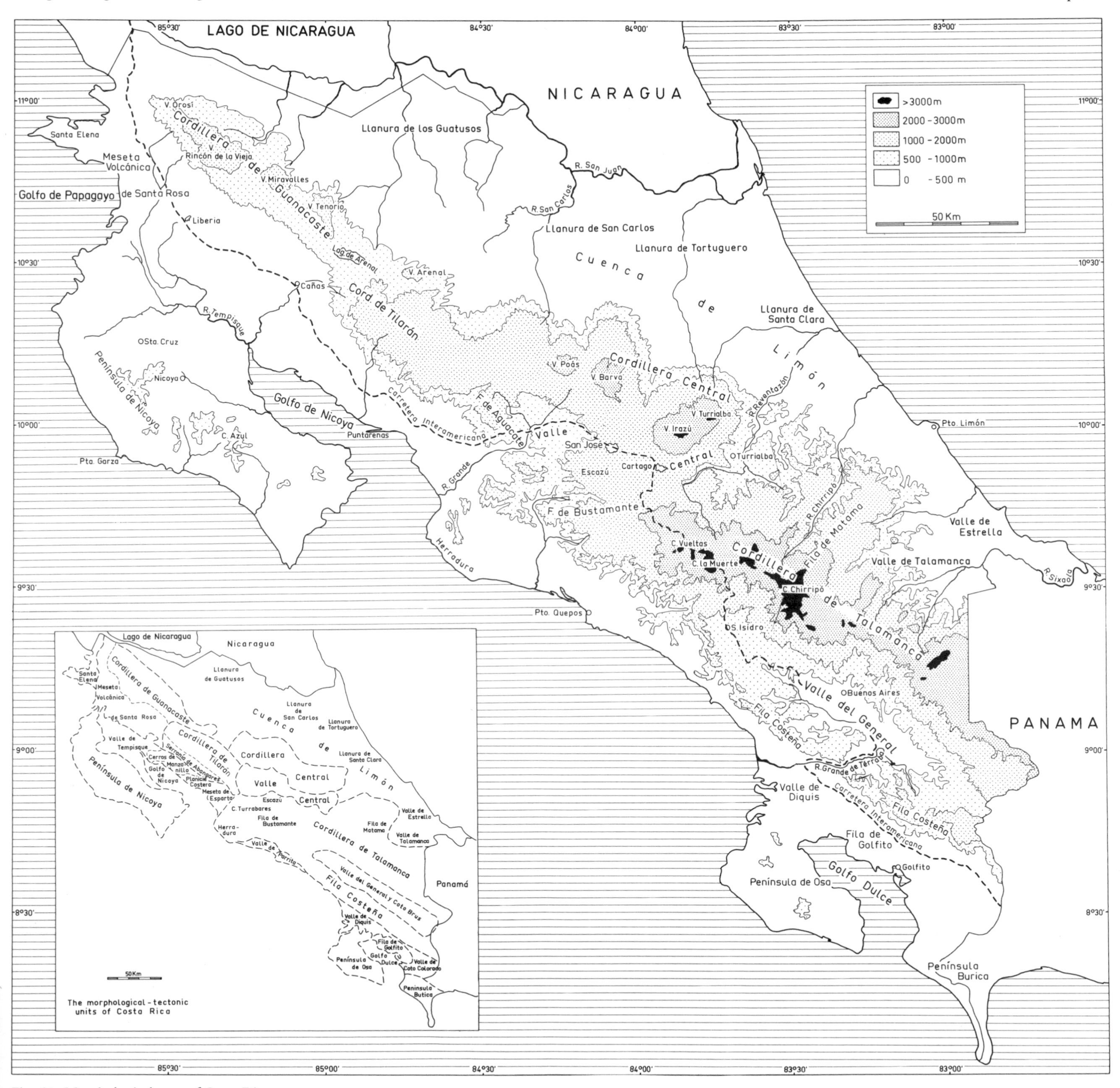

Fig. 65. Morphological map of Costa Rica.

The morphology of the Cordillera de Talamanca is dominated by the contrast between steeply incised valleys and extensive remains of flattish Pliocene forms at the crest; the Inter-american Highway follows this flattish relief over long distances.

The highest regions around the Cerro Chirripó (3,820 m) and the Cerro Kamuk (3,554 m) are formed by glacial erosion resulting from Late Ice Age glaciation (Fig. 67) (WEYL 1956, HASTENRATH 1973, BERGOEING 1978).

The Valle del General

To the southwest the Cordillera de Talamanca drops steeply away to an approximately 120 km long and about 10 km wide intramontane basin which is named according to the rivers draining from it as the Valle del General in the northwestern part and as the Valle de Coto Brus in the southeastern part. Both rivers join together in the Río Grande de Térraba. The basin contains thick accumulations of Pliocene and Pleistocene gravels and volcanoclastic rocks (HENNINGSEN 1966 a). Dissected by present rivers, the latter form extensive terraces which are covered by lateritic soils (MADRIGAL 1977). Extensive bauxite deposits have been detected in them (Chapter 9.15 and Fig. 195).

To the southwest the Valle del General rises gently to the Fila Costeña which is over 1,000 m high.

The Fila Costeña

The Fila Costeña is separated from the Cordillera de Talamanca by the Valle de General. It runs parallel to the Pacific coast from the valley of the Río Savegre up to the region of the Panamanian border where it slopes away and ends in the valley of the Río Chiriquí. The steep slope to the Pacific or to the Pacific coastal plains occurs in several NW/SE-striking step faults. The range is built up predominantly of marine Tertiary strata which are slightly folded and strike longitudinally along the axis of the range (HENNINGSEN 1966 a). Thick Middle to Upper Eocene shallow water limestones appear in the landscape from east of the Río Terraba up to and beyond the frontier with Panama, and they have earned the name **Fila de Cal** for this part of the range.

The Atlantic-Caribbean Lowland (Cuenca de Limón)

A large lowland region bounded by the volcanic chains, the Río San Juan and the Caribbean coast is located in northeastern Costa Rica. This region is the southeastern continuation of the Nicaragua Depression and, like this, it is a sedimentation basin dating back to the Early Tertiary; it bears the name of Limón Basin or Cuenca de Limón. Its surface is covered by alluvium from countless rivers, and at the foot of the volcanoes this alluvium takes the form of fans and lahar deposits. At Tortuguero some small Quaternary volcanic cones rise above the plain.

To the east of the Río Reventazón come a series of small mountain chains consisting of Tertiary sediments and exhibiting northwards overthrust folds while still incorporating some Pliocene material. Since this region promises to yield oil, detailed studies have been carried out here by the Compañia Petrolera de Costa Rica, which made available its unpublished results for the general geological map of Costa Rica. This region on the map therefore depicts far more structural details than those parts of the country which have been merely superficially explored (Fig. 66).

Fig. 68. Aerial photograph of the Santa Elena Peninsula, Costa Rica. The E/W-trending mountain chains follow the structure. In the middle of the picture is the peridotite/serpentinite massif, to the north (on the left of the photograph) come the Upper Cretaceous sediments, which are interrupted by a fault. In the background can be seen the volcanoes of the Cordillera de Guanacaste. Aerial photograph by the Instituto Geográfico Nacional de Costa Rica.

A number of individual regions in the lowland have been given their own name, and these are shown in the general map (Fig. 65).

The Pacific coastal region

The shape of the coast is fundamentally different in both parts of Central America. To the north the flat coast is hardly broken and the sole interruption is the shallow Fonseca Bay; on the other hand in the south, from Costa Rica to Panama, the coastline is extremely broken by peninsulas, bays and islands offshore with rocky beaches and cliffs, wave-cut benches and mountainous country which often approaches right up to the sea (WEYL 1969/70). This contrast continues in the marine area. A less broken shelf lies off northern Central America, and towards the ocean comes the Middle America Trench. The latter ends off northern Costa Rica. Off southern Costa Rica its place is taken by the Cocos Ridge with its uneven relief, its extensive submarine escarpments and seamounts.

In the **Pacific peninsulas** of Costa Rica later uplift has exposed Late Mesozoic ocean floor at altitudes of up to 1,000 m. The rocks of the ocean floor are designated, after the main occurrence on the **Nicoya Peninsula,** as the Nicoya Complex. The Complex will be dealt with in detail in the stratigraphic and tectonic section. To the southeast the Nicoya Complex extends in the Herradura Peninsula, near Quepos, in the Osa Peninsula, in the Fila de Golfito and in the Burica Peninsula.

The **Santa Elena Peninsula,** which is located in the extreme northwest, does not fit in at all with the structure of the rest of Costa Rica (Fig. 68). In contrast to the other NW/SE-striking structures this peninsula is composed of an E/W-striking peridotite-serpentinite massif which is transected by faults of the same orientation. Upper Cretaceous marine sediments, which also strike E/W, are accreted onto this massif at post-Eocene faults (DENGO 1962 b, Fig. 3). The pattern of the coastline with its almost submerged chains of mountains and island festoons indicates that recent subsidence took place in the vicinity of the peninsula, unless the explanation lies in eustatic changes in the sea level.

Depressions between the peninsulas and the mainland

Between the Nicoya Peninsula and the Cordillera de Guanacaste lies a broad depression, which DENGO (1962 b) called the **Valle de Tempisque** after the main river it contains, and which was divided by him into a number of sub-units. The depression continues to the SE in the **Gulf of Nicoya,** which was recognized as a subsidence area even by ROMANES (1912) and was interpreted by DENGO as the submerged part of the Tempisque Valley. A large number of submerged chains of hills emerge from the shallow Gulf in the form of elongated islands. Levelling surveys have revealed a recent tendency towards subsidence on the Puntarenas Peninsula (MIYAMURA 1975).

In the **Golfo Dulce** between Golfito and the Osa Peninsula it is likely that similar circumstances exist to those in the Gulf of Nicoya, however, little research has been carried out in this area. A basin located within the Gulf and extending to a depth of more than 200 m below sea level is very probable evidence of a strong subsidence tendency.

Between the peninsulas and on the parts of the peninsulas facing the mainland runs a series of **coastal plains** built up of alluvial deposits, and these form the area of the large banana plantations or of the follow-on crops. Towards the sea come extensive sandbars and behind them lie mangrove swamps (cf. WEYL 1969/70).

BERGOEING & BRENES (1978) proposed a slightly different morphological classification of Costa Rica which they at the same time explained in a very clear geomorphological map on the scale 1 : 1,000,000 (Mapa Geomorfológico de Costa Rica, San José 1978). They distinguish between:

"Llanuras de Inundación del Atlántico
Cordilleras volcánicas
Cuenca tectónica del río Tempisque
Serranías antiguas
Valle Central
Cordillera de Talamanca
Cuenca tectónica del río General
Areas litorales de reciente formación."

Their map skilfully combines such information as lithological units, morphology and — represented in different colours — their geological age. Their attempt to incorporate Quaternary deposits and terraces into the international stratigraphy is interesting, but proof has still to be supplied that this is correct.

4.12 Stratigraphic sequence and paleogeography

The sedimentary sequence of Costa Rica extends from the Upper Mesozoic to the Cenozoic. The oldest rocks are thick basaltic lavas frequently exhibiting pillow structures (Fig. 69), tuffs, sub-effusive basic igneous rocks and intercalated radiolarites, siliceous shales, siliceous limestones and heterogeneous breccias which are referred to as melange. They crop out on

Fig. 69. Pillow lava of tholeiitic basalt. Coast west of Playa El Coco, Nicoya, Costa Rica. Photo: R. WEYL.

the Pacific peninsulas of Nicoya, Herradura and Osa (cf. map, Fig. 66) and were studied systematically for the first time by DENGO (1962 a and b) on the Nicoya Peninsula, and he referred to them as the **Nicoya Complex** [4]. So far, however, despite many subsequent studies, there are still a great number of questions associated with the Nicoya Complex and its tectonic interpretation which have yet to be resolved or are under discussion.

Nevertheless, agreement seems to exist on the fact that the Nicoya Complex and analogous formations on the western edge of northern South America and in the Caribbean region

[4] Similar series are referred to as the "Basic Igneous Complex" in Panama and in the Western Cordilleras of Colombia and Ecuador.

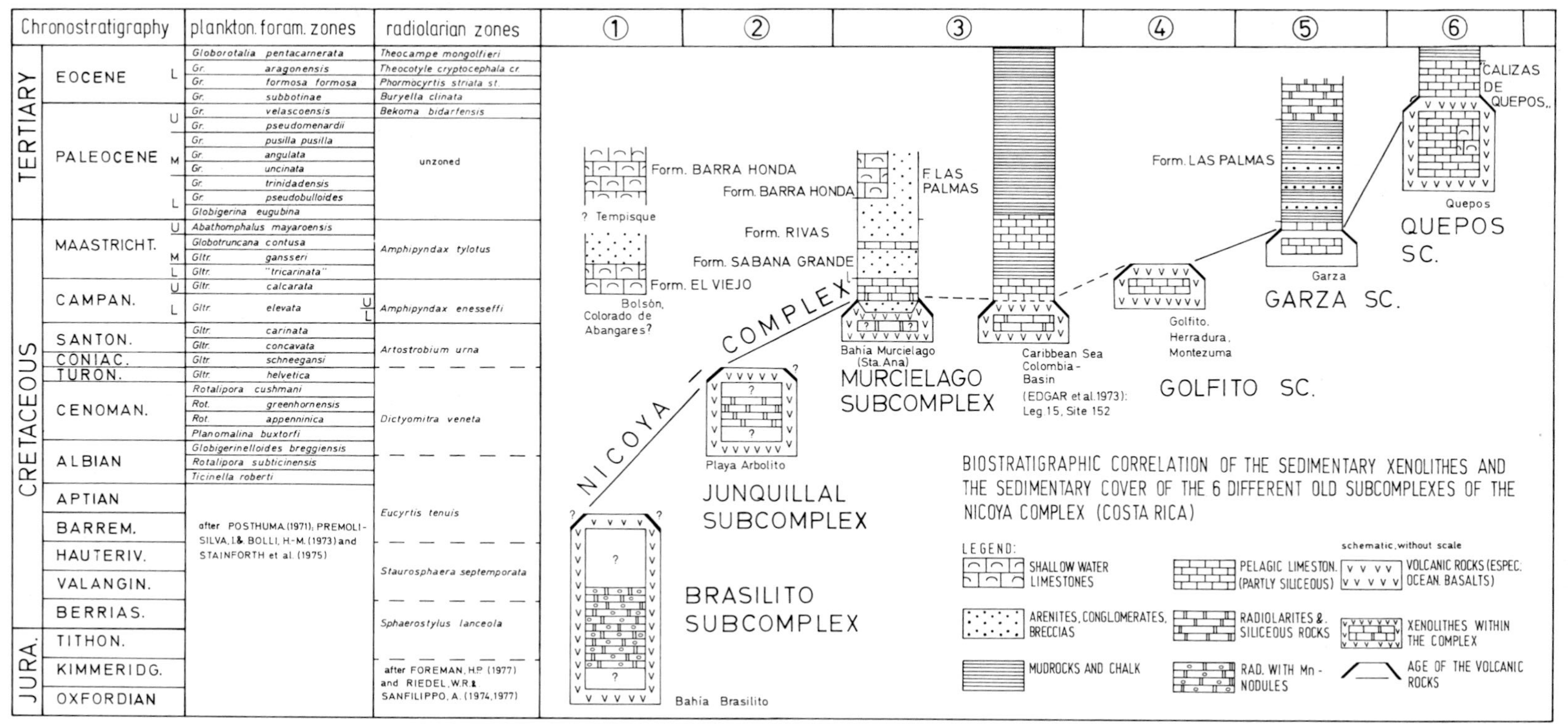

Fig. 70. Schematic representation of profiles together with age correlations of the different ancient subcomplexes of the Nicoya Complex in Costa Rica. The profile of site 152 (DSDP), situated south of the Nicaragua Bank, is included because this Caribbean lava sheet exactly fits the Murciélago Subcomplex. (From SCHMIDT-EFFING 1979 a and b).

can be interpreted as oceanic formations. Their age was put by DENGO at Cretaceous, possibly older. HENNINGSEN (in HENNINGSEN & WEYL 1967) narrowed the age range down to Campanian to Maestrichtian on the basis of some relatively sparse microfaunas. New finds of radiolarians are evidence, in PESSAGNO's view (in GALLI-OLIVIER 1977, 1979), of uppermost Jurassic in the case of the oldest rocks, while the most recent volcanic flows extend into the earliest Eocene. SCHMIDT-EFFING (1979) made a classification based on the radiolarian and foraminiferal fauna.

His classification was made possible by biostratigraphic dating of the overlying sediments and inclusions of sedimentary exotic blocks (xenoliths) within the basaltic sequences. According to SCHMIDT-EFFING the Nicoya Complex can now be divided into six subcomplexes of different ages (Fig. 70): "These subcomplexes have to be interpreted as being submarine, abyssal to bathyal sheet flows of mainly tholeiitic basalts up to several hundred meters in thickness. During these short-duration effusions, the sedimentary cover in the meantime deposited on the underlying effusives was more or less intensively reworked by these processes and deported within the new effusion body in the form of exotic blocks (xenoliths, volcanic melange), where some contact-metamorphic changes occurred."

SCHMIDT-EFFING (1979) distinguished between the following subcomplexes: [5]

1. The subcomplex of Brasilito containing radiolarite or jaspilite-xenoliths of the early Lower Cretaceous and Upper Jurassic (*Sphaerostylus lanceola* zone).
2. The subcomplex of Junquillal of post-Cenomanian to pre-Campanian age.
3. The subcomplex of Murciélago of about Santonian or lowermost Campanian age.
4. The subcomplex of Golfito with numerous xenoliths composed of siliceous limestones in which planktonic foraminifera and well-preserved radiolarians are found. The age of this subcomplex is post-Campanian, and it probably dates from the beginning of the Maestrichtian.
5. The subcomplex of Garza containing pelagic limestones rich in faunas of Middle Maestrichtian planktonic foraminifera; it is overlain by pelagic limestone of uppermost Maestrichtian.
6. The subcomplex of Quepos with xenoliths of Paleocene age and overlain by sediments of the Lower Eocene.

The recent discovery of Upper Albian ammonites (*Neokentroceras* sp.) within the volcano-sedimentary Nicoya Complex confirms the Lower Cretaceous age at least of a part of this series (AZÉMA et al. 1979). Fragments of *Pseudokossmaticeras* sp. from San Buenaventura de Abangares and Guanacaste permit part of the Rivas Formation to be assigned to the Campanian (SCHMIDT-EFFING 1975).

AZÉMA et al. (1979), in a paper published after my own manuscript was completed, describe microfaunas of Paleocene to Eocene age from Quepos and surroundings.

The basalts of the Nicoya Complex are overlain by or interlocked with a heterogeneous series of Cretaceous to Early Tertiary sediments which are much more complicated in structure than DENGO had been able to establish in his fundamental studies.

According to DENGO (1962 b) the Nicoya Complex is overlain by a clearly differentiated series of strata having the following formations (cf. Table 4 a):

---

[5] The following names have been selected according to type locality which is specified in Fig. 2 in SCHMIDT-EFFING (1979 a).

The discordantly superjacent **Sabana Grande Formation** dating from the Senonian or earlier and which contains siliceous shales, cherts and siliceous limestones. Above these, again discordantly, follows the clastic-tuffitic 1,800 m thick **Rivas Formation** which dates from about the Senonian. This is followed by the spatially limited **Barra Honda Formation** which is built up of shallow water limestones of Danian to Montian; and the last formation in the series is the sandy-silty **Las Palmas Formation** which is 900 m thick, dates from the Paleocene to Lower Eocene and is limited to the eastern tip of Nicoya.

Macro- and micropaleontologically substantiated profiling on the Nicoya Peninsula and the adjacent mainland by STIBANE et al. (1977) revealed some completely new aspects which necessitate a revision of previous views: Lithologically identical rocks, which DENGO grouped together in the above formations, have to be assigned to quite different ages. For example, cherts from the "Sabana Grande Formation" date from the Upper Jurassic to the Eocene; typical clastic series of the "Rivas Formation" can occur from the Campanian to the Eocene; shallow-water and reef limestones occur in some profiles in the Upper Campanian and are directly superjacent on the basalts of the Nicoya Complex, and they themselves are overlain by thick clastic deposits of the "Rivas" sediments.

It was further found that closely adjacent profiles display completely different series of strata both as regards facies and thickness: some of the strata are thin with shallow-water limestones and rudistan reefs, while the others are very thick with cherts at the base and clastic series with turbidites and olistostromes above. These findings led to the conclusion that considerable vertical block movements have taken place since the Campanian, and as a consequence the basins and ridges were moved closer together (Fig. 71).

On the basis of the findings which have been briefly described here, the development of the Nicoya Complex was traced by STIBANE et al. (1977), initially for the period from the Cam-

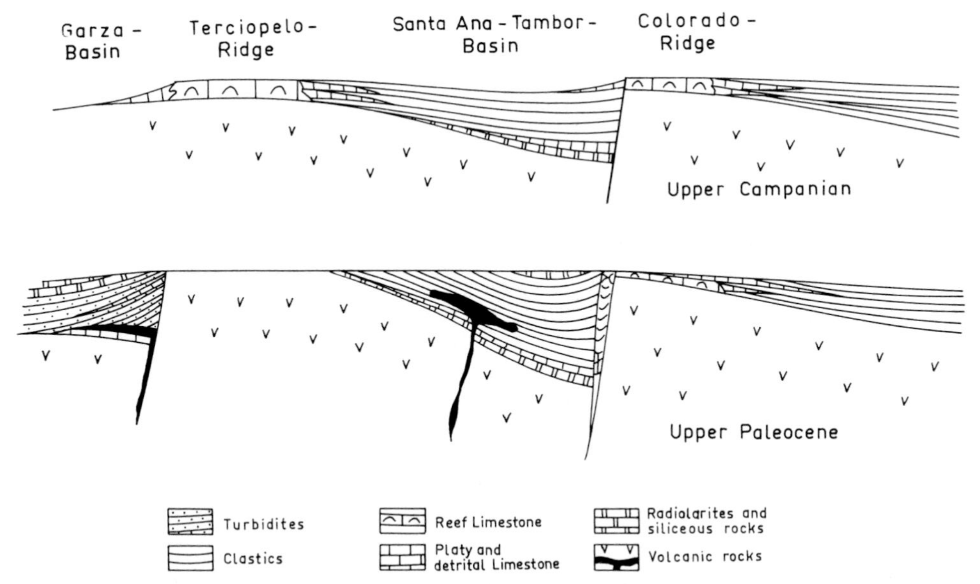

Fig. 71. Schematic cross-sections illustrating the development of Nicoya Peninsula. (From STIBANE et al. 1977, Fig. 6).

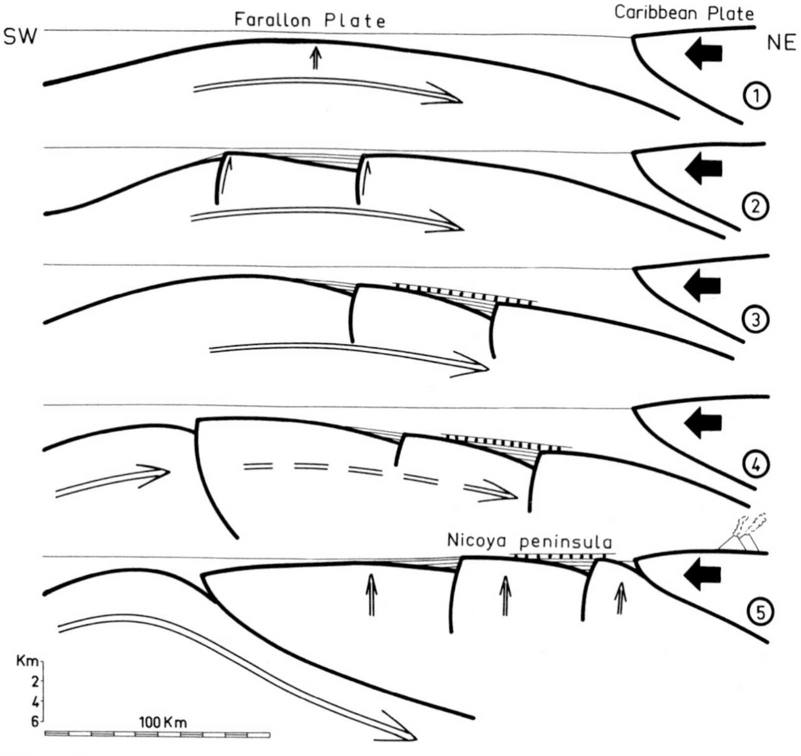

Fig. 72. Schematic cross-sections illustrating the development of the Cocos Plate in the region of Costa Rica. (From STIBANE et al. 1977, Fig. 7).

panian to the Tertiary and later in modified form by SCHMIDT-EFFING (1979) from the Upper Jurassic to the present day. Both interpretations were explained by the two authors in diagrammatic profiles which are reproduced in Figs. 72 and 73.

According to the first model (STIBANE et al. 1977) the floor of the Farallon Plate bulges as a result of crustal extension in the area of an Outer Gravity High (Fig. 72, 1, 2). This rejuvenated the topography and gave rise to clastic sedimentation from the uplifted blocks. Then shallow-water and reef limestones formed at the edges of these blocks. Further spreading caused the region of sedimentation to move towards the NE in the direction of a subduction zone located ahead of the Caribbean Plate. This caused subsidence below the CCD line and gave rise to renewed sedimentation of radiolarites (strong signature in Fig. 72) in the Paleocene and Eocene (Fig. 72, 3). New faults were opened up in the region of the Outer Gravity High (Fig. 72, 4), and these faults were then transformed into a second subduction zone which underlies the region of the present Nicoya Peninsula and caused the latter to rise to its present level in the Tertiary (Fig. 72, 5).

SCHMIDT-EFFING's model (1979 a, b) assumes a Jurassic ocean floor within the Pacific on which radiolarites were deposited and manganese nodules were formed (Fig. 73, Jurassic). Since the end of the Jurassic a volcanic aseismic "Nicoya-Azuero Ridge" has been developing on this crust and is moving in a northeastern direction (Fig. 73, Lower Cretaceous). Entirely

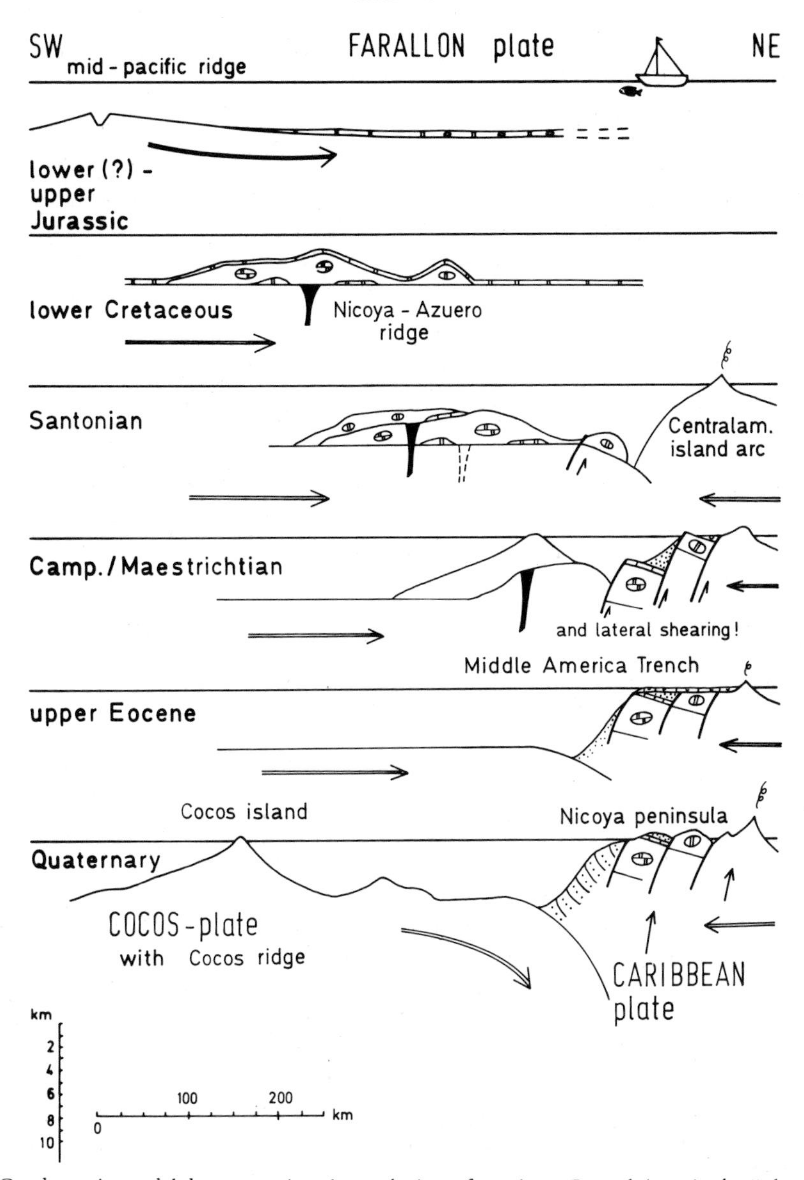

Fig. 73. Geodynamic model demonstrating the evolution of southern Central America by "plate boundary jumping". (From SCHMIDT-EFFING 1979 a and b).

independent of this ridge a subduction zone developed at about the beginning of the Upper Cretaceous in the area of the "Panama Strait", separating the Caribbean from the Farallon plate and connecting the ancient North and South American subduction zones (Fig. 73, Santonian).

The subduction zone is indicated by andesitic-dacitic volcanism in the Upper Cretaceous and Paleocene of Nicaragua, Costa Rica and Panama as well as by the start of sialic plutonic activity in Panama (Fig. 73, Santonian). During the Campanian the Nicoya-Azuero Ridge reached the subduction zone and caused parts of the crust to shear off in echelon fashion during the Campanian to Lower Eocene, and as a result they became welded to a "Nicaragua-Panama island arc" (Fig. 73, Camp./Maestrichtian). The associated vertical movements gave rise to a very pronounced facies change into thick clastic rocks on the one hand and shallow-water limestones on the other. At the same time, this explains the formation of a temporary land link between North and South America at the end of the Cretaceous — such a link must have existed, judging by the geographical distribution of animal species.

Basaltic volcanism on the still oceanic portions of the Nicoya-Azuero Ridge lasted up to the turn of the Paleocene/Eocene, while the ridge itself became partly linked with the Caribbean Plate and partly was subducted and thus disappeared (Fig. 73, Upper Eocene). Since the Eocene the rest of the Nicoya Complex has exhibited a platform character and is part of the Central American continent which was next to develop (Fig. 73, Quaternary). The reader is referred to SCHMIDT-EFFING's work (1979) for more details on this complicated process of formation.

Both models are in agreement in placing the formation of the Nicoya Complex in the region of the Pacific and in explaining its present position in terms of plate movements. Both models assume that individual blocks underwent considerable vertical movement during the Upper Cretaceous and Lower Tertiary, and this gave rise to very distinct facies differentiation. While STIBANE et al. regarded an Outer Gravity High far off the coast and a consequent uparching of the crust as the reasons for this faulting, SCHMIDT-EFFING seeks the reason in a subduction zone off a primary Upper Cretaceous island arc. The two models coincide again in the assumption that from the Eocene onwards the Nicoya Complex had by and large reached its present position in relation to the rest of Costa Rica and, apart from marginal ingression, remained part of the continent. It was recognized as such by DENGO (1962 b, Fig. 5).

According to GALLI-OLIVIER's conception (1977, 1979) the ophiolite suite — which is the name he gives to the Nicoya Complex — was formed in the period from the Tithonian to the Upper Santonian and was squeezed out by plate convergence in a relatively short period of time between the Upper Santonian and Lower Campanian. The following island-arc suite of plutonic and volcanic rocks as well as volcanogenic sediments first unconformably covered the ophiolite in early Campanian time. GALLI illustrates his conception of the actual structure of NW Costa Rica by the cross-section reproduced in Fig. 74. This conception does not, however, fit in with the view of STIBANE et al. (1977) and in particular of SCHMIDT-EFFING (1979).

Tertiary sediments of mainly marine origin with intercalated volcanic rocks and volcanoclastic deposits cover most of Costa Rica and probably also underlie the regions of Quaternary volcanism and alluvial deposits in the northern part of the country. During the fifties and sixties these sediments were studied in detail in connection with petroleum exploration work in the Limón Basin and at the northern foot of the Cordillera de Talamanca, but little information has been published on these studies (RIVIER 1973). DENGO (1962 b) also based his description of the Tertiary in Guanacaste province on studies made to determine the petroleum geology of the region. However, the stratigraphy and the lithological character of the Paleocene and Lower Eocene have been interpreted quite differently here by STIBANE et al. (1977). A study by ESCALANTE (1966) and a comprehensive report by KRUSHENSKY, MALAVASSI & CASTILLO (1976), which was preceded by mapping and description of several map sheets in this region (CASTILLO 1969, SANDOVAL 1971, KRUSHENSKY 1972, 1973), are avail-

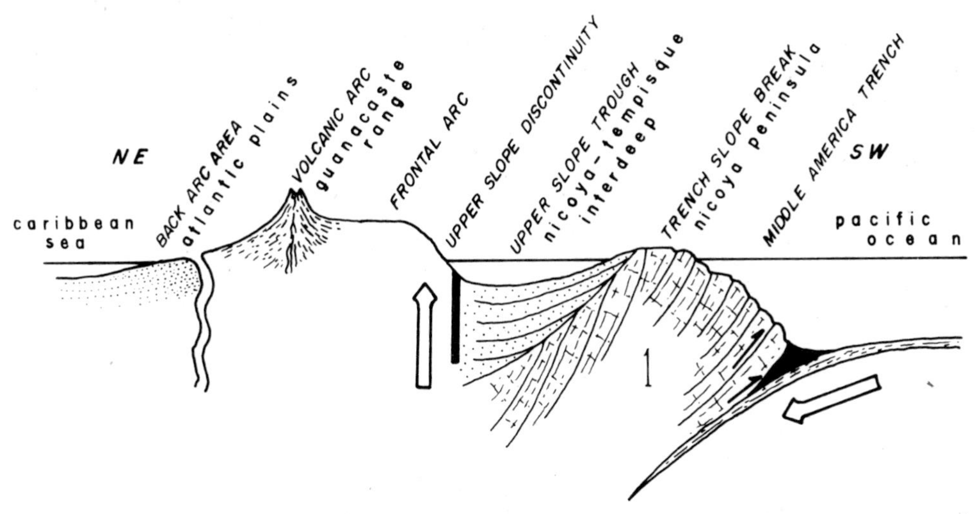

Fig. 74. Schematic cross-section through northwestern Costa Rica showing the accretionary prism and other tectonic belts. (From GALLI-OLIVIER 1979, Fig. 2).

able for the Valle Central and the adjacent northern part of the Cordillera de Talamanca. The Tertiary of the Fila Costeña was the subject of a study by HENNINGSEN (1966), and individual paleontological, stratigraphic and ecological studies have been conducted by WOODRING & MALAVASSI (1961), VAN DEN BOLD (1967/74), KRUCKOW (1974, 1976), TAYLOR (1973) and CARBALLO & FISCHER (1978). In addition, there are a number of unpublished "Licenciado" theses from the Escuela Centroamericana de Geología in San Pedro, which were made available to me in summary form [6].

DENGO (1962 b), HENNINGSEN (1966), WEYL (1971 c) and above all FISCHER and ESCALANTE correlated the Tertiary deposits in unpublished correlation charts which were kindly put at my disposal for me to evaluate. With modifications, they form the basis of Table 4 b. A somewhat different correlation chart published by Anonymous (1978) is reproduced in Table 4 a.

FISCHER and SCHMIDT-EFFING have undertaken to revise the Tertiary stratigraphy on the basis of the macro- and microfaunas, and this will almost certainly necessitate amendment of the following sections. What I have attempted to do is to demonstrate the extraordinary range of facies of the Tertiary deposits in the respective paleogeographic situation. I have based my statements to a large extent on a description given by RIVIER (1973), in which I was able to evaluate unpublished material gathered during petroleum exploration.

RIVIER assumes that during the **Paleocene to Middle Eocene** an ocean basin extended throughout the region of the present Cordillera de Talamanca in a northwest direction into the region of the Rivas anticline in Nicaragua and that large masses of sediments built up in this basin. Thus, in the Caribbean Lowland near the frontier with Panama, Upper Cretaceous to Paleocene occur in the form of 4,000 m thick volcanoclastic sediments which are followed without a break by a further > 2,000 m of Eocene deposits (**Río Lari Formation**) having

[6] I am grateful to Prof. FISCHER, Marburg, who taught for a long time at the Escuela, for letting me have this and other summaries of theses.

Table 4a. Stratigraphic correlation chart of Costa Rica. (From Anonymous 1978, Cuadro 1-1).

Serie	Edad	Nicoya y Guanacaste	Cerros de Escazú Montes del Aguacate y Talamanca
PLEISTOCENO		G. LIBERIA: riodacita 0,14 m.a.; riodacita 0,60 m.a.; andesita 0,80 m.a.; dacita 1,00 m.a.; dacita 1,60 m.a.; G. BAGACES	
	1,8		
PLIOCENO		INTRUSIVO GUA-CIMAL (Monzonita)	G. AGUACATE: diorita 2,3 m.a.; andesita 2,6 m.a.; andesita 3,2 m.a.; basalto 4,3 m.a.; basanita 4,6 m.a.
	5,0		
MIOCENO SUP		F. MONTEZUMA: limolitas, areniscas, conglomerados y areniscas 30 m	G. AGUACATE: diorita 8.5 m.a.; andesita 9,0 m.a.; granito 10,5 m.a.
	12,0		
MIOCENO MEDIO		F. PUNTA CARBALLO: areniscas finas calcáreas con megafósiles 200 m	F. CORIS; andesita 16,9 m.a.; F. SAN MIGUEL
	15,0		
MIOCENO INF			F. TURRUCARES
	22,5		
OLIGOCENO SUP		F. MASACHAPA: areniscas, limolitas, lutitas calcáreas 700 m	F. TERRABA
OLIGOCENO	33,0		
OLIGOCENO INF		?	
	39.0		?
EOCENO SUP		F. BRITO: conglomerados, areniscas gruesas, lutitas y calcarenitas, caliza arrecifal en Fila de Cal y Colorado, Gte. 2.400 m	
	44.0		
EOCENO MEDIO			
	50.0		
EOCENO INF			
	55.0		?
PALEOCENO SUP		F. LAS PALMAS: limolitas, lutitas, areniscas y calcarenitas 900 m	F. PACACUA
PALEOCENO	58.0		? Caliza Parritilla
PALEOCENO INF		F. BARRA HONDA: calizas arrecifales 300 m	
	65,0		
MAESTRICHTIANO		F. RIVAS: lutitas grises, arenisca y caliza 2.000 m	
	69,0		
CAMPANIANO			
SANTONIANO		F. SABANA GRANDE: caliza, silícea, radiolarita, pedernal 240 m	
	88,0	C. NICOYA: basaltos elipsoidales, gabros, diabasas y diorita	

F = Formation, G = Group, C = Complex.

Sureste de Costa Rica y oeste de Panamá	Fila costeña	Cuenca de Limón	Valle Central
F. ARMUELLES: conglomerado arenisca conglomerado arenisca conglomerado 1.150 m		Volcanismo no diferenciado; Limón caliza Molly Creek arenisca	Andesita 0,30 m.a. Ingnimbrita 0,77 m.a. Andesita 0,98 m.a. Andesita 1,67 m.a.
F. CHARCO AZUL	Arcillas rojas – lateritas F. PASO REAL: bloques volcánicos en cemento arenoso, areniscas tufáceas	F. SURETKA: arena, conglomerado–arcilla 300 m; RIO PEY 850 m	
F. BURICA	F. GATUN: areniscas tufáceas conglomerados, limolitas y lutitas	F. GATUN: lutitas, areniscas, calizas arcillosas 380 m; VENADO 300 m	
F. CURRE: conglomerado, arenisca			
F. TERRABA	F. TERRABA: lutitas negras calcáreas y calizas interestratificadas	F. USCARI: arenisca, limolitas, lutitas, calizas arcillosas 2.000 m	
		? F. DACLI: caliza, arenisca, conglomerado 1.500	
		F. SENOSRI: caliza, arenisca, conglomerado 850 m	
F. DAVID		CALIZA LAS ANIMAS 106 m	
F. BUCARO		F. TUIS: arenisca 850 m	
		?	
		RIO LARI 2.800 m Volcánico Caliza fosilífera Areniscas Sedimentos volcánicos Areniscas	
F. CHANGUINOLA: calizas, areniscas tobáceas, lavas intercaladas		F. CHANGUINOLA: tiene rocas volcánicas 1.200 m	
F. GOLFITO			
C. NICOYA			

Table 4b. Stratigraphic correlation chart for the Cenozoic of Costa Rica. Partially from unpublished data of ESCALANTE and FISCHER.

			SW Nicaragua	Guanacaste	Valle Central	SW Costa Rica	Limón Basin
QUATERNARY			Alluvium Volcan.; Las Sierras Group	Volcanics; Liberia Form.	Volcanics; Ujarrás Form.	Alluvium	Volc.; Limón Fm.; Molly Creek
TERTIARY	Plioc.	Upper	El Salto Formation 100m	Bagaces Form.; Monte Verde Form. 200m	Doán Form. 400–800m	Charco Azul Form.	Suretka Form.; Río Pey
		Lower			Aguacate Formation 900 m	Paso Real Form. 800m	Río Banano Form. 500m
	Miocene	Upper	El Fraile Form. 2500m; Tamarindo Group 900 m	? Monte Zuma Form. 30m ?; Aguacate Form. 800m ?	Coris Form. 400 m; Pacacua Form. max. 1200m; Turrúcares Form. 150m	Curré Formation 800m	Gatún Formation 1700m; Venado Form.
		Middle		Pta. Carballo Form. 200m	San Miguel Form. max 50m		
		Lower	Masachapa Formation 1500m		Térraba Formation 1300 m	Térraba Formation >1000 m	Uscari Formation 1700m
	Oligocene	Upper		Masachapa Form. 900 m			
		Middle					? Dacli Form. 1500m
		Lower					Senosri Formation 850m
	Eocene	Upper	Brito Formation 2550m	Brito Form. max. 2700 m	Las Animas; ? Parritilla Form.	Brito Form. 1100m; Fila de Cal; Damas	Las Animas Form.; Tuis Formation 850m
		Middle			?		
		Lower		Las Palmas Form. max. 900m			?
	Paleocene	Upper		Barra Honda Form. 180m			Río Lari Formation 2800m
		Lower					
CRETACEOUS			Rivas Formation 2800 m	Nicoya Complex		Nicoya Complex	Changuinola Formation 1200m

identical facies. In Guanacaste, on the other hand, and in particular on the Nicoya Peninsula, there is up to the Paleocene a distinct facies differentiation into shallow-water reef limestones (**Barra Honda Formation**) on the one hand and thick clastic sediments with turbidites and olistostromes (**Las Palmas Formation**) on the other. They are interpreted as being the result of considerable block movements along NW/SE-striking faults (STIBANE et al. 1977). So far no Paleocene to Lower Eocene deposits have been discovered in Central Costa Rica, so we can assume this was a continental region which was a source of sediment.

Middle to Upper Eocene have been described on the Pacific side of Costa Rica as the **Brito Formation,** which connects with the formation of the same name in Nicaragua. In the NW, in Guanacaste, it attains a thickness of around 1,600 m and of at least 1,000 m in the Fila Costeña. The Formation consists mainly of tuffitic arenites, and in the Fila Costeña the material for these was supplied, according to sedimentological evidence, from the SW, i.e. the present Pacific region (HENNINGSEN 1966 a – c). Widespread reef and foraminiferal limestones are intercalated and occur in great thicknesses, forming the topography, above all in the Fila de Cal in the eastern part of the Fila Costeña (see LOHMANN & BRINKMANN 1931, MALAVASSI 1961, WEYL 1961 a, DENGO 1962 b, HENNINGSEN 1966 a for lists of fauna).

There is obviously less faunal evidence for the Middle Eocene in the Limón Basin. The Upper Eocene is distinguished in the approximately 800 m thick **Tuis Formation.** RIVIER notes reef limestones occurring as intercalations in the region of the Río San Juan, i.e. in the extreme north of Costa Rica. In the central part of Costa Rica the mainly volcanoclastic and tuffitic **Pacacua Formation** is placed in the Upper Eocene without any biostratigraphic proof (KRUSHENSKY et al. 1976).

Altogether, the Upper Eocene is regarded as a period of low volcanic activity and mainly neritic sedimentation.

The **Oligocene** is again characterized by chiefly clastic-tuffitic sediments in which there is evidence, in the southern part, of a transition from purely marine to lagoonal facies. Various formations have been identified in the different regions; for example, the **Térraba Formation,** which is 1,300 to 2,000 m thick, is found in the Fila Costeña, in the Valle Central and in the northern part of the Cordillera de Talamanca (DENGO 1961, HENNINGSEN 1966) — this Formation extends from the Lower Oligocene into the Lower Miocene, and it was deposited in the Térraba Trough (Fig. 75). Isolated occurrences of presumably Oligocene age on the Nicoya Peninsula were associated by DENGO (1962 b) with the Nicaraguan **Masachapa Formation.** In the Limón Basin, a Lower Oligocene **Senosri Formation** (850 m thick), a Middle Oligocene **Dácli Formation** (1,500 m thick) and the **Uscari Formation** (1,700 m thick) extending from the Upper Oligocene to the Middle Miocene were identified and micropaleontologically attested in the course of petroleum drilling (BROWNE 1961, unpublished).

From the **Middle Oligocene** onwards a facies differentiation becomes apparent. In the central parts of Costa Rica the sea does not seem to have been as extensive as during the Eocene, so that probable coastlines running approximately parallel to the present margin of the mountain range were postulated on the western edge of the Limón Basin (BROWNE 1965, unpublished). The link with the Nicaragua Trough was probably restricted by uplift of a large island area in Guanacaste. In the region of the Fila Costeña dark, pyritiferous shales occur as deposits of a semi-euxinic basin which was separated from the Atlantic by initial uplift in the region of the later Cordillera de Talamanca. According to HENNINGSEN the clastic sedi-

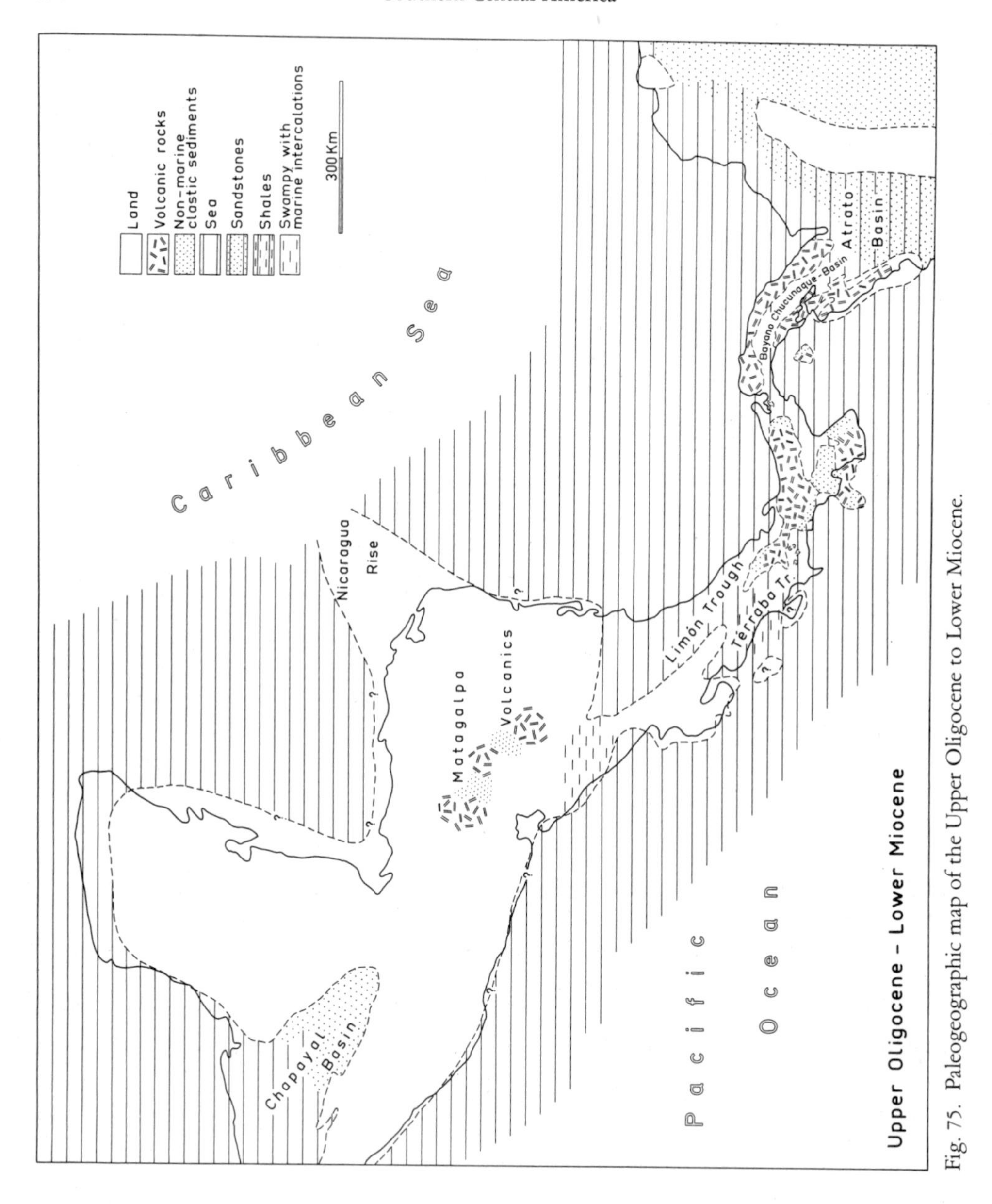

Fig. 75. Paleogeographic map of the Upper Oligocene to Lower Miocene.

mentary material in this basin came, as before, from a source region in the present Pacific Ocean.

RIVIER (1973) states that during the **Lower Miocene** the sedimentation conditions remained much the same as during the Upper Oligocene. Neritic facies predominated in the region of the Cordillera de Talamanca, and marlaceous-pelagic facies predominated in the Atlantic coastal region. Towards the end of the Lower Miocene the sea in fact became deep once more and pelagic microfaunas developed.

CARBALLO & FISCHER (1978) on the other hand refer to a change that took place in marine conditions between the Oligocene and the Lower Miocene. While relatively uniform neritic conditions prevailed in central and western Costa Rica during the Oligocene, the relief started to become much more pronounced during the Miocene, and sedimentation in shallow water predominated.

The **Middle Miocene** is by and large ushered in by a major discordance, and in some places parts of the Upper Oligocene to Lower Miocene Uscari Formation have been eroded away. The sediments are epineritic to paralic with intercalations of shallow-water limestones, reefs and lignites. This is evidence of the uplift of Costa Rica and in particular of the Cordillera de Talamanca, which contrasts with the considerable subsidence of the Limón Basin. Sandstones and siltstones of the **Gatún Formation** were deposited up to 1,700 m thick, together with carbonate and lignitic lenses in this basin. According to HENNINGSEN(1966 a, p. 20) tuffaceous arenites together with conglomerates and siltstones, which have yielded a few molluscan faunas and which are placed in the Middle to Upper Miocene as the **Gatún Formation** (Curré Formation), follow concordantly after the Térraba Formation in the region of the Fila Costeña as well. Small isolated occurrences of fossil-rich Middle Miocene are located on the eastern coast of the Gulf of Nicoya, where, following MACDONALD (1920) they were often described as the **Punta Carballo Formation** (DENGO 1962 b, p. 38).

A number of well-known deposits of fossiliferous Middle Miocene in the region of the Valle Central — for example the calcareous **sandstones of Turrucares** with their rich molluscan and foraminiferal fauna (WOODRING & MALAVASSI 1961) or the **limestones of San Miguel, Patarrá** and **La Chilena** (ESCALANCTE 1966), which were mined for cement production — are particularly important in terms of the paleogeography. They are evidence of an inter-oceanic link that existed in the Middle Miocene. Calcareous algae from the group of the lithothamnions, which were identified by GÓMEZ (1973) and KRUCKOW & GÓMEZ (1974), point to maximum water depths of 100 m in these marine straits. Their presumed course and their extent were discussed by KRUCKOW (1974), and his views have been incorporated in the paleogeographic map in Fig. 75.

CARBALLO & FISCHER (1978), in a monograph which did not appear until 1979, described the fauna and lithological make-up of the limestones of San Miguel. They refer to the limestones as the **San Miguel Formation** thus acknowledging the priority gained by HILL (1898) who gave them the name of "San Miguel Beds". Using the foraminifera as a guide, it is possible to deduce that the San Miguel Formation came into existence sometime between the zone of *Praeorbulina glomerosa* and *Globorotalia foshi robusta.* It thus belongs in the uppermost Lower Miocene to Middle Miocene. Judging by the fauna and the sediments, the deposition conditions were characterized by tropical temperatures, normal salinity, and by clean, turbulent water. It is assumed that the deposition area was a barrier of carbonate sediments that was unaffected by the mainland.

From the **Upper Miocene** onwards, marine deposits are limited to the coastal areas, and there is no evidence of an inter-oceanic link. On the Caribbean side the **Gatún Formation** extends into the Upper Miocene. Marine **Pliocene** has been described as the **Limón Formation** in the coastal area of Limón. Here the river and delta deposits interlock with lagoonal sediments and coral reefs which extend into the Pleistocene. On the Pacific side the **Montezuma Formation** on the southern tip of the Nicoya Peninsula extends probably into the Pliocene (DENGO 1961 b, p. 40). The Pliocene **Charco-Azul Formation** in western Panama stretches into the peninsulas of Burica and Osa. With its deep-water foraminiferal assemblages and a thickness of over 1,250 m, this formation indicates for the last time a fully marine milieu and severe subsidence in the coastal zone (cf. Panama).

In the interior of Costa Rica energetic uplift and vigorous magmatic activity are discernible from, at the latest, the Upper Miocene onwards. The 1,700 m thick conglomerates and sandstones of the **Suretka Formation**, in which individual shallow-water faunas appear only at the edges, lie as erosion products on the northern slopes of the Cordillera de Talamanca. The 800 – 900 m thick volcanoclastic **Paso Real Formation** (DENGO 1962 a, p. 149), which was also described by HENNINGSEN (1966 a, pp. 21/23) and also assumed to be Lower Pliocene, was deposited in the Valle del General — which can be regarded as the chronological extension of the Térraba Trough.

Thick Tertiary volcanic and volcanoclastic series of rocks, which were described under various formation names (**Aguacate, Doán, Pey** and **Cerros Curena Formations**) occur in the Valle Central, in the adjoining parts of the Cordillera de Talamanca and above all in the Cordillera de Tilarán. Because of the lack of fossils, their classification and stratigraphic position have still not been settled, and it will probably only be possible to do so with the aid of radiometric dating. Some indication of the age can, however, be derived from radiometric data of volcanic and associated plutonic rocks whose ages lie between 10.5 and 2.3 million years according to Table 4 a.

Very thick basic volcanic rocks occur over wide areas in the Fila de Aguacate and the Cordillera de Tilarán which adjoins to the northwest; these volcanics attracted attention early on as the country rock of countless gold deposits (ATWOOD 1882), but to this day they still raise unsolved questions. DENGO described them (1962 a) as the "**Aguacate volcanic series**" or (1962 b) as the "**Formación Aguacate**" and he associated it with similar rock series in the northern limbs of the Cordillera de Talamanca, such as the Fila de Puriscal and the Cerros Turrubares. DÓNDOLI had already at an earlier date (1954) distinguished an older andesitic-basaltic series of volcanic rocks in the eastern part of the Valle Central; this series was included by KRUSHENSKY, MALAVASSI & CASTILLO (1976) in the Aguacate Formation.

The age of the Aguacate Formation has so far only very vaguely been given as Mio-Pliocene on the basis of the relationship to other formation. Biostratigraphical or radiometric datings still have to be carried out. Indications of the age can be derived from the tilting of stratified sections of the formation, the discordantly overlying, very probably pre-Quaternary, volcanic rock series, the lack of volcanic landscape forms, the occurrence of flat relief on the top of the Cordillera de Tilarán and — a feature which was discovered for the first time by CHAVES & SÁENZ (1974) — from the intrusion of granitic plutonites into the Aguacate Formation. Provided that these are the same age as those of the Cordillera de Talamanca, it is more likely that this formation dates from the Miocene than from the Pliocene, and thus it moves closer in time to the series of "Lower Tertiary volcanic rocks" in the sense used by DENGO (1962 a). The petrological characteristics of the Aguacate Formation with its basic alkaline rocks will be examined in detail in the following section on magmatism.

CHAVES & SÁENZ (1974) have mapped the Aguacate Formation in the Cordillera de Tilarán and, on the basis of in-the-field findings, distinguished between an agglomerated-brecciated part and a basaltic part in the Formation; they also separated out a discordantly superjacent "**Formación Monteverde**" with andesitic lavas and rhyodacitic tuffs a good 200 m thick. Paleomagnetic measurements indicate that this clearly acid volcanism is Pliocene in age, and the authors wish to compare it with the Tertiary volcanism of northern Central America.

According to KRUSHENSKY, MALAVASSI & CASTILLO (1976) a clastic-volcanoclastic rock series with remains of indeterminate marine organisms overlies the Aguacate Formation in the eastern Valle Central; the series has been given the name **Doán Formation** and has been placed, with some reservation, in the Pliocene. According to the map and the cross-section the series is in intrusive contact in the valley of the Río Macho with the plutonic rocks of the Cordillera de Talamanca, is consequently older than these and, on the basis of their radiometrically determined ages (see below), it should be placed in the Miocene.

The combined **thickness of the Tertiary** material in the Limón Basin is more than 10,000 m, in the Térraba Trough approximately 4,000 m. It is difficult to estimate the thicknesses in the region of the Cordillera de Talamanca, but they are probably also in the order of several thousands of metres. It can be assumed that from the Middle Miocene onwards a central ridge was the region from which these thick sediments originated; the presence of this ridge is reflected above all in the coarse clastic Pliocene deposits of the Surekta and Paso Real Formations. In addition, the volcanic formations (Aguacate, Doán) of the central and northern parts of Costa Rica supplied copious amounts of detritus.

It is more difficult to determine the origin of the Eocene to Lower Oligocene sediments. Judging by detailed sedimentological studies carried out by HENNINGSEN (1966) in the Fila Costeña they consist almost exclusively of detritus of volcanic rocks of relatively uniform composition dating from the Upper Cretaceous to the Miocene. The distribution of heavy minerals and the orientation of individual grains and remains of fossils indicate that the material was transported from the south and southwest. The areas from which the material originated must therefore have been situated where the present-day Pacific Ocean is located. This assumption is confirmed by frequent remains of wood and charcoal.

Fine-grained tuffites, siliceous shales and tuffitic arenites predominate in the Cordillera de Talamanca, and these deposits may link up with those of the Fila Costeña. In the case of the Limón Basin with its particularly large thicknesses it seems on the other hand unlikely that the sediments originated in the area of the present-day Pacific. Instead, it is more probable that they came from a continental region located in the area of Rivas and Guanacaste (DENGO 1968, Fig. 11).

The **Pleistocene** is characterized by violent **volcanism** in the northern half of Costa Rica. Volcanic products in the form of lavas, incandescent tuffs and ashes and erosional volcanic products in the form of lahars and volcanoclastic deposits constitute the material from which the Cordillera Central and the Cordillera de Guanacaste and their adjacent volcanic plateaus are built up. They have been described under the names of various formations, one example of which is Irazú (see map, Fig. 135) together with the local names which are in current use (KRUSHENSKY 1972, 1973). The volcanoes are dealt with in detail in Chapter 5.4.

Fluviatile deposits, and in particular their terraces, are regarded by KRUCKOW (1974) as indicators of the recent abrupt uplift of the land. He also sees connections here with the recent sedimentation in the coastal region of which the Puntarenas Peninsula and the coast near Puerto Limón are examples. Further observations on the morphology and sedimentation in the Pacific coastal region are contained in WEYL (1969/70).

Due to the damming effect of lava flows, fresh-water basins with limnic sedimentation (**Ujarrás Formation,** ESCALANTE 1966, p. 62), in which diatomites may be imbedded, formed occasionally in the region of the Valle Central.

Quaternary moraines (WEYL 1956) in which HASTENRATH (1973) recently identified two to three stages of different altitude, weathering and vegetation density, are located in the highest section of the Cordillera de Talamanca, the massif of the Cerro Chirripó (3,820 m).

Lateritic weathering — dating certainly back to the Upper Tertiary — on nowadays dissected river terraces of the Valle del General led to the formation of mineable **bauxite deposits** which will be dealt with in detail in Chapter 9.

4.13 Magmatism

From the Upper Cretaceous to the Present, Costa Rica has been, almost without interruption, the scene of vigorous magmatic activity marked by a continuous rise in the SiO_2 content of the volcanic products and the development of alkaline rock types. This all reflects the conversion of oceanic to continental crust and the formation of the isthmic island arc (Fig. 76).

We have to thank DENGO (1962 a) for the first summary review of the magmatic development. DENGO's work was further expanded by WEYL (1969 c), PICHLER & WEYL (1975), BALLMANN (1976), AZAMBRE & TOURNON (1977), and BELLON & TOURNON (1978). The following phases of magmatic output have been identified:

Cretaceous to Lower Tertiary magmatism of the Nicoya Complex;
Lower to Middle Tertiary volcanism;
Late Miocene plutonism;
Pliocene volcanism;
Quaternary and Recent volcanism.

DE BOER (1979, p. 251), in a paper published after my own manuscript had been completed, distinguishes between eight successive structural-petrogenetic elements in Costa Rica.

The magmatism of the **Nicoya Complex** is dominated by basalts, often with pillow structures and mafic subvolcanic rocks. According to the main element and trace element analyses which have been carried out (WEYL 1969 c, Table 1, PICHLER & WEYL 1975, Tables 4 and 5), the chemical make-up of the complex corresponds to that of oceanic tholeiites which, together with intercalated eupelagic sediments, form a Cretaceous oceanic crust. They belong to the cover of submarine tholeiitic basalts which extends into the Antilles, covers the floor of the Caribbean (DONNELLY et al. 1973) and can be traced far to the south in the Cordillera Occidental of Colombia and Ecuador (GOOSSENS & ROSE 1973, PICHLER, STIBANE & WEYL 1974, GOOSSENS et al. 1977, GALLI-OLIVIER 1979).

The **peridotite-serpentinite massif** of the Santa Elena Peninsula forms part of the Nicoya Complex (HARRISON 1953, DENGO 1962 b). The parent rock was a lherzolite that is permeated by basic dike rocks such as metadolerites, amphibolites and rodingites (DE BOER 1979, p. 226), which indicate dynamic metamorphism of the ultrabasic rocks. According to DE BOER this is the oldest part of the Nicoya Complex. BELLON & TOURNON (1978, p. 995) give the radiometric age of an amphibole-gabbro as 88.8 ± 4.5 m.y. DENGO (1962 a) discovered that the ultrabasic material had been exposed to erosion even before the Senonian, and SCHMIDT-EFFING (1979) found that Campanian rudistan reefs had grown up on this material. DENGO and PICHLER & WEYL regard the peridotite-serpentinite massif as slightly metamorphosed mantle rock.

In a still unpublished study, AZÈMA & TOURNON (1979) describe the Santa Elena Peninsula as an overthrust block made up of material from the upper mantle and consisting of the following units:

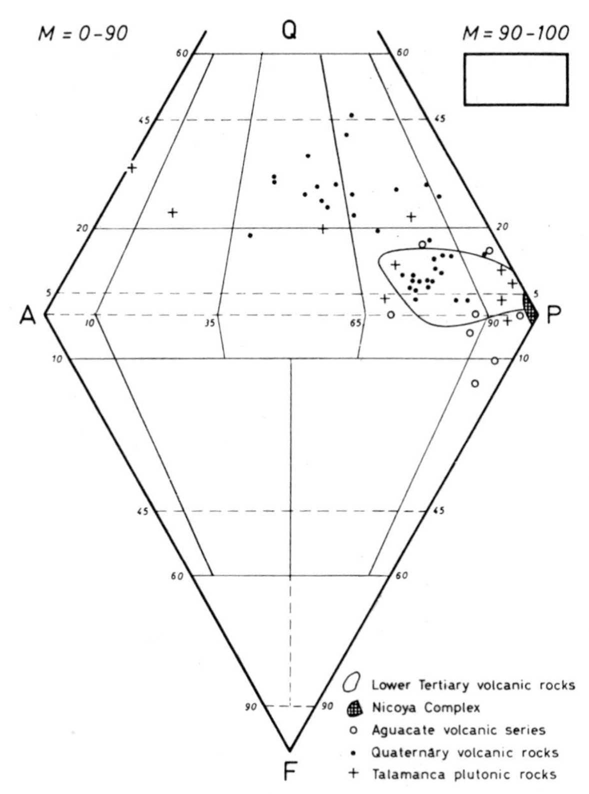

Fig. 76. Distribution of the magmatic series of Costa Rica in the STRECKEISEN double-triangle. (From PICHLER & WEYL 1975, Fig. 6).

A lower unit composed of basic volcano-sedimentary deposits; chaotic megabreccias; an upper unit formed by basic and ultrabasic rocks.

Emplacement occurred before the Upper Campanian and lasted up to the Maestrichtian. The units are covered by Upper Cretaceous to Lower Tertiary sediments and Quaternary ignimbrites. Together with DENGO (1962 a) and WEYL (1969/70), the authors believe there is a possible connection with the Clipperton Fracture Zone.

When studying the magmatic development of Costa Rica, it should be borne in mind that, according to the concepts of plate tectonics, which are expounded in Section 4.12, the Nicoya Complex was formed in the Pacific, far from its present position, and it is consequently, to some extent, an allochthonous element in the structure of the country. In contrast, the subsequent magmatic phases took place autochthonously in the region of the island and continental arc which formed during the Tertiary. However, the position of this arc in the geographic co-ordinate system need not necessarily have coincided with its present position.

Lower to Middle Tertiary volcanic rocks are widely distributed within the marine sediments of Costa Rica. DENGO (1962 a, pp. 142/143) and HENNINGSEN (1966 a, pp. 25/27) name a number of examples which probably belong chiefly to the Eocene but cannot yet be

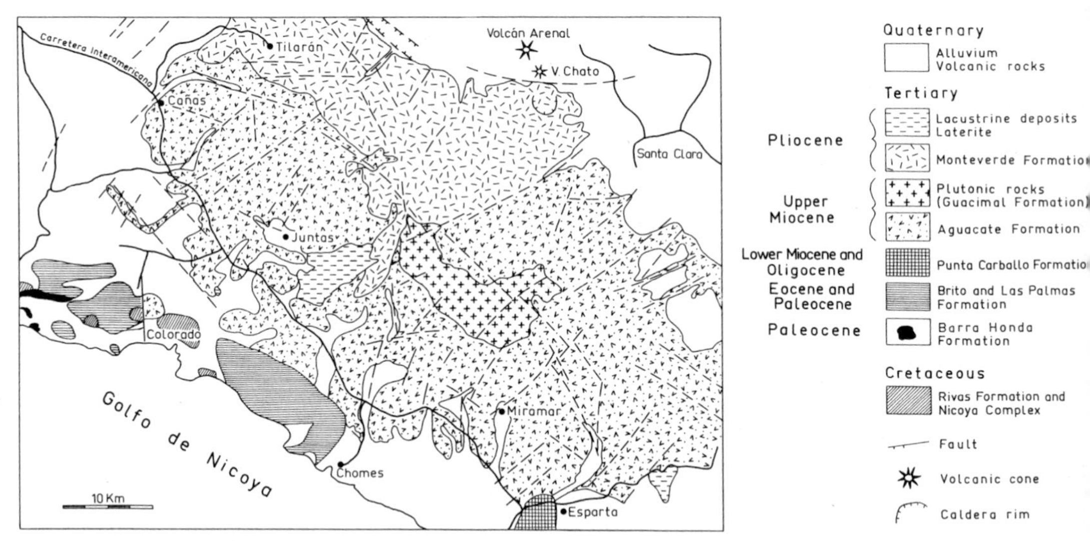

Fig. 77. Geological map of the Cordillera de Tilarán, Costa Rica. (Redrawn from CHAVES & SÁENZ 1974).

dated with certainty. Petrographic data are available above all on the western Cordillera de Talamanca (PICHLER & WEYL 1975) where dark *quartz-andesites* [7] and *quartz-latiandesites* crop out along the Carretera Interamericana in conjunction with tuffitic arenites which date presumably from the Miocene. The SiO_2 values between 50.62 and 57.52% clearly distinguish this mass of rocks from the basalts of the Nicoya Complex (Fig. 76), and there is evidence of an incipient trend towards an increase in the silica levels of the magmas which indicate the formation of an island arc. BELLON & TOURNON (1978, p. 956) distinguish two phases of andesitic series within the Cordillera de Talamanca which they date radiometrically as 16.9 ± 2.5 and 8.3 ± 0.4 m.y.

Apart from this, alkali rocks undersaturated with SiO_2 which DENGO (1962 a) separated out as the **Aguacate Formation** and placed in the Mio/Pliocene, crop out in the Fila de Aguacate and in the Fila de Puriscal. He regarded the Aguacate Formation as younger than the plutonism of the Cordillera de Talamanca. However, in the course of mapping the mining area of Abangares and Miramar, CHAVES & SÁENZ (1974) discovered a large granitic pluton which intrudes into the Aguacate Formation at that point; the Aguacate Formation must therefore be placed before the plutonism, and CHAVES & SÁENZ put it provisionally in the Middle Miocene (Fig. 77). New radiometric data indicate the age of a hypersthene-andesite as 3.2 ± 0.2 m.y. and a subvolcanic diorite as 2.1 ± 0.1 m.y. (BELLON & TOURNON 1978, p. 956).

The typical rocks of the Aguacate Formation are *olivine-nepheline-tephrite* (TOURNON 1973, Table 1, PICHLER & WEYL 1975, Table 2), *nepheline-bearing mugearites, olivine mugearites* and *olivine hawaiites.* In addition, however, *quartz-andesites* and *quartz-latiandesites* also occur. A more detailed examination of these rocks, determination of their ages and an explanation of

[7] Rock definitions based on the STRECKEISEN (1967) method using the RITTMANN (1973) norms are indicated in italics (see also p. 177/178).

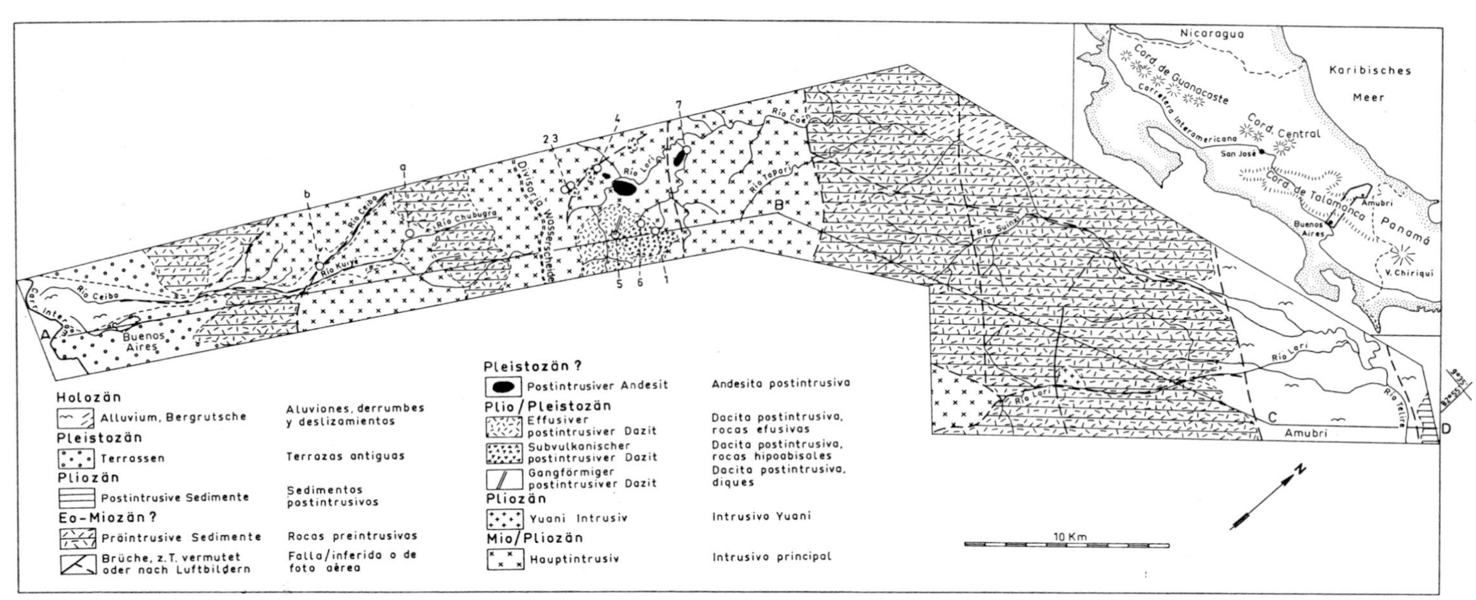

Fig. 78. Geological map of a section across the eastern Cordillera de Talamanca. (From BALLMANN 1976, Fig. 1). 1 – 7 mark the positions of post-intrusive volcanics, a and b of intrusives.

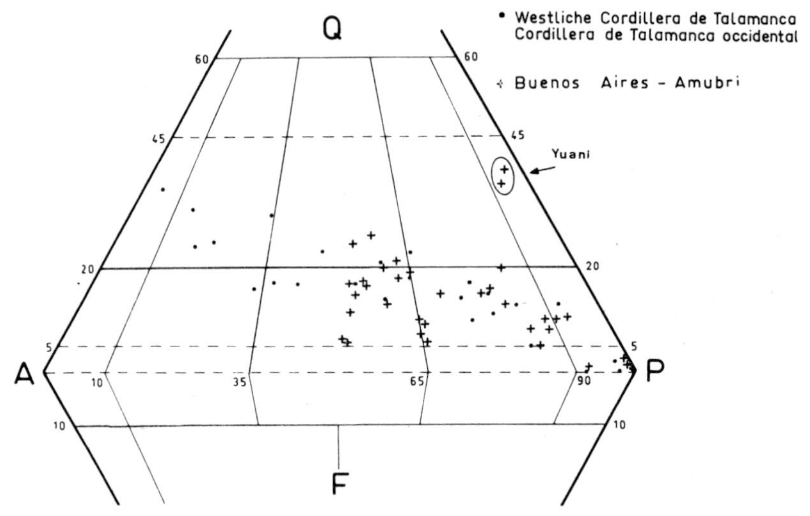

Fig. 79. Distribution of intrusives from the Cordillera de Talamanca in the STRECKEISEN double-triangle. According to their modes. (From BALLMANN 1976, Fig. 2).

the relationships between the rocks would be desirable and could be helpful in settling the problem of the conditions under which they originated: Were they produced by oceanic islands, with which the chemical properties would seem to coincide, or were they generated by volcanic activity in the hinterland of a paleo-subduction zone, like the occurrence of Quaternary alkaline rocks in the Atlantic region of Costa Rica?

Miocene plutonism played a major part in the structure of the Cordillera de Talamanca and has recently also been identified in the Cordillera de Tilarán. Its distribution is only very diagrammatically represented on the geological map of Costa Rica (Fig. 66); a more detailed picture was revealed by cross-sections along the Carretera Interamericana near División (WEYL 1957, Figs. 4 and 7) and in the eastern Cordillera. These cross-sections were recorded in the course of an exploration program conducted by the United Nations (BALLMANN 1976) along a traverse between Buenos Aires and Amubri (Fig. 78).

The plutonism was placed by DENGO (1962 a) in the Lower Miocene and by PICHLER & WEYL (1973) in the Upper Miocene. On the basis of some radiometric dating, however, the age was found to be Lowermost Upper Miocene [8]:

Gabbrodiorite, División:	10.3 ± 0.5 m.y.	(according to DE BOER in BALLMANN 1976, p. 507)
Gabbro, Tapantí:	11.4 – 11.5 m.y.	(according to a paper by SUTTER 1977)
Monzodiorite, San Isidro:	9.6 m.y.	(according to a paper by SUTTER 1977)
"Granitoide", northern Cord. de Talamanca	10.1 ± 0.5 m.y. 8.5 ± 0.4 m.y.	(BELLON & TOURNON 1978, p. 956).

There is a great variety in the intrusive rocks but quartz-monzonitic rocks prevail (Fig. 79). According to WEYL (1957) and BALLMANN (1976) gabbros, gabbrodiorites, diori-

[8] Anonymous (1978, Table 1-1) gives without any details the following data: granite 10.5, diorite 8.5, diorite 2.3 million years.

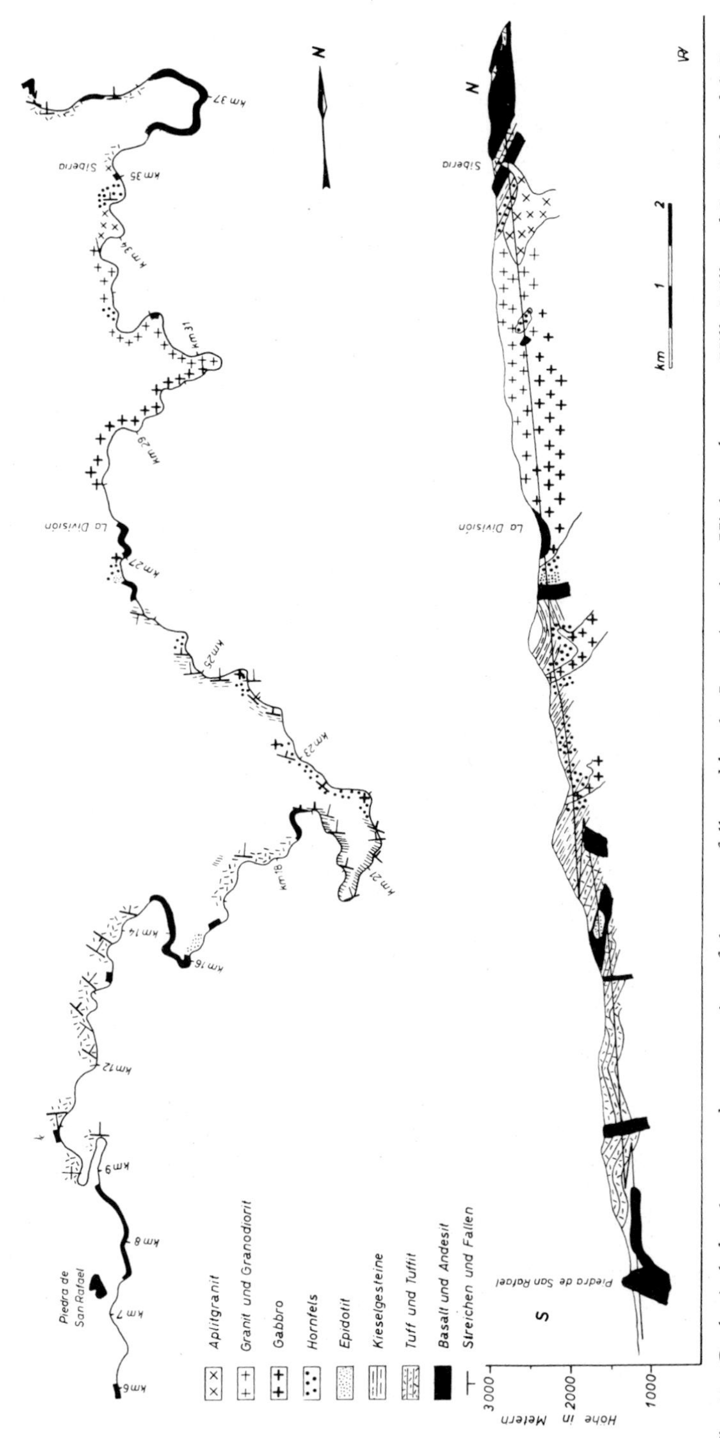

Fig. 80. Geological sketch map and cross-section of the route followed by the Inter-American Highway between Villa Mills and San Isidro del General. Cordillera de Talamanca, Costa Rica. (From WEYL 1957a, Fig. 4).

tes, quartziferous diorites, quartziferous monzodiorites, quartziferous monzonites, granodiorites, granites and alkaline rocks (aplite granites) were encountered. The cross-section at División revealed a vertical differentiation into granites and aplite granites in the upper part and gabbros in the lower part (Fig. 80).

Twelve chemical analyses (WEYL 1961 a, Table 11 and WILLIAMS & MCBIRNEY 1969, Table 1) also proved the existence of a broad scatter in the SiO_2 content between 46.2 and 72.5%, although the basic rocks appear to be over-represented in the rocks which were selected for analysis. The serial indexes (Sigma according to RITTMANN) indicate a medium to weak calc-alkaline character of the material and DENGO (1962 a, p. 146) had already referred to a weak calc-alkaline character of some intrusions.

Altogether the plutonism represents an advanced stage in the transformation of oceanic into continental crust and can probably be interpreted as the outcome of violent subduction between the Cocos and Caribbean plates in the Upper Tertiary.

Pliocene volcanism is claimed to have been found in various parts of Costa Rica, but it seems to have been less important compared with the Lower to Middle Tertiary and the subsequent Quaternary volcanism. Phenobasalts, which were described by DENGO (1962 a, pp. 148/49), occur within the Suretka Formation in the Limón Basin on the Río Pey. In addition, dike swarms of phenobasalts intrude into the sediments of the Uscari and Gatún Formation (DENGO, p. 153), the age of which has not been precisely determined. DENGO thinks that they are contemporaneous and interprets them as concomitants of the tension events occurring in the Upper Miocene/Pliocene.

Volcanism of alkaline character commenced in the Pliocene in the lower reaches of the Río Reventazón with the production of nephelinites, basanites, basalts and teschenites. Radiometric ages are 5.1 ± 0.75 and 4.6 ± 0.25 m.y. according to AZAMBRE & TOURNON (1977) and BELLON & TOURNON (1978). ROBIN & TOURNON (1978) report on further finds of alkaline rocks from the Caribbean side of Costa Rica.

According to DENGO (1962 a, p. 149) and HENNINGSEN (1966, pp. 27/28) basic pyroclastics and lavas, which on the basis of the mineral contents were described as leukoandesite basalts and basalts, are intercalated into the Pliocene Paso Real Formation of the Valle del General.

BERGOEING et al. (1979), in a paper published after my own manuscript was completed, describe plutonic rocks of Miocene to Pliocene age and volcanic rocks of Pliocene to Quaternary age from the Fila Costeña.

In the eastern Cordillera de Talamanca these obviously basic rocks are contrasted with porphyric dacites and quartz-andesites which, according to BALLMANN (1976, pp. 507 – 510), are younger than the intrusive rocks at that point and thus date from the Pliocene to Pleistocene. In material terms the *quartz-andesites* are similar to the Lower Tertiary volcanics of the Cordillera de Talamanca. The chemistry of the *dacites* is somewhere between that of the *quartz-latiandesitic* flows of the Quaternary Cordillera Central and that of the *dacitic-rhyodacitic* ignimbrites of Guanacaste.

SUTTER (paper presented in 1977) gives the radiometric age of a phenobasalt from the Valle Central as 5.7 m.y. and BELLON & TOURNON (1978, p. 958) date an andesitic breccia from San Vito as 2.6 ± 0.4 m.y.

The very powerful **Quaternary volcanism** of the Cordillera Central and of the Cordillera de Guanacaste of Costa Rica is discussed together with the other Quaternary volcanism of Central America in Chapter 5.4. From the material standpoint it represents the continuation

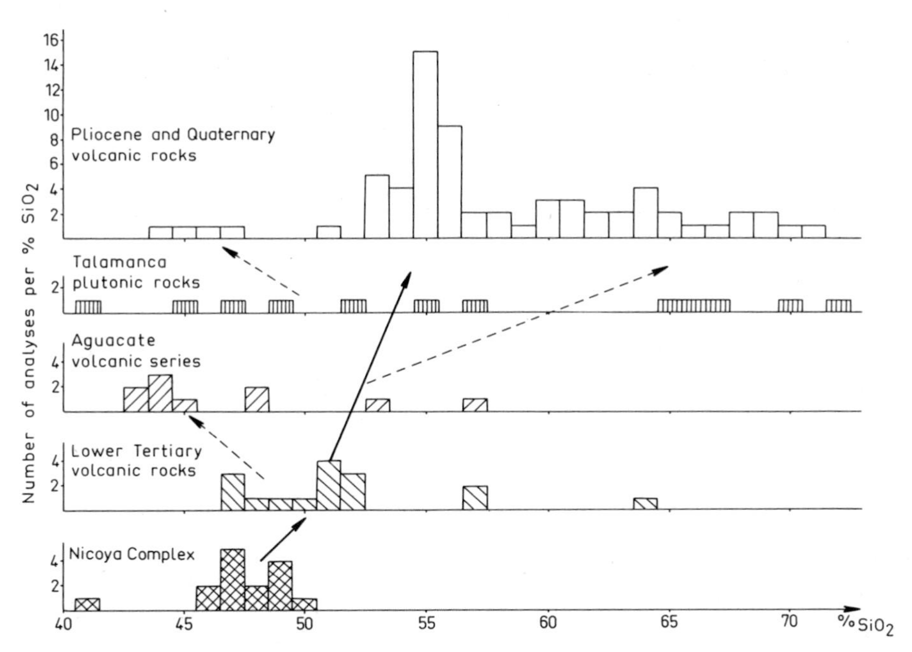

Fig. 81. Histogram of all available chemical analyses of the volcanic and plutonic rocks from Costa Rica showing the crustal evolution of the area.

and for the time being the conclusion in the development from mafic to sialic magmatism in southern Central America. However, according to recent discoveries by TOURNON (1973) and AZAMBRE & TOURNON (1977) the predominance of sialic calc-alkaline flows in the main chains of the volcanoes contrasts with an alkaline province in the Atlantic slopes and in the Limón Basin. BELLON & TOURNON (1978, p. 957) give the radiometric age of a basanite from Colorado as 1.2 ± 0.4 m.y.

The magmatic development of Costa Rica can be summarized graphically in a diagram showing the increase in the silica level in the main flows and subsidiary flows up to the level of undersaturated alkaline rocks (Fig. 81). In addition to the development of the sediments from deep-sea to shallow-water and continental deposits, this also reflects the development into an island arc and into the present continental crust.

4.14 Structure

Very different structural elements collide and to some extent overlap each other in Costa Rica. They stand out clearly in the general geological map (Fig. 66), and they are reflected to a large extent in the morphological classification of the country, which was discussed at the beginning. The various structures are as follows:

1. The regions of the **Nicoya Complex** along the Pacific coast;
2. in the northwestern part of the country, the Tertiary volcanic and plutonic rocks of the **Cordillera de Tilarán** which adjoin the Nicoya Complex;

3. above and alongside the foregoing units come the Quaternary volcanic chains of the **Cordillera de Guanacaste** and the **Cordillera Central,** which are elements of the Central American Volcanic Chain;
4. the southern boundary of the volcanic chain is formed by the transverse depression of the **Valle Central,** which is also probably a transverse fault zone;
5. the **Cordillera de Talamanca,** which adjoins to the southwest and which is composed of Tertiary sediments, volcanic and plutonic rocks;
6. the similarly structured **Fila Costeña** (=Pacific Coastal Range) which splits off from the Cordillera de Talamanca;
7. the intramontane **basin of the Río General** and **Río Coto Brus** between the two mountain chains;
8. the **Limón Basin,** which is all that is left today of a large Tertiary basin, located in front of the mountains to the northeast.

1. Of the regions that make up the **Nicoya Complex,** the Nicoya Peninsula was mapped in the fifties by the geologists of the Cia. Petrolera de Costa Rica. DENGO (1962 b) published the 1 : 300,000 scale map with an explanatory text, and this proved to be an important basis for progress in acquiring knowledge about the geology of southern Central America. The map is unfortunately out of print; a simplified re-drawn version is reproduced in Fig. 82. New conceptions about the structure and facies of the Nicoya Complex will certainly necessitate some changes to the map as well; however, before this can be done some very accurate special mapping and biostratigraphical classification work must be carried out.

DENGO (1961) has supplied some observations on the Herradura Peninsula opposite Nicoya, and HENNINGSEN(1966 a, pp. 43 – 49) and AZÉMA et al. (1979) have provided data on the region around Quepos and Golfito. The Osa Peninsula has so far not been studied in detail. The following paragraphs relate therefore mainly to the Nicoya Peninsula.

All researchers regard the Nicoya Complex as oceanic crust which, according to the current theory, owes its structure and position at the edge of the continent to plate tectonic events. However, views diverge greatly on the events which led to this state. These views are discussed in Section 4.12.

Some of the rocks in the Nicoya Complex have undergone severe deformation. The basalts contain zones of crush breccias. Chloritisation and foliation, such as occur at the Punta Matapalo on the Osa Peninsula, indicate deformation in the area of the greenschist-facies. The sediments which are intercalated in the igneous rocks have undergone intensive local folding or have been tilted up to the vertical position. Where such structures could be identified, they have different directions of strike: SW/NE in the northwestern part of the Nicoya Peninsula or also NW/SE as can be seen on aerial photographs of the hinterland of the Bahía de Tamarindo. Elsewhere, however, the basalts, in particular the pillow basalts, seem to be unfaulted so that even the hyaloclastic rocks lying between them occur in the form of glass.

DENGO (1962 b, p. 67) stressed that when the geological survey was made of Guanacaste, no detailed studies were made of the Nicoya Complex, but he assumed that it may have been deformed several times. In DENGO's opinion a discordance lies between the Nicoya Complex and the overlying sedimentary strata commencing with the Campanian Sabana Grande Formation.

In contrast, STIBANE et al. (1977) regard the fold structures, in particular of the siliceous sediments, as the results of synsedimentary compression. They see the latter as following

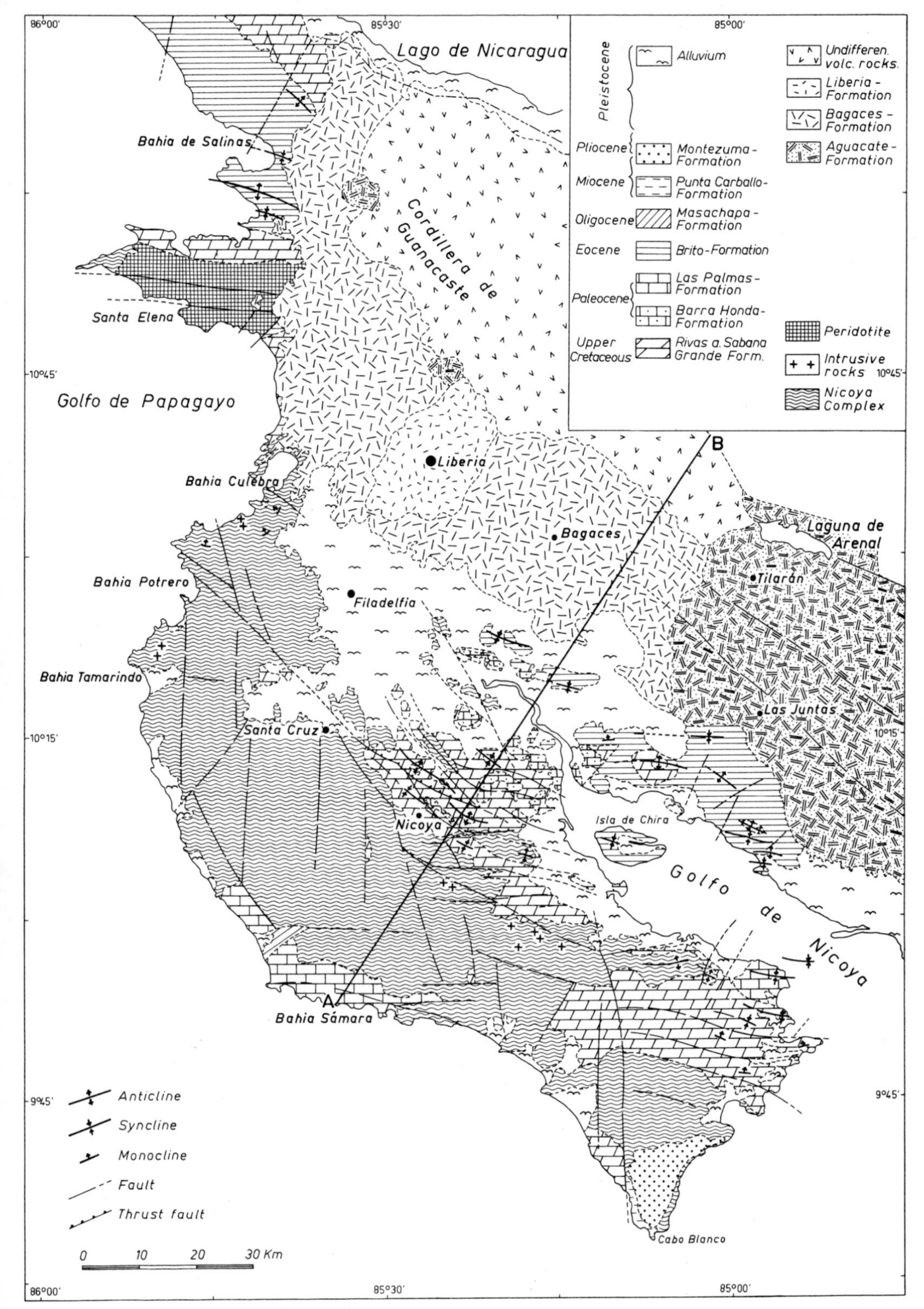

Fig. 82. Geological map of Guanacaste province, Costa Rica. (Simplified redrawing from DENGO 1962b).

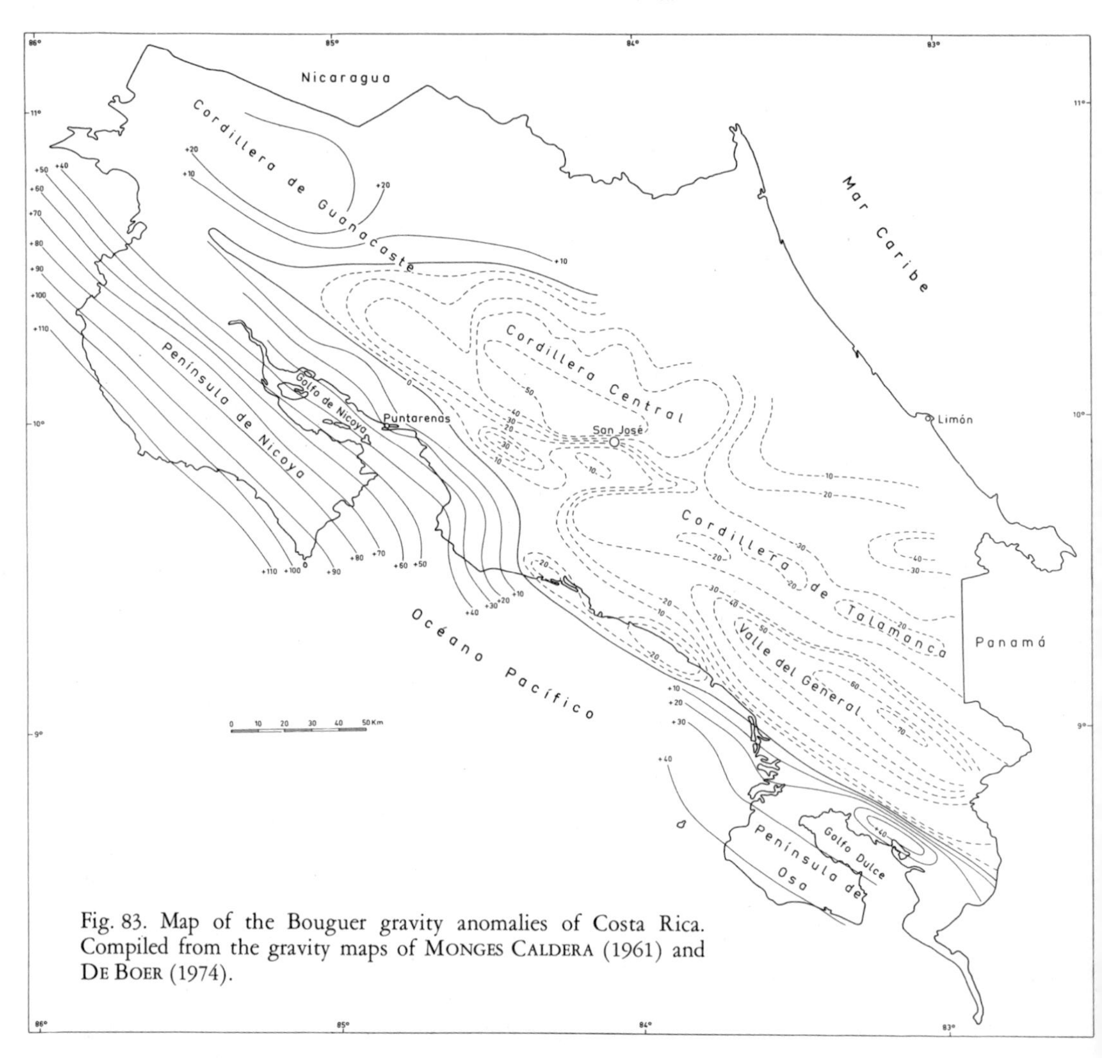

Fig. 83. Map of the Bouguer gravity anomalies of Costa Rica. Compiled from the gravity maps of MONGES CALDERA (1961) and DE BOER (1974).

from vertical block movements which led to the formation of the facies change and of the submarine relief detected by them.

The regions of the Nicoya Complex stand out in the gravity map through the rapid rise in the positive Bouguer anomalies (Fig. 83). A rapid drop towards negative anomalies in the central part of Costa Rica marks a rapid submergence or cessation of the Nicoya Complex which is confirmed in the continental crustal profile below the Cordilleras of Guanacaste. A corresponding process is the submergence of the Nicoya Complex beneath the thicker series of Upper Cretaceous-Tertiary sediments in the **Tempisque Basin**. Mapping carried out here by the Cia. Petrolera de Costa Rica revealed vigorous block folding striking in general NW/SE (Fig. 84). This folding also dominates the sediments in the eastern part towards Nicoya, which crop out in the shelf in the form of an extremely broken coastline with islands ahead of it.

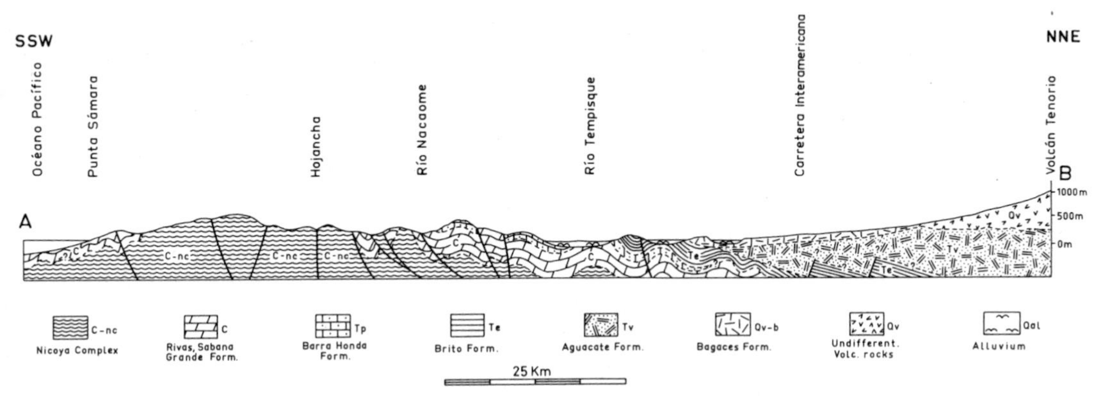

Fig. 84. Geological cross-section through the Nicoya Peninsula, the Tempisque Basin and the southern rim of the Cordillera de Guanacaste (Guanacaste Volcanic Chain). (Redrawn from DENGO 1962b).

At this point STIBANE et al. (1977, p. 347) would like to explain "the thick folded clastic rocks of the Rivas and Las Palmas Formations in the SE of the Peninsula in terms of gravitational slip into a subsiding basin" (this is a free quotation from their work). In their opinion large faults such as the "Falla de Tempisque" and the "Falla de Samara" are decisive structures which led to the formation of basins and uplifted blocks and thus to very extensive slide phenomena. The relationship with plate movement over and beyond an Outer Gravity High has already been desribed in the section on stratigraphy.

With its E/W-striking peridotite/serpentinite and its likewise E/W-striking faults within and along its northern boundary, the **Peninsula of Santa Elena** is something of a foreign body in the structure of Costa Rica (Fig. 68). The peridotite is interpreted as mantle rock (DENGO 1962 a and b, WEYL 1969 c), which started to erode as early as in the Campanian. The first person to study this area, namely HARRISON (1956) and also after him DENGO, considered the possibility that this peridotite and its accompanying structures constitute the continuation of the Clipperton Fault Zone of the eastern Pacific and also the boundary between the continental crust of northern Central America, including the Nicaragua Rise, and the oceanic crust of the southern part of Central America together with the Colombia Basin. The possible objection that the Clipperton Fault Zone ends at the Middle American Trench does not seem very convincing to me, because the latter probably developed as a result of Tertiary subduction quite some time after emplacement of the peridotite. Judging by the paleogeographic situation of the Upper Cretaceous, the peridotite belongs at all events in the region of the ocean floor of that time, and the structures of the Santa Elena Peninsula could be regarded as relics of a crustal condition completely different from that prevailing today.

2. The mountain chains which are grouped together as the **Cordillera de Tilarán** are built up from magmatic rocks which have long been separately identified as the Aguacate Formation but which were not described in detail until CHAVES & SÁENZ (1974). These authors' discovery of a large intrusive mass about 200 km^2 in extent (Fig. 77), whose rock type can be compared with the intrusive rocks of the Cordillera de Talamanca, was particularly important. CHAVES & SÁENZ date the Aguacate Formation as Miocene. They share the opinion already advanced earlier by the present author (1961 a, p. 76) that the Cordillera de Tilarán represents the northwestern continuation of the Cordillera de Talamanca.

The crest of the range, which rises to 1,400 m, is covered by remains of a peneplain and the range is bounded on both sides by NW/SE-striking fault scarps. The course of these scarps can be followed over a distance of 170 km so that they are given their own names, namely "Falla del Río Cañas" and "Falla de Las Juntas" in DENGO's general map of Guanacaste. CHAVES & SÁENZ discovered further E/W-, SW/NE- and SE/NW-striking fault systems, of which the SW/NE system is the main site of mineralization. Altogether, the Cordillera de Tilarán is a horst mountain bounded on both sides by faults, and it submerges in the northwest into the Depression of Tilarán.

3. The **Cordillera de Guanacaste** and the **Cordillera Central** are dealt with in Chapter 5.4 as parts of the Quaternary volcanic chain of Central America. The linear configuration of the volcanoes in the Cordillera de Guanacaste along a NW/SE-strike would seem to indicate a similarly trending fault system linking up with the faults which bound the Cordillera de Tilarán. In the Cordillera Central the arrangement of the volcanoes indicates two obliquely intersecting fault systems, namely a WNW/ESE-striking system with the volcanoes Viejo, Poás and Barba, and a NE/SW-striking system with the volcanoes Irazú and Turrialba. Further fractures are suspected in the basement of the Cordillera Central, and it is along these that the basement, which is built up of Tertiary volcanics, breaks away to the Atlantic Lowland.

In the gravity map the Cordillera Central stands out because of a strong negative Bouguer anomaly (Fig. 83), which is probably caused by the flows of lighter specific gravity from the large volcanoes. The anomaly crops out towards the northwest and is replaced in the Cordillera de Guanacaste by a slight positive anomaly.

According to seismic measurements carried out in the region of the volcano Arenal, the basement of the volcanic chains consists of continental crust with the following structure (according to MATUMOTO et al. 1977):

Layer	P-wave Velocity (km/sec)	Crossover (S-P) Time (sec)	Thickness (km)
1	5.1	7.5 (+ 1.0)	8.2 (+ 1.1)
2	6.2	16.0 (+ 1.0)	12.9 (+ 0.8)
3	6.6	26 (+ 4)	22.3 (+ 5.4)
4	7.9		

This finding is all the more remarkable since the oceanic crust of Nicoya is located in the immediate vicinity. The transition of both types of crust is characterized by steep gradients in the Bouguer anomalies.

4. The **Valle Central** is situated between the volcanoes of the Cordillera Central in the north and the scarp of the Cordillera de Talamanca in the south. Earlier authors compared it with intramontane depressions of the North American Cordilleras (HILL 1898) or regarded it as a fault graben (SCHAUFELBERGER 1935). If, however, the volcanoes of the Cordillera Central were not there, the Valle Central would not be an upland basin but merely a broad step in the foreland of the Cordillera de Talamanca.

The basement of the Valle Central consists of Quaternary river gravels, lake deposits, unconsolidated volcanic masses and lavas which, commencing with the (?) Pliocene Doán Formation, are discordantly superjacent on the Tertiary. The basement is dominated by block folding with a mainly WNW/ESE-strike in the folds and faults running mainly at right angles to them (KRUSHENSKY et al. 1976, Fig. 2). A diagrammatic cross-section through the eastern Valle Central by ESCALANTE (Fig. 85) reveals the block folding and the discordance over the Miocene Aguacate Formation. According to KRUSHENSKY et al. (1976, p. 132) a further discordance is located between the (?) Upper Eocene Pacacua Formation and the Oligocene/Miocene Térraba Formation.

The marine Miocene of the Valle Central is nowadays situated between 600 and 1,890 m above sea level. KRUCKOW (1974, 1976) gives the rate of uplift as between 0.040 and 0.095 mm per annum; however, the terraces of the Río Barranca, which were surveyed by him, indicate periods of vigorous uplift and relative quiescence. Based on levelling surveys carried out along the route followed by the railway between Puntarenas, San José and Puerto Limón, MIYAMURA (1975) calculates that the present rate of uplift is 1 – 2 mm per annum; however, the measurements are obviously greatly affected by local movements.

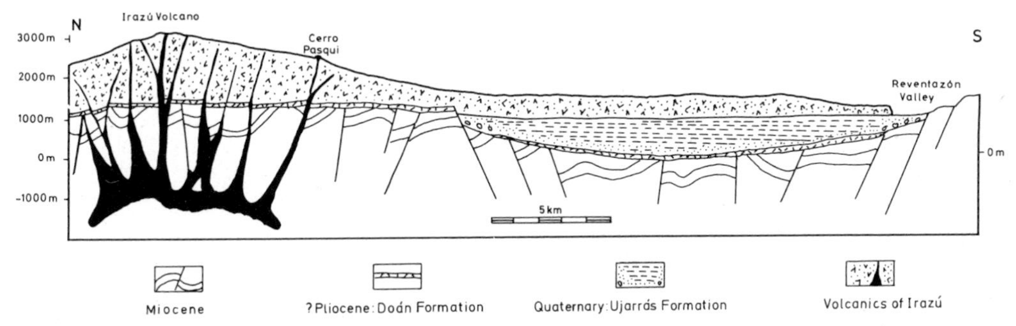

Fig. 85. Geological cross-section through the eastern Valle Central. (Simplified redrawing from ESCALANTE 1966, Fig. 4).

In addition to these details it should not be forgotten that the Valle Central separates the chain of young volcanoes of Central America from the Tertiary mountain range of the Isthmus, if we disregard its branches in the Cordillera de Tilarán. This boundary zone is seismically active and was the site of many serious earthquakes (GRASES 1975, Mapa 7). The last interoceanic link in the region of Costa Rica ran through this zone during the Miocene. Its western extension takes in the end of the Nicoya Peninsula with E/W-striking structures, and the Middle America Trench ends in the oceanic region where it is replaced by the Cocos Ridge.

5. In the southern part of the Valle Central, along a series of faults with considerable throw, rises the **Cordillera de Talamanca,** the highest and largest mountain range in Costa Rica, which continues into the Cordillera Central of western Panama. Data on its structure are given in WEYL (1957 and 1961 a) so that here we will merely summarize the fundamental facts and incorporate new knowledge that has been obtained.

In the western part of the range, and in particular on the northern flank, Tertiary sediments and intercalated volcanics predominate. According to our present knowledge they are flat-bedded or slightly inclined. Measurements carried out along the drop of the Carretera Interamericana between División and the basin of the Río General revealed slight folding (Fig. 80), into which intrusive rocks have penetrated. In the eastern part, on the other hand, along the traverse surveyed by BALLMANN (1976), by and large only horizontal strata or strata distorted by faults were found. Plutonic rocks are obviously much more in evidence in the eastern part of the range than in the west and north (Fig. 78).

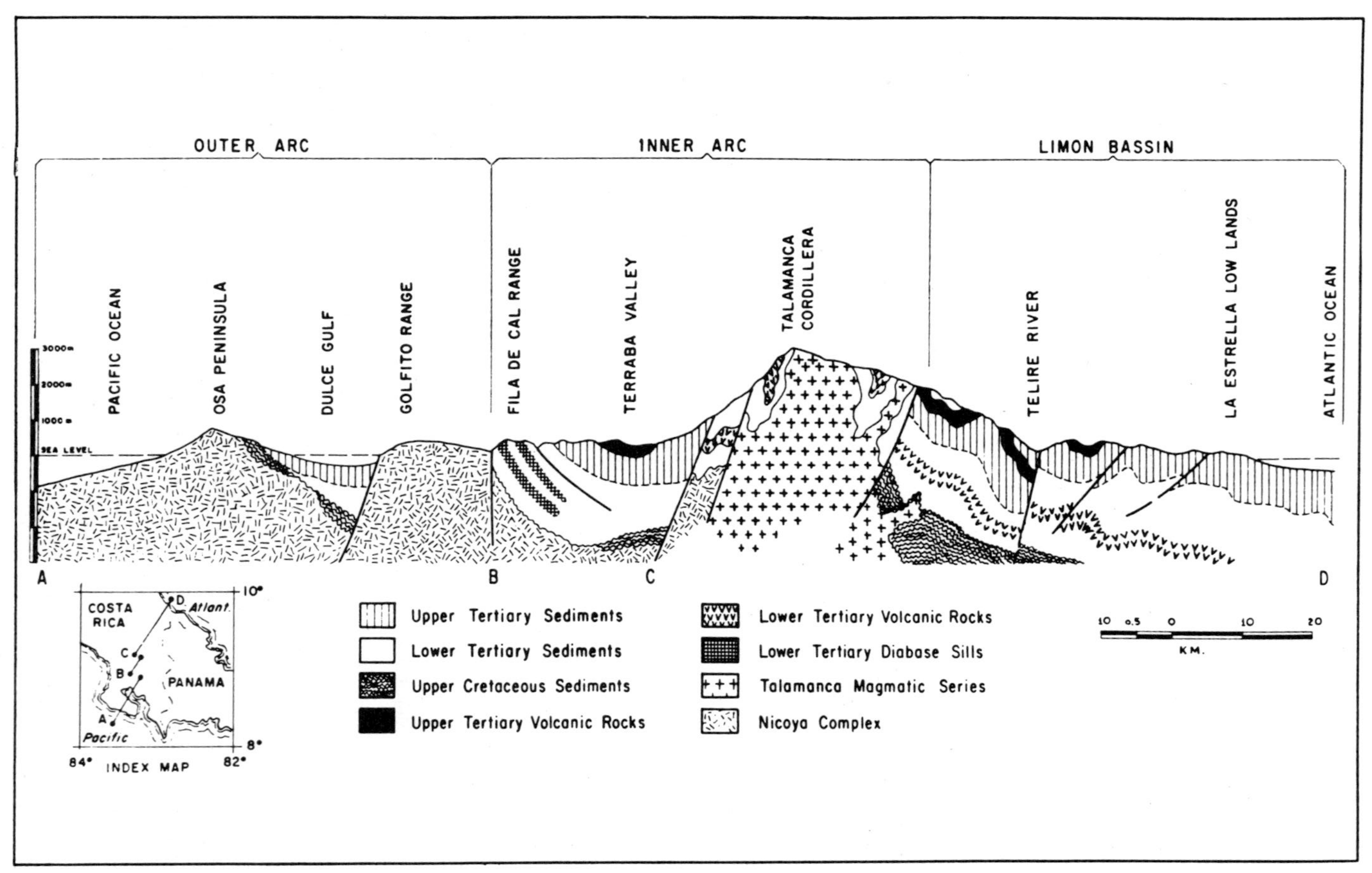

Fig. 86. Schematic cross-section through southeastern Costa Rica. (From DENGO 1962a, Fig. 2).

Faults, even if so far they can only be indicated in a very general way (cf. map, Fig. 66), are more important with regard to the present pattern of the Cordillera. The faults probably run mainly with the strike of the morphological longitudinal axis and at right angles to it. Corresponding fault and joint systems were detected on aerial photographs in the region of Cerro Chirripó (WEYL 1957, Fig. 13, 1961 a, Fig. 29). The steep southern slope of the range to the Valle del General must be regarded as a stepped escarpment and was shown as such in the cross-section given by DENGO (Fig. 86). On the northern side, on the other hand, the Cordillera slopes down relatively gently to its foreland which is made up of Tertiary sediments. This is presumably the result of tilting towards the southwest. The Cordillera de Talamanca is, after all, to be regarded as a block mountain intersected by plutons in which folding plays only a secondary role.

The violent young uplift of the range is expressed in the high-lying remains of flattish relief (WEYL 1957 a, pp. 185 – 196) and in the deep valleys incised into this relief, as well as in the Tertiary (probably Miocene) shallow-water sediments on the mountain crest of Alto de la Asunción at an altitude of 3,400 m. KRUCKOW (1974, p. 252) assumed a rate of uplift of around 0.15 mm per annum, assuming continuous uplift. However, it has already been discussed in connection with the river terraces that the uplift was more probably discontinuous.

A moderate negative Bouguer gravity anomaly follows the course of the range. Because of this anomaly the uplift of the range can be interpreted as an isostatic compensation.

6. According to HENNINGSEN (1966 a, pp. 28 – 33) only weak folding with mainly NW/SE-trending strata prevails in the **Fila Costeña** as well. The intensity of folding seems to increase in the stratigraphically deeper stockworks, but a discordance can only be detected between the Miocene Gatún and the Pliocene Paso Real Formation. The morphology of the Fila Costeña is much more strongly influenced by faults which trend NW/SE and give rise to a steep, step-like escarpment of the range to the Pacific. These faults are less pronounced on the northern side in the more gentle slope down to the Valle del General. Thus the Fila Costeña appears to be a fault block tilted to the south, which is smaller than but structurally similar to the Cordillera de Talamanca and which has only weak internal folding (cf. Figs. 86 and 87).

7. The broad **valley of the Río General** and the **Río Coto Brus** is covered by the Pliocene Paso Real Formation and by fluviatile Quaternary. In the general geological map the Pliocene is depicted as being very limited in extent, but both DENGO (1962 a, Fig. 4) and HENNINGSEN (1966 a, Fig. 6) show it extending over a great distance longitudinally through the basin. The strong negative Bouguer anomaly (Fig. 83) accompanying it indicates very thick Tertiary material. The valley thus nowadays appears to be a chronological continuation of the Tertiary Térraba Trough. However, the deep incising of the Quaternary river terraces indicates that the region is undergoing uplift, even though to a lesser extent than the mountain ranges bordering it on both sides (MADRIGAL 1977).

8. The **Limón Basin** also developed out of the Tertiary Limón Trough which, however, extended further into the present slope of the Cordillera de Talamanca and into the Valle Central. Petroleum exploration carried out in the fifties and sixties revealed a strong folding with overthrust folds, a structural style which continues into the northern coastal region of Panama (cf. Fig. 88). Unfortunately no details have been published but the most important structures were incorporated into the general geological map of Costa Rica. A structural sketch of the coastal region of Limón is given in the first (German) edition, Fig. 30 (WEYL 1961 a).

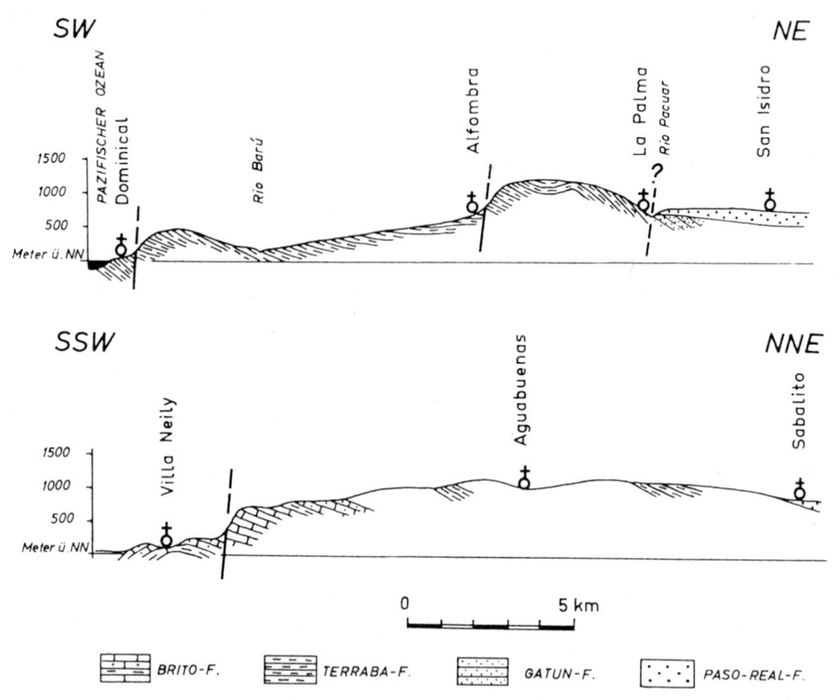

Fig. 87. Cross-sections through the northwestern (upper) and southeastern Fila Costeña, Costa Rica. Exaggeration 2×. (From HENNINGSEN 1966a, Fig. 5).

According to data supplied by RIVIER (1973, p. 151) a widespread clear discordance occurs between the Uscari and Gatún Formations, i.e. at the base of the Middle Miocene and, according to older sources (WEYL 1961 a, Table 5), further discordances follow between the Gatún and the Pliocene Suretka Formations and between the latter and the Quaternary. In view of the uncertain classification of many of the identified formations, it is still rather unclear to what extent these discordances can be correlated with those of the Valle Central.

Many people have attempted to arrive at an **overall view of the tectonics of Costa Rica;** the theories reflect not only the respective knowledge but also the fundamental conception of tectonics held by the authors and prevalent at the times when they were writing. While SAPPER (1937, pp. 120 – 123) obviously set out to collect facts from which he concluded, among other things, that the geological history of Costa Rica followed a different course from that of Panama, STILLE (1940, pp. 321 – 326) was mainly concerned with placing the development in the pattern of tectonical phases which he had traced all over the world. I myself (1957, p. 197, 1967, p. 371) attempted to show that the development of Costa Rica, starting from the Cordillera de Talamanca, was a tectonic magmatic cycle as defined by STILLE, RITTMANN and others. The thick Tertiary deposits with their basic volcanism were interpreted as geosynclinal formations. An orogenesis commenced in the Miocene with only weak folding but vigorous granitic plutonic activity. This was followed by a period of relative calm and peneplanation which was interrupted by terminal faulting and (subsequent) volcanism. The structure of the orogen was characterized by predominant fracture tectonics, gentle folding and vigorous magmatism.

DENGO (1962 – 1973) commented in a series of papers on the structure and on the tectonic development of Costa Rica, and he included southern Nicaragua and western Panama in a "Southern Central American orogen". Following on from the subdivision of the circumpacific belt as conceived by GUTENBERG & RICHTER (1954), he distinguishes between an outer arc, later called a Pacific arc, incorporating the regions of the Nicoya Complex in Costa Rica and the southern coast of Panama, an inner arc with the Cordillera de Talamanca and its continuation into Panama, as well as the volcanic mountains of Costa Rica, and the Caribbean Basins of Nicaragua/Limón, Bocas del Toro, Darién and Chocó. Thus structurally and developmentally related regions are united over a large area, but nothing is said about the conditions leading to their formation.

By and large following the model of an orogenic cycle, DENGO distinguishes chronologically a prototectonic phase as the time when the Nicoya Complex formed; the latter is taken to be a submarine ridge in the Pacific surmounted by a volcanic island arc. The ridge is interpreted as the possible reaction of an impact between Caribbean and Pacific crust, thereby predicting the theories of modern plate tectonics. The following orogenic phase covers the period from the Upper Cretaceous to the Late Tertiary and is characterized by the formation of troughs with thick sedimentation, folding of the Laramide orogeny, Late Tertiary plutonism and the uplift of young mountain ranges. The cycle is concluded by a post-orogenic phase with the final vertical movements and the build-up of the Quaternary volcanic mountains.

Since this conception was first drafted, the theories of plate tectonics have undergone explosive development, particularly in the Central American-Caribbean region (MALFAIT & DINKELMAN 1972, HEATH & VAN ANDEL 1973, HEY 1977, HEY et al. 1977).

In a new edition of his synthesis (1973, p. 44) DENGO has, however, deliberately refrained from commenting on this, because the time is still not ripe for a model that can satisfactorily explain all phenomena. In the meantime, however, further knowledge has been obtained so that an attempt may be made to interpret certain aspects of the tectonics of Costa Rica in accordance with the plate tectonics model:

The oceanic crust of the Nicoya Complex formed at the time when the two American plates drifted apart (LADD 1976). Rifts or transform faults may therefore exist in the ultrabasic rocks and faults of the Peninsula of Santa Elena. The subsequent development of the Nicoya Peninsula from the Campanian to the Paleocene is explained by STIBANE et al. (1977) who hypothesize plate movement via an Outer Gravity High. Evidence of the resulting subduction should be sought in the northeast. The sedimentary troughs of Costa Rica, which deepened considerably during the Tertiary, could thus be understood as the reaction of the crust to the subduction process taking place beneath them. Initial island arc volcanism is indicated by the quartz-andesitic volcanic rocks of the "basement" of eastern Panama (GOOSSENS et al. 1977, cf. chapter on Panama).

Deformation of the Tertiary sediments (discordances in the Miocene) and the violent Upper Miocene plutonic activity can be regarded as signs of accelerated subduction under an island arc which formed in Costa Rica and Panama. Subsequent uplift of the mountain range in the eastern part of Costa Rica and in the western part of Panama can be interpreted as the isostatic counterweight at the end of the subduction phase in this part of Central America; the cessation of subduction is also indicated by the decline in magmatism and in reduced earthquake activity (cf. the development diagrams drawn up by MALFAIT & DINKELMAN 1972, Fig. 5 and HEATH & VAN ANDEL 1973, Fig. 7, and our Fig. 178). To the north of the Valle

Central, earthquakes, volcanism and the Middle America Trench are all evidence that plate subduction is continuing right up to the present day.

DE BOER (1979), in a paper published after my own manuscript was completed, makes use of volcanological-petrographical observations, paleomagnetic measurements and microtectonic analyses carried out in the area of the peninsulas of Santa Elena and Nicoya, to attempt an interpretation of the structure of Costa Rica in terms of plate tectonics. Unfortunately, he was unable to take account of the facts reported by STIBANE et al. (1977) and SCHMIDT-EFFING (1979) which cast doubt in particular on his dates. DE BOER summarized his theories as follows:

"The basement of the Costa Rican outer arc consists of two major complexes. The older is composed of peridotite-serpentinite, pillow lava and radiolarite; the younger is made up of gabbrodiorite, pillow lava, pyroclastic rocks and siliceous limestone.

The older pillow lavas are believed to be oceanic crust generated along the north-south-spreading Carnegie Ridge during the late Coniacian. The younger lavas flowed from fissures along a west-northwest-trending volcanic belt (Culebra arc) which developed in this crust during Early to Middle Campanian time, when it collided with the Chortis block. Paleomagnetic evidence suggests that the older sequence originated in the Southern Hemisphere, and the younger in the Northern.

During the Paleocene, the crust fragmented and separated into the Caribbean and Cocos plates, probably as a result of the outer arc escaping the tectonic influence of the Carnegie Ridge and entering that of the ancestral East Pacific Rise. This fragmentation resulted in the formation of two parallel volcanic belts (San Antonio and Cachimbas arcs) in the inner deep (Tempisque Valley), which remained active throughout the Eocene. It is postulated that subduction of the Cocos beneath the Caribbean plate was initiated during Oligocene time and resulted in the formation of yet another volcanic belt (Tilarán-Talamanca arc). The outer arc was uplifted, folded, and thrust southwestward. The resulting pattern shows a gradual clockwise rotation west to northwest, and northeastward migration of the volcanic arcs through time. Aeromagnetic and tectonic data indicate that differential uplift and later gravitational décollement of the sedimentary rock blanket characterize the tectonic deformation of singular volcanic belts, and that tectonic overprinting is usually restricted to one major phase."

4.2 Panama

4.21 Introduction and general morphology

Right up until the first half of this century, geological research in Panama was by and large restricted to the Canal Zone, which was, however, very thoroughly studied. The petroleum exploration activity which then commenced yielded mostly unpublished data, although TERRY's summary description (1956) emerged to give the first general appraisal of the area. Fresh impetus was given by the explorations carried out for non-ferrous metals, in particular copper, in the sixties by the United Nations and paralleled by studies conducted by private companies. This work yielded general maps of large areas in the western, central and eastern parts of Panama, and the accompanying texts are available in the form of unpublished reports. Further impetus came from the attempt to find a route for the second inter-oceanic canal in the region of the border between Panama and Colombia; surveying of the route went hand in hand with geological studies of the region.

On the basis of this work the Dirección General de Recursos Minerales was able, in 1976, to put out a new general geological map of the country, which represented a considerable advance over TERRY's map and his interpretation.

In the most recent phase of research petrological, geochemical and deposit-related studies have been carried out on the widely-distributed igneous rocks in the country; these studies are closely connected with the search for and discovery of copper deposits.

The fundamentals of the topography of Panama are depicted in simplified form in the map in Fig. 89. The most important elements are as follows:

In the west there is a central mountain range which constitutes the direct continuation of the Cordillera de Talamanca of Costa Rica. It is called the **Cordillera Central** and sometimes also the **Serranía de Tabasara.** In the west the Cordillera rises to altitudes of 2,500 m while towards the east it slopes slowly away towards the depression of the Canal Zone. The mountain range, which is built up chiefly of Tertiary volcanics and their detritus, is surmounted by the **volcanoes** El Barú (=Chiriquí) at 3,475 m, Cerro Santiago at 2,826 m and El Valle at 1,173 m. Ahead of the Cordillera Central lies flatter hilly country and a coastal plain of varying width. In the north it is broken up by the Laguna de Chiriquí and the islands of Bocas del Toro. On the Pacific side the mountainous peninsulas of Burica, Soná and Azuero continue the highly broken character of the Pacific coast of Costa Rica. Between them the submerged, funnel-shaped mouths of rivers with mangrove swamps reach far into the mainland.

To the east of the Canal Zone lie mountain ranges and a continental watershed with the **Serranía de San Blas** and the **Serranía del Darién** only barely 15 km from the Caribbean coast. To the south of the narrow mountain chains extend the **basins** of the Río Bayano, the Río Chucunaque and the Río Tuira. These rivers, too, possess extremely broken mouths which extend deep inland. On the Pacific side, small fault block mountains rise out of the lowland to heights of 934 m (Serranía de Majé) and 1,581 m (Serranía del Sapo).

The morphological features of Panama, which have been described in very broad outline here, reflect to a large extent the geological structure, as is easy to see from the general geological map in Fig. 90. We will return to this in more detail in Section 4.24.

TERRY (1956) made some important observations on the morphology of Panama which are quoted here from the first (German) edition of this book. He distinguished between a mature stage of relief in the east and a youthful stage of relief in the west with central Panama between the two.

Eastern Panama is characterized by wide, flat valley floors with meandering rivers, which are accompanied by arc-shaped lakes. The valley-floor divide, for example between the river systems of the Río Bayano and the Río Chucunaque, is so flat that it cannot be seen from the air, and the inhabitants of the country cross it in their boats. The planation of the broad lowland areas of eastern Panama took place after the folding of the Tertiary including the Pliocene Chucunaque Formation, whose uplifted strata have been cut off. The lower position of the peneplain indicates that tectonic repose has reigned since its formation. The extensive submerged river mouths of eastern Panama, particularly in the region of the Golfo de San Miguel, owe their formation to the post-glacial rise in the sea level. The morphological findings are confirmed by slight seismic activity.

Extensive remains of a peneplain landscape are found at about 1,300 m on the horsts of the Serranía de Pirre and between the Balsa and Sambú Basins. According to TERRY the process

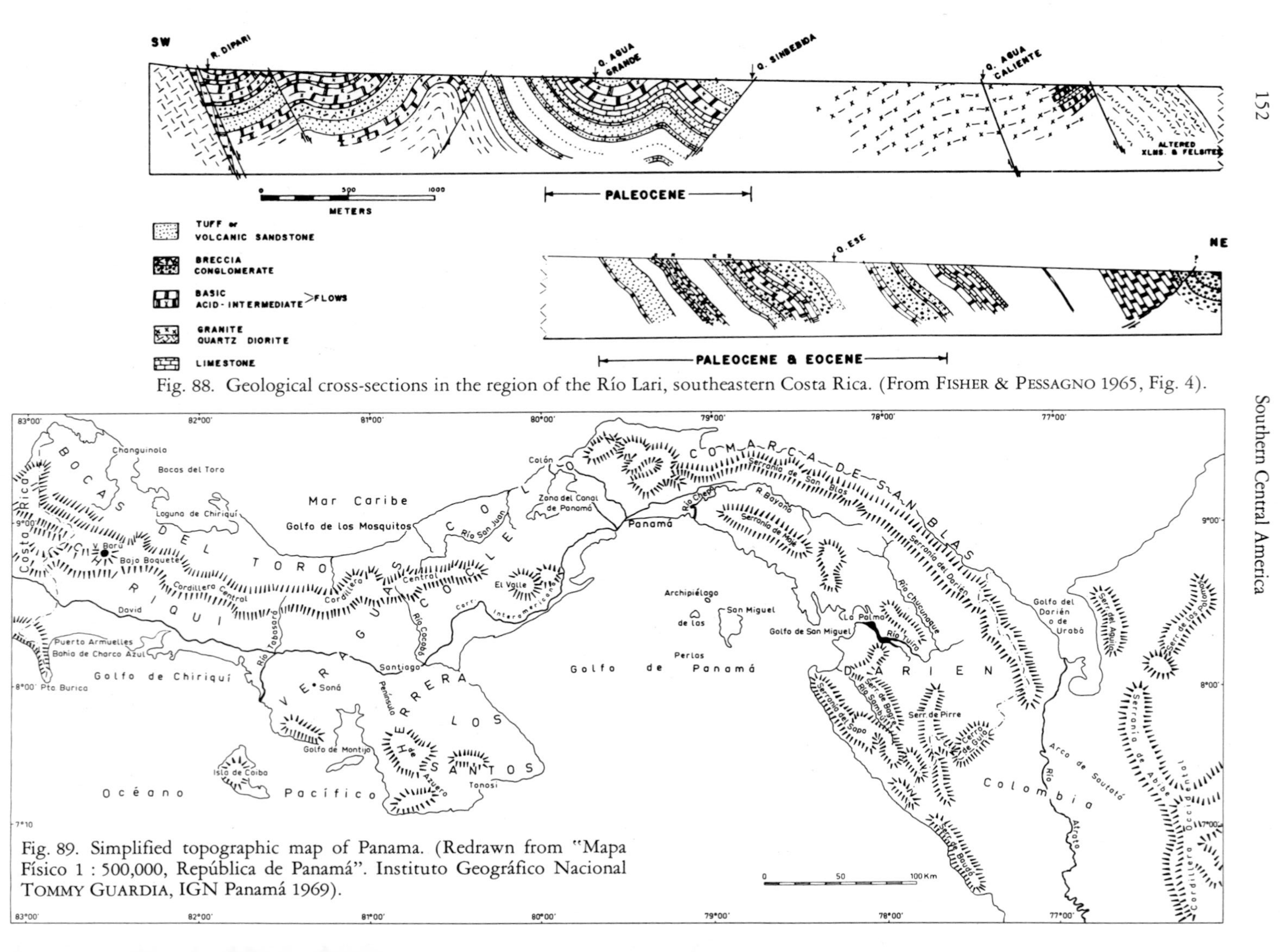

Fig. 88. Geological cross-sections in the region of the Río Lari, southeastern Costa Rica. (From FISHER & PESSAGNO 1965, Fig. 4).

Fig. 89. Simplified topographic map of Panama. (Redrawn from "Mapa Físico 1 : 500,000, República de Panamá". Instituto Geográfico Nacional TOMMY GUARDIA, IGN Panamá 1969).

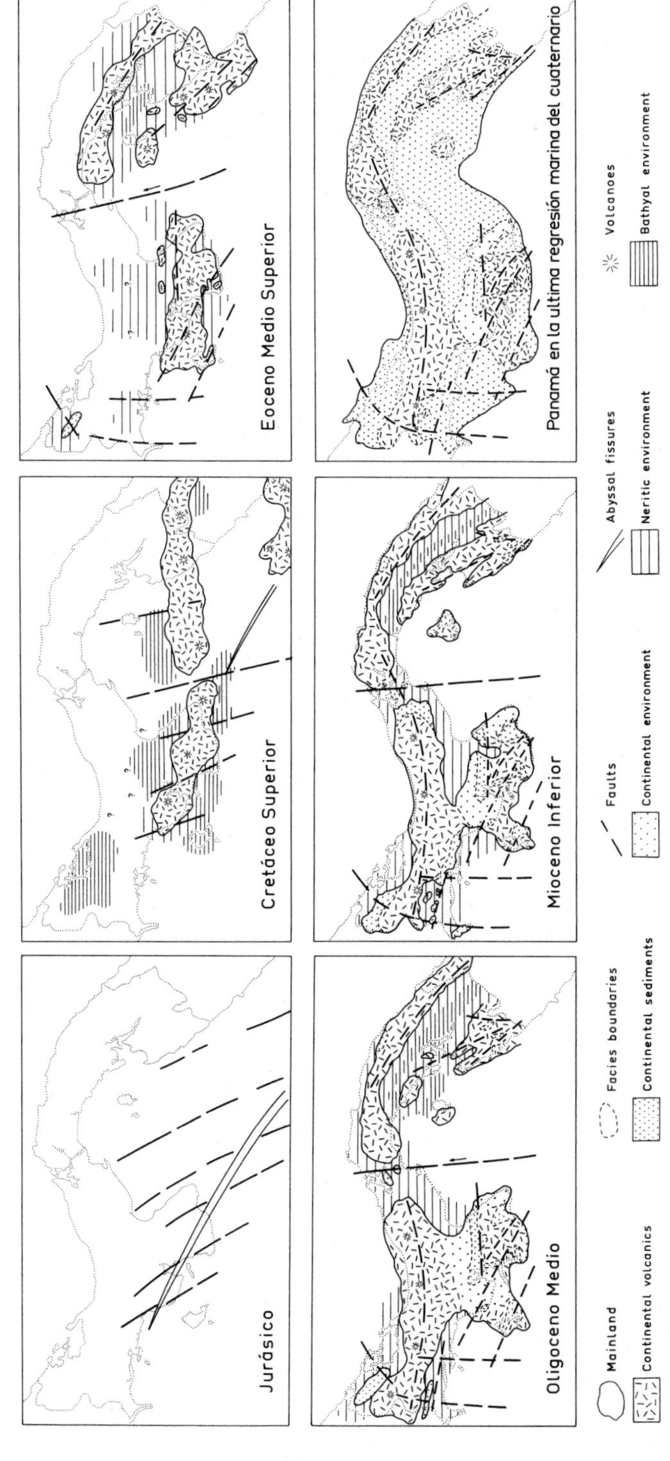

Fig. 91. Paleogeographic maps of Panama. (Redrawn from Recchi 1975).

of peneplanation commenced after the pre-Gatún folding, i.e. in the Middle Miocene. The present altitude is the result of vertical displacement of the fault blocks.

A slightly younger stage in the erosion cycle is found in central Panama. The rivers are consequent, and their courses are fairly straight. Submerged river mouths occur only in a limited area on the Pacific side and they end at the plain of Santiago. In some regions, for example on the Azuero Peninsula, traces of recent faulting are evident. El Valle, Cerro Quema and San Miguel de la Borda are young volcanoes.

In western Panama the central highlands, with peaks rising to more than 2,000 m, predominate. Youthful relief is evident and traces of flat relief are left at high altitude. Near the western boundary, on the Pacific slope, TERRY observed an extensive area of flat relief from 3,000 to 4,000 feet high between Cañas Gordas and Breñon. The relief slopes away gently to the southwest towards the depression between the highlands and the peninsula of Burica. The latter is also covered by areas dipping to the north, and it can be interpreted as a tilted block. The landscape of western Panama is dominated by the massive isolated volcanic cone of Chiriquí.

4.22 Stratigraphic sequence and paleogeography

The **basic to ultrabasic volcanics** which crop out in the southeastern part of the Azuero Peninsula and on the Soná Peninsula (DEL GIUDICE & RECCHI 1969) are regarded as the oldest rocks. They comprise picrites, pyroxenites, melagabbros, gabbros, basalts, often with a pillow structure, and agglomerates. Greenschists with chlorite and actinolite which are found at certain points could be a slightly metamorphosed facies of this ultrabasic series, particularly since the chemical make-up is the same. It is not possible to give any precise indication of the age. RECCHI (1975) placed them, with some reservation, in the Jurassic and interpreted them as submarine flows from an oceanic fissure (Fig. 91).

Up until a short time ago only very sparse and vague information was available about the **Cretaceous** in Panama (TERRY 1956, p. 29, WOODRING 1957, p. 13), but in recent years some very important data have been obtained (FISHER & PESSAGNO 1965, DEL GIUDICE & RECCHI 1969, BANDY 1970, BANDY & CASEY 1973, CASE et al. 1971, CASE 1974).

According to these authors the **Basic Igneous Complex,** which is composed of basic volcanic rocks, as well as the Nicoya Complex of Costa Rica, belong to the Cretaceous and some parts probably also belong to the lowermost Tertiary. The Basic Igneous Complex crops out on the Azuero Peninsula, in the vicinity of the Gulf of San Miguel, and above all in the Caribbean coastal ranges of eastern Panama. Boreholes which were sunk in the Serranía del Darién and on the Gulf of San Miguel in the course of prospecting a possible canal route penetrated up to more than 400 m of basalts, tuffs and agglomerates. They are overlain by highly deformed cherts and fine-bedded, siliceous sediments which contain large quantities of radiolarians and sometimes foraminifera. The species *Dictyomitra torquata* FOREMAN is the index species for the **Campanian** (BANDY & CASEY 1973, p. 3083). The rocks may be compared, both lithologically and on the basis of the organisms they contain, with similar material from deep-sea drillings carried out in the Caribbean and the Pacific, and the Basic Igneous Complex can be assigned to the second, likewise Campanian, reflector of the deep-sea basins (cf. MOORE & FAHLQUIST 1976).

Some samples containing large quantities of Campanian radiolarians and Maestrichtian foraminifera were also found when surveying the Colombian Canal Route 25 near the Panamanian border, and they confirm the theory of an Upper Cretaceous abyssal ocean floor (BANDY 1970, Fig. 3).

In the Canal Zone highly altered volcanic rocks and tuffs of the Basic Igneous Complex are overlain by the Eocene Gatuncillo Formation so that only a pre-Eocene age can be detected here.

Small occurrences of dacites, which are found on the Azuero Peninsula beneath fossiliferous Upper Cretaceous limestones of the **Ocú Formation,** can be placed with reservation in the Cretaceous. They can be regarded as the roof facies or effusive facies of a 69 ± 10 m.y. old quartz-diorite occurrence which is also overlain by the limestones and which covers large areas on the Azuero Peninsula (cf. map Fig. 93).

A small occurrence of quartz sandstones near the farm of La Pitalosa is interpreted as the local erosion product of the quartz-diorite. Near Ocú the sandstones are overlain by well stratified fossiliferous limestones alternating with pelites, tuffs and limestones. The fauna contains

Globotruncana lapperenti
Globotruncana ventricosa
Globotruncana contusa
Globigerinelloides multispina?
Gümbelina spec.,

which are evidence of Upper Campanian to Maestrichtian. Similar limestones discovered in the region of Río Changuinola will be examined in detail in the following, but the material on Azuero is probably much less thick. At all events, the limestones indicate that for a short period of time the sea ingressed in the area of the peninsula.

A more complete cross-section through the Upper Cretaceous from the upper reaches of the **Río Changuinola** in the area of the frontier between Panama and Costa Rica has been described by FISHER & PESSAGNO (1965). The dense limestones are for the most part biomicritic and pelmicritic and they contain a rich, exclusively planktonic foraminiferal fauna which is evidence of **Upper Campanian to Maestrichtian.** Syngenetic swirls and glide structures occur in the limestone beds. The interbedded tuffs and tuffitic sandstones contain feldspar and quartz in a glass-rich matrix whose refraction index indicates andesitic to dacitic chemical composition. This is matched by the composition of lava beds, in which labradorite, green pyroxene and corroded quartzes occur as phenocrysts in a matrix of glass-containing feldspar and quartz microlites. The rocks are designated "vitrophyric dacites".

FISHER & PESSAGNO interpret the **Changuinola Formation** as a facies of the open sea with medium to abyssal depth. The volcanics are regarded as an indication of an oceanic volcanic chain; the acid character of the volcanic products probably corresponds on the one hand to the dacites of Azuero and on the other differs clearly from the basic volcanism of the Basic Igneous Complex.

On the basis of these findings it is justified to assume that large areas of Panama, in particular the eastern section, were the floor of a deep sea during the Upper Cretaceous, where basaltic volcanism predominated and abyssal sediments were deposited. On Azuero and in the northwestern part of the country, on the other hand, acid volcanics and quartz-dioritic intrusions indicate other conditions, and along with FISHER & PESSAGNO it is possible to imagine the existence of volcanic islands surrounded by open sea. RECCHI (1975), too, comes to the conclusion that an initial volcanic island group formed in the Panama region during the Up-

Table 5. Stratigraphic correlation chart of Panama. (From Informe técnico 1, Proyecto Minero, Fase II,

PERIODO	EPOCA	NIVELES	SHELTON 1952		TERRY 1956		WOODRING 1957	I. O. C. S. 1968		
			Región Central del Darién		Este de Panamá		Lago Gatún, Litoral Pacífico, Zóna del Canal	Noroeste del Darién		Región Pirr
			Formación	Litología	Formación	Litología	Formación	Formación	Litología	Formació
TERCIARIO	PLIOCENO	SUP.	CHUCUNAQUE	Lutitas, Areniscas, Conglomerados	CHUCUNAQUE	areniscas, lutitas arenosas, lutitas, foraminiferas				CHUCUNAQU
		MEDIO								
		INF.					CHAGRES			
	MIOCENO	SUP.	PUCRO	Areniscas, lutitas, calizas, arenosas				PUCRO		
		MEDIO	GATUN INFERIOR	Areniscas, arcillas lutitas arenosas, Areniscas, conglo-merádicas	PUCRO / GATUN INFERIOR	areniscas calcáreas calizas, areniscas, lutitas, conglomerados	GATUN	GATUN INFERIOR	Areniscas	GATUN
		INF.	AGUAGUA	conglomerados / lutitas, areniscas, material carbonoso	AGUAGUA	Lutitas, calizas	LADO PACIFICO / PANAMA / CUCARACHA / CULEBRA			
	OLIGOCENO	SUP.			ARUSA	Marmol, cenizas	superior: CAIMITO	AGUAGUA	Tobas, Areniscas tobáceas	TOPALIZA
		MEDIO	ARUSA	Lutitas calcáreas, calizas, tobas			inferior: Miembro ? marino / ? BOHIO / Obispo	ARUSA		
		INF.	CLARITA	Calizas, lutitas, areniscas, conglomerados		Intrusiones básicas				
	EOCENO	SUP.	CORCONA	areniscas, lutitas calcáreas, calizas	?		GATUNCILLO	CLARITA	Calizas arenosas, areniscas	CORCON
		MEDIO	?	Aglomerado, areniscas, lutitas, calizas						LUTITICO - ARENACEA
		INF.	?	Aglomerados				?	Conglomerado basal	
	PALEOCENO									
CRETACICO		SUP.				Intrusiones hipabisales				
			COMPLEJO DEL BASAMENTO	Aglomerados de rocas igneas básicas	COMPLEJO DEL BASAMENTO					

per Cretaceous (Fig. 91). This theory fits in with the development of the magmatism in the direction of a calc-alkaline island arc-series (cf. Chapter 4.23).

The **Tertiary of Panama** contains thick sequences of marine to terrestrial, mostly volcanically influenced sediments, as well as thick terrestrial volcanoclastic series and covers of basic, intermediate and acid flows and ignimbrites. Its development varies greatly in the individual regions of Panama so that positive stratigraphic correlation does not always seem possible, and the classification varies from region to region, as is evident from Table 5 which is taken from Informe técnico No. 1 of the Proyecto Minero, Fase II, Panamá, 1972.

In the northwest the development links up with that of Costa Rica and is predominantly marine. Terrestrial volcanic and plutonic formations predominate in central Panama. The region of the Canal Zone is an area of transition to the again mainly marine conditions in eastern Panama, which in turn are transitional to the corresponding series in Colombia.

In **western Panama** the Tertiary crops out in the northern and southern slopes of the central mountain range, while in the interior of the country it is covered by the volcanic rocks from the volcano of Chiriquí. The development corresponds by and large to that of the adjoining boundary region of Costa Rica. The distribution of the individual Tertiary formations 'anama 1972).

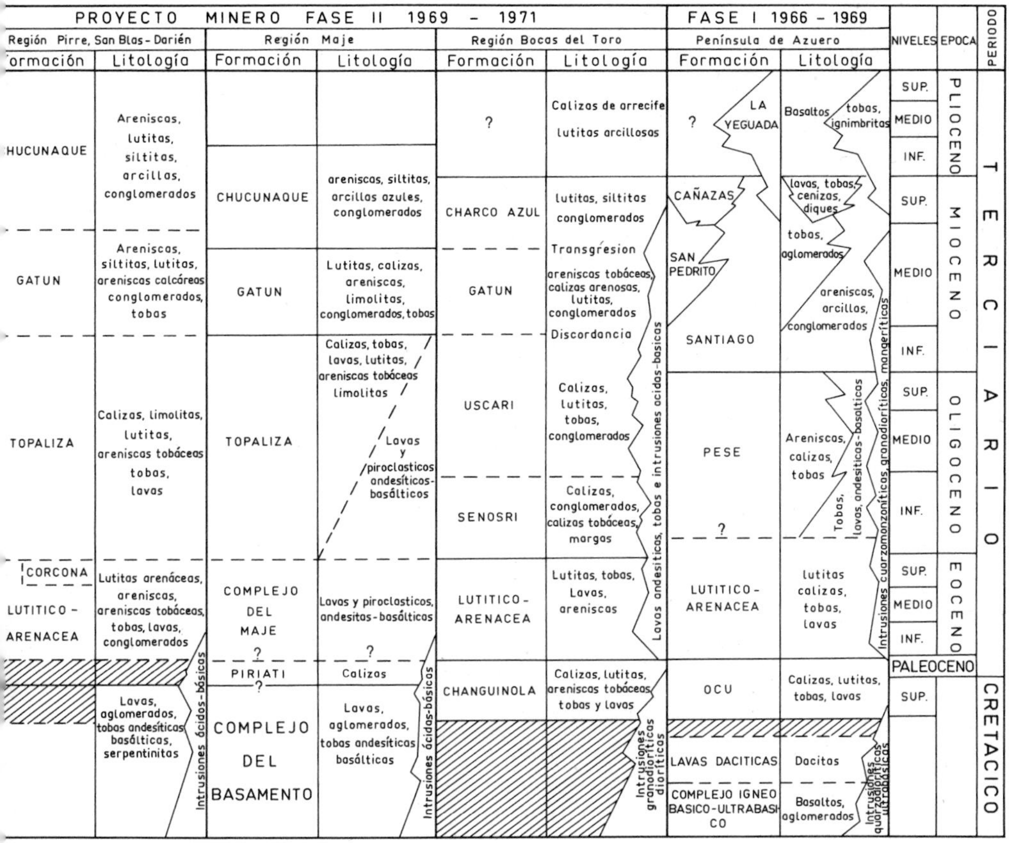

PROYECTO MINERO FASE II 1969 – 1971						FASE I 1966 – 1969				
Región Pirre, San Blas - Darién		Región Maje		Región Bocas del Toro		Península de Azuero		NIVELES	EPOCA	PERIODO
Formación	Litología	Formación	Litología	Formación	Litología	Formación	Litología			
CHUCUNAQUE	Areniscas, lutitas, siltitas, arcillas, conglomerados			?	Calizas de arrecife lutitas arcillosas	? LA YEGUADA	Basaltos, tobas, ignimbritas	SUP. MEDIO INF.	PLIOCENO	TERCIARIO
		CHUCUNAQUE	areniscas, siltitas, arcillas azules, conglomerados	CHARCO AZUL	lutitas, siltitas conglomerados	CAÑAZAS	lavas, tobas, cenizas, diques; tobas, aglomerados	SUP.	MIOCENO	
GATUN	Areniscas, siltitas, lutitas, areniscas calcáreas conglomerados, tobas	GATUN	Lutitas, calizas, areniscas, limolitas, conglomerados, tobas	GATUN	Transgresión; areniscas tobáceas, calizas arenosas, lutitas, conglomerados	SAN PEDRITO; SANTIAGO	areniscas, arcillas, conglomerados	MEDIO		
TOPALIZA	Calizas, limolitas, lutitas, areniscas tobáceas tobas, lavas	TOPALIZA	Calizas, tobas, lavas, lutitas, areniscas tobáceas limolitas; Lavas y piroclasticos andesíticos-basálticos	Discordancia; USCARI	Calizas, lutitas, tobas, conglomerados	PESE	Areniscas, calizas, tobas	INF.		
								SUP. MEDIO	OLIGOCENO	
				SENOSRI	Calizas, conglomerados, calizas tobáceas, margas	?	Tobas, lavas, andesíticas-basálticas	INF.		
CORCONA; LUTITICO - ARENACEA	Lutitas arenáceas, areniscas, areniscas tobáceas, tobas, lavas, conglomerados	COMPLEJO DEL MAJE	Lavas y piroclasticos, andesitas - basálticas	LUTITICO - ARENACEA	Lutitas, tobas, Lavas, areniscas	LUTITICO - ARENACEA	lutitas calizas, tobas, lavas	SUP. MEDIO INF.	EOCENO	
		? PIRIATI ?	? Calizas	CHANGUINOLA	Calizas, lutitas, areniscas tobáceas tobas y lavas	OCU	Calizas, lutitas, tobas, lavas		PALEOCENO	
	Lavas, aglomerados, tobas andesíticas basálticas, serpentinitas	COMPLEJO DEL BASAMENTO	Lavas, aglomerados, tobas andesíticas basálticas			LAVAS DACITICAS; COMPLEJO IGNEO BASICO-ULTRABASICO	Dacitas; Basaltos, aglomerados	SUP.		CRETACICO

Intrusiones ácidas-básicas (Región Pirre; Región Maje). Lavas andesíticas, tobas e intrusiones ácidas-básicas; Intrusiones granodioríticas dioríticas (Región Bocas del Toro). Intrusiones cuarzomonzoníticas, granodioríticas, mangeríticas; Intrusiones cuarzodioríticas ultrabásicas (Península de Azuero).

was established in detail by the mapping of the Proyecto Minero (Fig. 92), however, the project did not allow for a more detailed study so that the data given in the report were drawn from TERRY (1956).

The sequence begins with the **Eocene Búcaru Formation** which appears in Table 5 as the "Formación lutítico-arenácea". On the southern side the name "Formación David" is also used. The formation consists of pelites, sandstones, limestones and intercalated pyroclastics. Orbitoid limestones are known from the Burica Peninsula. The **Lower Oligocene Sinosri Formation** [9] consists of consolidated agglomerates and clays containing marine molluscs and Lepidocyclinae. It accompanies the anticlines of Sinosri and Changuinola. The **Middle Oligocene to Lower Miocene Uscari Formation** consists of shales over 1,000 m thick which were the subject of petroleum exploration activity. Tuffitic and arenitic material is included in them. A rich molluscan fauna together with the contemporaneous fauna of Costa Rica was reported by OLSSON (1922), and a similarly rich foraminiferal fauna was reported by TERRY (1956, pp. 41/42). The **Middle Miocene Gatún Formation** consists of tuffitic sandstones,

[9] The spelling "Senosri" also seems to be accepted usage.

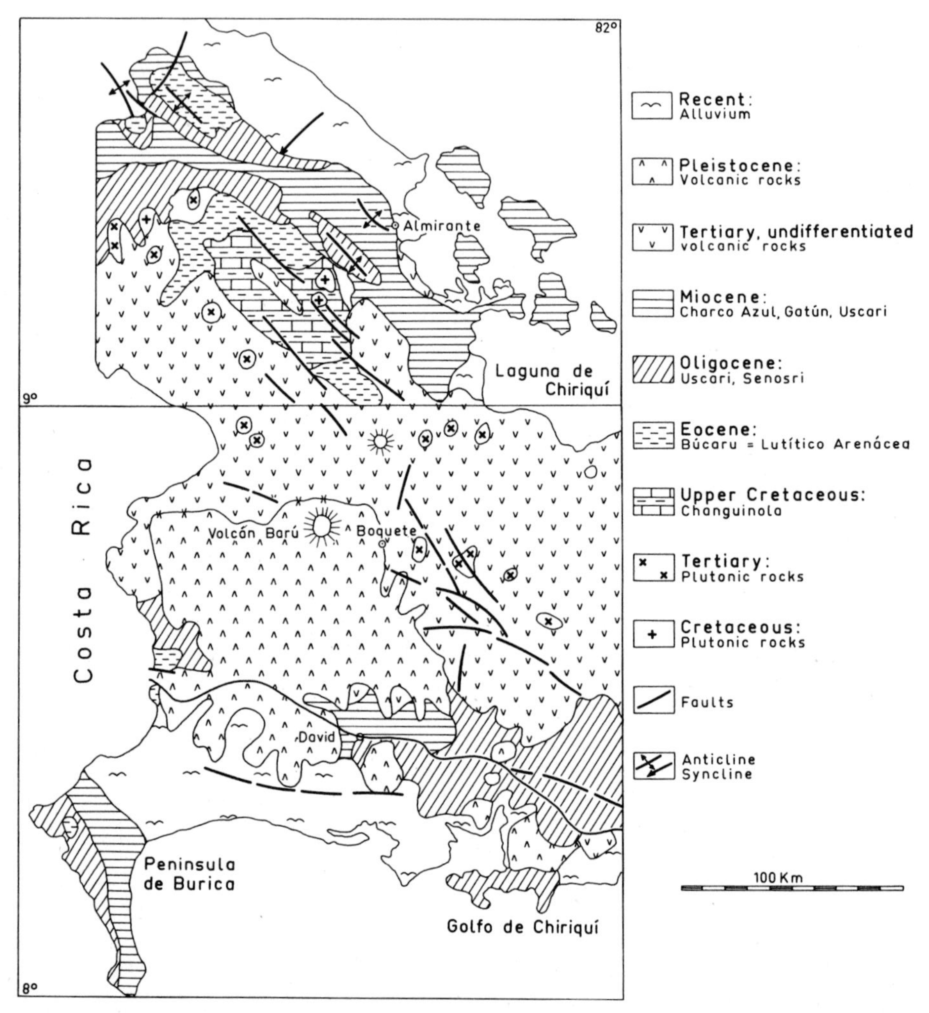

Fig. 92. Geological map of western Panama. (Redrawn from Anonymous 1972d).

sandy limestones, pelites and conglomerates. Terrestrial influences are detectable in the form of impure coals and lignites.

The **Upper Miocene** has not been more accurately delineated, but it is present and is contained partly in the **Charco Azul Formation** of the Burica Peninsula which is normally placed in the **Pliocene.** This formation transgressively overlies older Tertiary and commences with an approximately 200 m thick basic conglomerate. The total thickness was determined to be 2,500 to 2,800 m in a borehole sunk near Puerto Armuelles; in the lower strata the Miocene and in the upper strata the Pliocene each account, without any doubt, for about 800 to 900 m. With the exception of the basic conglomerate, the sediments are fine-grained and, in agreement with the microfauna, indicate that deep water existed in the sedimentation region. No such corresponding strata have so far been found in the interior of the province of Chiriquí. On the Atlantic side the Pliocene is represented by conglomerates in the Talamanca val-

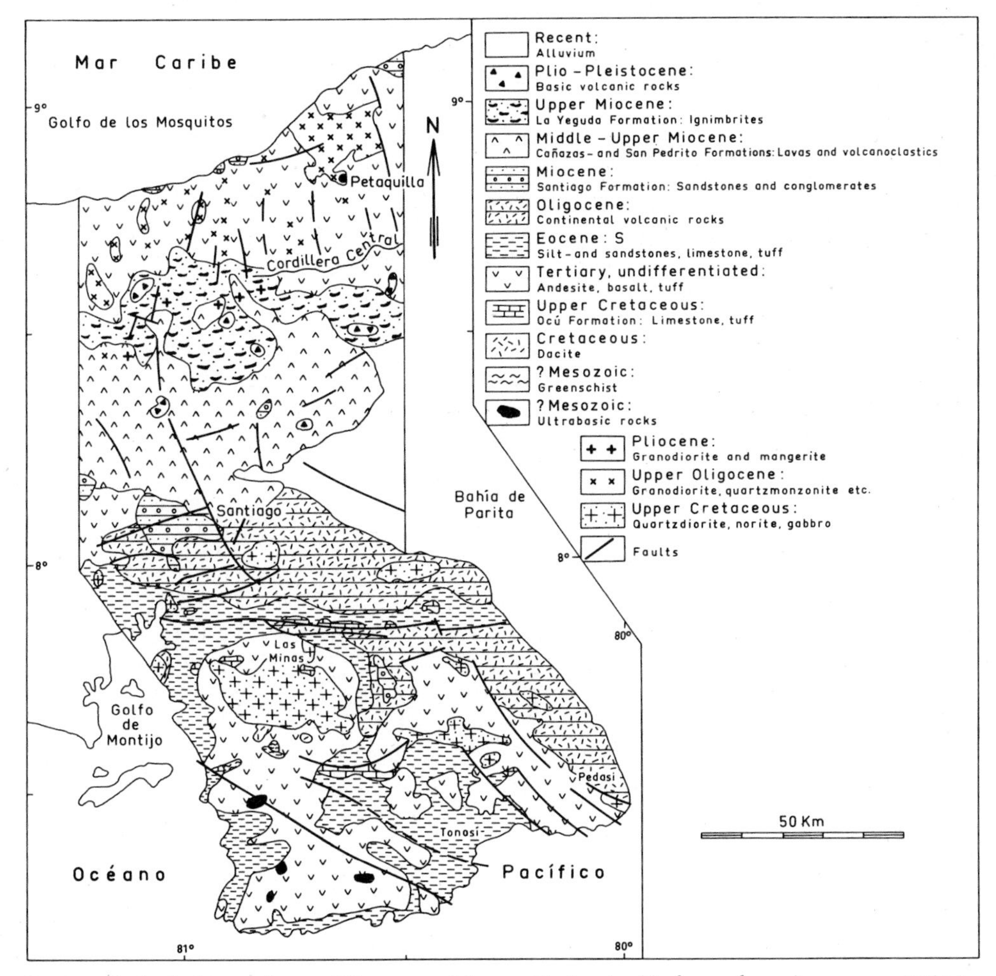

Fig. 93. Geological map of central Panama and Azuero Peninsula. (Redrawn from Anonymous 1969c).

ley and in neighbouring regions of the province of Bocas del Toro as well as by coral limestones and marls in the coastal region.

Pleistocene conglomerates over 200 m thick are located several tens of metres above sea level near Puerto Armuelles and indicate the mobility of this region which is permeated by faults and frequently struck by earthquakes.

In **central Panama,** the Azuero Peninsula and the regions adjoining to the north between 80° 30′ and 81° 30′ (map, Fig. 93), the following Tertiary sequence, which was not previously known, was revealed by the studies carried out for the Proyecto Minero de Azuero, and it is described on the basis of the unpublished report prepared by DEL GIUDICE & RECCHI (1969). The classification and correlation with other regions are apparent from Table 5.

Eocene-Oligocene ("Formación lutítico arenácea"): Pelites, tuffs, hornblende andesites in the lower sections, andesitic-basaltic intercalations in the entire series. In the top, reef limestones with

Lepidocyclina pustulosa toberli
Lepidocyclina chaperi
Asterocyclina spec.
Operculinoides spec.
Lepidocyclina spec.
Globigerina spec.
Lithotamnium spec.
Cubanaster acunai
and unspecified corals.

While TERRY (1956) believed that the marine sediments were limited to a trench and thus to the depression of the Río Tonosi, widespread finds of marine Eocene to Oligocene indicate a much wider marine distribution (Fig. 91).

Oligocene. Oligocene sediments occur in two facies: in neritic sandstones ("areniscas de Pesé") and the limestones of El Barro on the one hand and in terrestrial tuffs and tuffites with local intercalations of marine sediments on the other. The sequence reflects a regression which is accompanied by violent volcanism. The volcanic products are phenoandesites and phenobasalts and their pyroclastics, and in the north pheno-olivine basalts occur (Fig. 91).

Miocene. Sandstones with argillaceous intercalations and conglomerates are designated the "Formación Santiago". TERRY (1956) and OLSSON (1942) had already described marine faunas from this formation. The Middle to Upper Miocene "Formación San Pedrito" is composed of continental tuffs, agglomerates, phenobasalts and phenoandesites with frequent occurrences of petrified wood. The (?) Upper Miocene "Formación Cañazas" consists of a large cover of andesitic-basaltic lavas, a large number of veins which pass through the San Pedrito Formation, and some "laccoliths". Radiometric dating (K/Ar) yielded an age of 17.7 ± 0.6 m.y.

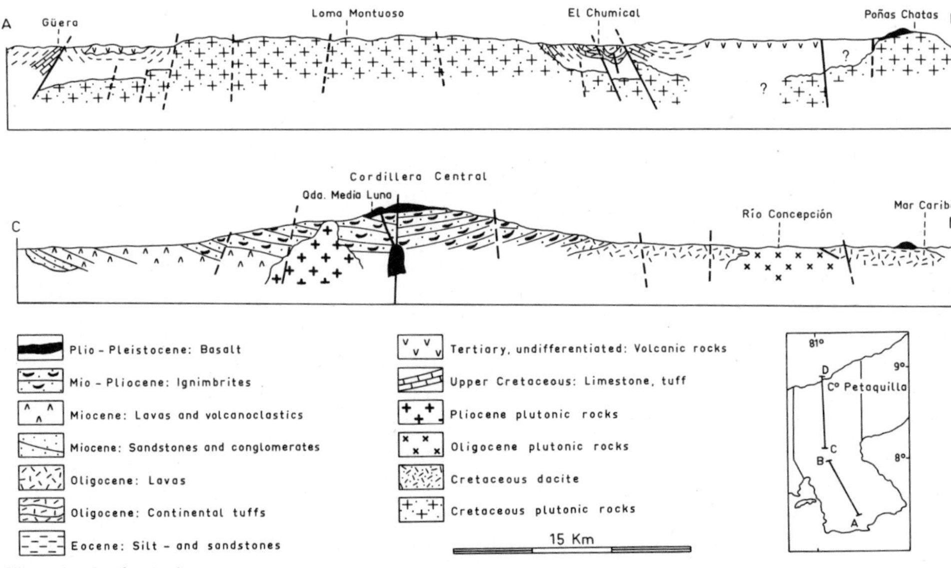

Fig. 94. Geological cross-sections through central Panama. (Redrawn from Anonymous 1969c).

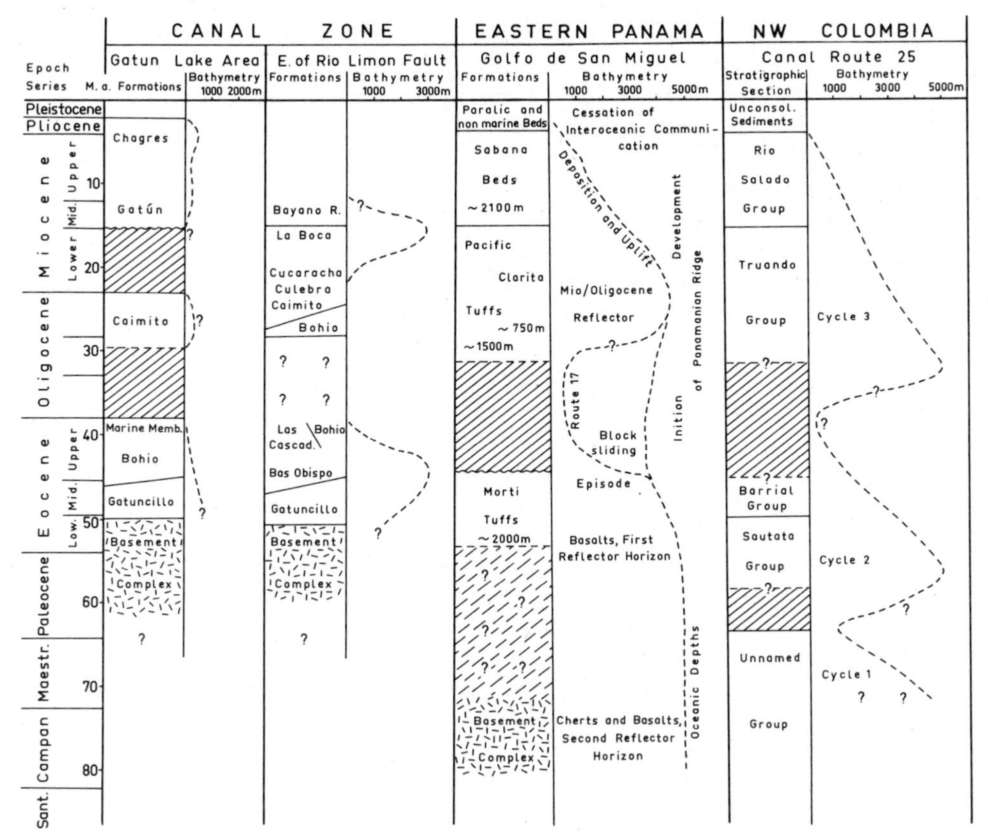

Fig. 95. Correlation of geological formations and depositional-tectonic models, central and eastern Panama. (Redrawn from BANDY 1970, Fig. 2 and BANDY & CASEY 1973, Fig. 3).

Mio/Pliocene. Tuffs and thick covers of phenodacitic to phenorhyolitic ignimbrites, which were designated the "Formación La Yeguada", are widespread in the Cordillera Central. From the brief description these must be ignimbrites of various colour and exhibiting varying degrees of welding, and their thick beds with their steep edges determine the pattern of the landscape in the Cordilleras. K/Ar dating yielded an age of 12.6 ± 0.8 m.y.

Miocene to Recent basalts are superjacent on the ignimbrites and their age was determined to be 10.3 ± 0.4 m.y. Quite young olivine basalts overlie the alluvia of the Río Concepción, and in the Cordillera Central two small cones of basaltic scoriae were found.

The cross-section C – D in Fig. 94 shows the sequence of Tertiary volcanic formations which were fractured by faults and dissected by intrusive bodies.

The Tertiary sequence of strata in the **region of the Canal Zone** has been described in detail by WOODRING & THOMPSON (1949), JONES (1950) and WOODRING (1957), and the rich molluscan fauna has been dealt with in monographs by WOODRING (up to 1973). The classification is described in detail in Fig. 95 and in Table 5 it is compared with the other Tertiary regions of Panama. Since the stratigraphy was described in detail in the first (German) edition of this book, it is sufficient here to refer back to that work (WEYL 1961 a, pp. 86–90). The work conducted by BANDY (1970) and VAN DEN BOLD (1972) yielded new conceptions about the facies development. According to BANDY, different tendencies exist on both sides of the faults which pass through the Canal Zone with an ENE-strike: As a general rule subsidence and thus temporary deposition in a bathyal to abyssal milieu prevailed on the eastern side of the faults, in particular of the Limón Fault, while on the western side terrestrial to neritic conditions existed. It is possible to distinguish clearly between an Eocene and a Miocene deposition cycle.

Within the Lower Miocene VAN DEN BOLD (1972 b) was able, on the basis of the ostracod fauna, to distinguish between brackish water, shallow water and open sea as the deposition milieu; these series also include the terrestrial-volcanogenic rocks of the Cucaracha, Bas Obispo and Pedro Miguel Formations (Fig. 96).

The discovery of a mammalian fauna in the Lower Miocene Cucaracha Formation is important; the fauna contains ungulates of the following genera: *Diceratherium, Anchitherium, Archaeohippus, Mericochoerus* and *Brachycrus* (WITHMORE & STEWART 1965). The fauna is closely related to North American ungulate faunas so that it must have been possible for the animals to migrate via Central America into the Canal Zone. This corresponds to the paleogeographical pattern of the Lower Miocene (Figs. 75 and 91).

In contrast to the neritic and shallow-water sediments of the Canal Zone, **eastern Panama** contains thick series of abyssal deposits which accumulated in basins between the rising blocks of the Basic Igneous Complex. The basins are as follows: Along the middle axis lies the **Chucunaque-Bayano Basin** as an extension of the Colombian Atrato Basin, and in the Pacific coastal region comes the **Sambú Basin** which continues in the Gulf of San Miguel. The up to more than 5,000 m thick Tertiary sediments stand out in the gravity map because

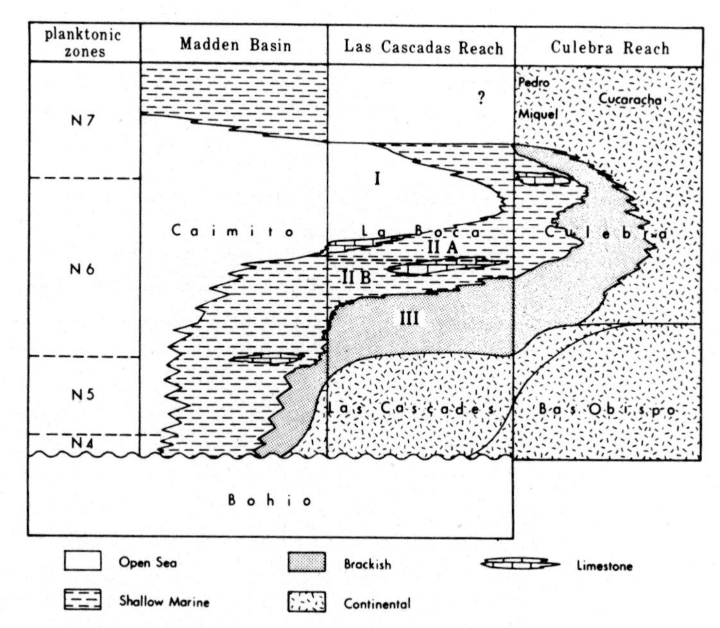

Fig. 96. Correlation and facies of the Caimito, La Boca and Culebra Formations. (From VAN DEN BOLD 1972b, Fig. 5).

of negative Bouguer anomalies, while the regions of the uplifted zones of the Basic Igneous Complex, in particular in the Serranía del Darién, are zones of extreme excess gravity.

TERRY's (1956) description of the stratigraphic sequence has been summarized in detail in WEYL (1961 a, pp. 91 – 92) (cf. Table 5). WING & MCDONALD (1973, p. 829) provided the following revised summary of the stratigraphic sequence in the most important basins of eastern Panama:

Age	Formation	Thickness	Description
		Tuira-Chucunaque Sub-basin	
Early Pliocene-late Miocene	Chucunaque		Sandstone and shale containing foraminiferal fauna; occupies narrow inner trough of SD-VII
Middle early Oligocene	Pucro	610 m	Calcareous sandstone and thin limestone beds
	Gatún	1,070 m	Shale, underlain by conglomerate, sandstone, and thin limestone lenses
Early Miocene	Aguagua	610 – 1,370 m	Benthonitic and bituminous shale, limestone, and sandstone
Late Oligocene	Arusa	460 m 310 m	Massive, brownish, tuffaceous marlstone White-weathering, tuffaceous, massive mudstone
Middle early Oligocene	Clarita	300 m	Tuffaceous carbonate rocks (mostly limestone) and sandy limestone
Late Eocene	Corcona	350 – 400 m	Shale, sandy, marine, calcareous; massive agglomerate, conglomerate, limestone (includes volcanic material)
Middle early Eocene			Chert (possibly equivalent to Quayaquil chert of Ecuador), limited distribution (Pacific Hills area)
Pre-late Eocene	Basement		Andesite flows; crystalline rocks (somewhat silicic on Caribbean side)
		Bayano Sub-basin	
Late Miocene		915 – 1,220 m	Soft, argillaceous siltstone, Foraminifera
Middle Miocene		460 m	Reef-type nearshore limestone beds, 15 km long; present on south side of SD-III syncline
Early Miocene			Sandstone and siltstone
Late Oligocene			Tuffs, conglomerates, calcareous and tuffaceous sandstone, and tuffaceous siltstone
Middle early Oligocene			Limestone (calcareous algae, Foraminifera)
Late Eocene			Sandy shale, agglomerate, conglomerate, tuffaceous and calcareous sandstone, tuffaceous siltstone
Pre-late Eocene	Basement		Andesite flows; crystalline rocks (silicic on Caribbean side; mafic on Pacific side)

On the basis of the lithology, the radiolarian fauna and the foraminiferal fauna, BANDY (1970) and BANDY & CASEY (1973) developed a theory regarding the geological formations and the crustal movements depicted in them; the results of this theory are summarized in the diagrams in Fig. 95.

According to this figure, the **Eocene** is superjacent on the oceanic rocks of the Basic Igneous Complex and there are no signs of a transgression. The tuffitic Eocene deposits, which are about 2,000 m thick, are described as "Morti Tuffs". They commence with cherts rich in radiolarians and deep-water assemblages of foraminifera, while towards the upper strata the number of shallow-water components gradually increases, and along Canal Route 17 they extend into the neritic milieu. At the Gulf of San Miguel the sedimentation region remains, on the other hand, in the abyssal range. Basalt flows in the sediments are correlated with the upper seismic reflector horizon in the oceanic basins of the Caribbean and of the Pacific.

The **Lower Oligocene** is characterized along Canal Route 17 by marine shallow-water and freshwater deposits and by an erosion gap. A shallowing of the sea floor, although much less pronounced, is also recorded in the Gulf of San Miguel. In the **Middle Oligocene** a new cycle with an abyssal milieu began, and this is reflected in the radiolarian-rich and foraminifera-deficient sediments. Up to the end of the **Miocene** the sea floor became shallower, probably mainly because of a great amount of infilling, until paralic and limnic conditions set in during the Pliocene. The final interruption of the inter-oceanic communication is placed in the period around 3 m.y. ago.

Biogeographic studies of coiling direction in the planktonic foraminiferal genus *Pulleniatina* from paleomagnetically dated piston cores in the Atlantic, Pacific, and Indian Oceans suggested that the marine connection across the Isthmus of Panama was severed at least by 3.5 m.y. ago (SAITO 1976). In a later study, KEIGWIN Jr. (1978) has compared the ranges and coiling direction preferences of *Neogloboquadrina pachyderma* at different times in the early Pliocene of the Panama, Colombian and Venezuelan Basins. The disappearance of this foraminiferal assemblage from the Panama Basin at 3.6 m.y. B.P. may mean that the Central American seaway was closed at this time. The disappearance of *Pulleniatina* in the Colombian Basin was closer to 3.1 m.y. ago. This suggests the possibility of surface-water communication between the Atlantic and Pacific until that time and indicates that the Panamanian land bridge was completed by 3.1 m.y. B.P. Following KEIGWIN (1978, p. 633) "therefore it appears that there is a range of suggested minimum dates for the closure from 3.6 to 3.1 m.y. B.P.".

These data are in accordance with KEIGWIN (1976), who found out that the late Miocene in the Panama Basin is marked by a decrease in warm-water foraminiferal species and an increase in the *Neogloboquadrina* group. He postulated the gradual uplift of the Isthmus of Panama and the resulting decreased inflow of warm, saline water from the Atlantic Ocean as the causal mechanism for this faunal change.

Along Canal Route 25 in Colombian territory, a similar cyclical development to that in eastern Panama is found (Fig. 95).

The paleogeographical development of Panama was summarized by RECCHI (1975) in six sketch maps (Fig. 91), which were drawn without taking account of plate movements.

4.23 Magmatism

Since the time of the Upper Cretaceous, Panama, just like Costa Rica, has been the scene of violent magmatic events in the course of which the oceanic crusts with basaltic volcanic rocks were converted into a continental-type crust or island arc-type crust with sialic magmatic rocks. The mapping work carried out by the United Nations (Anonymous 1969 c, DEL GIUDICE & RECCHI 1969) and works by FERENČIĆ (1971), CASE (1974), GOOSSENS et al. and

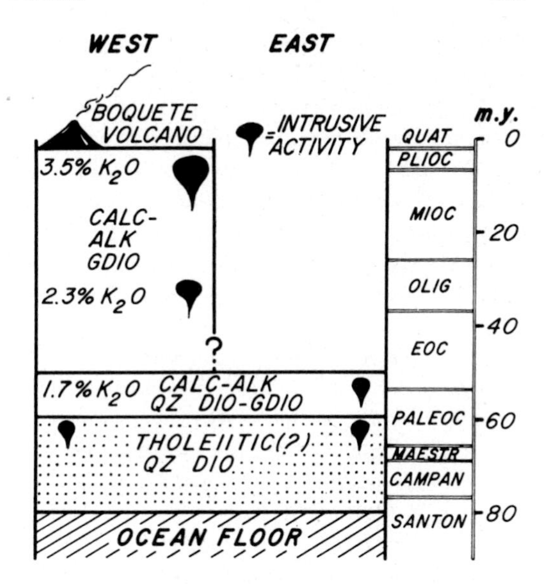

Fig. 97. Schematic illustration of the evolution of magmatism in Panama. (From KESLER et al. 1977, Fig. 7).

Table 6. Statistical comparison of geochemistry of Central American tholeiitic rocks.

	1	2	3	4
Oxides (wt. %)				
SiO_2	53.1	49.8	51.6	49.6
Al_2O_3	15.1	14.9	15.9	16.0
FeO	10.1	11.8	9.5	11.5
MgO	5.7	7.7	6.7	7.8
CaO	10.6	10.9	11.7	11.3
Na_2O	2.8	2.5	2.4	2.8
K_2O	0.6	0.4	0.4	0.2
TiO_2	1.4	1.2	0.8	1.4
Elements (ppm)				
Cu	215	133	—	77
Zn	92	121	—	—
Ni	95	54	30	97
Co	78	43	—	32
Cr	202	51	50	297
Rb	30	38	5	10
Sr	154	113	200	130
Zr	93	116	70	95
V	—	—	270	202
Ba	145	—	75	14
No. of analyses	12	9	—	—

1. Average of Panamanian Basic Igneous Complex.
2. Average of the Nicoya Complex, Costa Rica.
3. Average island-arc tholeiitic basalt.
4. Average ocean floor tholeiite.

From GOOSSENS et al. 1977, Table 4.

KESLER et al. (1977) reveal the fundamental principles of this development, even if many areas remain unexplored and countless questions unanswered (Fig. 97).

The development begins with the output of the tholeiitic basalts of the **Basic Igneous Complex,** which builds up, particularly in eastern Panama, the mountain ranges of San Blas and the upthrown blocks between the Gulf of San Miguel and Colombia. The occurrence in association with eupelagic sediments of the Upper Cretaceous, and a chemical composition which corresponds to the tholeiites of oceanic ridges and incipient island arcs, permit us to interpret these basalts or basaltic andesites [10] as formations of oceanic crust. The average chemical values are given in Table 6 after GOOSSENS et al. (1977, Table 4); individual values from twelve analyses conducted in the vicinity of the Gulf of San Miguel are incorporated in the diagrams in Figs. 98 to 100. According to the chemical composition (Table 6) the rocks around the Gulf of San Miguel are clearly more acid than those of Nicoya, and they represent a transition to the island arc series.

The Basic Igneous Complex can be traced along the western edge of the Andes to Ecuador and it corresponds to the Nicoya Complex of Costa Rica (PICHLER et al. 1974, GOOSSENS et al. 1977). DONNELLY et al. (1973) grouped it together with basalts from deep-sea drillings in the Caribbean and from the Antilles and spoke of a "truly gigantic igneous province", which owes its origin to one of the greatest magmatic events of the Phanerozoic.

The emplacement of **quartz-dioritic intrusions** commenced in the Maestrichtian, lasted into the Eocene and was replaced from the Oligocene onwards by granodioritic rocks. Volcanic activity commenced in the Canal Zone in the Oligocene and lasted in the western half of Panama into the Late Tertiary and, with the volcanoes of Barú, Santiago and El Valle into the Quaternary. In the eastern part, on the other hand, the magmatic activity died out as early as in the Lower Tertiary. A diagram of the magmatic activity is reproduced in Fig. 97 after KESLER et al. (1977), and the radiometric dates compiled in Table 7 were taken from the same authors.

The **volcanic** sequence is best known through the mapping activity of the United Nations carried out in central Panama north of the Azuero Peninsula (DEL GIUDICE & RECCHI 1969, FERENČIĆ 1971). It commences with the marine volcanoclastic San Pedrito Formation and merges in the Miocene into the terrestrial andesitic rocks of the Cañazas Formation. It is followed by the ignimbrites of the Mio/Pliocene La Yeguada Formation, which in turn is overlain by Pliocene basalts. ISSIGONIS (1973) discovered similar sequences in the western continuation of the Cordillera Central (Serranía de Tabasara) on the Cerro Colorado.

The **plutonic sequence** was studied in detail, in particular with regard to the mining of copper, and the results were summarized by KESLER et al. (1977). They discovered an old formation phase with quartz-dioritic intrusions and a young phase with granodioritic intrusions. The older phase includes the plutonic rocks of Azuero, Cerro Azul and Río Pito, while the

[10] To avoid misunderstandings, the designations of the rocks used in the original works will be used here, too. Calculation of the standards and definition in accordance with STRECKEISEN/RITTMANN's method yielded the following definitions:

In the Basic Igneous Complex:	5 *mela quartz-andesites,* 3 *quartz-andesites,* 2 *quartz-tholeiite basalts,* 1 *quartz-tholeiite hawaiite,* 1 *olivine-bearing latiandesite.*
In the quartz-diorite group:	5 *tonalites,* 1 *leuco quartz-norite,* 1 *mela-tonalite.*
In the granodiorite group:	5 *granodiorites,* 1 *leuco quartz-diorite,* 1 *trondjemite,* 1 *leuco quartz-monzonite,* 1 *monzogranite,* 1 *quartz-monzodiorite,* 1 *leuco quartz-diorite,* 1 *quartz-bearing anorthosite.*

Table 7. Radiometric ages for magmatic rocks from Panama. (From KESLER et al. 1977).

Volcanic rocks		
Basalt, Veraguas	whole rock	10.3 ± 0.4 m.y.
Ignimbrite, Yeguada Formation, Veraguas	biotite	12.6 ± 0.8 m.y.
Basalt, Cañazas Formation, Veraguas	whole rock	17.5 ± 0.6 m.y.
Granodiorite group		
Granodiorite, Cerro Colorado	biotite	3.34 ± 0.05 m.y.
Monzonite porphyrite, Serranía de Tabasará	hornblende	7.3 ± 1.16 m.y.
Diorite, Petaquilla	hornblende	32.6 ± 1.16 m.y.
Quartz monzonite, Petaquilla	hornblende	36.41 ± 2.06 m.y.*
	feldspar	28.98 ± 0.35 m.y.
Quartz-diorite group		
Quartz-diorite, Río Pito	hornblende	48.45 ± 0.55 m.y.
	feldspar	49.23 ± 0.57 m.y.
? Intrusion, Azuero	hornblende	53.00 ± 3.0 m.y.
Quartz-diorite, Cerro Azul	hornblende	61.58 ± 0.7 m.y.
	feldspar	51.11 ± 0.58 m.y.**
Quartz-diorite, Azuero	hornblende	64.87 ± 1.34 m.y.
	feldspar	52.58 ± 0.63 m.y.**
Quartz-diorite, Azuero		69.0 ± 10.0 m.y.

* probably air argon contamination
** probably partial postcrystallization argon loss

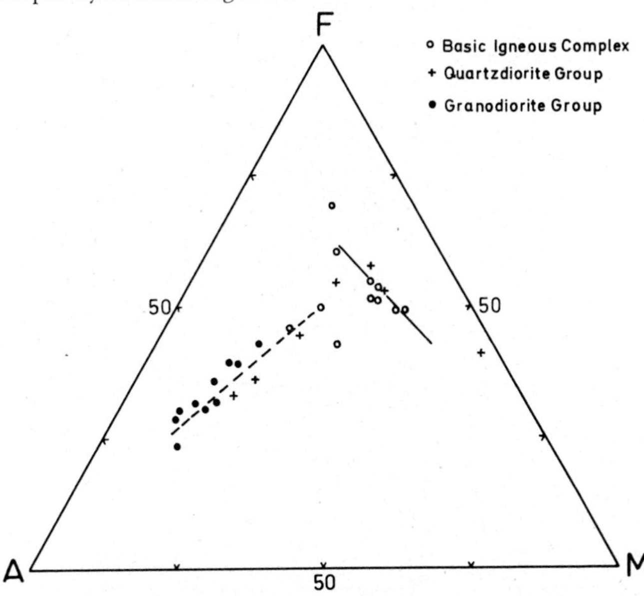

Fig. 98. AFM plot of chemical analyses of Panamanian magmatic rocks. Unbroken line illustrates possible tholeiitic trend, dotted line the calc-alkaline trend. (Data from GOOSSENS et al. 1977 and KESLER et al. 1977).

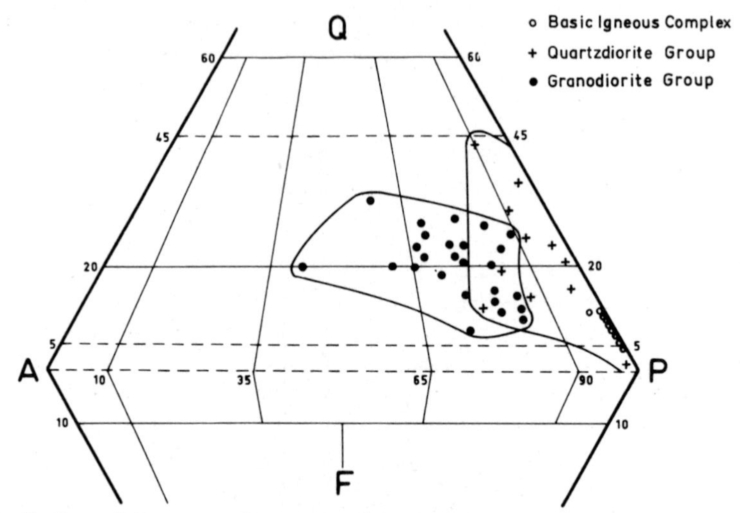

Fig. 99. Distribution of the magmatic groups of Panama in the STRECKEISEN double-triangle. Basic Igneous Complex according to the RITTMANN norm, plutonic rocks by modal composition from KESLER et al. (1977, Fig. 2).

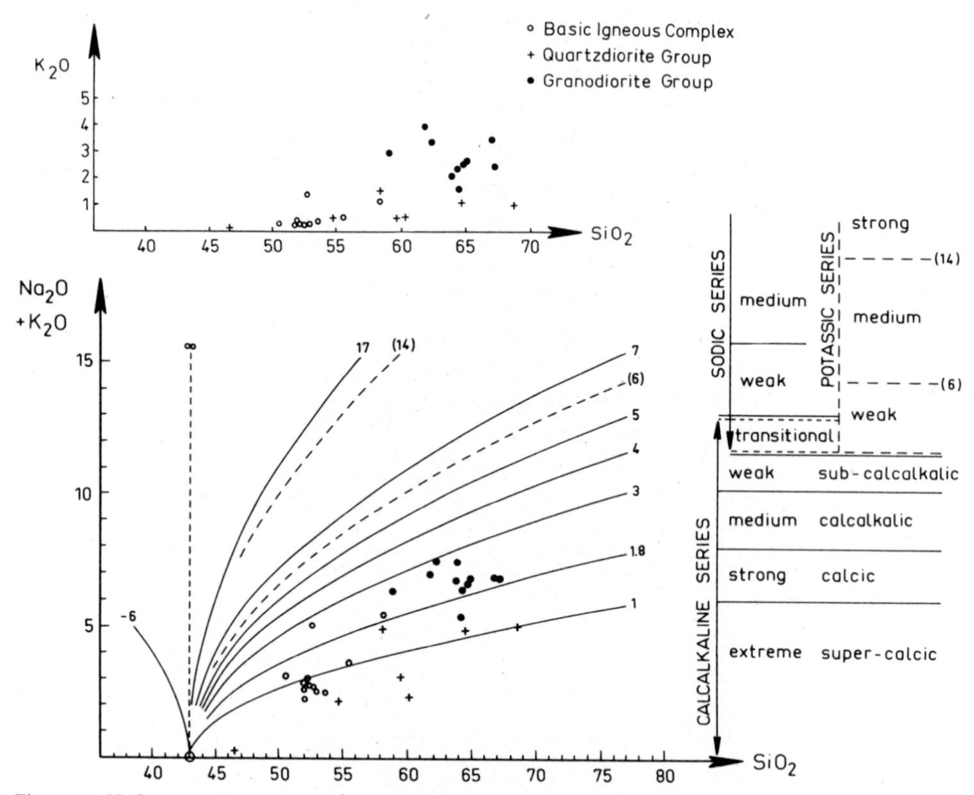

Fig. 100. K_2O over SiO_2 plot and serial index (alkali versus SiO_2) of the Panamanian magmatic groups. (Data from GOOSSENS et al. 1977 and KESLER et al. 1977).

younger takes in the granodioritic intrusions of Petaquilla, Cerro Colorado, Charcha and Bocas del Toro in western Panama and of the Río Guayabo in eastern Panama. The copper deposits which are discussed in detail in Chapter 9 are linked with these formations.

KESLER et al. (1977) confirm the differences in the material of both series by carrying out a large number of chemical analyses, including determination of trace elements. The older sequence, for which they introduce the designation "tholeiitic group" to characterize it, still exhibits close material links with the "Basic Igneous Complex". The younger sequence, which is also designated the "calc-alkaline group", corresponds on the other hand in its chemical composition to the island arc magmatism. Some of the chemical data are summarized in the diagrams in Figs. 98 to 100 together with those of the Basic Igneous Complex, and they show the trend of the magmatism of the ocean floor through to the continental type of island arc type magmatism. The series in the Río Pito region tend more towards the calc-alkaline than towards the tholeiitic group, which also makes it apparent why they belong to the copper-bearing regions. Within the calc-alkaline group the potassium content increases in the course of time.

The average values for the trace elements are listed in Table 8, and they also show the intermediate position of the "tholeiitic group".

KESLER et al. (1977) see correlations between the magmatic and tectonic development of Panama which followed different courses in the eastern and western parts of the country. The older series is still connected with the Basic Igneous Complex and could be regarded as an initial stage of an island-arc-type magmatism. The younger series, which commences with the Eocene and gradually shifts towards the west, is regarded as being related to subduction. Agreement seems to exist here with the spreading rate of the Cocos and Nazca plates. The extinction of magmatism in the eastern part of Panama confirms the pattern of plate movements which MALFAIT explains as being caused by a slowing down and cessation of subduction of the Nazca Plate at the end of the Miocene (MALFAIT & DINKELMAN 1972, VAN ANDEL & HEATH 1973, LONSDALE & KLITGORD 1978. Cf. also Chapter 8 and Figs. 178, 179).

Table 8. Average trace element abundances in the three magmatic groups of Panama. (After KESLER et al. 1977, Table 3 and GOOSSENS et al. 1977, Table 4).

	Basic Igneous Complex	Tholeiitic intrusive group	Calc-alkaline intrusive group
Cu	215	102	111
Zn	92	91	63
Pb	—	5	4
Cr	202	65	14
Ni	95	30	7
Co	78	40	12
V	258/3/2	277	151

The samples from Río Pito are contained in the "calc-alkaline group". All data are in ppm.

4.24 Structure

Panama is more than 700 km long from border to border, and within this distance the individual regions of the country exhibit widely differing structures. Detailed information on the structure is available from the western boundary area, from the central part of the country including the Azuero Peninsula, from the Canal Zone and from the eastern regions.

In the **western frontier region,** on the northern side of the Cordillera Central, which here is sometimes also called the Cordillera de Talamanca, the NW/SE-trending folds and faults of Costa Rica continue (see map, Fig. 92). Cretaceous or Lower Tertiary crops out in the core of the anticlines and Upper Tertiary in the synclines. A great number of faults interrupt the folds along the strike (see cross-section, Fig. 88). The anticlines of Sinosri and Changuinola stand out.

According to TERRY (1956, pp. 76 – 78) the majority of the anticlines on the Caribbean side are asymmetrical and their steep flanks face north where they slope steeply and where also it is suspected that overthrusting occurred. This suspicion was confirmed by drilling carried out on the Isla Colón. In the Sinosri anticline the strata on the northern flank are either vertical or overtilted, and on the southern flank they dip at an angle of 40 – 45° to the SE. The smaller Jarkin anticline which follows inland has a similar structure.

Thus, after all, a combination of folding and faulting exists here in the eastern continuation of the Limón Trough.

Tertiary volcanics, which are intersected by a large number of small intrusions, are located on the northern flank of the watershed in the Cordillera Central, a range which rises to over 3,000 m. The southern flank is covered by Quaternary volcanic products from the volcano of Barú (=Chiriquí). Beneath them, at the southern foot of the range, highly disturbed Tertiary occurs again, and TERRY discovered fault throws of up to 2,000 m in it. As far as is known, in this region the folds strike NW/SE.

The **structure of central Panama** with the peninsula of Azuero as shown in the maps prepared by the Proyecto Minero de Azuero is much different from the way TERRY (1956) conceived it (see map, Fig. 93 and cross-sections, Fig. 94). According to DEL GIUDICE & RECCHI (1969) folds play only a secondary role here in the predominating continental lavas, agglomerates and ignimbrites. On the other hand, monoclinal blocks bounded by faults predominate. To the south of the Azuero Peninsula the faults strike NNW to NW. To the north comes the region of major intrusions which is bounded in the north as in the south by large E/W-striking faults. Along their flanks the Cretaceous strata of the Ocú Formation crop out. The faults permeate the entire Azuero Peninsula from the Golfo de Montijo to the Bahía de Parita. Faults have also been detected within the intrusive bodies (Fig. 94). A fault block structure with NW to E/W-striking faults and monoclinal orientation prevails in the relatively low and flat region around Santiago, where continental volcanics of the Oligocene and Miocene are found. The central mountain zone is built up of unfolded Tertiary lavas and in particular of the thick ignimbrite covers of the La Yeguada Formation, which are crowned by Miocene to Pleistocene basalts. The uplifting of the Cordillera is primarily due to the accumulation of volcanic products. As far as can be seen, they dip at a shallow angle on both sides towards the crest of the Cordillera. From aerial photographs it has been possible to detect a number of faults striking in different directions. N/S-striking faults predominate at the northern foot of the Cordillera, and TERRY believed that he could trace some of them through to the Pacific

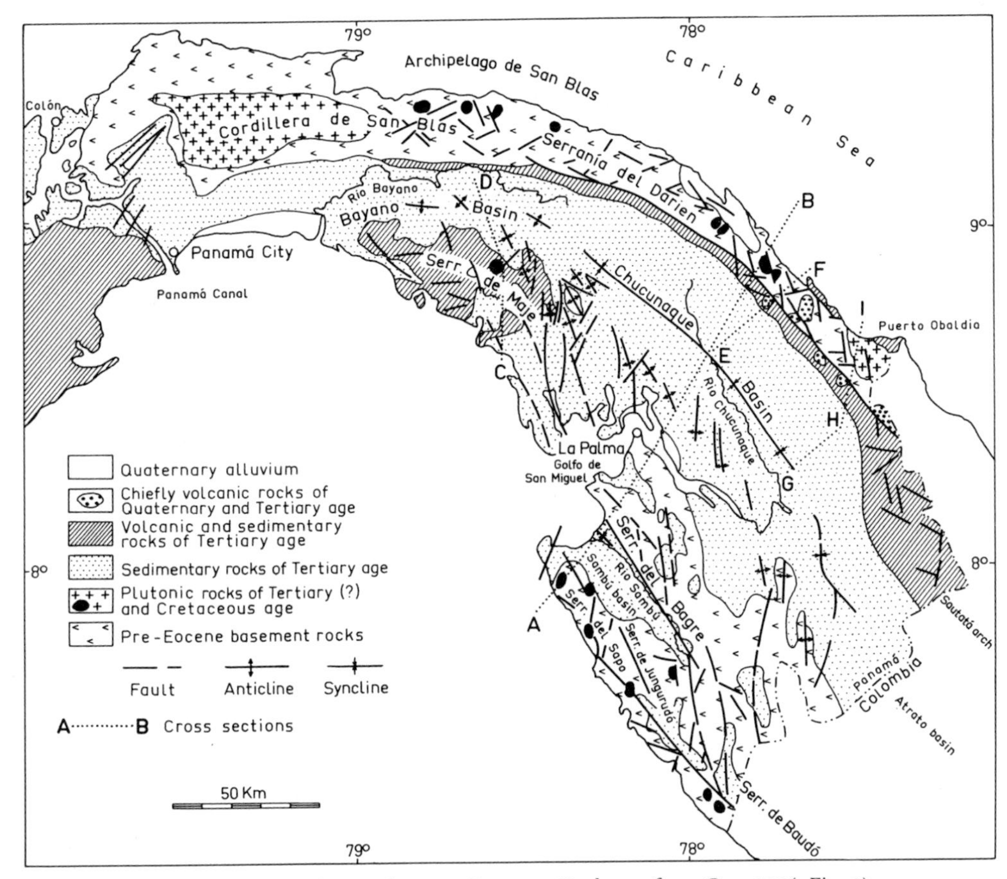

Fig. 101. Generalized structural map of eastern Panama. (Redrawn from CASE 1974, Fig. 1).

coast. DEL GIUDICE and RECCHI also talk of a trans-isthmic fault which is clearly recognizable in aerial photographs.

If the results of the Proyecto Minero Azuero are compared with TERRY's structural map of Panama (1956, Fig. 4), it is found that the course of the axes of the anticlines as assumed by TERRY in central Panama is evidently not confirmed.

The **region of the Canal Zone** is dominated by pronounced fracture tectonics. JONES (1950) has described the latter in detail for the Gatún region and has mapped the faults from aerial photographs and from field observations (cf. also WOODRING's map 1957, in WEYL 1961 a, Fig. 36). In one of these fault systems Lake Madden is situated in a narrow Tertiary graben in the middle of the Basic Igneous Complex. The Quebrancha syncline running in a N/S direction to the west of this is also accompanied by faults.

The structure of **eastern Panama** is completely different; here, upthrown mountain blocks of the oceanic Basic Igneous Complex alternate with the basins of thick, mainly marine Tertiary sediments which are embedded in them (see map, Fig. 101 and cross-sections, Fig. 102). Here, surveying of a possible canal route and petroleum exploration incorporating gravimetric measurements and radar imaging of the inaccessible regions have yielded a much

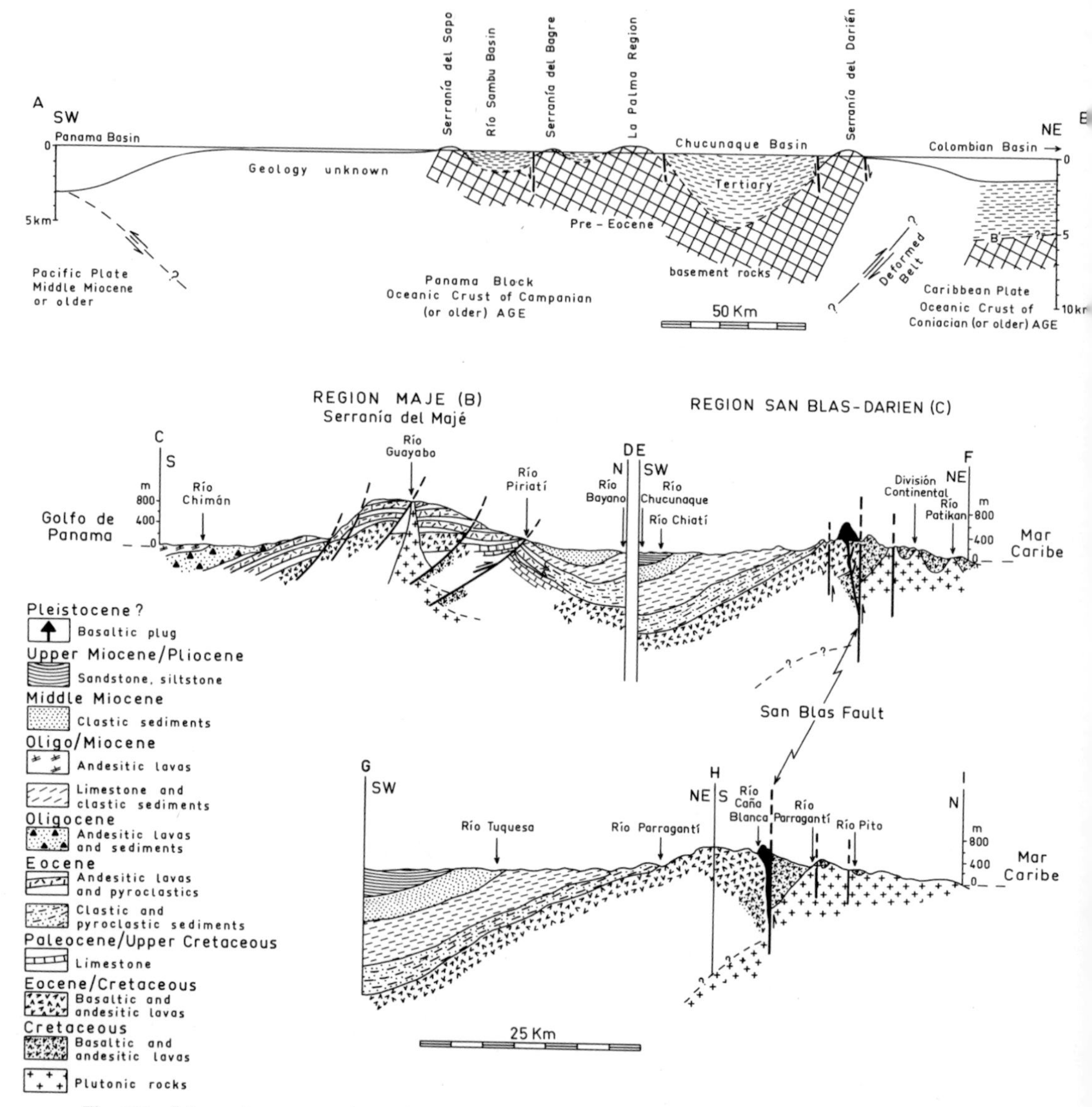

Fig. 102. Schematic cross-sections through eastern Panama. See also Fig. 101. (A-B from CASE 1974, Fig. 3; C-D, E-F, G-H-I from Anonymous 1972c).

clearer picture (H. C. MACDONALD 1969, WING 1970, CASE et al. 1971, WING & MACDONALD 1973, CASE 1974) than TERRY (1956). Additional data derived mainly from exploration of ore deposits were provided by the Proyecto Minero (Informe Técnico 4, 1972).

Altogether the structures form one or several arches convex towards the north, which are also reflected in the gravity anomaly map (Fig. 103). Their individual elements from N to S are:

The **Cordillera** (or **Serranía**) **de San Blas** and the **Serranía del Darién,** which are made up of Cretaceous or pre-Eocene basic volcanics with large and small Cretaceous and Tertiary intrusions, of which that of the Río Pito was studied in detail because of its copper mineralization (Proyecto Minero, Fase II, Informe Técnico 4, 1972). The range is regarded as being formed by elongated horst mountains of the Basic Igneous Complex, which are intersected by countless faults that have been detected mainly on radar images. To the south they are separated from the large Tertiary basins by the **San Blas Fault** (Falla de San Blas).

From west to east these basins are divided up into the following sub-basins: The **Bayano Basin** which extends from the Bahía de Panamá to the transverse structure of the Cañazas Platform. The Bayano Basin is a large semi-graben with a central synclinal trough which submerges towards the SSW and continues under the broad shelf of the Gulf of Panama. The basin filling ranges from the Eocene to the Pleistocene. On the other side of the Cañazas Platform the large **Chucunaque-Tuira Basin** runs in an SSE direction; this basin is an asymmetrical syncline filled with 4,600 m of sediments dating from the Eocene to the Holocene.

This basin is in turn subdivided by a transverse fault into sub-basins. The basin is accompanied on the SW side by some anticlines running at an acute angle to it, in whose core the Basic Igneous Complex crops out. A series of **horst-anticlinal structures** in the form of low mountain chains, such as the **Serranía de Bagre,** the **Serranía del Sapo** and the **Serranía de Pirre,** follow towards the Gulf of Panama.

The Basic Igneous Complex crops out in them as well. Smaller synclines and the **Sambú Basin,** which comes to the surface in the Gulf of San Miguel, are interposed between these mountains. The eastern flank of the Sambú Basin is formed by a fault which can be traced beyond the Gulf of San Miguel. Details of its structure are given in the work by MACDONALD (1969), WING (1970) and WING & MACDONALD (1973), which unfortunately were not available to me in their entirety. Therefore only the main features of the highly differentiated network of folds and faults which was recognized by them could be incorporated, after WING & MACDONALD (1973, Fig. 7), in the maps in Figs. 90 and 101.

In the gravity map (CASE 1974) the outcrops of the Basic Igneous Complex are indicated by positive Bouguer anomalies up to more than 125 mgal, and the basins are identified by low-gravity values (Fig. 103). A continuation of the Basic Igneous Complex is indicated by a gravity high in the Gulf of Panama, and this is followed to the south by a narrow zone – convex to the north – of strong negative free-air anomalies which have to be interpreted as basins containing young unconsolidated sediments. Similarly, a narrow zone of negative free-air anomalies, which can be taken as indicative of a trough filled with Tertiary sediments, runs off the northern coast of Panama. The gravity data appear to indicate that a large strike-slip fault runs along the eastern coast of Azuero and another runs in the Canal Zone (Fig. 103).

On the basis of the geological and gravimetric findings, CASE (1974) comes to the logical conclusion that eastern Panama is not underlain by continental crust but is composed of an uplifted block of oceanic crust. He discusses four possible means by which this came about: Panama as part of the Atlantic Plate, as part of the Caribbean Plate, as part of the Pacific Plate or as an island arc and spreading centre. The latter interpretation seems to him to agree best with the findings, and it is also supported by more recent petrological data:

"In a context of plate motions, the convex northward shape of eastern Panama, as well as the deformed sedimentary basin along the north margin revealed by gravity and seismic surveys, suggests that the Caribbean plate has moved relatively southwestward beneath eastern

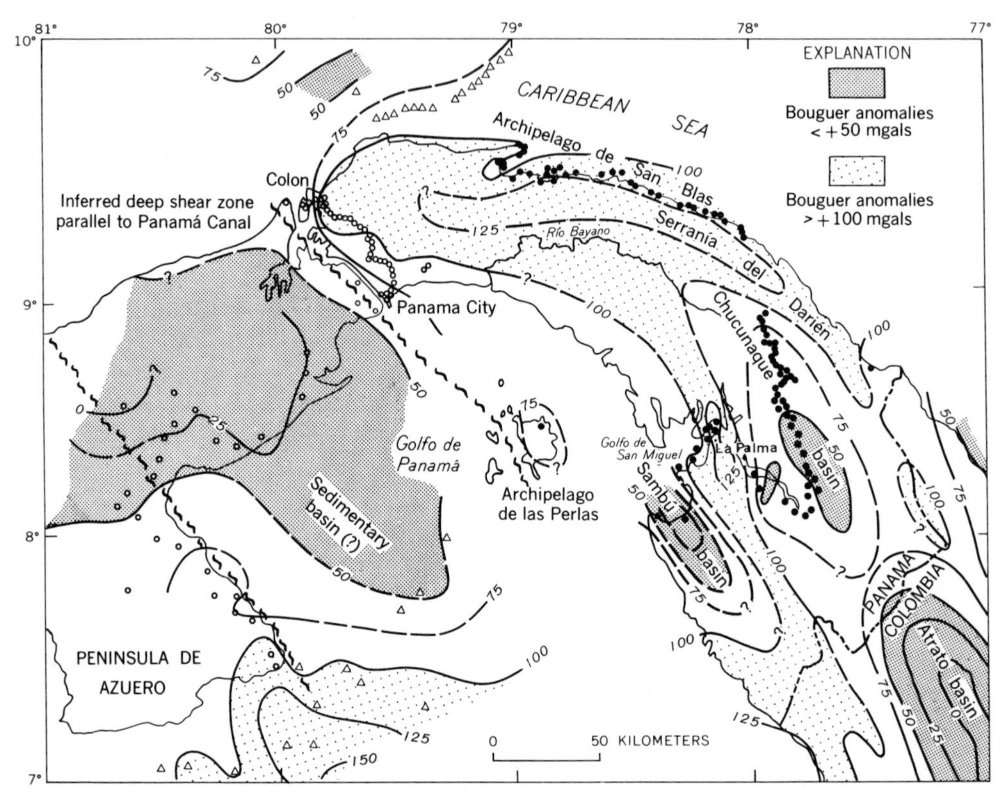

Fig. 103. Simple Bouguer anomaly map of central and eastern Panama. Circles and triangles = measuring stations, ∼ ∼ = supposed wrench faults. (From CASE 1974, Fig. 2).

Panama. Morphology of western Panama and Costa Rica suggests the opposite sense of relative movement: northeastward flow of the Pacific plate beneath the arc. Why two arcs might fortuitously be joined to form the isthmus is unclear, but examples of such arc-arc juxtapositions, separated by transform faults, are known in the western Pacific (WILSON 1965, LEPICHON 1968). HAYES & EWING (1970, p. 57) have stated that Panama is a zone separating two buried trench structures that lie opposite one another: one on the Pacific side and one on the Atlantic side.

Various investigations (LLOYD 1963, p. 99, DENGO 1968), on the other hand, have proposed that the sigmoidal shape of the isthmus results from relative compressive movements between nuclear Central America and South America" (CASE 1974, p. 648).

Thus, according to CASE, it is probable that the isthmus was uplifted as a type of wedge, either by episodic countermovements between the Caribbean and Pacific plates, or as a result of differences in the rates of movement between the two plates moving in the same direction. The cross-section in Fig. 102 A – B explains this interpretation put forward by CASE.

Recent plate tectonic theories (LONSDALE & KLITGORD 1978) relate the uplifting of the oceanic crust to obduction in the Early Tertiary. The reaction of the crust to subduction of the Cocos and Nazca plates beneath the Caribbean Plate can probably be seen in the development of the magmatism of Panama and of its Tertiary sedimentation basins. See Chapter 8 for more on this.

The **link between the Panamanian structures and those of South America** remained unexplained for a long time, because there was as good as no geological information available on this area, and the topographical maps were inaccurate. The first people to provide information on this topic were HUBACH (1930) and TROLL (1930), and their findings have to a large extent been confirmed by more recent works. In the Loma del Cuchillo, in the middle of the Atrato Basin, TROLL found evidence of a link between the Serranía del Darién and the Colombian Cordillera Occidental; he recognized the independence of the depression of the Gulf of Urabá between two branches of virgation of the Cordillera Occidental and he recognized that there was a connection between the Serranía de Baudó and the Serranía del Sapo; and he even traced this Coastal Cordillera through to the Pacific peninsulas of the Isthmus. From this he concluded that a link existed between the Andes and the mountains of the Central American Isthmus.

This conclusion was to a large extent confirmed, although from quite different standpoints, mainly by the works of CASE et al. (1971) and CASE (1974), who drew on preliminary work carried out by JACOBS et al. (1963). It has been proved by many findings within the Atrato Basin on the one hand and by a strongly positive anomaly on the other, that the Serranía del Darién definitely continues in the basement of the depression towards the SSE. This ridge is nowadays referred to as the "Sautatá Arch" or "Arco de Sautatá", and within it the Loma del Cuchillo discovered by HUBACH and TROLL constitutes the largest outcrop of the Basic Igneous Complex. This Sautatá Arch separates the Atrato Basin of Colombia from the Urabá-Sinú Basin. The Atrato Basin continues in the Tuira-Chucunaque Basin of Panama. And finally, the series of the oceanic Basic Igneous Complex run along the Pacific coastal range from Ecuador via Colombia through to the Pacific peninsulas of Panama and Costa Rica (PICHLER, STIBANE & WEYL 1974, GOOSSENS et al. 1977).

5. The volcanic ranges and plateaus

5.1 Introduction

The volcanic ranges and plateaus of Central America stretch in an unbroken system south-eastwards from the Mexican frontier to the central part of Costa Rica. Here the system is disrupted by the Cordillera de Talamanca and it continues on into Panama in the form of isolated volcanoes. While the historically active volcanoes or those volcanoes which became extinct only during the Quaternary run parallel to the Pacific coast and reflect the subduction of the Cocos Plate beneath the Caribbean Plate, a region of Tertiary volcanism extends to the north and northeast of this series of volcanoes. It commences in the west, in Guatemala, where it is relatively narrow, then it broadens considerably in Honduras and El Salvador and finally it extends in Nicaragua to the edge of the Caribbean Coastal Plain. It ends in the south at the Nicaragua Depression and thus belongs to the northern part of Central America (Fig. 104).

The volcanism of this region is closely linked with plutonism which, however, extends back chronologically to the Cretaceous and is most widely distributed to the north of the volcanic region.

Tertiary and Quaternary volcanism and the plutonism mentioned above will be discussed in this Chapter. The older, Mesozoic volcanism, which still has not been researched in great detail, was discussed in connection with the stratigraphy of northern Central America. The quite different magmatic sequence of southern Central America has also been discussed in connection with the development of Costa Rica and Panama.

Since the first (German) edition of this book appeared, our knowledge of the volcanic region defined above has increased decisively and the new data have been incorporated into the plate tectonic theories. PICHLER & WEYL (1973) summarized the results with special emphasis on the petrochemical aspects. This description must be updated and modified in the light of later works.

One problem that arose in the course of preparing the following description concerns the **petrographic designation** of the volcanic products. The designation is based on widely varying criteria such as field observation, microscopic determination of the mineral content and definition on the basis of the chemical characteristics, for which in turn various norms (CIPW and RITTMANN norm) were used. It is not possible to achieve uniformity among the designations of the rocks when they are arrived at by such different methods. Therefore, as a rule, the designations employed by the authors are used in the text. Where possible, definitions were based on the STRECKEISEN (1967) method using the RITTMANN norms (RITT-

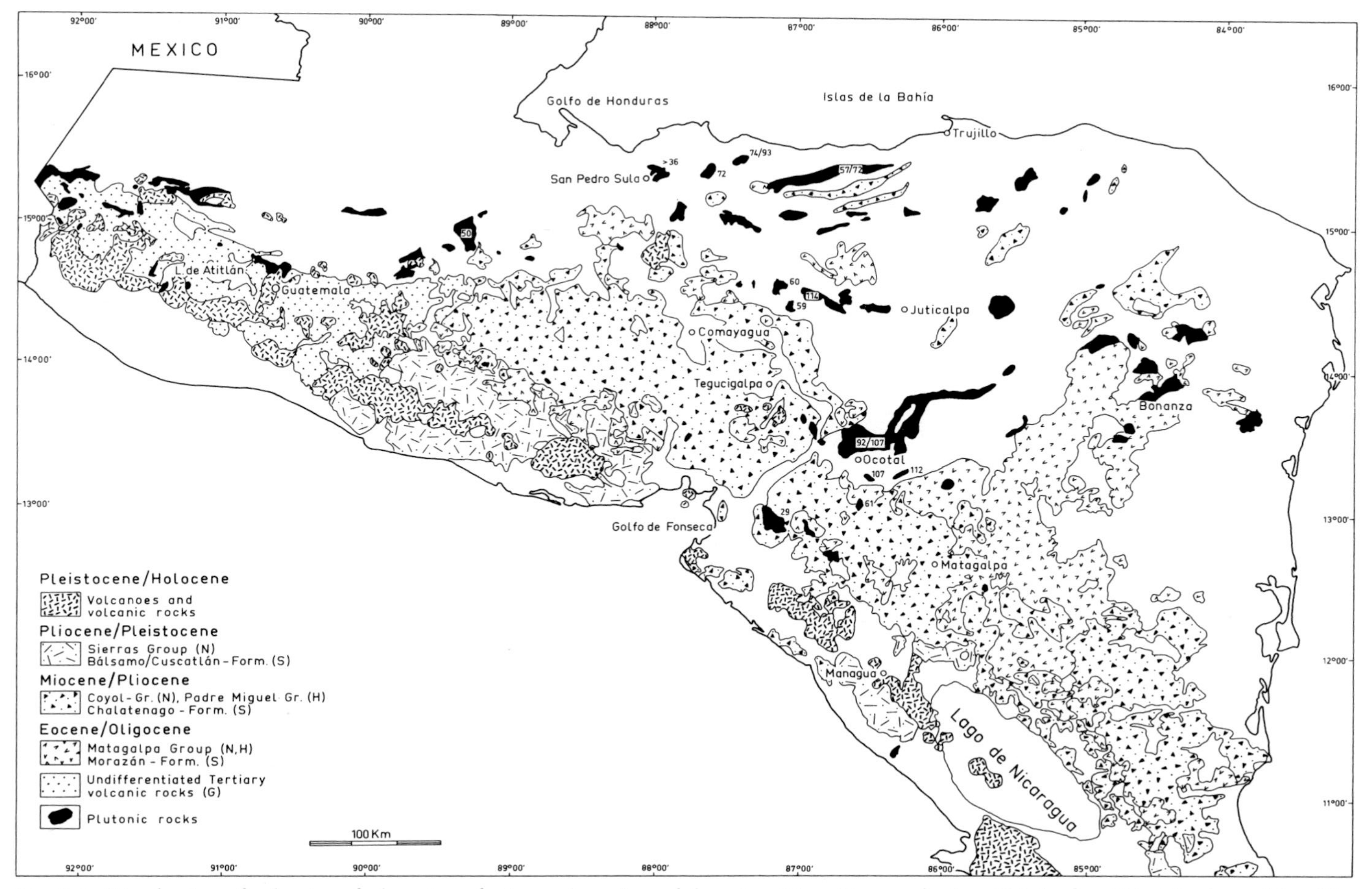

Fig. 104. Distribution of volcanic and plutonic rocks in northern Central America. (G = Guatemala, S = El Salvador, H = Honduras, N = Nicaragua).

MANN 1973) and these designations are indicated in italics. The same remarks hold true in particular for the summary tables and diagrams and for the products of the Quaternary volcanoes. I am grateful to my colleagues Prof. Dr. PICHLER and Dr. STENGELIN in Tübingen for calculating the norms. I am also grateful to Dr. PICHLER for reviewing this chapter.

5.2 Tertiary volcanic area

Tertiary volcanic rocks cover most of the highlands of Nicaragua, wide areas of central and southern Honduras, northern El Salvador and a narrow belt of Guatemala (see general map, Fig. 104, and the geological maps for the individual countries). The existence of this volcanic area has long been known; SAPPER (1937, Fig. 2) was able to define its extent accurately in broad terms. He mentioned free-flowing eruptions, mainly of andesites, also of basalts, rhyolites, melaphyres and porphyries, but he stressed the entirely inadequate knowledge of the Upper Tertiary formations which, by his estimate, cover approximately 170,000 square kilometres, i.e. 22% of the entire area of Central America.

SONDER (in BURRI & SONDER 1936) described the landscape of the volcanic area in Nicaragua in extremely vivid terms, as follows:

"The inner highlands form a unique structure. If one travels on the new road from Managua to Matagalpa, one has the impression at many points of being in a sedimentary tableland in which horizontally stratified hard beds alternate with soft strata (see Figs. 105 and 106). However, the immediate environment shows that the steeply sloping steps of the valley walls consist of lava flows usually several metres thick while the less resistant zones in between contain volcanic ashes, breccias, etc. The breccias frequently contain large quantities of volcanic glass. The lava flows, as well as the breccias and tuffs, are composed of dacitic, andesitic and basaltic lavas. The entire structure shows that copious lava flows alternated with violent explosive activity over long periods of time" (p. 41).

The rocks collected by SONDER were described by BURRI as liparites, andesites and olivine basalts. He referred several times to their hyaline base and their fluidal texture but he was unable at that time to recognize that this is a typical structure of ignimbrites, which, together with lavas and other pyroclastic rocks, dominate the volcanic areas of Central America. Our knowledge of this fact is derived mainly from the systematic studies carried out by MCBIRNEY & WILLIAMS (1965) and WILLIAMS & MCBIRNEY (1969). These works were followed by a large number of detailed studies concentrating mainly on the stratigraphy but also on petrographic and chemical aspects, and a large number of radiometric datings were carried out. Unfortunately these studies also led to the establishment of regionally different "Formations" and "Groups", which do not always necessarily coincide with the known facts. REYNOLDS (1977) therefore attempted a correlation which will be dealt with further below.

In **Nicaragua,** MCBIRNEY & WILLIAMS (1965, pp. 15 – 24) recognized two series of volcanic products which differ clearly from each other both with regard to the rock inventory and topographically. The older series, consisting of basalts, andesites, dacites and their pyroclastics, was grouped together by the authors along with some underlying and intercalated freshwater sediments as the "**Matagalpa volcanic series**" and placed in the Oligocene and Miocene. According to radiometric datings of 17.2 and 19.1 m.y., the upper parts of this series extend into the Lower Miocene. The Matagalpa series covers mainly the northern part of the central highlands with an uneven mountainous and hilly terrain.

Fig. 105. Rim of the volcanic plateau of central Nicaragua near San Isidro, Inter-American Highway. Photo: R. WEYL.

Fig. 106. Weathering forms of horizontally bedded ignimbrite. Road from Matagalpa to Sébaco, central Nicaragua. (Erroneously described as phenoandesite in the first (German) edition, Fig. 41). Photo: R. WEYL.

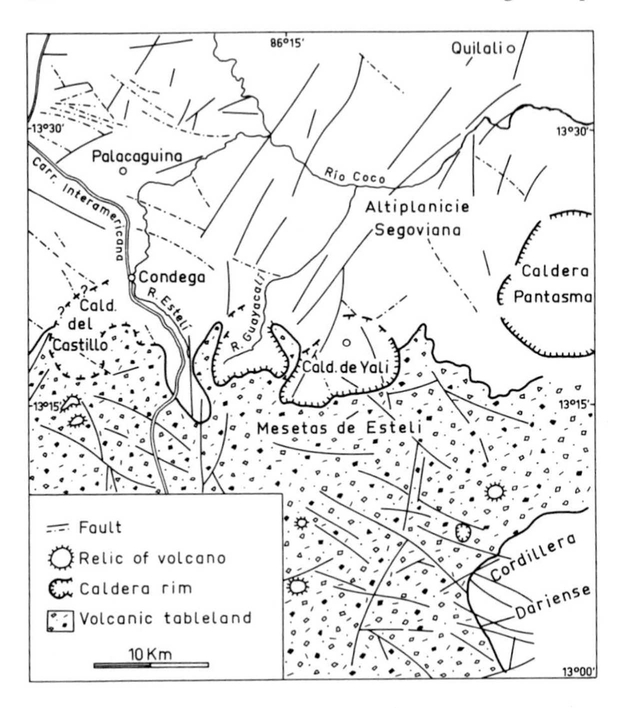

Fig. 107. Northern rim of the ignimbrite plateau of central Nicaragua. (Redrawn from GARAYAR 1971, Fig. 6).

In sharp contrast to this are the flat mesas and plateaus with steep scarps which are built up of **acid ignimbrites** with intercalated tuffites and basic lavas: "The ignimbrites, basic lavas, and interbedded sediments seem to fill embayments between higher, peninsula-like ridges, and in places rounded hills of seemingly older andesite rise like islands surrounded by a sea of ignimbrite" (MCBIRNEY & WILLIAMS 1965, p. 20). From the air MCBIRNEY & WILLIAMS were able to distinguish clearly the extent of these ignimbrites (1965, Map 1). They link up to the southeast with the Matagalpa Formation, extend to the edge of the Nicaragua Depression and then crop out again in the coastal region in the Tamarindo Formation (see the cross-section in Fig. 60). As a rule the ignimbrites overlie the Matagalpa series but also interlock with the uppermost strata and they were classified in the Miocene and Pliocene.

HODGSON & FERREY (1971), GARAYAR (1971), HODGSON (1971) and Anonymous (1972 b) drew up a detailed division of the volcanics into the **Matagalpa and Coyol Group** (Table 3) and they described a multiple alternation of basic, intermediate and acid ignimbritic (ash flow) products whose stratigraphic sequence agrees well with a number of radiometric datings:

Upper Coyol Group	Upper series of lavas and agglomerates with basalts and andesitic basalts (6.65 m.y.) Ash flows with rhyolitic ignimbrites (12.0 – 13.1 m.y.)
slight angular discordance	
Lower Coyol Group	Ash flows with dacitic and, at their base, andesitic ignimbrites (13.8 – 14.9 m.y.) Lower series of lavas and agglomerates with andesites predominating (12.8 – 15.2 m.y.)
slight angular discordance	
Matagalpa Group	Andesites and basalts with secondary acid volcanics

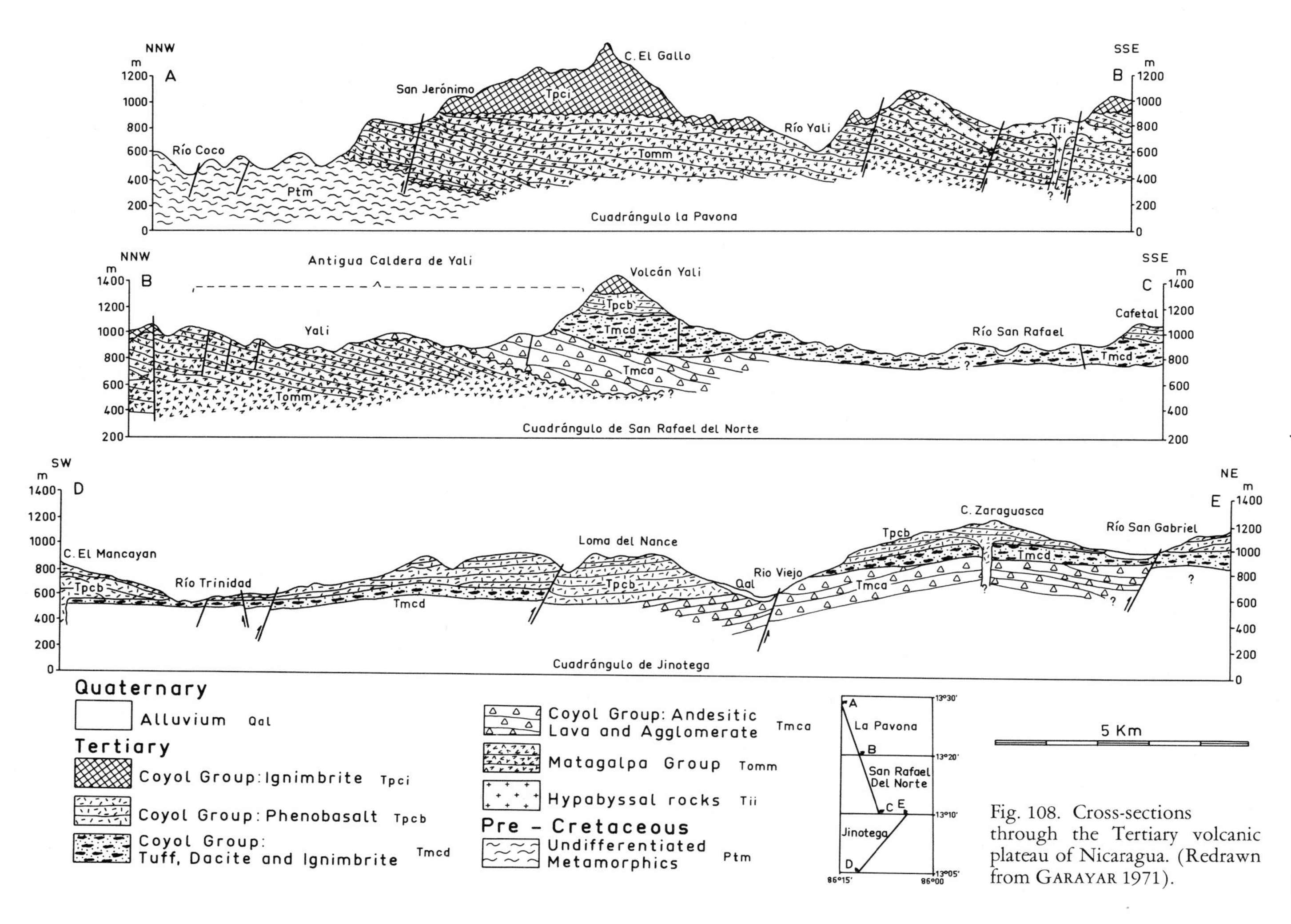

Fig. 108. Cross-sections through the Tertiary volcanic plateau of Nicaragua. (Redrawn from GARAYAR 1971).

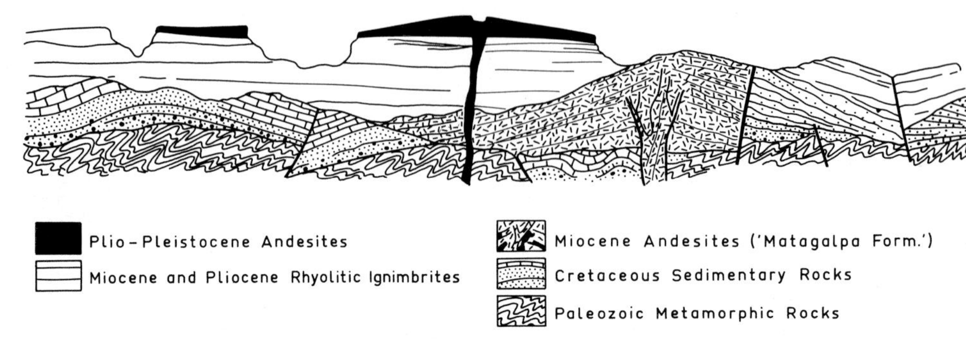

Fig. 109. Schematic section showing the relations of the Tertiary volcanic series and underlying basement in central Honduras. (Slightly modified from McBIRNEY & WEILL 1966, Fig. 4).

According to this the Coyol Group should be placed in the Upper Miocene. Its equivalent in the Pacific coastal region is the petrographically similar Tamarindo Group. The Matagalpa Group was placed in the Eocene to Miocene [11].

Several large calderas, whose shapes can still be recognized in the landscape, are regarded as the eruption centres of the ignimbrites. They are concentrated on the northern edge of the ignimbrite plateau (Fig. 107). GARAYAR, in numerous profiles, has shown the sequence of volcanic series among themselves and together with approximately contemporaneous plutonites, and three of these profiles have been redrawn in Fig. 108. The exaggeration of the elevation, which is necessary for this purpose, does not correctly depict the plateau character of the landscape (see Fig. 105).

The studies performed by GARAYAR and Anonymous contain a wealth of individual descriptions and many chemical rock analyses. The petrology will be discussed in detail later on in connection with the other areas of Tertiary volcanism.

In **Honduras,** the last country of Central America which was studied by WILLIAMS & MCBIRNEY (1969) for its volcanism, the sequence of volcanic products is in principle the same as in Nicaragua. MCBIRNEY & WEILL (1966) summarized them in a schematic cross-section which is reproduced, with slight modifications, in Fig. 109. According to this cross-section the older Matagalpa Formation and the younger ignimbrites are sharply separated from each other. In more recent works the latter are called the **Padre Miguel Group,** a name which was introduced in southeastern Guatemala by BURKART (1965).

The Matagalpa Formation consists chiefly of andesitic and basaltic lavas and their pyroclastics together with a few acid intercalations. It crops out in separate areas which were described by WILLIAMS & MCBIRNEY (see map of Honduras, Fig. 58). Volcanogenic sediments, which grade into the red beds of the Subinal Formation, are intercalated in the volcanics. Radiometric dating of 28 m.y. (EVERETT & FAKUNDINY 1976, p. 34) fits in with the Oligocene age of the Matagalpa Formation.

The Formation is overlain by younger ignimbrites with radiometric ages of 19.1, 18.0, 16.8 and 15.0 m.y., which indicates that they were produced in the Lower Miocene. The ignimbri-

[11] There is a discrepancy between Table 3, which is reproduced from the original, and the classifications made here, because in Table 3 the Miocene/Pliocene boundary is still shown at 13 m.y., whereas here, based on a more recent assessment (BERGGREN 1972), it is drawn at 5 m.y.

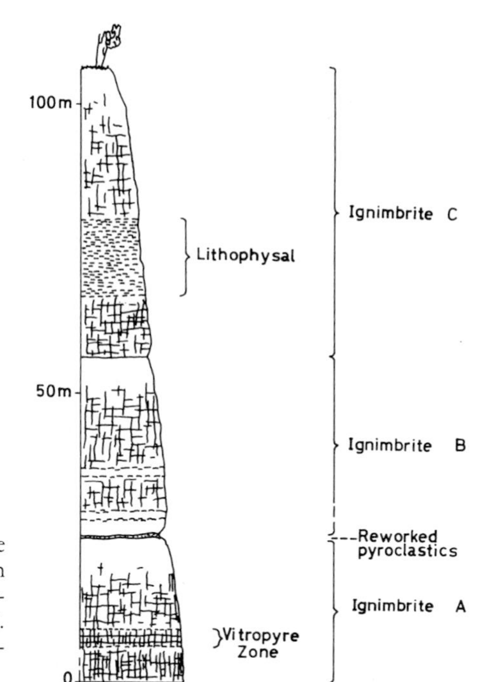

Fig. 110. Section through ignimbrites of the Padre Miguel Group. Zambrano Plateau, Inter-American Highway between Tegucigalpa and Comayagua. (Redrawn from "Trip log, Post-Congress Excursion". IV. Reunión de Geólogos Centroamericanos, Tegucigalpa 1974).

Fig. 111. Vitrophyric zone of ignimbrite of the Padre Miguel Group overlain by chemically altered ignimbrite. Inter-American Highway at southern rim of the Comayagua Graben. Photo: R. WEYL.

tes in the region of the Fonseca Gulf, which extend into the Pliocene, are regarded as younger. The ignimbrites were extruded over a flat relief and nowadays cover the greater part of the central highlands of Honduras. It is cautiously estimated that they originally covered an area of 50,000 km^2. Their thickness fluctuates between 300 and approximately 900 m so that their volume can be estimated at something in excess of 5,000 km^3; the volume of all Central American ignimbrites was put at more than 10,000 km^3 (PUSHKAR et al. 1972, p. 269). The ignimbrites probably erupted from fissures; however, no detailed evidence regarding the eruption centres is available.

The ignimbrites are for the most part weakly to moderately welded and thus belong to the "Sillar type". However, welded layers having the appearance of macroscopically homogeneous glass also occur (Figs. 110 and 111). The red colouration is assumed to have been caused by fumarole activity and the green and yellow colouration by post-volcanic conversion with the formation of clay minerals.

The predominant rocks are *rhyolitic* and *rhyodacitic* ignimbrites, while *dacitic* and *andesitic* types are less frequent. The volcanic eruption was concluded by locally limited basaltic flows which, however, already date from the Quaternary.

CURRAN & MACDONALD (provisional communication 1977) recognized a very differentiated sequence in the Siguatepeque region of central Honduras:

50 m	of unnamed basalt
250 m	Cerro Verde Formation with 30 m of Porterillos rhyolite and 30 m of Siguatepeque olivine tholeiite
200 m	Guique Formation with rhyolitic and andesitic ignimbrites, tuffs and volcanoclastic sandstones
100 m	Ocote Arrancado Formation with erosional remains of a rhyolite about 30 m.y. old
200 m	Matagalpa Formation with andesitic, highly jointed lavas.

The alternation of mafic and felsic products corresponds to a similar sequence in the Coyol Group of Nicaragua, and it raises the question of the origin of such different magmas.

The alternation of mafic and felsic products prevails also in the western highlands of Guatemala. Here, KILBURG (1978), on the Quadrangle Concepción Tutuapa, found on the base of the volcanic pile approximately 500 m of felsic material comprising pumice, perlite, rhyolite and ash flow tuffs. Overlying the basal unit are about 300 m of mafic volcanics consisting of porphyritic basalt and andesite, lahars and minor scoria. Above are distinctive series of leucocratic crystal tuffs that have a combined thickness of approximately 400 m. The highest unit in the suite consists of lahars and porphyritic andesite and basalt having a thickness of about 300 m. Chemically these rocks are calc-alkaline in character. The volcanic series is believed to have originated from two paleovolcanic vents in the southern part of the quadrangle.

In the surroundings of Tegucigalpa ignimbrite is extracted as a building stone in many small quarries and it is used in particular in public buildings such as the airport. WILLIAMS & MCBIRNEY pointed out that the Maya pyramids of Copán were built of ignimbrites which were quarried in the immediate vicinity. The particularly ornamental treatment of the stelae of Copán, which evolved right up to the stage of three-dimensional openwork sculpture, could only have been developed in ignimbrite, which is easy to work and can be quarried in large blocks (Fig. 112).

To the south the Tertiary volcanic covers of Honduras overlap into the northern part of **El Salvador.** They were mapped and classified (cf. map of El Salvador, Fig. 150) by the Geologi-

Fig. 112. Mayan statue and pyramid built of ignimbrite. Ruins of Copán. Photo: R. WEYL.

sche Mission der Bundesanstalt für Bodenforschung (=Geological Mission of the Federal Institute for Geosciences and Natural Resources in Hannover, Federal Republic of Germany). As in the other countries, an older, predominantly intermediate to basic and a younger, predominantly acid, ignimbritic sequence of products, which were designated the **Morazán** or **Chalatenango Formation** respectively, were distinguished and placed with reservation in the Oligocene to Miocene (see also WIESEMANN 1975). Since more details are still forthcoming from the Mission, I must refer the reader to the description of ignimbrites in WEYL (1961 b) and to the chemical data given in WEYL (1966 a).

In central and southern El Salvador extensive basic to intermediate products follow the **Bálsamo Formation,** and basic to acid products come after the **Cuscatlán Formation** — these products can be placed in the Pliocene to Pleistocene. The basic Bálsamo Formation is thus intercalated in the acid products of the Chalatenango and Cuscatlán Formations and divides these up. It is still not possible to say to what extent the sequence corresponds to that of the Coyol Group in Nicaragua. (Details on the geology of El Salvador are given in Section 5.51).

In **Guatemala** Tertiary volcanics cover a broad strip of the highlands between the central mountain chains and the range of Quaternary volcanoes, which are superimposed on the southern edge of the highlands. WILLIAMS (1960) gave a summary of this region, which was followed by mapping activity by US universities (Index map, Fig. 197 and general geological map of Guatemala, Fig. 43). WILLIAMS detected a regional division into three areas: in the west andesitic and basaltic flows predominate; in the central part mainly very thick glowing tuffs occur together with fine-grained white tuffs and ignimbrites with intercalated limnic fluviatile tuffites; in the east, acid lavas and glowing tuffs predominate. The material erupted from fissures, some of which can be detected in a series of rhyolitic to rhyodacitic domes of the Tecum Uman Ridge. The age of the eruptive products can be given as mainly Pliocene merely on the basis of the few diatom floras. WILLIAMS stressed the contrast between these fissure eruptions, which are typical for the Tertiary not simply of Central America but also of Mexico and the USA, and the great central eruptions in the uppermost Tertiary and the Quaternary.

Later mapping work carried out by Dartmouth College, Hanover, N.H. (USA) (Director R. STOIBER) in southeastern Guatemala covered a cross-section through the Tertiary volcanic series, and REYNOLDS (1977) evaluated this cross-section to arrive at a summary description. He was able to distinguish between three eruptive sequences which were referred to as the "Lower, Middle and Upper Group" and these may be characterized as follows:

Lower Group: This comprises a 600 – 800 m thick sequence of intermediate to acid tuffs, ignimbrites and lava flows. It can be subdivided into a lower formation of ignimbrites and lavas of dacitic composition and an upper formation of rhyolitic to rhyodacitic lavas, tuffs and ignimbrites. On the basis of the radiometric age of 15.7 ± 0.6 m.y. of a granite intruding into the oldest tuffs and from the dating of a tuff at 16 m.y., the group can be assigned to the Lower Miocene.

Middle Group: This consists of at least 800 m of mainly massive andesitic lava covers with secondary basalts and dacites as well as pyroclastic rocks. The age can be given merely on the basis of the position as Upper Miocene to Pliocene.

Upper Group: This consists of subjacent pumice tuffs of dacitic to rhyolitic composition, a middle series of lavas of the "classic basalt-andesite-rhyolite-association" (REYNOLDS, p. 23) and overlying basalts and andesites. K/Ar dating carried out on biotites of these lavas yielded an age of 4.0 ± 0.2 m.y. so that the group can be assigned to the uppermost Miocene to Pliocene.

REYNOLDS (1977) is of the opinion that five lithostratigraphic units can be traced through most of the Central American volcanic covers, although the units are referred to by a number of local names. His attempt to correlate these units is convincing, particularly since he does not introduce any new names but sticks to the old ones which come mainly from El Salvador (WIESEMANN 1975).

1. **Matagalpa Group:** The type locality is the vicinity of Matagalpa, Nicaragua (MCBIRNEY & WILLIAMS 1965). It comprises several cycles of andesites, basalts and acid volcanics. Radiometric data indicate that the eruption commenced in the Paleocene and ended in the Lower Miocene. The Morazán Formation in El Salvador is regarded as equivalent.

2. **Subinal and Totogalpa Formations:** The type locality is near El Progreso in the Motagua Valley, Guatemala (HIRSCHMAN 1963) and Totogalpa in NW Nicaragua (ZOPPIS &

DEL GIUDICE 1960). Continental volcanogenic red beds of still uncertain age interlock with the volcanics.

3. **Chalatenango Formation:** The type locality is Chalatenango in northern El Salvador (WIESEMANN 1975). The formation constitutes the main phase of the acid ignimbritic eruption. Equivalent formations are the lower formation of the Padre Miguel Group of Guatemala and Honduras (BURKART 1965) or the Lower Group in the REYNOLDS sequence (1977) as well as the middle formation of the upper Coyol Group in Nicaragua (Anonymous 1972 b). On the basis of radiometric dating the main eruption can be placed in the Middle to Upper Miocene.

4. **Bálsamo Formation:** The type locality is the Balsam Range back from the Pacific coast of El Salvador. It consists of basaltic, andesitic and dacitic lavas, particularly in the coastal region of El Salvador, and it extends into Nicaragua and Guatemala at the base and on the coastal side of the Quaternary volcanic chain. Equivalent formations are the Middle Group in the REYNOLDS sequence (1977) in Guatemala and the uppermost lavas of the Coyol Group in Nicaragua. The age is Miocene to Pliocene.

5. **Cuscatlán Formation:** The type locality is Cuscatlán in El Salvador (WIESEMANN 1975). The formation consists of acid tuffs and tuffites overlain by acid and basic lavas. Equivalent formations are the upper sections of the Padre Miguel Group (BURKART et al. 1973) to the north in Guatemala or the Upper Group identified by REYNOLDS (1977), and possibly also acid tuffs in central Honduras, but there are no equivalents in Nicaragua. The age is Pliocene to Pleistocene.

SUTTER (1977, provisional communication) feels that another way to arrive at an overall classification of the Tertiary and Quaternary volcanic activity of Central America is to evaluate as many radiometric datings as possible, and this has so far resulted in the following pattern:

Around 30 m.y. BP: Commencement of violent volcanic activity.

30 – 10 m.y.: Eruption of andesitic-basaltic lavas of the Matagalpa Group with the climax occurring between 16 and 12 m.y.

30 – 9 m.y.: Production of ignimbrite covers with the climax occurring around 9 m.y.

10 – 5 m.y.: Production of basalt covers over the ignimbrites and of countless lava covers which are taken to be Quaternary.

1.5 m.y.: Basal lava covers from volcanoes of various sizes.

According to this, ignimbrites were produced during the entire period of eruption, whereas the andesitic-basaltic magmas were erupted in a more cyclical manner and peaked around 14 m.y. At all events, basic and acid magmas alternated many times and the process seems to have been regional in character.

The **petrographic definition** of the Tertiary volcanic rocks was based partly on macroscopic findings and partly on the microscopically determined mineral contents as well as partly on the basis of chemical analyses and the norms calculated therefrom; these norms are vital for arriving at a positive definition, particularly in the case of rocks with a glassy base. A large number of chemical rock analyses, reported mainly by WILLIAMS & MCBIRNEY (1969) and Anonymous (1972 b), enable us to define and characterize 75 rocks uniformly in accordance with the RITTMANN norm and the STRECKEISEN procedure (1967):

	Ignimbrites	Others
alkali rhyolite	6	–
rhyolite	12	1
rhyodacite	17	3
dacite	5	3
plagidacite	–	2
quartz-alkali trachyte	1	–
quartz-latite	–	2
quartz-latiandesite	1	12
quartz-andesite	–	6
andesite	–	2
olivine andesite	–	1
quartz tholeiite	–	1

According to this table the *quartz-latiandesites* clearly predominate among the lavas, while basalts are lacking according to the chosen definition. With few exceptions the ignimbrites are *rhyolitic, rhyodacitic* or *alkali-rhyolitic*. "Andesitic" ignimbrites, which are mentioned in the literature, can be assigned, according to the norms, to the quartziferous rocks.

A review of the SiO_2 content (Fig. 113) reveals a clear predominance of 52 – 56% SiO_2 in the lavas, and a clear predominance of SiO_2 values in excess of 70% in the ignimbrites. Values around 60% SiO_2 are less common and thus permit both groups of rocks to be distinguished from each other, even though SiO_2 contents up to 70% may occur in the lavas. In the STRECKEISEN double-triangle and in the serial index plot (alkali versus SiO_2) (RITTMANN's sigma values) the ignimbrites link up directly, however, with the lavas; they represent the continuation of the lavas in the more acid range. These groups of rocks belong in the calc-alkaline series with a strong to moderate serial character. In this respect they are similar to the outputs of the Quaternary volcanoes (Fig. 114 and 115).

The formation of approximately 10,000 km^3 of acid magmas, which built up the ignimbrite covers of Central America, is a petrogenetic problem which still has not been solved, and it is one which applies to all large ignimbrite areas such as, for example, the Tertiary "Central Province" of Mexican volcanism with its much larger ignimbrite covers (GUNN & MOOSER 1971). The discussion on the **origin of the Central American ignimbrites** in fact reveals that, as more data are gathered, so a solution to the problem becomes even more difficult to find.

MCBIRNEY & WEILL came to the following unequivocal conclusion (1966, pp. 12/13), in particular on the basis of the Sr^{87}/Sr^{86} ratios: "We believe that the combined evidence of geo-

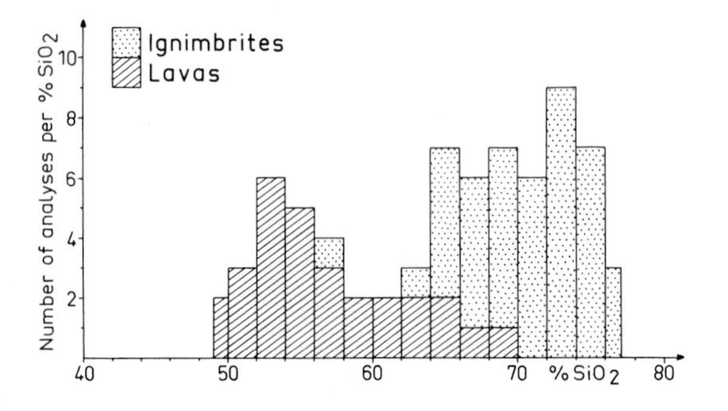

Fig. 113. Histogram of the SiO_2 content of Central American Tertiary lavas and ignimbrites.

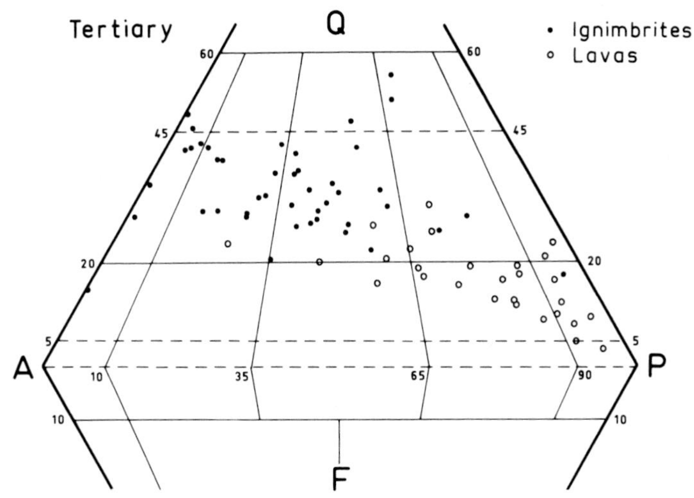

Fig. 114. Distribution of Central American Tertiary lavas and ignimbrites in the STRECKEISEN double-triangle.

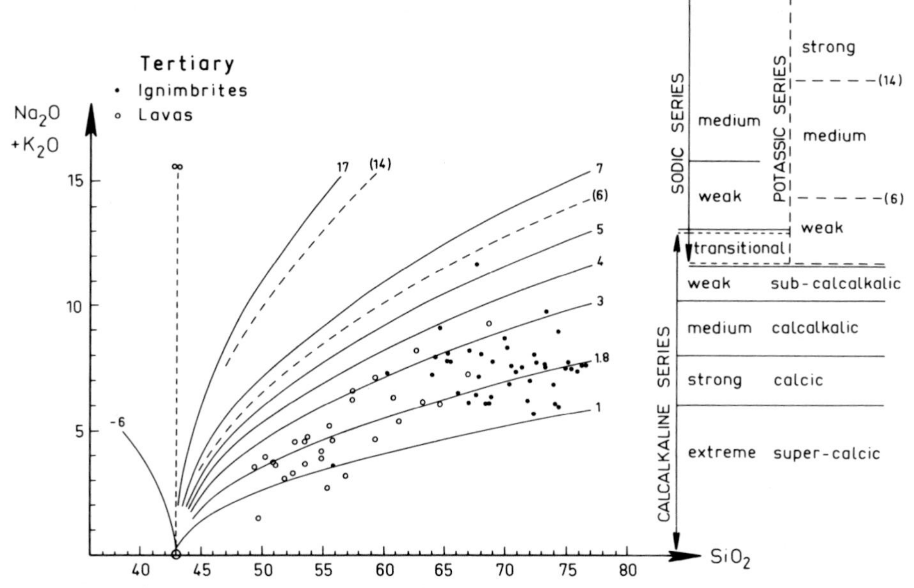

Fig. 115. Serial index plot (alkali versus SiO_2) of Central American Tertiary lavas and ignimbrites.

logic occurrence, relative volumes, melting relations, and isotopic compositions is consistent with derivation of the rhyolite ignimbrites from a zone of widespread fusion of the crust."

On the basis of petrological, geochemical and experimental data and considerations, WILLIAMS & MCBIRNEY (1969, p. 58) stated: "The fact that ignimbrites have high Sr^{87}/Sr^{86} ratios may indicate that they are contaminated magmas derived from more primitive sources, or that they are produced by anatectic processes in which only part of the radiogenic strontium is inherited in the rhyolitic magma."

PUSHKAR et al. (1972) developed a series of models, such as partial or complete dissolution of crustal rocks, assimilation of crustal material in closed or open systems, and came to the following conclusion: "These considerations leave us without a satisfactory explanation for the almost world-wide restriction of voluminous rhyolites and granites to the continents. Even though this problem is presently unresolved, any search for the source of siliceous magmas should give at least as much attention to the mantle beneath the continents as to the overlying sialic crust."

Following on from the above-quoted works, PICHLER & WEYL (1973, pp. 388 – 389) arrived at the following opinion: The Sr^{87}/Sr^{86} ratios found by PUSHKAR et al. (1972, Table 2) in the ignimbrites of Central America fluctuate between 0.7035 and 0.7175; they thus overlap the values of crustal rocks which would be considered as the starting material for an anatectic magma. Eight of the fifteen values are below 0.7050 and thus correspond to those of Quaternary "andesitic" calc-alkaline lavas. Furthermore, they are below the average Sr^{87}/Sr^{86} ratio in the sialic crust but above that for the material in the upper mantle. These mean values of the isotope ratios thus fit best with the concept of magma formation in the deeper crustal regions and/or in the Low Velocity Zone of the upper mantle in which a relatively low Sr^{87}/Sr^{86} ratio can be assumed. However, those ignimbrites in which the Sr^{87}/Sr^{86} ratio is greater than 0.7050, which corresponds to sialic crustal material, are regarded as anatectic products of higher sections of crust or must at least be strongly influenced by assimilation of crustal material. The obviously widespread alternation of basic and acid flows could be explained by magma being formed at different depths.

It is still an unsolved question why partially dissolved xenoliths are lacking here, a fact to which WILLIAMS & MCBIRNEY (1969, p. 56) refer. And above all it still has to be explained what geotectonic impulse led to the formation of large quantities of rhyolitic magmas; the impulse could have been in the form of regional burials of crustal material, although there are no indications that such a process occurred. If, however, one regards the volcanic covers as a magma flow arising from subduction at the boundary of the Cocos and Caribbean plates, then their aerial extent requires a special explanation, because this differs clearly from the more linear volcanism of the Quaternary. REYNOLDS regards the boundaries of the youngest Tertiary flows in El Salvador, i.e. the Cuscatlán Formation, as the beginning of a segmentation of subduction, which STOIBER & CARR (1974) described for the Quaternary volcanism.

A **shifting of the volcanic activity** from the Upper Tertiary to the present in the direction of the present Pacific border is worth noting. In Nicaragua the Matagalpa Group was mainly concentrated in the northern central highlands. The younger Miocene-Pliocene Coyol Group is, on the other hand, clearly shifted southwestwards. On its southwestern edge, or on the northeastern edge of the Nicaragua Depression, is located the series of extinct Pliocene-Pleistocene volcanoes. Further to the southwest comes the chain of active volcanoes in the interior of the Nicaragua Depression.

In Honduras a regional demarcation of the Tertiary volcanics is not so obvious, yet according to the general map of the country large occurrences of the Oligocene Matagalpa Formation stretch further north than the extensive covers of the Middle to Upper Miocene Padre Miguel Group. In contrast, the shift is clearer to the south in El Salvador, where the oldest (Oligocene?) Morazán Formation predominates in the northern part of the country. The somewhat younger (Miocene) Chalatenango Formation covers approximately the same area. It is followed by the Pliocene-Pleistocene Cuscatlán Formation with the ruins of the volcanoes of the Bálsamo Formation, both of which extend into the Pacific coastal region. They are surmounted by the most recent, sometimes still active, volcanoes of the San Salvador Forma tion.

The southward shift of volcanic activity is also very clear in Guatemala, where the Tertiary volcanic regions connect up in broad belts with the central crystalline rock on which are directly superimposed the Quaternary and active volcanoes where the terrain slopes down to the Pacific Coastal Plain.

If the formation of magma in the subduction zone between the Caribbean Plate and the Cocos Plate is regarded as the important, though not the only, source of volcanic activity, then the displacement of the volcanic activity must reflect a shift in the subduction zone, which has resulted in a southwards-oriented accretion of the Caribbean Plate in the Central American continent. Such assumptions could perhaps be confirmed by other characteristics of subduction zones, such as the position of a foredeep which, in keeping with the present distance, would have had to be situated about 200 km ahead of each volcanic region. It might be possible to prove that this is the case in the last sections of the Nicaragua Trough, which was situated at a corresponding distance from the region of the Matagalpa volcanoes. It should be possible to find similar clues in the thickness of the crust and the thickness of the sediment within the shelf off the remaining parts of Central America. The evidence is slowly coming to light in the sediment troughs detected by seismic reflection and by drilling in the shelf region of Guatemala and Nicaragua.

5.3 Plutonism in northern Central America

Plutonic rocks have long been known in the region of northern Central America; already SAPPER (1937) indicated a large number of occurrences in his maps; however, it was not possible to give more than a general definition of them and to arrive at approximate ages. ROBERTS & IRVING (1957, p. 26) stated more precisely that small intrusions break through the Cretaceous sediments in Central Guatemala, Honduras, Nicaragua and El Salvador and consequently must date from the Late Cretaceous to Upper Tertiary. They also recognized their importance for the process of mineralization in the Central American gold and silver deposits. The present author has compiled further data on this subject (1961 a, pp. 113 – 115). WILLIAMS & MCBIRNEY (1969, pp. 6 – 14) provided a more recent summary and published a series of chemical rock analyses, which are contained in the diagrams in Fig. 119. Since that time countless individual studies have been carried out, even though the statement still stands that: "Much more radiometric and field work must be done before a clear picture can be drawn of the plutonic history of Central America".

Nevertheless, a large number of **radiometric datings** combined with field studies, in particular those carried out by US universities, permit us to draw some conclusions about the age of intrusion of the plutonic rocks. Since the data are scattered and recorded in literature which is sometimes difficult to obtain, those sources which were accessible have been summarized in Table 9 and Fig. 116. Fig. 104 contains a regional review of the larger plutons and — where known — their age. Only the data for the Cretaceous-Tertiary intrusions have been included; for the Paleozoic intrusions the reader is referred to p. 29. It was not, however, possible to include in the map a number of small and very small stocks which are intruded into the Tertiary volcanics.

The review shows that the young intrusions commenced in the Jurassic and accumulated above all in the Upper Cretaceous and lowest Tertiary. This intrusion phase is connected

Table 9. Radiometric ages of Cretaceous and Tertiary plutonic rocks of northern Central America.

Guatemala		
Chiquimula pluton, isochrone of main rock sequence (CLEMONS & LONG 1971, p. 2738)		50.0 ± 5.0 m.y.
Chiquimula pluton, biotite-whole rock isochrone (CLEMONS & LONG 1971, p. 2739)		95.0 ± 1.0 m.y. 83.9 ± 1.7 m.y. 84.6 ± 1.7 m.y.
Granite, Guijor, Chiquimula (LEVY 1970, Anexo 1 – 1)	biotite	84.0 ± 3.0 m.y.
Trondhjemite, near Chinautla (WILLIAMS 1960, p. 5)		92 m.y.
Gneissoid biotite-rich granite or monzonite, Río Hondo Power Plant, north side of Motagua Valley (WILLIAMS & McBIRNEY 1969, p. 6)		99.7 m.y.
Hornblende-biotite mafic quartz-diorite, Santiago Bay, Lake Atitlán (probably reheated) (WILLIAMS & McBIRNEY 1969, p. 8)		13.8 m.y.
Hornblende-biotite quartz-diorite, San Juan, Lake Atitlán (probably reheated) (WILLIAMS & McBIRNEY 1969, p. 8)		9.0 m.y.
Hornblende-biotite quartz-monzonite, San Juan Lucas Toliman (probably reheated) (WILLIAMS & McBIRNEY 1969, p. 8)		8.5 m.y.
Biotite and muscovite from pegmatite dike, Chuacús Series, El Chol (McBIRNEY 1963, p. 228)		66.0 m.y.
Granite, south of La Fragua (LEVY 1970, Anexo 1 – 1)	biotite	24.0 ± 0.5 m.y.
Tonalite, Los Tablones, Zacapa (LEVY 1970, Anexo 1 – 1)	biotite	22.0 m.y.
Tonalite, Los Tablones, Zacapa (LEVY 1970, Anexo 1 – 1)	hornblende	35.0 m.y.
Mylonitic biotite granite, Río Hondo (LEVY 1970, Anexo 1 – 2)		99.7 m.y.
Quartz-biotite monzonite, Cerro Zacapa (LEVY 1970, Anexo 1 – 2)		19.7 m.y.
Hornblende-biotite granite, Cerro Redondo, Santa Rosa Quadrangle (REYNOLDS 1977, p. 25)		15.7 ± 0.6 m.y.
Honduras		
Granodiorite, San Pedro Sula (HORNE et al. 1976 c, p. 586)	biotite	35.9 ± 0.7 m.y.
Tonalite, Mezapa pluton, Cordillera Nombre de Diós (HORNE et al. 1976 c, p. 586)	biotite	71.8 ± 1.4 m.y.
Augite tonalite, Tela batholite, Cordillera Nombre de Diós (HORNE et al. 1976 c, p. 586)	biotite biotite hornblende	80.5 ± 1.5 m.y. 73.9 ± 1.5 m.y. 93.3 ± 1.9 m.y.
Tonalite, Las Mangas batholite, Cordillera Nombre de Diós (HORNE et al. 1976 c, p. 586)	muscovite	57.3 ± 1.1 m.y.
Tonalite, Piedras Negras, Cordillera Nombre de Diós (HORNE et al. 1976 c, p. 586)	biotite hornblende	56.8 ± 1.1 m.y. 72.2 ± 1.5 m.y.
Granodiorite, Minas de Oro stock (HORNE et al. 1976 b, p. 94)	biotite hornblende	58.2 ± 1.8 m.y. 60.6 ± 1.3 m.y.
Porphyritic dacite, San Francisco stock (HORNE et al. 1976 b, p. 94)	biotite	58.6 ± 0.7 m.y.
Adamellite, San Ignacio stock (HORNE et al. 1976 b, p. 96)	biotite	114.0 ± 1.8 m.y.
Biotite granite, San Pedro-Pto. Cortéz (LEVY 1970, Anexo 1 – 2)		27.0 ± 2.7 m.y.

Table 9 continued

Nicaragua		
Granodiorite, Dipilto pluton (ZOPPIS BRACCI 1961, p. 43)		83.0 ± 3.0 m.y.
Granodiorite, Dipilto pluton (Anonymous 1972 b, p. 89)		107.4 ± 1.2 m.y.
Biotite granite, Cerro Sumús stock (Anonymous 1972 b, p. 89)	biotite	107.5 ± 2.2 m.y. 107.3 ± 2.1 m.y.
Hornblende diorite, San Juan del Río Coco stock (Anonymous 1972 b, p. 90)	hornblende	112.4 ± 1.5 m.y.
Pyroxene diorite, Cerro el Diablo stock (Anonymous 1972 b, p. 90)	plagioclase	60.8 ± 1.0 m.y.
Granodiorite, Santo Tomás del Nance stock (Anonymous 1972 b, p. 90)		28.6 ± 0.4 m.y.

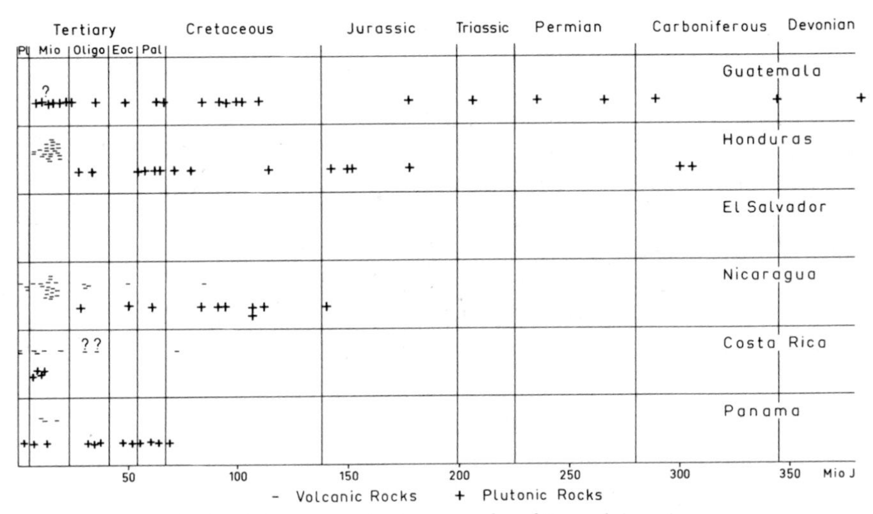

Fig. 116. Radiometrically dated plutonic and volcanic rocks of Central America.

with the Laramide orogenesis in northern Central America. A second main phase occurs in the Late Tertiary and is concentrated above all in the regions of Tertiary volcanism. The radiometric dates are rather scant; however, the contacts unequivocally indicate a date in the Late Tertiary. The intrusions in question are usually the very near-surface intrusions which are important for the process of mineralization and to which reference is made in Chapter 9. It is logical to assume a genetic link with the volcanic magmas, particularly as the chemistry of the two types of rock is very similar (see Figs. 114, 115, 119).

The individual descriptions reveal a wide range of variation in the intrusive rocks which may be illustrated by describing two typical plutons.

One particularly fine example is the **intrusion of San Juan del Río Coco** in NW Nicaragua (GARAYAR & VIRAMONTE 1971, 1973). Rectangular in shape, it extends longitudinally for 13 km and covers an area of about 52 km². From NE to SW the following rocks occur:

hornblende-peridotites, hornblendites, hornblende-gabbrodiorites, hornblende-diorites, biotite-tonalites, biotite-granodiorites and muscovite-granodiorites; their modes are listed in the following table.

The modes of rocks in the San Juan del Río Coco intrusion. After GARAYAR & VIRAMONTE 1971/1973.

	1	2	3	4	5	6	7
Quartz	–	–	·	5	20	25	25
Potassium feldspar	–	–	·	·	5	25	40
Plagioclase	–	2	25	45	60	40	20
Muscovite	–	–	·	·	·	·	13
Biotite	–	–	·	·	13	10	·
Hornblende	45	75	60	35	2	·	·
Pyroxene	30	20	10	10	·	·	·
Olivine	20	–	·	·	·	·	·
Pyrrhotine	5	5	5	·	·	·	·
Titanite	·	·	·	5	·	·	·
Garnet	–	–	·	·	·	·	3

1. hornblende-peridotite, 2. hornblendite, 3. hornblende-gabbrodiorite, 4. hornblende-diorite, 5. biotite-tonalite, 6. biotite-granodiorite, 7. muscovite-granodiorite.

The fabric of the peridotites is described as poikilitic with large hornblendes. They replace the pyroxene. The sometimes high percentage of pyrrhotine is worth noting. In acid units the potassium feldspar increases while replacing the plagioclase, and biotite replaces the hornblende. Both according to the mineral content as well as to six reported chemical analy-

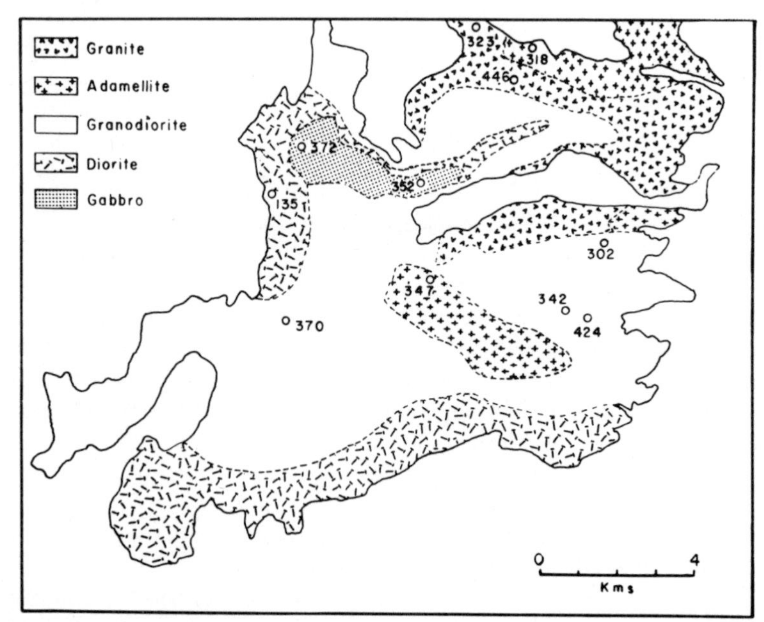

Fig. 117. Map of the southern part of the Chiquimula pluton showing the approximate distribution of rock types. (From CLEMONS & LONG 1971, Fig. 4).

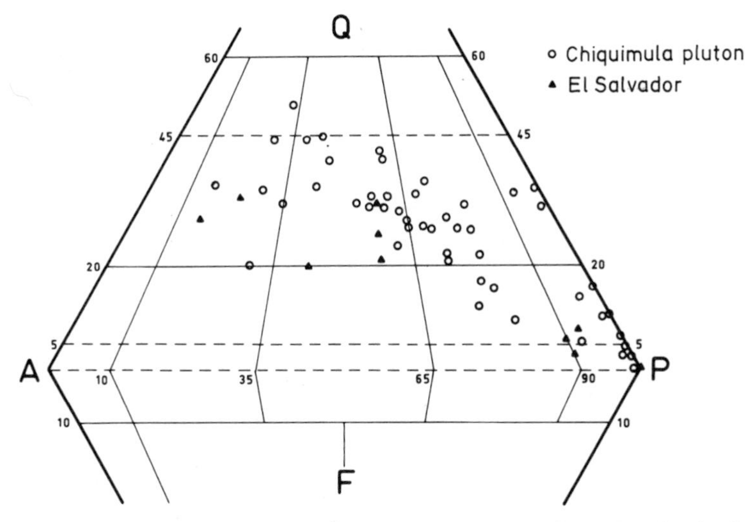

Fig. 118. STRECKEISEN double-triangle of the modal mineral content of the Chiquimula pluton and plutonic rocks of El Salvador. (Data from CLEMONS & LONG 1971 and WEYL 1961a).

ses, there is a sharp division between the peridotites to diorites at 55.0 – 56.7% SiO_2 on the one hand and the tonalites and granodiorites at 69.5 – 74.1% SiO_2 on the other.

The intrusion, with its association of rocks which was hitherto unknown in Central America, is regarded as the result of gravitational crystallization differentiation in a tubular magmatic body which is now horizontally oriented due to subsequent tectonic tilting. The age of the hornblende-diorite has been determined as 112 ± 1.5 m.y. (Anonymous 1972 b, p. 89).

Very much larger, but not differentiated through to peridotites, is the **Chiquimula pluton,** which covers approximately 300 km^2, south of the Motagua Valley in Guatemala (CLEMONS & LONG 1971). In the outcrops, which are regarded as relatively young, granodiorite is the predominant rock while in deeper incisions gabbro occurs. The more acid rocks prevail in the eastern part of the pluton (map, Fig. 117), which may have been tilted here after its intrusion. The range of variation in the rocks was confirmed by a large number of modes (Fig. 118). The variation is ascribed to a combination of differentiation and assimilation. The age of the pluton was placed at 50 ± 5 m.y. by Rb/Sr determination, and this agrees with field observations; the age of biotites from the granites and adamellites on the other hand was put at 95 ± 1 m.y., which may indicate earlier intrusion of these parts of the pluton.

The range of variation in the intrusive rocks in northern El Salvador is also very large, and their modal values have been included in the diagram in Fig. 118. Although they have not been dated radiometrically, their age was recognized as Tertiary by GREBE (1961) and the present author (1961 a), and this was confirmed by the mapping carried out by the German Geological Mission (WIESEMANN 1975). This fits in with the radiometrically determined age of 15.7 ± 0.6 m.y. for a biotite granite from Cerro Redondo (Santa Rosa Quadrangle) in eastern Guatemala (REYNOLDS 1977, p. 25).

The relatively few **chemical analyses** of plutonic rocks in northern Central America (WEYL 1961 a, Table 11, Nos. 1 – 3; WILLIAMS & MCBIRNEY 1969, Table 1, Nos. 1 – 3; Anonymous 1972 b, Table IV – 8 C, Nos. C 43 – 45) provide a provisional summary and indi-

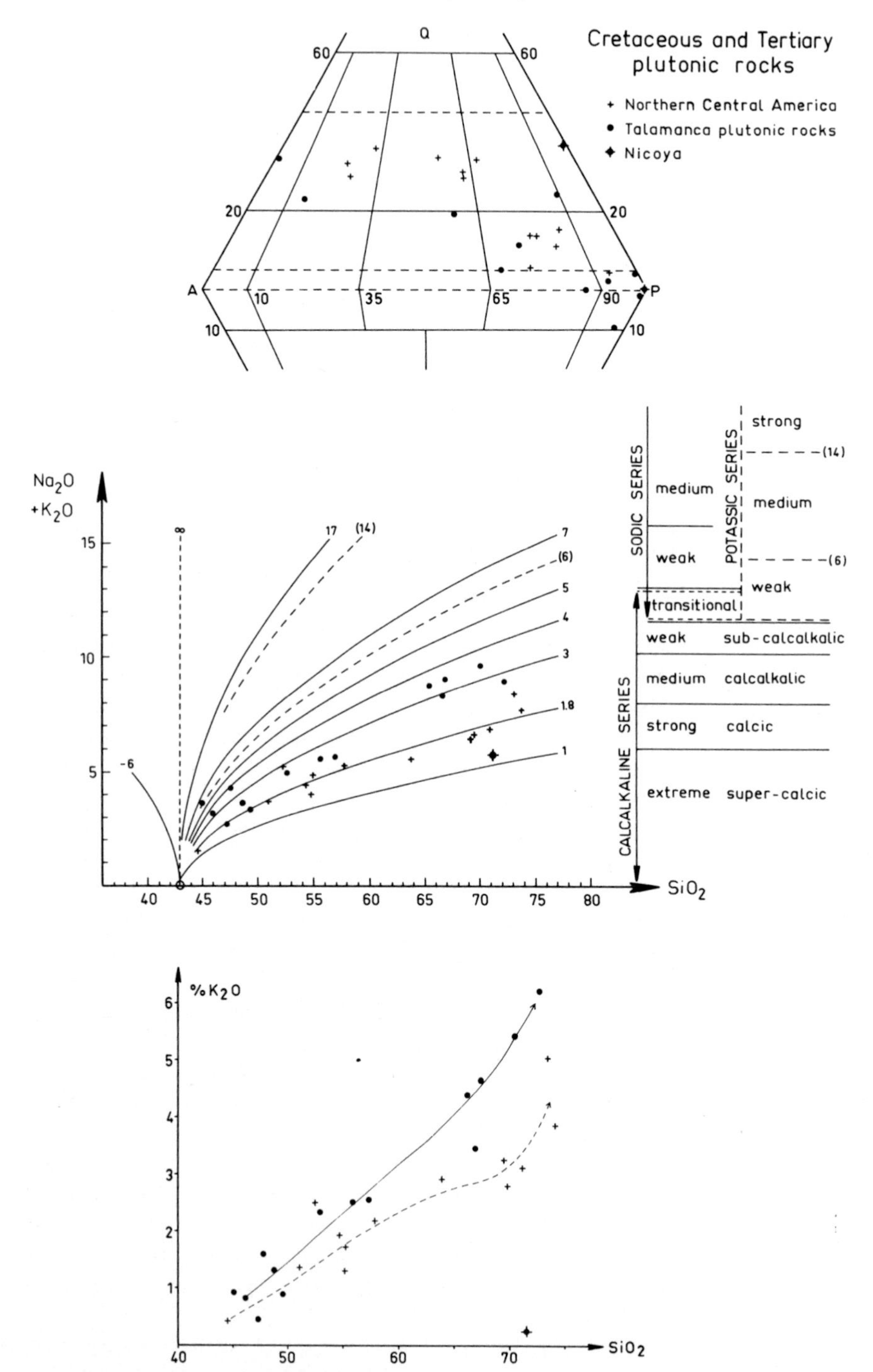

Fig. 119. Distribution of Central American plutonic rocks in the STRECKEISEN double-triangle, serial index (alkali versus SiO_2), and K_2O versus SiO_2 plot.

cate the strong to medium calc-alkalic character of the plutonic rocks. In contrast to them the Tertiary plutonic rocks of the Cordillera de Talamanca of Costa Rica have a much higher alkali and, in particular, potassium content (Fig. 119).

WILLIAMS & MCBIRNEY (1969, p. 6) were the first to point out that the Cretaceous-Laramide plutons are located in the northern part of the Tertiary volcanic plateau, while the Tertiary plutons are located within this region so that a shift in the magmatic activity towards the southwest must have occurred. This was confirmed by Anonymous (1972 b, p. 96) for the Nicaraguan region, and the data which have become known since that time (see Table 9) indicate that such a shift took place from the northeast during the Cretaceous and lowermost Tertiary towards the southwest in the Upper Tertiary. This coincides with the displacement of the volcanic activity in the direction of the present Pacific border (Fig. 104).

If it is assumed that the plutonic magmas formed as a result of plate subduction, this would be evidence of a shift in the corresponding paleo-subductions. However, it is a questionable practice to line up the volcanoes and plutonic rocks of the geological past in a diagram and to deduce corresponding subductions from this. In this respect the otherwise informative maps prepared by MALFAIT & DINKELMAN (1972) appear very hypothetical.

5.4 Quaternary volcanism

The geological structure and topography of the Pacific border of Central America is molded to a large extent by young and in many cases still violently active volcanoes. The volcanic chain stretches 1,100 km almost without interruption from the border between Mexico and Guatemala to the volcano Turrialba in Costa Rica and from there, with larger intervals of a good 300 km between the individual volcanoes, through to and into Panama. According to SAPPER (1913, pp. 163/164) 101 first-order volcanoes, i.e. more or less independent volcanic edifices, belong to the Central American volcanic chain. Their number increases considerably when smaller and extinct volcanoes are taken into account. The list prepared by BOHNENBERGER (1978) indicates the following numbers of volcanoes of various size and age in the countries of Central America:

Guatemala	288
El Salvador	180
Honduras	18
Nicaragua	58
Costa Rica	30
Panama	8

The young and, in particular, the active volcanoes attracted attention even at the time of the Spanish Conquest. Systematic research on them commenced around the middle of the 19th century with the main interest initially being focussed on their position, structure, the type of activity and the recorded history of their activity. The works by DOLLFUS & MONT-SERRAT (1868), MORITZ WAGNER (1870), MONTESUSS DE BALLORE (1888), VON SEEBACH (1892), SAPPER and TERMER characterize this period of research which was in a way concluded by the comprehensive descriptions provided by SAPPER (1913, 1937). PUTNAM (1926), VON WOLFF (1929) and BURRI & SONDER (1936) concentrated on petrological questions. The study of Central American volcanoes was given an important impetus after the Second World War by

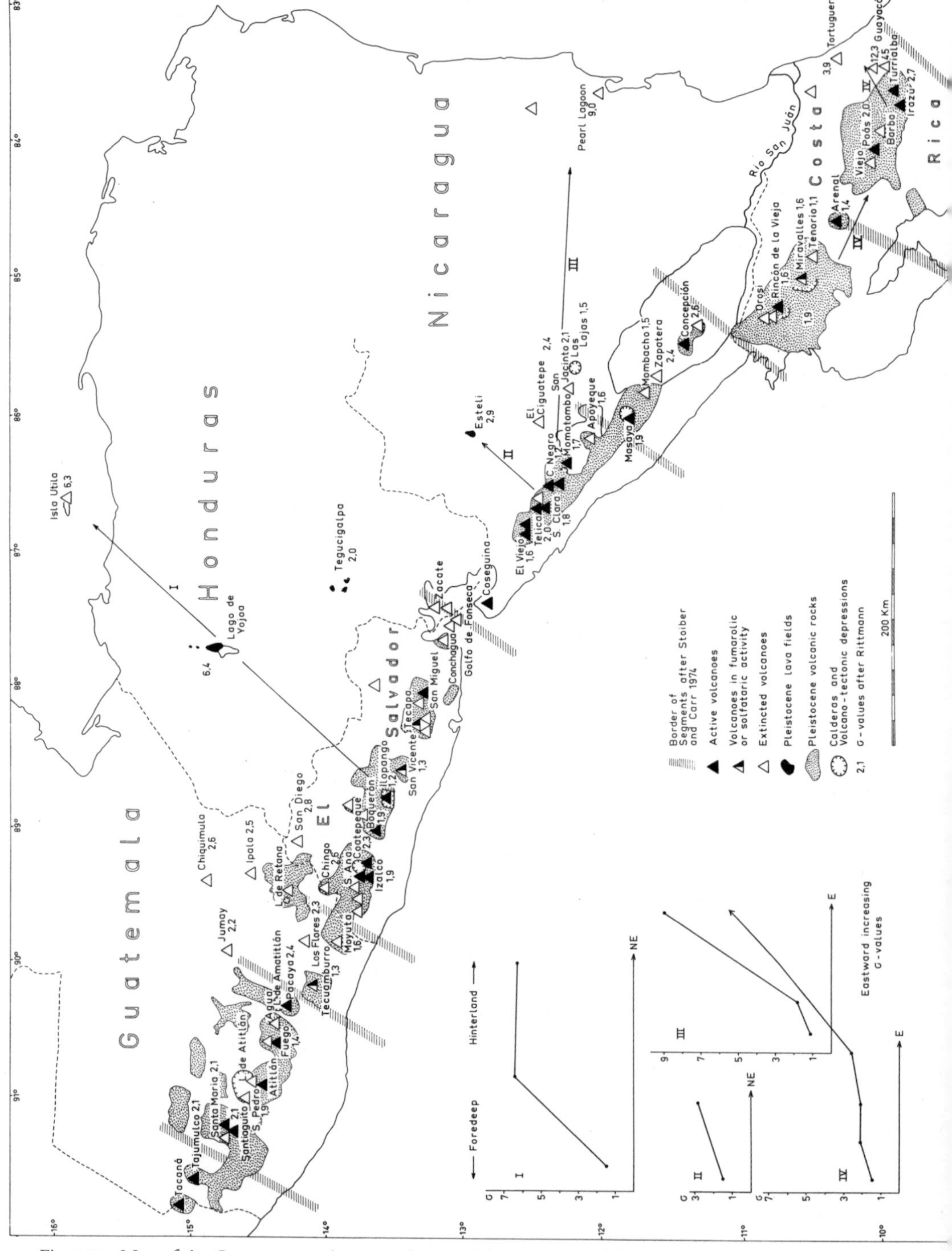

Fig. 120. Map of the Quaternary volcanoes of Central America. (Modified from PICHLER & WEYL 1973, Fig. 9)

the founding of research establishments in El Salvador which produced the works by MEYER-ABICH (1952 – 1956), WILLIAMS (1953 – 1956) and WEYL (1953 – 1957), so that for a time the Salvadorian section of the volcanic zone was one of the best researched. During this period the "Catalogue of the Active Volcanoes of the World . . . P. VI, Central America, 1958", which was prepared by MOOSER, MEYER-ABICH & MCBIRNEY, also appeared.

The works by WILLIAMS, MCBIRNEY and co-workers (1963 – 1969) denote a decisive step towards modern volcanological research in Central America; these authors travelled systematically through one Central American country after the other describing and recording the volcanic phenomena, including the Tertiary volcanism which up until that point had been very badly neglected, in a series of monographs. These studies were then continued on the active volcanoes right through to the present day by STOIBER and ROSE Jr. and their pupils, who concentrated in particular on the fumarolic activity, while at the same time a number of other papers on most of the volcanic regions in Central America have appeared since 1960. Petrological-chemical problems and links with the tectonics of the Pacific border or with plate tectonics have received increasing attention (MOLNAR & SYKES 1969, MCBIRNEY 1969, PICHLER & WEYL 1973, CARR & STOIBER 1974, CARR et al. 1979).

Other important volcanic events and research results can only be briefly dealt with in this book, and in the general map in Fig. 120 only the largest volcanoes could be mentioned by name. The data on the number of **active volcanoes** fluctuate according to the various definitions of volcanic activity. VON SEEBACH (1892) had identified 26 as active, SAPPER named 26 active volcanoes and 7 volcanoes in the fumarole or solfatara stage. The 1958 catalogue of volcanoes names 31 volcanoes which have been known to erupt, while 8 are in the solfatara or fumarole stage. Our list records in each case the most recent violent manifestation of volcanic activity. According to this list, 27 volcanoes were active in historical times and produced ash or lava, and a further 9 volcanoes are in the fumarole stage. Central America is thus one of the most active volcanic regions on earth.

SAPPER (1913) reported in detail on the historically recorded eruptions of Central American volcanoes, and he pointed out that despite the gaps in the reports, in particular from early historical times, the activity is characterized by pronounced periodicity. MCBIRNEY (1956) had also drawn attention to the extremely pronounced periodicity in Nicaragua. The third decade of the 16th century was probably a period of particularly strong activity, and during the 19th and 20th centuries the pattern of activity for the various decades was as follows:

Number of active volcanoes

Decade	Number	Decade	Number
1801 – 1810	1	1891 – 1900	2
1811 – 1820	1	1901 – 1910	7
1821 – 1830	5	1911 – 1920	7
1831 – 1840	4	1921 – 1930	7
1841 – 1850	6	1931 – 1940	4
1851 – 1860	9	1941 – 1950	4
1861 – 1870	9	1951 – 1960	8
1871 – 1880	6	1961 – 1970	14
1881 – 1890	5		

SAPPER (1897) made some remarkable discoveries about the spatial arrangement of the volcanoes, and these data have recently been returned to by STOIBER & CARR (1974) or CARR & STOIBER (1977) who interpreted them in connection with seismic activity and plate tectonics: According to these authors the Central American volcanoes are not arranged in an

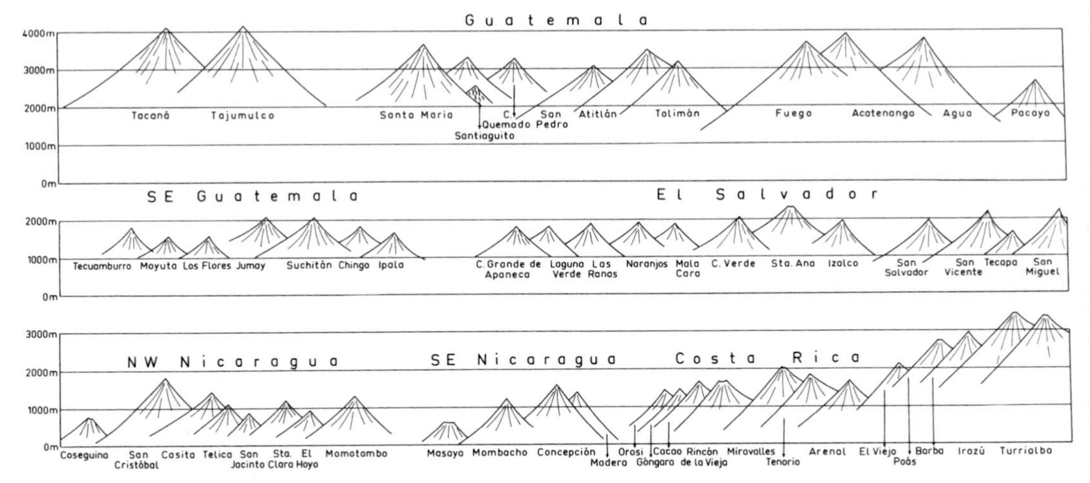

Fig. 121. Schematic diagram showing the elevation of groups of Central American volcanoes.

unbroken chain (SAPPER talks of a longitudinal fissure) but are distributed over a number of short separate chains which are offset in relation to each other. STOIBER and CARR see this as evidence of **plate segments**, with different angles of dip of the subduction zone, which are offset in relation to each other in the crust by transverse faulting. They distinguished at first 7 and later 8 such segments whose more or less wide boundaries are incorporated in the map in Fig. 120.

The individual segments are distinguished by the following features

1. Types of volcanic eruption. The segment in central Guatemala is particularly active, while the neighbouring segments have so far only exhibited solfataral activity. With the exception of Arenal, nuées ardentes are known only from central Guatemala. Salvador and Nicaragua combine violent explosive eruptions with the production of large lava flows, while the volcanoes of Costa Rica, with the exception of Arenal, have not produced any lava in historical times but have merely erupted steam and ash from deep central craters.

2. Form and structure of the volcanic centres – these are very similar within each segment but differ from the neighbouring segment. This is clearly expressed in, among other things, the heights of the volcanic edifices and the heights of the volcanoes above sea level (Fig. 121). Many of the large volcanoes of Guatemala are arranged on transverse fissures, while those in eastern Guatemala and El Salvador are arranged annularly. The volcanoes of Nicaragua and of the Cordillera de Guanacaste in Costa Rica form chains of individual volcanoes, while those of the Cordillera Central in Costa Rica are joined in a coherent mountain range.

3. Volumes of the recent volcanic centres and the ratio of volume to segment length differ distinctly (STOIBER & CARR 1974, Tab. 1).

4. The chemistry of the volcanic products reveals differences from segment to segment, and these will be dealt with in connection with the discussion of the petrology.

5. Large complex volcanic edifices (Santa Maria, Pacaya) have built up on the segment boundaries, presumably as a result of easy ascent paths for the magma.

According to CARR et al. (1979): "Recent volcanic rocks in Central America occur in three structural settings. The first in a narrow zone of volcanoes, the volcanic front, where large volumes of calc-alkaline rocks are erupted. Most volcanoes in this zone are basaltic to andesitic composite cones in lineaments parallel to the strike of the seismic zone and clearly related to it. The second structural setting is behind the volcanic front and above transverse structures where the strike or dip of the seismic zone abruptly changes. Here quartz- and olivine-normative basalts are common and nepheline-normative basalts are present. Occasional rhyolites make the distribution of lavas distinctly bimodal. Most volcanoes in this structural setting are small monogenetic shields and cinder cones forming clusters located 25 – 100 km or more landward from the volcanic front. The third structural setting is far behind the volcanic front. Here strongly undersaturated lavas are erupted from volcanoes having no relation to the structure of the inclined seismic zone" (p. 387).

In the following list of larger volcanoes the subdivision into segments is followed.

List of the important volcanoes of Central America

The data are taken from the following summary works and from the works listed for the individual volcanoes or groups of volcanoes:

SAPPER (1913) for the entire region
MEYER-ABICH (1956) for Guatemala and El Salvador
MOOSER, MEYER-ABICH & MCBIRNEY (1958) for the entire region
BOHNENBERGER, BENGOECHÉA, DÓNDOLI & MARROQUIN (1966) for the entire region
BOHNENBERGER (1969) for Guatemala
MARTÍNEZ & VIRAMONTE (1971, 1973) for Nicaragua
CARR & STOIBER (1974) for the entire region
BOHNENBERGER (1978) for the entire region
Nombres Geográficos de Costa Rica, Vol. II. Oronomía. Instituto Geográfico Nacional, San José, (undated) for Costa Rica
MCBIRNEY (unpublished) [12] for the entire region

The geographical co-ordinates were rounded up to full minutes. Differing height data are given in brackets. The designations of the types of volcano are taken for the most part from the catalogue of volcanoes (MOOSER, MEYER-ABICH & MCBIRNEY 1958).

The data for the volcanoes are arranged according to the following pattern: Name (synonyms), latitude and longitude, height above sea level, height of volcanic edifice.

The type of volcano, the volcanic activity, and the main rock types are quoted from the authors, and the definition according to RITTMANN/STRECKEISEN is shown in italics.

The average SiO_2 content and the average serial index (Sigma RITTMANN).

Recent references.

Western Guatemala

Tacaná 15°08′ N, 92°07′ W. Height 4,092/2,300 m.
Strato volcano with caldera rings and a small culminating plug. Sporadic eruptions of steam and ash, solfataric activity.
Hornblende-hypersthene andesite (MOOSER et al. 1958).
No chemical analyses.
MOOSER et al. (1958, pp. 36, 38).

Tajumulco 15°03′ N, 91°54′ W. Height 4,220/1,200 m.
Compound strato volcano built on high plateau. Weak fumarolic activity and explosions in 1821 (?) and 1863.
Andesite, basaltic andesite (MCBIRNEY unpubl.).

[12] I am particularly grateful to Prof. MCBIRNEY for letting me have the data from a large number of unpublished chemical analyses.

Quartz-latiandesite, leuco quartz-latiandesite.
Average SiO_2 : 57.5%. Average serial index 2.07.
MEYER-ABICH (1958, pp. 40 – 41).

Central Guatemala

Siete Orejas 14°49′ N, 91°37′ W. Height 3,370/1,100 m
Eroded strato volcano. Short lived but cataclysmic outburst of Vulcanian and Krakatoan type between 5,000 and 10,000 years ago.
Rhyodacitic pumice.
Average SiO_2: 71.08%. Average serial index: 1.50.
WILLIAMS (1960, p. 33, 36).

Santa Maria (Figs. 122, 136) 14°45′ N, 91°33′ W. Height 3,772/1,500 m.
Large strato volcano with explosion crater on SW slope and lava dome Santiaguito. Explosions 1902 – 1911, then formation of lava dome Santiaguito.
Basaltic andesite (ROSE et al. 1977), hornblende augite hypersthene basalt (MCBIRNEY unpubl.).
Andesite, quartz-andesite, leuco quartz-andesite, leuco plagidacite.
Average SiO_2 values: 52.5% and 64.2%. Average serial index: 2.14.
BONIS (1965), ROSE (1972 – 1974), ROSE, STOIBER & BONIS (1970), ROSE et al. (1977), STOIBER & ROSE (1969).

Santiaguito (Fig. 122) 14°44′ N, 91°34′ W. Height 2,520/250 m.
Complex lava dome. In 1922, an endogenous volcanic dome began to grow in the 1902 explosion crater of Santa Maria volcano. Extrusion was accompanied by pyroclastic eruptions, lava flows, and strong fumarolic activity.
Tridymite-bearing hyperstehene dacite (ROSE 1972), porphyritic dacite (ROSE et al. 1977).
Leuco quartz-andesite.
Average SiO_2: 63.5%. Average serial index: 2.08.
ROSE (1972 – 1974), ROSE, STOIBER & BONIS (1970), ROSE et al. (1977), STOIBER & ROSE (1969, 1970), STOIBER et al. (1971).

Cerro Quemado (Volcán de Quezaltenango) 14°48′ N, 91°31′ W.
Height 3,370/1,100 m.
Large, complex lava dome. Lava effusion in 1785.
Hornblende-mica andesite and hornblende-andesitic pumice (MEYER-ABICH 1958, p. 48).
No chemical analyses.
BONIS (1965), GALL (1966).

Lago de Atitlán (Fig. 123) 14°40′ – 14°45′ N, 91°05′ – 91°16′ W.
Lake level 1,500 m.
Cauldron formed by collapse as a result of subterranean withdrawal of magma.
WILLIAMS (1960, p. 29 – 32).

San Pedro 14°39′ N, 91°16′ W. Height 3,020/1,500 m.
Strato volcano with parasitic cones. Extinct.
Hypersthene-augite andesite (MCBIRNEY unpubl.), pyroxene andesite (BOHNENBERGER 1978).
Leuco quartz-latiandesite, rhyodacitic pumice.
Average SiO_2 values: 58.72%, 69.85%. Average serial index 1.91.
WILLIAMS (1960, pp. 43 – 44).

Atitlán (Suchiltepequez) 14°35′ N, 91°11′ W. Height 3,537/2,000 m.
Strato volcano, twin of Tolimán volcano. Ten explosions recorded in 19th century, fumaroles.
Andesite (MCBIRNEY unpubl.).
Leuco quartz-andesite.
SiO_2: 54.43%. Serial index 1.80.
WILLIAMS (1960, p. 45).

Tolimán 14°37′ N, 91°11′ W. Height 3,158/1,550 m.
Strato volcano, twin of Atitlán volcano. Fumaroles during 1870 – 1892.
Andesite (MCBIRNEY unpubl.).
Leuco quartz-andesite.
SiO_2: 59.40%. Serial index 2.00.
WILLIAMS (1960, pp. 44 – 45).

Fig. 122. Santa Maria Volcano with Santiaguito Dome. Photo: TERMER 1923.

Fig. 123. Lake Atitlán with Toliman and Atitlán Volcanoes, Guatemala. (From WEYL 1961a, Fig. 47).

Fuego (Fig. 137) 14°29′ N, 90°53′ W. Height 3,763/2,400 m.
Large strato volcano, twin of Acatenango volcano. The most active volcano in Guatemala, more than 50 eruptions since 1524. Tephra eruptions and lava flows. In 1971 strongest eruption since 1900.
Basaltic ash (ROSE et al. 1974), andesitic basalt (MCBIRNEY unpubl.).
Andesitic ash, *andesite.*
Average SiO_2: 51.1%. Average serial index 1.35.
TERMER (1964), BONIS & SALAZAR (1974), ROSE, BONIS, STOIBER, KELLER & BICKFORD (1974), CRAFFORD (1975), ROSE & STOIBER (ed.) (1976), DAVIES et al. (1978), ROSE et al. (1978).

Acatenango (Fig. 45) 14° 30′ N, 90°53′ W. Height 3,976/2,500 m.
Large strato volcano, twin of Fuego volcano. Explosions during 1924 – 25, and 1926 – 27, 1972.
Andesite, basaltic andesite (MCBIRNEY unpubl.).
Leuco quartz-latiandesite, quartz-andesite.
Average SiO_2: 57.72%. Average serial index 1.99.
CARR & STOIBER (1974, p. 328).

Agua (Fig. 45) 14°28′ N, 90°45′ W. Height 3,766/2,400 m.
Large strato volcano. Extinct, mud flow in 1541.
Hypersthene-augite andesite (MCBIRNEY unpubl.).
Quartz-latiandesite.
Average SiO_2: 58.30%. Average serial index: 2.30.
WILLIAMS (1960, pp. 47 – 48).

Pacaya (Fig. 124) 14°23′ N, 90°36′ W. Height 2,552/1,000 m.
Complex strato volcano. Frequent explosions during 1565 – 1846. Active lava and ash eruptions from 1965 to the present.
Andesitic basalt (1961 lava), rhyodacitic dome (MCBIRNEY unpubl.), olivine basalt (ROSE 1967).
Andesite, rhyodacite.
Average SiO_2 values: 50.86, 70.30%. Average serial index: 2.39.
WILLIAMS (1960, pp. 48 – 54), TERMER (1964), BOHNENBERGER (1967 a), ROSE (1967), STOIBER & ROSE (1970), EGGERS (1972, 1978), EGGERS et al. (1976).

Eastern Guatemala (Figs. 40 and 125)

Tecuamburro 14°10′ N, 90°25′ W. Height 1,840/700 m.
Complex eroded strato volcano. Solfatara (Azufral).
Andesite, basalt (CARR 1974).
Plagidacite, leuco quartz-andesite.
Average SiO_2: 60.10%. Average serial index: 1.34.
MEYER-ABICH (1958, p. 67), CARR (1974, pp. 120 – 121).

Laguna de Ayarza 14°25′ N, 90°07′ W. Height 1,409/300 m.
Twin caldera. Plinian eruption and collapse between 5,000 and 10,000 years ago.
Rhyolite, rhyolitic-rhyodacitic pumice (WILLIAMS 1960, MCBIRNEY unpubl.).
Rhyodacite.
Average SiO_2: 70.10%. Average serial index: 1.98.
WILLIAMS (1960, pp. 54 – 57).

Moyuta 14°02′ N, 90°06′ W. Height 1,662/600 m.
Strato volcano with 3 lava cones. Extinct.
Andesite (WILLIAMS et al. 1964, CARR 1974).
Leuco quartz-latiandesite, plagidacite, dacite.
Average SiO_2 : 60.91%. Average serial index: 1.58.
WILLIAMS et al. (1964, pp. 28 – 29), CARR (1974, pp. 118 – 119).

Jumay 14°42′ N, 90°00′ W. Height 2,176/600 m.
Strato volcano. Extinct.
Porphyritic olivine-augite basalt (WILLIAMS et al. 1964), andesite (CARR 1974).
Andesite, quartz-latiandesite.
Average SiO_2: 54.80%. Average serial index: 2.17.
WILLIAMS et al. (1964, p. 25), CARR (1974, pp. 112 – 113).

Los Flores 14°18′ N, 90°00′ W. Height 1,600/600 m.
Strato volcano. Extinct.

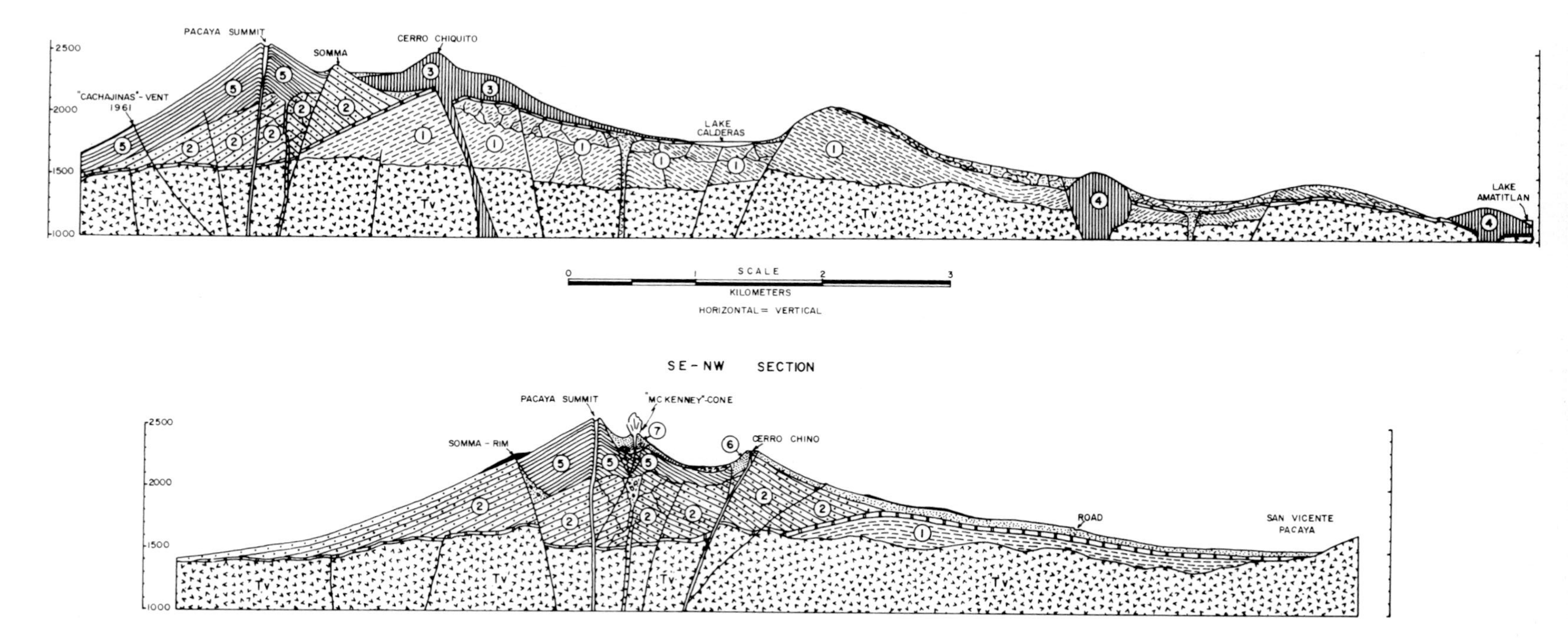

Fig. 124. Hypothetical sections through the Pacaya volcanic complex, Guatemala. (From BOHNENBERGER 1967 a).

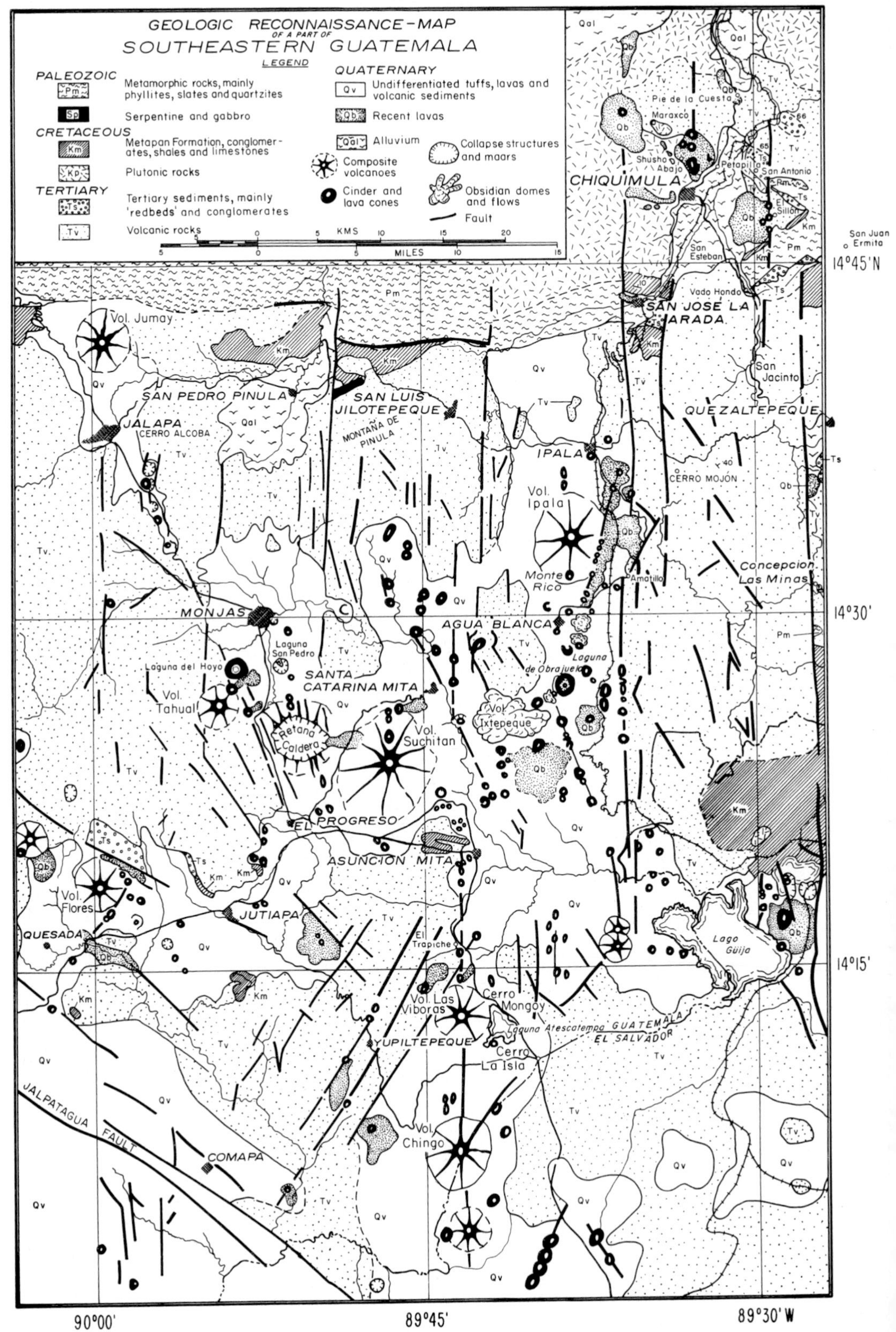

Fig. 125. Geological reconnaissance map of southeastern Guatemala. (From WILLIAMS, MCBIRNEY & DENGO 1964, Fig. 2).

Vitrophyric olivine-augite basalt (WILLIAMS et al. 1964), basalt (CARR 1974).
Andesite.
Average SiO_2: 53.20%. Average serial index: 2.27.
WILLIAMS et al. (1964, p. 24), CARR (1974, pp. 136 – 137).

Laguna de Retana 14°25′ N, 89°51′ W. Height 1,040/200 m.
Caldera. Extinct.
Olivine basalt, dacite (WILLIAMS et al. 1964), dacite (MCBIRNEY, unpubl.).
Dacite.
Average SiO_2 (dacite): 64.78%. Average serial index: 2.39.
WILLIAMS et al. (1964, pp. 29 – 32).

Suchitán 14°24′ N, 89°47′ W. Height 2,042/1,200 m.
Large eroded strato volcano. Extinct.
Basalt (WILLIAMS et al.), dacite (CARR 1974).
Leuco quartz-latiandesite.
Average SiO_2: 61.24%. Average serial index: 2.74.
WILLIAMS et al. (1964, pp. 25 – 26), CARR (1974, pp. 112 – 113).

Chingo 14°07′ N, 89°44′ W. Height 1,775/900 m.
Strato volcano. Extinct.
Olivine-augite basalt (WILLIAMS et al. 1964), basalt (CARR 1974).
Andesite, olivine andesite.
Average SiO_2: 52.60%. Average serial index: 2.57.
WILLIAMS et al. (1964, p. 29), CARR (1974, pp. 124 – 125).

Ixtepeque 14°25′ N, 89°41′ W. Height 1,292/500 m.
Large dome of obsidian. Extinct.
Rhyolite, black obsidian (CARR 1974).
Rhyolite.
Average SiO_2: 74.20%. Average serial index: 2.20.
WILLIAMS et al. (1964, pp. 38 – 39), CARR (1974, pp. 122 – 123).

Laguna de Obrajuela 14° 27′ N, 89° 39′ W. Height 878/100 m.
Large crater in dome. Extinct.
Pumiceous rhyolitic obsidian (WILLIAMS et al. 1964), basalt (CARR 1974), andesitic basalt (MCBIRNEY unpubl.).
Olivine andesite, quartz-latiandesite, rhyolite.
Average SiO_2 values: 54.15%, 71.25%. Average serial index: 2.04.
WILLIAMS et al. (1964, p. 43), CARR (1974, pp. 130 – 131).

Cerro Tempisque 14°30′ N, 89°38′ W. Height 1,050/150 m.
Lava dome. Extinct.
Rhyolitic obsidian (WILLIAMS et al. 1964, CARR 1974).
Rhyolitic obsidian.
Average SiO_2: 74.50%. Average serial index: 2.13.
WILLIAMS et al. (1964, p. 36), CARR (1974, pp. 122 – 123).

Ipala 14°33′ N, 80°38′ W. Height 1,650/800 m.
Large strato volcano with large crater lake. Extinct.
Olivine basalt (WILLIAMS et al. 1964), basalt (CARR 1974).
Tholeiitic hawaiite.
Average SiO_2: 51.43%. Average serial index: 2.51.
WILLIAMS et al. (1964, pp. 27 – 28), CARR (1974, pp. 132 – 133).

El Salvador

Sierra de Apaneca (Fig. 126)

Cerro Grande de Apaneca 13°50′ N, 89°49′ W. Height 1,854/900 m.
Strato volcano. Extinct.
Basalt (CARR 1974).
Andesite.
SiO_2: 52.00%. Serial index: 2.02.
CARR (1974, pp. 116 – 117).

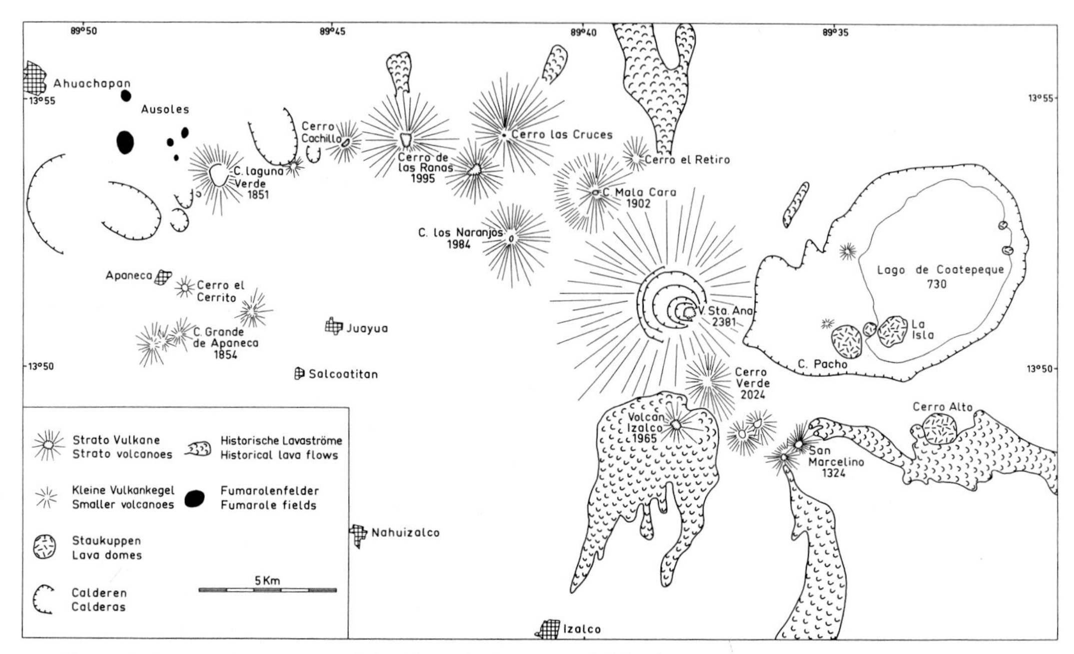

Fig. 126. Reconnaissance map of the Sierra de Apaneca, El Salvador.

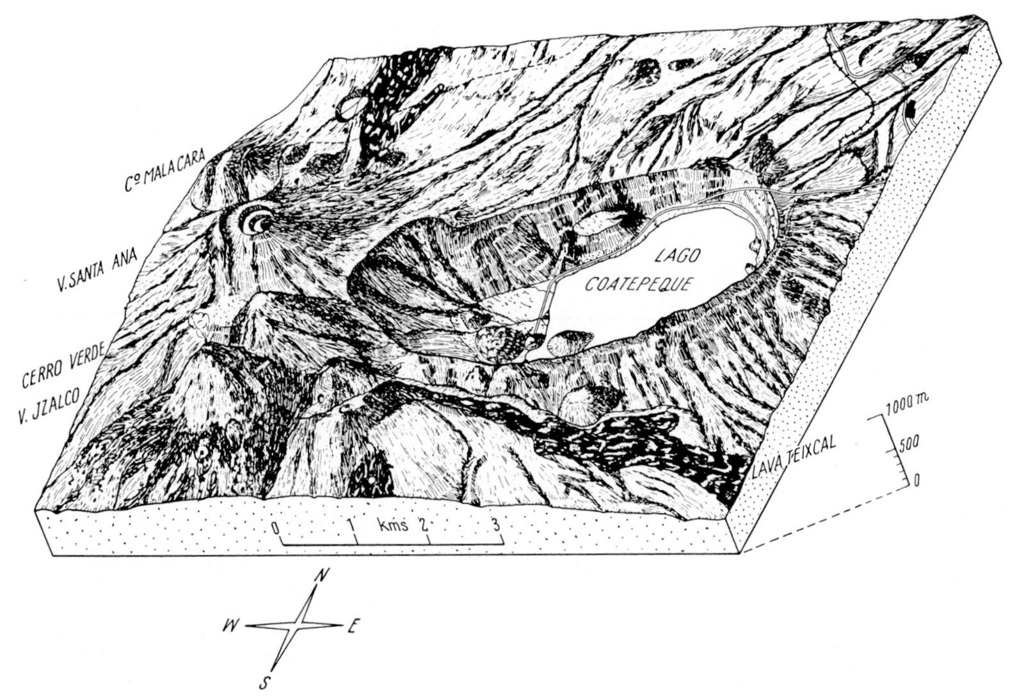

Fig. 127. Diagram of the Santa Ana and Izalco Volcanoes in the Lago de Coatepeque area of El Salvador. (From Meyer-Abich 1958, Fig. 16a).

Laguna Verde 13°54′ N, 89°47′ W. Height 1,851/900 m.
Strato volcano. Extinct.
Andesite (CARR 1974).
Quartz-latiandesite.
SiO_2: 57.80%. Serial index: 2.15.
CARR (1974, pp. 116 – 117).

Cerro Los Naranjos 13°52′ N, 89°41′ W. Height 1,984/900 m.
Strato volcano. Extinct.
Andesite, basalt (CARR 1974).
Andesite.
Average SiO_2: 53.10%. Average serial index: 2.37.
CARR (1974, pp. 116 – 117).

Santa Ana (Ilamatepec) (Fig. 127) 13°51′ N, 89°38′ W. Height 2,381/1,800 m.
Strato volcano. Ash eruptions in historic time, fumarolic activity recorded until 1955.
Olivine basalt, labradorite andesite (WEYL 1955), basalt, andesite (CARR 1974).
Quartz-andesite, leuco quartz-andesite.
Average SiO_2 values: 53.52%, 61.40%. Average serial index: 2.00.
WEYL (1955, p. 17), MEYER-ABICH (1956, pp. 44 – 46, 1958, pp. 71 – 74), CARR (1974, pp. 114 – 115), KILBRIDGE (1979).

Izalco (Figs. 127, 140, 141) 13°49′ N, 89°38′ W. Height 1,965/650 m.
Strato volcano. Formed in historic time since 1770 by very frequent eruptions.
Basalt, olivine basalt (ROSE & STOIBER 1969).
Andesite, quartz-andesite.
Average SiO_2: 52.61%. Average serial index: 1.87.
ROSE & STOIBER (1969), STOIBER et al. (1975), KILBRIDGE (1979).

Cerro Verde (Fig. 127) 13°50′ N, 89°37′ W. Height 2.024/600 m.
Strato volcano. Extinct.
Basalt (WEYL 1955), basalt (CARR 1974).
Tholeiitic hawaiite.
SiO_2: 53.05%. Serial index: 2.13.
WEYL (1955 p. 17), CARR (1974, pp. 114 – 115).

Lago de Coatepeque (Figs. 127, 142) 13°50′ – 13°53′ N, 89°31′ – 89°37′ W. Lake level 730 m.
Cauldron subsidence with small lava domes. Late Pleistocene Plinian eruption and collapse.
Original volcano: pyroxene andesite, dacite, olivine-bearing basalt and basaltic andesite. Explosion products: dacitic pumice (WILLIAMS & MEYER-ABICH 1955), rhyolitic and rhyodacitic pumice (MEYER 1964), Cerro Pacho lava dome: rhyolite (CARR 1974).
Quartz-latiandesite, rhyolite, rhyodacitic and *dacitic* pumice.
Average SiO_2: 54.10%, pumice 65.82%, Average serial index: 1.60, pumice 2.35.
WILLIAMS & MEYER-ABICH (1955, pp. 11 – 17), MEYER (1964), CARR (1974, pp. 114 – 115).

San Marcelino (Cerro Chino) 13°48′ N, 89°35′ W. Height 1,324/200 m.
Cinder cone with lava flows. Eruptions in 17th century and (?) 1722.
Labradorite andesite (WEYL 1955), andesite (CARR 1974).
Quartz-latiandesite.
SiO_2: 55.20%. Serial index: 2.75.
WEYL (1955, p. 17), MEYER-ABICH (1958, pp. 82 – 83), CARR (1974, pp. 114 – 115).

San Salvador (Boquerón, Quezaltepeque) (Fig. 148) 13°44′ N, 89°17′ W. Height 1,967/1,250 m.
Strato volcano. Several eruptions in historic time, lava flow and formation of a small cinder cone (Boqueroncito) in 1917.
Labradorite-andesite, basalt (WEYL 1955), andesite (MEYER-ABICH 1958), cristobalite-bearing andesite, hypersthene-augite andesite (MCBIRNEY unpubl.), calc-alkaline basalt and andesite, tholeiitic andesite (FAIRBROTHERS et al. 1978).
Quartz-latiandesite, leuco quartz-latiandesite, quartz-andesite, quartz-bearing andesite, quartz-tholeiite mugearite.
SiO_2: 51.7 – 59.0%. Average serial index: 2.35.
WEYL (1955, pp. 19 – 21), MEYER-ABICH (1958, pp. 83 – 88), FAIRBROTHERS et al. (1978).

Lago de Ilopango with Islas Quemadas 13°40′ N, 89°03′ W. Lake level 438 m.
Volcano-tectonic depression with lava dome. Plinian eruptions and collapse in Pleistocene and Holocene. Eruption of Islas Quemadas dome in 1879 – 1880.
Hornblende-hypersthene dacite (WEYL 1955), dacite pumice and dacite lava (WILLIAMS & MEYER-ABICH 1955).
Dacite, rhyodacite.
Average SiO_2: pumice 67.58%, dome 66.08%. Average serial index: pumice 1.48, dome 1.07.
WILLIAMS & MEYER-ABICH (1955, pp. 24 – 40), WEYL (1955, 1961 a, pp. 130 – 132), SHEETS (1976, 1977, 1979), STEEN-MCINTYRE (1976), TROTTER (1977).

San Vicente (Chichontepeque) 13°77′ N, 88°51′ W. Height 2,173/1,800 m.
Strato volcano. Fumaroles and solfataras.
Labradorite andesite (WEYL 1955), hypersthene-augite andesite (MCBIRNEY unpubl.).
Leuco quartz-andesite.
SiO_2: 59.65%. Serial index: 1.34.
WEYL (1955, pp. 23 – 24), MEYER-ABICH (1958, pp. 92 – 94).

Tecapa 13°30′ N, 88°33′ W. Height 1,592/600 m.
Strato volcano with crater lake. Fumaroles (El Tronador) and solfataras (Laguna de Alegría).
Basalt, olivine basalt (WEYL 1955), hypersthene basalt (MEYER-ABICH 1958), pyroxene basalt (MCBIRNEY unpubl.).
Olivine-bearing andesite.
SiO_2: 49.26%. Serial index 2.37.
WEYL (1955, p. 25), MEYER-ABICH (1958, pp. 94 – 95).

Usulután 13°25′ N, 88°28′ W. Height 1,453/1,200 m.
Strato volcano. Extinct.
Olivine-augite basalt (MCBIRNEY unpubl.).
Olivine andesite.
SiO_2: 48.54%. Serial index: 2.10
SAPPER (1913, p. 56).

Chinameca 13°28′ N, 88°20′ W. Height 1,228/500 m.
Strato volcano with caldera. Fumaroles and solfataras (Ausoles de Chinameca).
Pyroxene andesite (MCBIRNEY unpubl.).
Quartz-latiandesite.
SiO_2: 58.36%. Serial index: 1.79.
MEYER-ABICH (1958, p. 96).

San Miguel (Chaparrastique) 13°26′ N, 88°16′ W. Height 2,132/1,900 m.
Strato volcano. Frequent eruptions recorded since 1586.
Olivine-augite basalt, basaltic andesite (MEYER-ABICH 1958).
Andesite.
SiO_2: 51.01%, 47.20%. Serial index: 1.94.
MEYER-ABICH (1958, pp. 97 – 102), TAYLOR & STOIBER (1973), CARR & STOIBER (1974).

Conchagua 13°17′ N, 87°50′ W. Height 1,250/1,250 m.
Strato volcano. Extinct. Record of eruptions doubtful.
Olivine basalt (PUTNAM 1926), hypersthene andesite (MEYER-ABICH 1958), basalt (MCBIRNEY unpubl.).
Quartz-andesite.
SiO_2: 52.31%. Serial index: 1.50.
PUTNAM (1926, p. 817), MEYER-ABICH (1958, pp. 102 – 104).

San Diego 14°17′ N, 89° 29′ W. Height 780/300 m.
Cinder and lava cone. Extinct.
Olivine basalt (WEYL 1955), basalt (CARR 1974).
Tholeiitic olivine hawaiite.
Average SiO_2: 50.50%. Average serial index: 2.78.
WEYL (1955, pp. 15 – 16), CARR (1974, pp. 128 – 129).

Guazapa 13°54′ N, 89°07′ W. Height 1,438/1,000 m.
Eroded large strato volcano. Extinct.
Olivine basalt (BOHNENBERGER 1978).

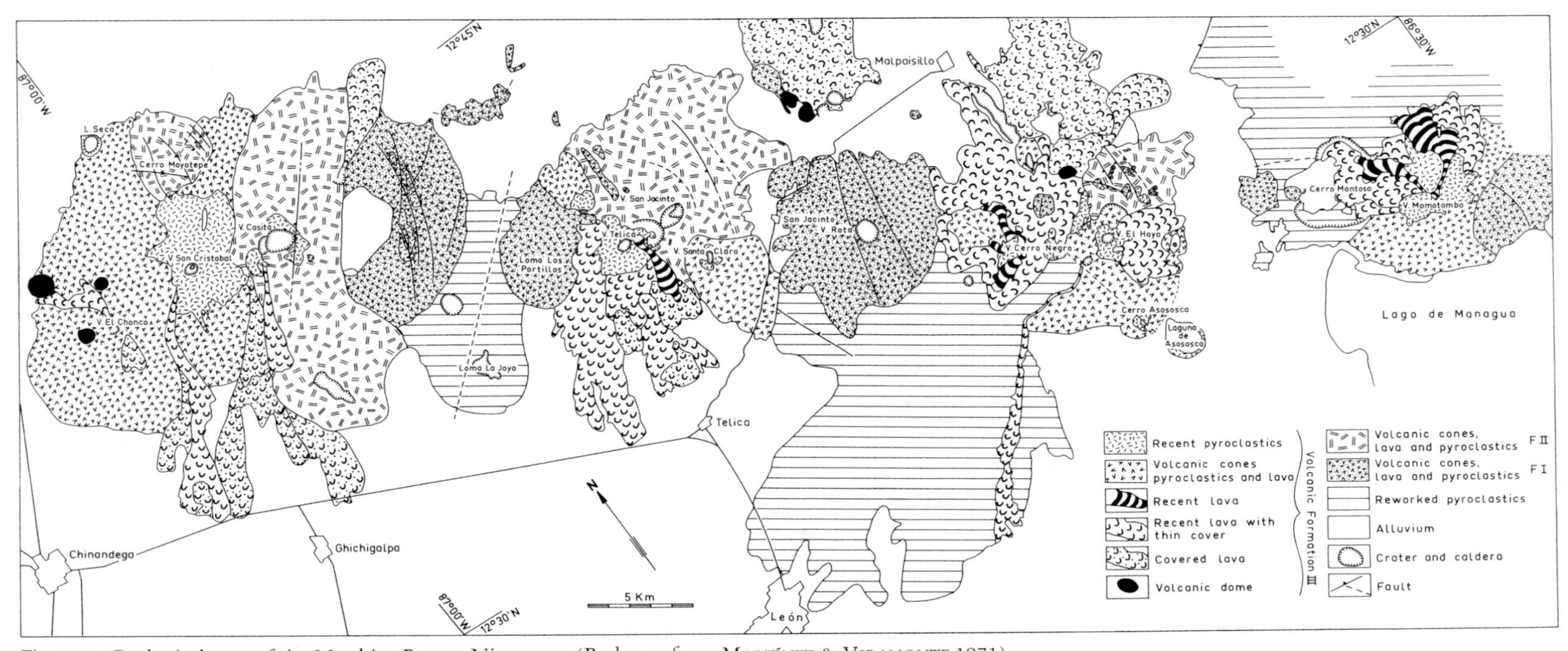

Fig. 128. Geological map of the Marabios Range, Nicaragua. (Redrawn from MARTÍNEZ & VIRAMONTE 1971).

Cacaguatique 13°45′ N, 88°12′ W. Height 1,500/1,000 m.
Large eroded strato volcano. Extinct.
Pyroxene basalt (BOHNENBERGER 1978).

Northwestern Nicaragua

Coseguina 12°59′ N, 87°34′ W. Height 859/859 m.
"Basaltic" shield with large central collapse crater. Great explosion of Krakatoan type in 1835.
Pyroxene basalt, olivine-bearing pyroxene basalt, andesitic bomb (WILLIAMS 1952)
Leuco quartz-andesitic bomb.
SiO_2: 58.84%. Serial index: 1.60
WILLIAMS (1952), MCBIRNEY (1958, pp. 109 – 111).

Cordillera de los Marabios (Fig. 128)

San Cristobal (El Viejo) 12°42′ N, 87°01′ W. Height 1,745/1,700 m.
Composite strato volcano. Ash ejection in 1971, strong gas emission in 1977.
Basalt, olivine basalt (MCBIRNEY & WILLIAMS 1965, MARTÍNEZ & VIRAMONTE 1971), olivine-augite andesite (MCBIRNEY unpubl.).
Andesite.
SiO_2: 50.75%. Serial index: 1.59.
MCBIRNEY & WILLIAMS (1965, pp. 29 – 30), MARTÍNEZ & VIRAMONTE (1971, p. 5).
Casita (Chichigalpa) 12°41′ N, 86°59′ W. Height 1,405/1,400 m.
Strato volcano. Active in 16th century.
Basalt, andesitic basalt (MCBIRNEY 1958), augite basalt (MARTÍNEZ & VIRAMONTE 1971), hypersthene-augite andesite (MCBIRNEY unpubl.).
Quartz-andesite, dacitic pumice.
SiO_2: 53.96%. Serial index: 1.93.
MCBIRNEY (1958, p. 114), MCBIRNEY & WILLIAMS (1965, p. 30), MARTÍNEZ & VIRAMONTE (1971, p. 5).
Telica 12°36′ N, 86°52′ W. Height 1,010/100 m.
Strato volcano. Active in 19th century, ash ejection in 1965 – 68.
Basalt, olivine basalt (BURRI & SONDER 1936, MCBIRNEY & WILLIAMS 1965, MARTÍNEZ & VIRAMONTE 1971).
Quartz-latiandesite.
SiO_2: 47.91%, 53.92%. Serial index: 1.62.
BURRI & SONDER (1936), MCBIRNEY & WILLIAMS (1965, p. 30), MARTÍNEZ & VIRAMONTE (1971, p. 6), CARR & STOIBER (1974, p. 329).
Santa Clara 12°34′ N, 86°49′ W. Height 834/800 m.
Strato volcano. Record of smoke eruption in 16th century doubtful.
Porphyritic basalt (MCBIRNEY & WILLIAMS 1965).
Andesite.
SiO_2: 48.36%. Serial index: 1.42.
MCBIRNEY (1958, pp. 117 – 118, Fig. 26), MCBIRNEY & WILLIAMS (1965, p. 38).
San Jacinto 12°37′ N, 86°49′ W. Height 900/800 m.
Strato volcano. Extinct.
Augite basalt, vitrophyric andesite (MARTÍNEZ & VIRAMONTE 1971).
Leuco andesite, dacite, rhyodacite.
SiO_2: 54.10%, 65.00%, 68.50%. Serial indices: 3.15, 1.64, 1.63.
MCBIRNEY (1958) (as synonym of Santa Clara volcano), MARTÍNEZ & VIRAMONTE (1971, p. 21).
Cerro Negro (El Nuevo, Las Pilas in early works) (Fig. 129) 12°31′ N, 86°42′ W. Height 675/500 m.
Cinder cone with lava fields. Born in 1850, erupted 6 times in 1867 – 1971.
Olivine basalt (BURRI & SONDER 1936, MARTÍNEZ & VIRAMONTE 1971), basalt (ROSE et al. 1974).
Olivine-bearing andesite, andesite, quartz-andesite, dacite.
SiO_2: 49.84 – 63.53%. Serial indices: 0.91 – 1.88.
BURRI & SONDER (1936), MCBIRNEY & WILLIAMS (1965, p. 31), VIRAMONTE & DI SCALA (1970), MARTÍNEZ & VIRAMONTE (1971, p. 20), VIRAMONTE (1973), STOIBER & ROSE (1973), TAYLOR & STOIBER (1973), ROSE et al. (1974).

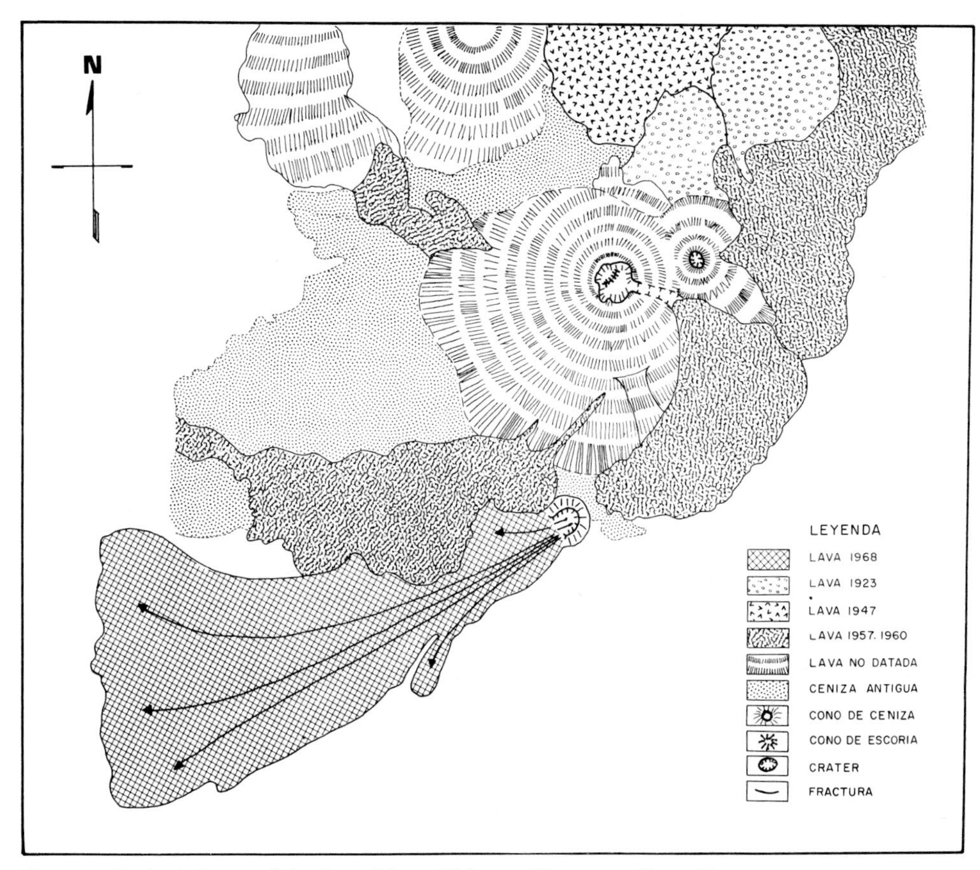

Fig. 129. Geological map of the Cerro Negro Volcano, Nicaragua. (From VIRAMONTE 1973).

El Hoyo (Las Pilas) 12°29′ N, 86°40 – 42′ W. Height 938/840 m.
Composite strato volcano. Eruptions in 1952 – 1954, fumaroles in 1977.
Basaltic lavas and pyroclastics, no analyses.
MCBIRNEY (1958, p. 123).

Laguna de Asososca 12°26′ N, 86°40′ W. Lake level 80 m.
Explosion-maar. Extinct.
Olivine basalt, augite-hypersthene andesite (BURRI & SONDER 1936).
Quartz-andesite, olivine-bearing andesite.
SiO_2: 47.18%, 55.95%. Serial indices: 0.92, 1.09.
BURRI & SONDER (1936).

Momotombo (Fig. 130) 12°25′ N, 86°32′ W. Height 1,191/1,900 m.
Strato volcano. Explosive eruptions since 16th century. Geothermal energy in exploration.
Hypersthene-olivine basalt (BURRI & SONDER 1936), dacite pumice (MCBIRNEY unpubl.).
Quartz-andesite, dacite pumice.
SiO_2: 52.90%, 66.47%. Average serial index: 1.72.
BURRI & SONDER (1936), MCBIRNEY & WILLIAMS (1965, pp. 31 – 32).

Apoyeque (Laguna de Apoyeque) 12°15′ N, 86°20′ W. Height 420/380 m.
Strato volcano with great collapse crater. Extinct.
Porphyritic dacite lava, dacite pumice (MCBIRNEY unpubl.).
Leuco quartz-latiandesite, dacite pumice.

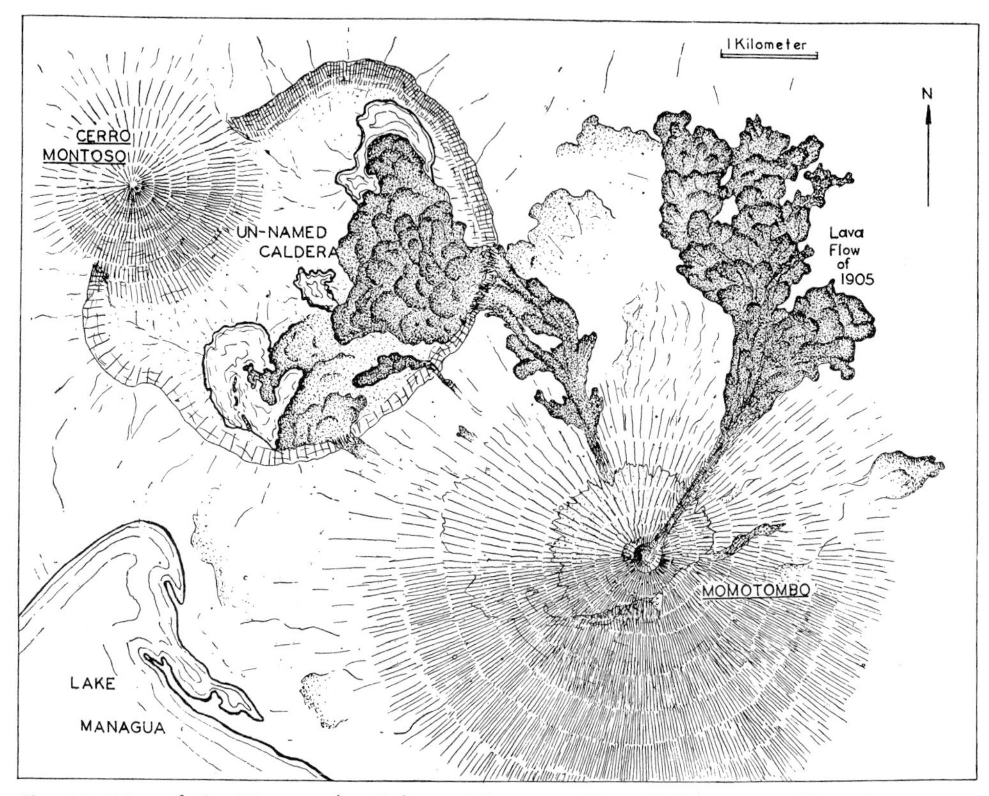

Fig. 130. Map of the Momotombo Volcano, Nicaragua. (From McBIRNEY 1958, Fig. 28).

Average SiO_2: 63.90%. Average serial index: 1.55.
McBIRNEY & WILLIAMS (1965, p. 32).

Southeastern Nicaragua (Fig. 131)

Masaya (Santiago, Nindiri, San Pedro) (Fig. 132) 11°57′ N, 86°09′ W. Height 635/430 m.
Caldera containing composite volcano. Frequent eruptions in historic time. Intra-caldera flows and intermittent appearance of lava lake (1970 – 77).
Olivine-augite basalt (McBIRNEY & WILLIAMS 1956), olivine basalt, augite-olivine basalt, hypersthene dacite pumice (UI 1973).
Tholeiitic basalt, tholeiitic hawaiite, olivine andesite, andesite, quartz-latiandesite, latiandesite.
Average SiO_2: 50.09%. Average serial index: 1.93.
McBIRNEY (1956), McBIRNEY & WILLIAMS (1965, p. 34), UI (1973), CARR & STOIBER (1974).

Apoyo 11°55′ N, 86°01′ W. Height 468/400 m.
Large caldera with lake.
Iddingsite-augite basalt, vitrophyric pyroxene dacite (McBIRNEY & WILLIAMS 1965), porphyritic olivine-bearing hypersthene-augite basaltic andesite, porphyritic augite-olivine basalt (UI 1973).
Andesite, quartz-andesite, leuco quartz-andesite, quartz-latiandesite, dacite, rhyodacite.
Averages SiO_2: *Andesite* 48.25%, *quartz-andesite* 53.90%, *dacite* 66.60%. Average serial index: 1.68.
McBIRNEY & WILLIAMS (1965, p. 61), UI (1973), McBIRNEY (unpubl.).

Mombacho 11°50′ N, 85°59′ W. Height 1,345/1,300 m.
Strato volcano. Active in 1560 and 1850.

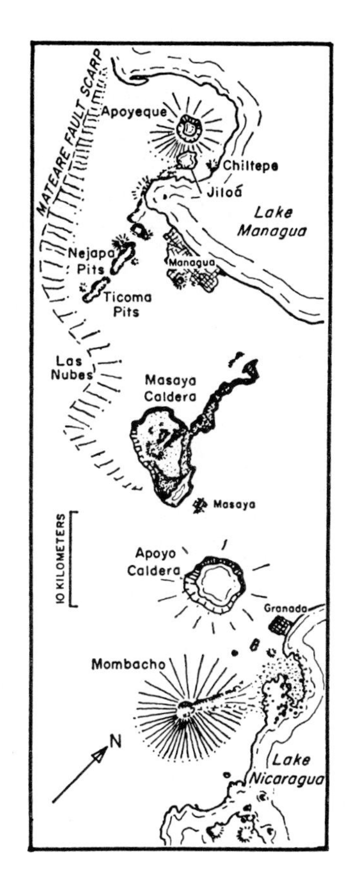

Fig. 131. Volcanoes between Lake Managua and Lake Nicaragua. (From McBIRNEY & WILLIAMS 1965, Fig. 7).

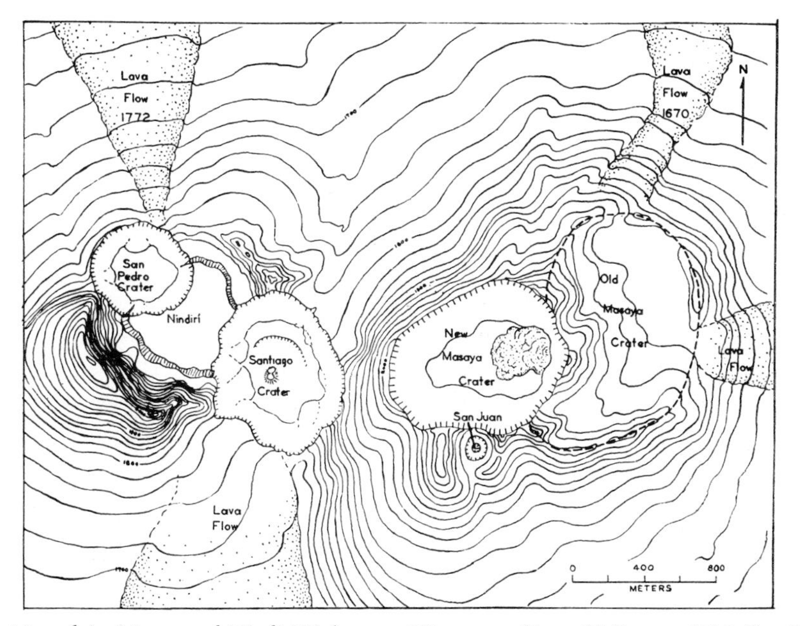

Fig. 132. Map of the Masaya and Nindiri Volcanoes, Nicaragua. (From McBIRNEY 1956, Fig. 2).

Fig. 133. Aerial photo of the Rincón de la Vieja Volcano, Cordillera de Guanacaste, Costa Rica. (Aerial photo by Instituto Geográfico Nacional de Costa Rica).

Olivine-bearing andesitic basalt, olivine-bearing vitrophyric hypersthene-augite andesite, dacite (McBirney & Williams 1965), basalt, andesitic basalt (McBirney unpubl.).
Tholeiitic basalt, quartz-andesite, dacite.
SiO_2: 47.92%, 55.69%, 66.76%. Average serial index: 1.50.
McBirney & Williams (1965, p. 34 – 35, 61).

Concepción (Omotepe) 11°32′ N, 85°37′ W. Height 1,610/1,600 m.
Strato volcano. Frequent gas and ash eruptions since 18th century, lava flow in 1957, ash eruptions in 1977.
Andesite, andesitic basalt, dacite (McBirney & Williams 1965).
Latiandesite, quartz-latiandesite, leuco quartz-latiandesite, dacitic pumice.
SiO_2: 51.90%, 57.90%, 64.06%. Average serial index: 2.53.
McBirney & Williams (1965, pp. 35 – 36, 38).

Northwestern Costa Rica

Cordillera de Guanacaste (Figs. 65, 66)

Orosi 10°59′ N, 85°28′ W. Height 1,440/1,100 m.
Strato volcano. Extinct.
Hypersthene-augite andesite (McBirney unpubl.).
Leuco quartz-andesite.
SiO_2: 50.17%. Serial index: 1.22.

Góngora 10°58′ N, 85°28′ W. Height 1,686/1,300 m.
Strato volcano. Extinct.
No data.

Cacao 10°56′ N, 85°27′ W. Height 1,659/1,200 m.
Strato volcano. Extinct.
No data.

Fig. 134. Aerial photograph of the Arenal Volcano, Costa Rica. In the centre of the picture is the lateral crater which formed in 1968, and below that the lava flows of the years 1968 to 1975. (Aerial photograph by the Instituto Geográfico Nacional de Costa Rica, 9.4.1978).

Rincón de la Vieja (Fig. 133) 10°49′ N, 85°20′ W. Height 1,806/1,600 m.
Composite large strato volcano with 6 craters. Sporadic ash eruptions in 1860, 1863, 1966 – 1967, 1969.
Dacite, labradorite rhyodacite (WEYL 1969), hypersthene-augite-andesite (MCBIRNEY unpubl.).
Quartz-latiandesite.
Average SiO_2: 57.20%. Average serial index: 1.58.
P. FERNÁNDEZ (1966, 1967, 1968), WEYL (1969 c, p. 438).

Miravalles 10°45′ N, 85°09′ W. Height 2,028/1,600 m.
Composite strato volcano. Fumaroles (Los Hornillos).
Hypersthene-augite andesite (MCBIRNEY unpubl.).
Quartz-latiandesite.
SiO_2: 58.87%. Serial index: 1.58.
SAPPER (1913, pp. 110 – 111, 159).

Tenorio 10°40′ N, 85°01′ W. Height 1,916/1,500 m.
Strato volcano. Extinct.
Olivine-bearing hypersthene-augite andesite (MCBIRNEY unpubl.).
Leuco quartz-andesite.
SiO_2: 58.77%. Serial index: 1.13.

Arenal (Fig. 134) 10°28′ N, 84°42′ W. Height 1,633/1,000 m.
Strato volcano. Active in 15th century, strong eruptions since 1968, glowing avalanches, ash and lava eruptions.
Andesite (MELSON & SÁENZ 1974).
Quartz-andesite, leuco quartz-andesite.
Average SiO_2: 55.00%. Average serial index: 1.38.
CHAVES (1969), MELSON & SÁENZ (1968, 1974), MINAKAMI & UTIBORI (1969), FUDALI & MELSON (1972), MATUMOTO & UMAÑA (1975), Anonymous (1974 d).

Fila de Aguacate

Cerro Pelón (882 m)
Cerro Tinajita (925 m)
Cerro Mondongo (1,020 m)
Between
9°59′ and 10°01′ N,
84°30′ and 84°31′ W.
Presumably extinct and eroded volcanic edifices.
Latite, dacite, ignimbrites.
BERGOEING, MALAVASSI & PROTTI (1979).

Cordillera Central (Figs. 65, 66)

Viejo 10°15′ N, 84°19′ W. Height 2,060/1,000 m.
Composite strato volcano. Extinct.
No data.

Póas 10°12′ N, 84°14′ W. Height 2,704/1,700 m.
Composite strato volcano with collapse crater. Ash and steam eruptions since 1838.
Quartz-latiandesite, rhyodacite.
Average SiO_2: 55.80%. Average serial index: 1.99.
KRUSHENSKY & ESCALANTE (1967), P. FERNÁNDEZ (1968), BENETT & RACCINICHI (1978).

Barba (Barva) 10°08′ N, 84°06′ W. Height 2,906/1,800 m.
Composite strato volcano. Ash eruption in 1867.
Hypersthene-augite andesite (MCBIRNEY unpubl.).
Quartz-latiandesite.
WILLIAMS (1952, pp. 163 – 164).

Irazú (Volcán de Cartago) (Fig. 135) 9°59′ N, 83°51′ W. Height 3,432/2,000 m.
Composite large strato volcano. Ash eruptions in 1920 and 1963 – 1965.
Quartz-latiandesite, quarz-latite, dacite.

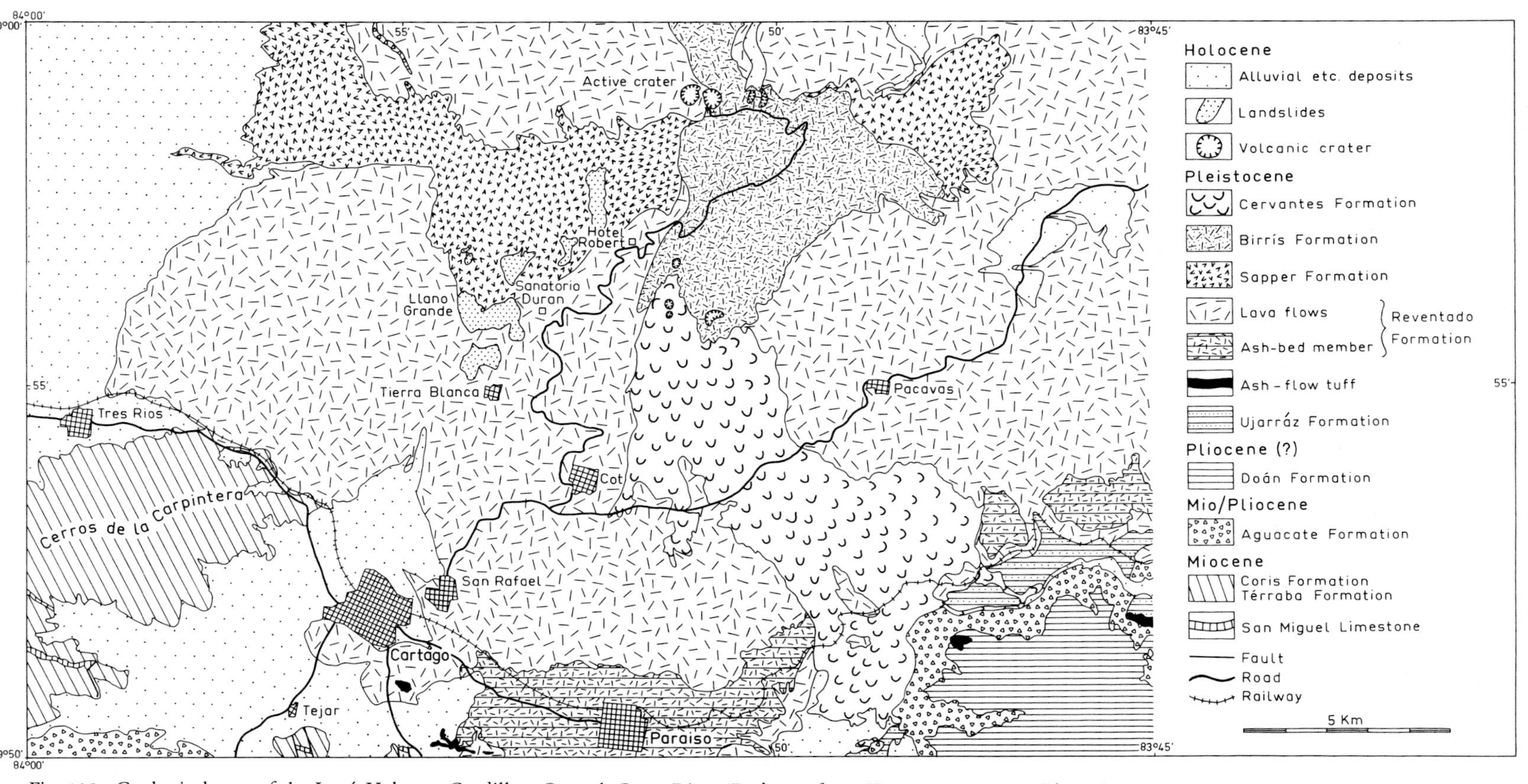

Fig. 135. Geological map of the Irazú Volcano, Cordillera Central, Costa Rica. (Redrawn from KRUSHENSKY 1972, Plate 2).

Average SiO_2: 56.00%. Average serial index: 2.70.
MURATA, DÓNDOLI & SÁENZ (1966), KRUSHENSKY & ESCALANTE (1967), WALDRON (1967), KRUSHENSKY (1972, 1973), KESEL (1973), ALLEGRE & CONDMINES (1976).
Turrialba 10°01′ N, 83°46′ W. Height 3,339/1,900 m.
Composite large strato volcano. Active in 18th and 19th century.
Basalt (PUTNAM 1926), hypersthene-olivine augite basalt (MCBIRNEY unpubl.).
Andesite, rhyodacite.
Average SiO_2: 51.88%. Average serial index: 2.44.
PUTNAM (1926, p. 817), MCBIRNEY (1958, pp. 144 – 146).

Panama (Figs. 89, 90)

Chiriquí (Barú) 8°49′ N, 82°33′ W. Height 3,475/2,500 m.
Large strato volcano. Thermal springs (Cerro Pando and Los Hornitos).
Andesite, basalt (SAPPER 1913).
MÉRIDA (1973).
El Valle 8°35′ N, 80°10′ W. Height 1,173/1,000 m.
Shield volcano with caldera.
No data.

Volcanoes in the hinterland of the Pacific Chain

Honduras (Figs. 58, 120)

Volcanoes on the high plateau near Tegucigalpa.
Basaltic shields and detached caps.
Quaternary, partly (?) Tertiary.
Olivine-bearing tholeiite basalt (WILLIAMS & MCBIRNEY 1969).
Tholeiitic hawaiite, tholeiitic mugearite.
Average SiO_2: 51.12%. Average serial index: 2.03.
WILLIAMS & MCBIRNEY (1969, pp. 65 – 70), Mapa Geológico de Honduras 1 : 50,000, Hoja 2757 IG San Buenaventura, Hoja 2756 IIG Tegucigalpa, Hoja 2758 IVG Zambrano.
Lake Yojoa volcanic field
Extinct scoria cones, collapse pits, and lava fields.
Olivine basalt, trachyte, ferro-augite trachyte, ferro augite-fayalite trachyte (WILLIAMS & MCBIRNEY 1969), olivine-augite basalt, olivine-titanaugite basalt with 5 – 10% nepheline normative (MERTZMAN 1976).
Alkali olivine hawaiite, latite, trachyte.
Average SiO_2: 46.98%, 59.35%. Average serial index: 6.56.
WILLIAMS & MCBIRNEY (1969, pp. 70 – 71), MERTZMAN (1976).
Isla Utila (Gulf of Honduras) 16°05′ N, 86°43′ W. Height 50 m.
Small lava and tuff cones. Extinct.
Olivine-augite alkaline basalt, sideromelane tuff (WILLIAMS & MCBIRNEY 1969).
Alkalic olivine hawaiite.
SiO_2: 47.31%. Serial index 6.23.
MCBIRNEY & BASS (1969 a), WILLIAMS & MCBIRNEY (1969, pp. 71 – 72).

Nicaragua (Figs. 62, 120)

Volcanoes near the northeastern side of the Nicaragua Depression
Cerro el Ciguatepec 12°32′ N, 86°09′ W. Height 603/500 m.
Lava and scoria cone. Extinct.
Porphyritic olivine-augite basalt (MCBIRNEY & WILLIAMS 1965).
Qaurtz-tholeiite hawaiite.
SiO_2: 51.75%. Serial index: 1.38.
MCBIRNEY & WILLIAMS (1965, p. 25).
Cerro San Jacinto 12°22′ N, 86°01′ W. Height 300/200 m.
Small strato volcano. Extinct.

Basalt, pyroxene andesite, pyroxene dacite (McBirney & Williams 1965).
Quartz-latiandesite.
SiO_2: 58.95%. Serial index: 2.09.
McBirney & Williams (1965, p. 25 – 26).

Las Lajas Caldera 12°18′ N, 85°44′ W. Height 686/400 m.
Broad lava shield with large caldera. Extinct.
Olivine-augite-pigeonite basalt, andesitic basalt, dacite (McBirney & Williams 1965).
Andesite, quartz-andesite, dacite.
SiO_2: 50.40%, 55.10%, 62.96%. Average serial index: 1.55.
McBirney & Williams (1965, pp. 26 – 28).

Pearl Lagoon, Atlantic Coastal Plain. Approximately 12°42′ N, 83°45′ W. Height approximately 110 – 130 m.
Small isolated cluster of cones with well-defined craters. Extinct.
Scoriaceous alkali basalt (McBirney & Williams 1965).
Alkali-olivine mugearite.
SiO_2: 45.89%. Serial index: 9.04.
McBirney & Williams (1965, p. 28).

Costa Rica (Figs. 66, 120)

Cerro Tortuguero 10°25′ N, 83°32′ W. Height 119/119 m.
Small cone of lapillis and bombs. Extinct.
Olivine-augite basalt (Tournon 1973).
Olivine hawaiite.
SiO_2: 46.78%. Serial index 3.94.
Tournon (1973, pp. 140 – 147).

The following data relate chiefly to volcanic events in the last two centuries and to new research results, where they were available.

The **volcanic chain of central Guatemala,** with its more than 3,000 m high mountains and the cauldron subsidences of the Lago de Amatitlán and Atitlán at their feet, forms a landscape of extraordinary beauty (Fig. 123). In addition, the chain includes three of the most active volcanoes in Central America, namely Santa Maria/Santiaguito, Fuego and Pacaya. Paleomagnetic measurements, which can be roughly correlated with data from cores in the Gulf

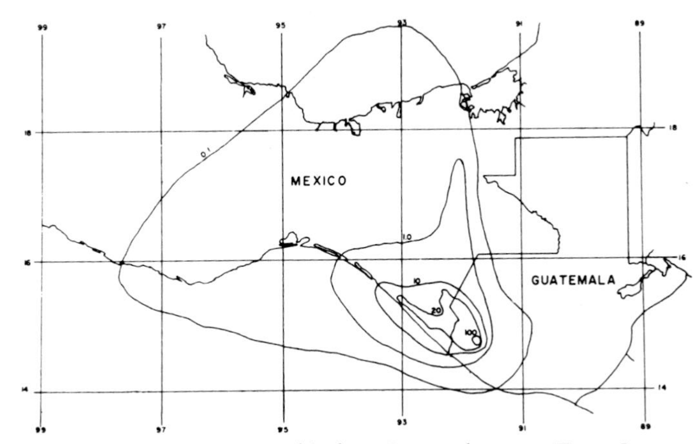

Fig. 136. Isopach map of Santa Maria's ash blanket, Guatemala 1902. (From Sapper 1905 and Rose 1972, Fig. 2).

of Mexico, and lake sediments near Tlapacoya, Mexico, suggest that the **Santa Maria Volcano** was probably built up over a period of about 30,000 years. The volcano had been in repose for at least several hundred years prior to 1902. In 1902, a catastrophic eruption out of the SW flank of Santa Maria produced an estimated 5.5 km^3 of pyroclastic dacite debris (Fig. 136). In 1922, **Santiaguito,** an endogenous volcanic dome, started to grow in the explosion crater (Fig. 122) and continues to grow to this day. Brief lava flows and small eruptions of nuées ardentes as well as intensive fumarolic activity are linked with this process.

Both the eruption of Santa Maria in 1922 and the extrusion of Santiaguito sparked a number of studies, in particular those by ROSE and STOIBER. In 1977, ROSE et al. summarized their geological, paleomagnetic, petrological and geochemical studies in a description of the development of Santa Maria: "The basaltic andesites and associated pyroclastics, which make up this volcanic succession and comprise the latter 40% of the volcano's eruptive output, show systematic increases in SiO_2 , K_2O, Rb, and Zr, and decreases in MgO and CaO, up the succession. These evolutionary changes anticipate the dacites which first appeared after a period of repose in Santa Maria, first as pyroclastic debris in the catastrophic flank eruption of 1902, and later as the endogenous dacite dome of Santiaguito which appeared in 1922 and has continued to grow to the present day. The increasing effects of crystal fractionation in the latter part of Santa Maria's history, as reflected in the compositional changes in the basaltic andesites, and its culmination in the Santiaguito dacites, is thought to be due to the inhibiting effect which increasing cone height has had on the volcano's eruptive frequency through time, so progressively increasing the lengths of the repose periods when fractionation could pro-

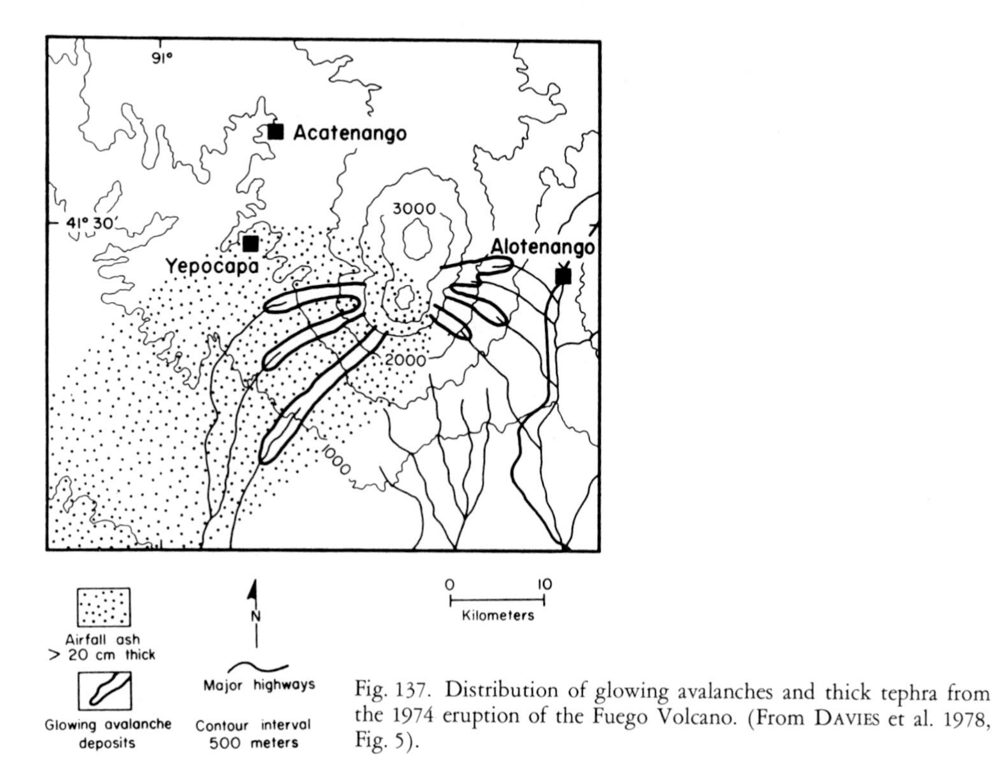

Fig. 137. Distribution of glowing avalanches and thick tephra from the 1974 eruption of the Fuego Volcano. (From DAVIES et al. 1978, Fig. 5).

ceed undisturbed in the top of the feeding magma column. Data from other high composite cones in Guatemala suggest that 3,500 – 4,200 metres represents a threshold cone height above which basaltic andesite is not erupted in this region, and it is believed that this limitation led to the historically documented dormancy of Santa Maria during which the Santiaguito dacites evolved."

While it is true that geological studies were caried out in connection with the engineering of a hydro-electric installation in the region of the **Lago de Atitlán** (Fig. 123), over which tower the volcanoes of Toliman, Atitlán and San Pedro, no new publications have appeared since WILLIAMS (1960, pp. 16 – 19). The N 10° E-orientation of Atitlán and Toliman is of interest with regard to the tectonic situation of the volcanoes.

The volcanoes of **Fuego** and **Acatenango** are oriented in the same direction; Acatenango experienced an ash eruption in 1972 and Fuego has had several violent ash eruptions since 1962. The most violent of the latter occurred in September 1971, when an ash layer up to 30 cm thick covered the immediate surroundings. The ash layer was still 1 cm thick up to 160 km west of the volcano. Glowing avalanches filled barrancos up to 20 m deep. The ashes were studied by ROSE et al. (1974), and the socio-economic consequences of the eruption, particularly on agriculture, were examined by BONIS & SALAZAR (1974). Further eruptions of glowing avalanches took place in 1974, and they were the subject of a detailed analysis by DAVIES et al. (1978) (Fig. 137).

Thick covers of **dacitic and rhyodacitic pumice** (Fig. 138), which was deposited sometimes as tephra and sometimes in the form of glowing avalanches and was then in some cases re-distributed by fluviatile action, are characteristic of the volcanic highlands of Guatemala and the intramontane basins that they contain. WILLIAMS (1960) provided a first general de-

Fig. 138. Roadside exposure of unbedded pumice (glowing avalanche deposits). Road from Paztún to Panajachel in the highlands of Guatemala. Photo: R. WEYL.

scription of the distribution and formation of these pyroclastic rocks, most of which he regarded as having been brought from elsewhere, although their origin could not be satisfactorily settled. Initial ^{14}C datings on charcoal embedded in the rocks gave ages of at least 16,000 or 2,340 years. A later dating (BONIS et al. 1966) yielded a figure of 31,000 or 35,000 years.

On the basis of megascopic characteristics, the proportion of basic phenocrysts and the stratigraphic position, detailed studies by KOCH & MCLEAN (1975) distinguish between 26 Pleistocene tephra layers, 4 ash-flow tuffs, 4 layers of fluviolacustrine sediments and countless buried soils. It was found that the tephra originated from the volcanoes of Pacaya, Agua, Acatenango/Fuego, the region of the Lago de Atitlán and the Laguna de Ayarza (Fig. 139).

The thickness of the entire sequence fluctuates between more than 30 m in the highlands and in the vicinity of the source volcanoes and over 100 m in the basins. The individual tephra layers cover 600 to 7,500 km², and the largest incandescent tuff unit extends over at least 16,000 km². It extends into the region of Cobán and Huehuetenango. The tephra was for the most part carried northwards, i.e. counter to the prevailing wind direction. This coincides with the northward transportation of the Coatepeque pumices of El Salvador, which was discovered by MEYER (1964) (Fig. 142), and it may be possible to find the explanation in the corresponding direction of the upper atmosphere winds. Such an interpretation is supported by the transportation of the Santa Maria ashes in 1902 and by observations made of ash eruptions from Fuego and Poás which extended very high into the atmosphere. At lower altitudes their ash was transported southwards by the trade winds while the direction was reversed at higher altitudes.

The age of the eruptions has been determined by recent ^{14}C datings on carbon and K/Ar datings on hornblendes, biotites and sanidines as between 40,000 years and 1,84 m.y. KOCH & MCLEAN point out that the most recent ash flow is only 40,000 years old and that the recurrence interval of such eruptions is not known. A repeat of this eruptive event is therefore entirely possible and it would have catastrophic consequences.

Ashes and pumice ejecta from Central American volcanoes have been detected in deep-sea drilling operations far out into the Pacific Ocean, and the largest occurrences cover areas of 300,000 and 400,000 km² (BOWLES et al. 1973). Most, if not all, of the ash layers are not older than 300,000 years. On the basis of their trace element content, individual layers may be assigned to eruptions in Guatemala (Tecpán) and El Salvador (Ilopango).

DONLEY (1971) published a geomorphological study of the pumice landscapes of Guatemala.

The **Pacaya** is a complex, severely faulted volcanic edifice to the south of Guatemala City and thus in the area of the N/S-trending Guatemala Graben. Andesitic to rhyolitic domes, andesitic strato volcanoes, basaltic cinder cones and thick layers of dacitic pumice are involved in its structure (EGGERS 1972). Two profiles taken from a survey by BOHNENBERGER (1967 a) in Fig. 124 and the map sheet for Amatitlán (2059 IIG) explain the structure. A period of violent activity with the production of olivine-bearing porphyric andesitic/basaltic lavas (BOHNENBERGER et al. 1966, p. 30) commenced in 1961. In 1962 the western flank below the summit collapsed. A new crater formed in the subsidence and its eruptions were of the Strombolian type, at first violent, later moderating, and continuing at that level through to the present day.

The basin of the **Lago de Amatitlán**, which lies ahead of Pacaya to the north, must be interpreted as a volcano-tectonic depression, which is probably related with the structure of the edifice of Pacaya.

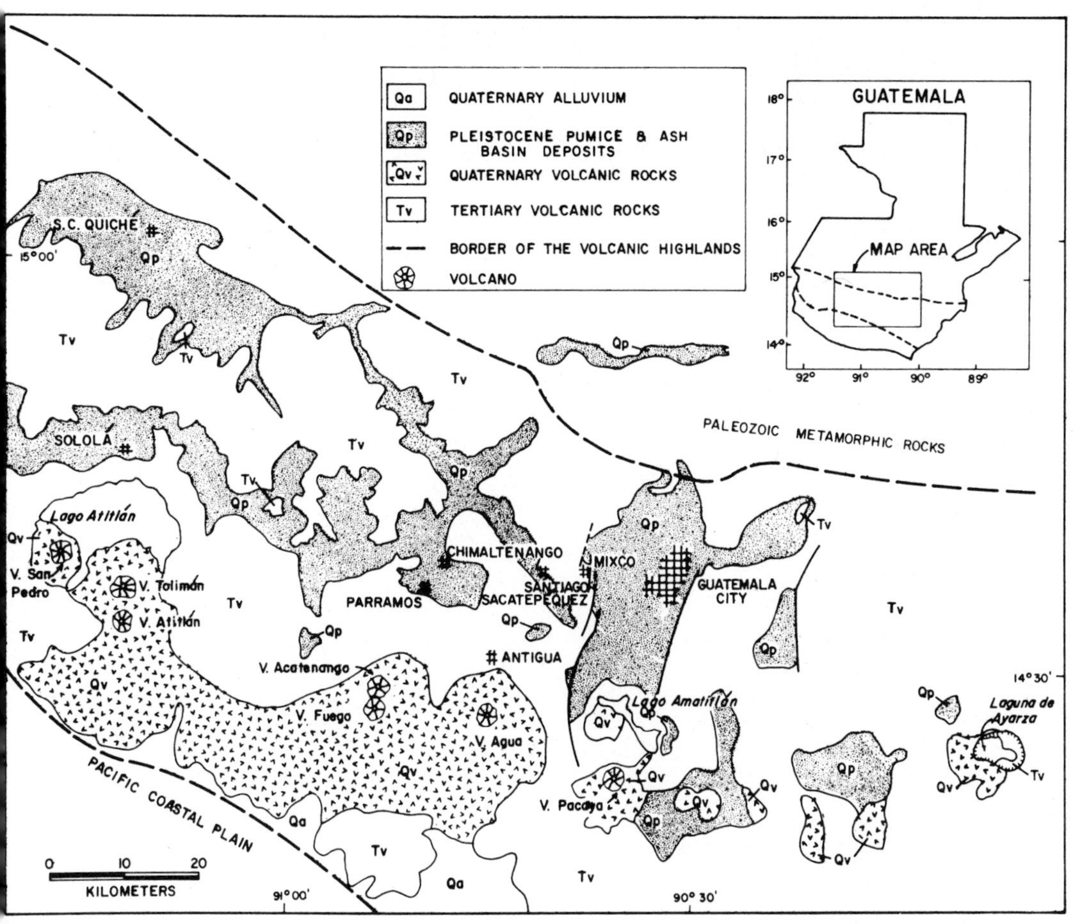

Fig. 139. Generalized geological map of the volcanic highlands of Guatemala. (From KOCH & MCLEAN 1975, Fig. 1).

The Quaternary volcanism of **southeastern Guatemala** differs noticeably from that of the highland (WILLIAMS, MCBIRNEY & DENGO 1964, CARR 1974). Active volcanoes are lacking. The large extinguished volcanoes consists mainly of phenobasalts, small basaltic scoria cones are frequent, but pumice is rare. In its place we find flows and domes of rhyolitic obsidian, which in turn is lacking in the highlands of Guatemala. The eruption centres are linked to N/S-trending faults (Fig. 125) in contrast to the usual arrangement of the volcanoes along WNW/ESE-striking lines. The association of basaltic and rhyolitic products is unique in Central America and it is regarded as a late eruption stage with extreme fractional crystallization differentiation in basaltic magma chambers.

The volcanoes, which are arranged in a semicircular pattern in the western part of **El Salvador,** are grouped together as the **Sierra de Apaneca** (Fig. 126). WILLIAMS & MEYER-ABICH (1955) and KILBRIDGE (1979) gave a detailed description of them. The historically attested eruptive activity has shifted to the eastern section of the group of volcanoes where, in

Fig. 140. Eruption phases of the Izalco Volcano, El Salvador, from its birth in 1770 up to 1968. Black areas indicate times of recorded eruptions. (From ROSE & STOIBER 1969, Fig. 1).

particular, Izalco — which erupted in 1770 — has been especially active (Fig. 140). Following a period of repose which commenced in 1957, a brief lava eruption occurred on the south-eastern side in October/November 1966, and this event was described by ROSE & STOIBER (1969). They noted that fumaroles in the crater region cooled down at an annual rate of 18 °C before and 45 °C following the eruption, and that their content of F and Cl rapidly declined between 1965 and 1974 (Fig. 141). The lavas from Izalco are *andesites* and *quartz-bearing andesites* [13].

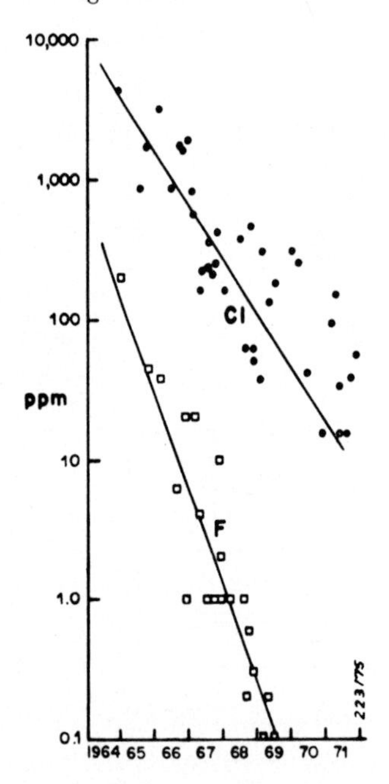

Fig. 141. Plot showing the changes in the Cl and F concentrations in condensates from the fumaroles of Izalco, from 1965–1972. (From STOIBER et al. 1975, Fig. 10).

[13] In a recently submitted M. A. thesis KILBRIDGE (1979) summarizes the magmatic evolution of Izalco Volcano and its relation to the Santa Ana Complex: "Major and trace element chemistry, petrography, least squares mixing analysis and trace element calculations of Santa Ana Complex lava suggest that they are products of shallow level crystal fractionation. Thirty-one basalts and andesites from Izalco and Santa Ana volcanoes were analysed for 9 major elements, total water and 7 trace elements. The Santa Ana basaltic parent magma has a higher concentration of K_2O, TiO_2, Ba, Sr, Zr and Rb than the Izalco parent. The volcanoes fractionate similar proportions of plagioclase, augite, olivine, magnetite and apatite. Silica concentration in Izalco decreases with time. This trend may be the result of tapping successively deeper levels in a zoned magma chamber, with silica higher toward the top of the chamber. Periodic influx of basaltic magma which mixes with leftover fractionated magma in the chamber which then erupts is an alternative mechanism. The two parent magmas for Izalco and Santa Ana volcanoes may be sequential melts from the same source."

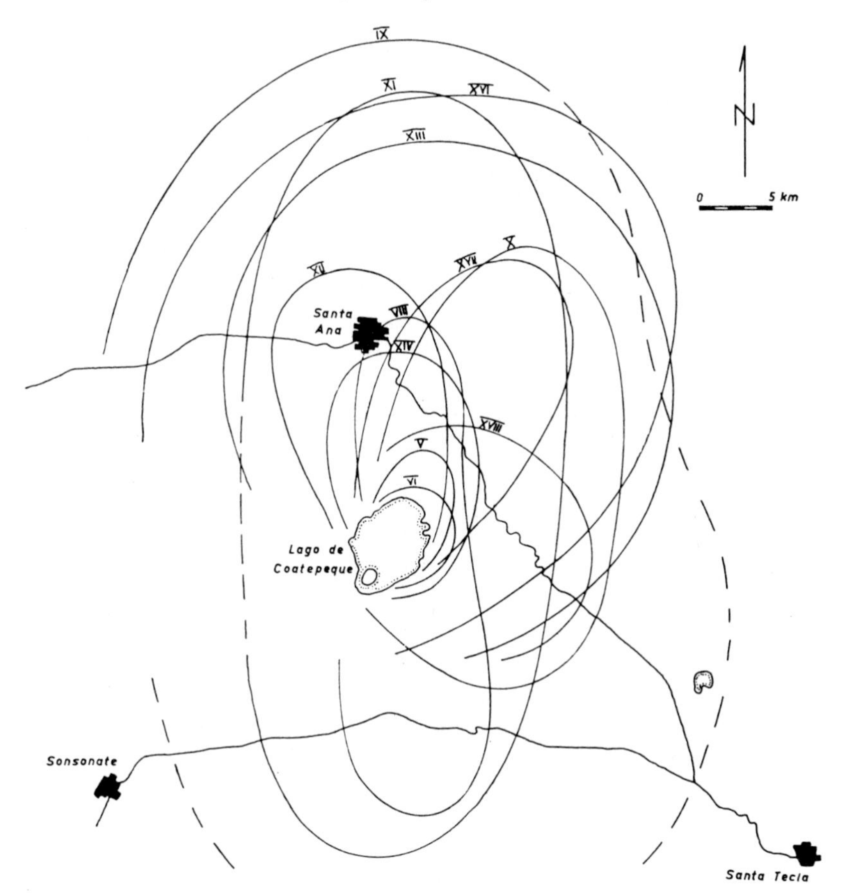

Fig. 142. 3m-isopach map of the pumice blankets on Coatepeque, El Salvador. (From MEYER 1964, Fig. 8).

The fumarole field "Salitral", from which steam has been extracted for several years to generate electrical power, is situated on the northern flank of the Sierra de Apaneca (see Chapter 9.2).

On the northern side of the volcano of Santa Ana lies the large caldera of the **Lago de Coatepeque,** whose border is built up of a series of basic rocks (Fig. 127). This series is overlain by thick *rhyodacitic* to *rhyolitic* pumices, which were laid down as tephra and from nuées ardentes, and their age was determined by MEYER (1964) at around 45,000 years. On the basis of the analysis of heavy minerals the same author was able to classify the pumices stratigraphically, to identify the predominantly northwards and eastwards oriented tephra fans (Fig. 142) and to separate them from the pumices of the Empalizada Volcano to the west. A quantitative calculation revealed that the volume of pumice masses that have been erupted is in the order of 80 km^3, which is sufficient to explain the caldera collapse with a mass deficit of approximately 50 km^3. Following the collapse, small basic ash and lava cones and acid domes were extruded. Some warm springs at the base of the caldera are the last vestiges of volcanic activity.

As a young volcano-tectonic depression, the basin of the **Lago de Ilopango,** which is situated immediately to the east of the capital city, has been the object of much attention since the journey made by DOLLFUS & MONT-SERRAT (1868), and it has been re-interpreted in modern terms by WILLIAMS & MEYER-ABICH (1953, 1955); this interpretation was summarized in detail in the first (German) edition of this book. Gigantic eruptions of *dacitic* pumice resulted in the gradual collapse of the depression. The pumices, which are referred to as "Tierra blanca" in local parlance, are separated by soil horizons and differ from each other through their contents of basic silicates (hornblende, orthoaugite, clinoaugite) (WEYL 1954).

The uppermost layers of the Tierra blanca cover countless remains of the protoclassic Maya civilization which point to dense settlement of the region. Archaeological investigations have dated these horizons at about 2,000 years BP. They were destroyed around 260 A.D. by a catastrophic eruption of Ilopango, and the survivors were forced to move north. This catastrophe explains the sudden appearance of styles of artifact in the northern Maya regions which previously had been associated with the region of El Salvador (SHEETS 1976, 1977, 1979, TROTTER 1977).

Field studies and petrographic investigations (STEEN-MCINTYRE 1976) revealed that the eruption began with ash flows which spread over a distance of 40 km. These were quickly followed by eruptions of airborne tephra which extended into the western part of El Salvador and destroyed a region about 3,000 km^2 in extent. Both units have petrographic properties similar to the lower Ilopango pumice and different from the Coatepeque.

The eruptions of the Ilopango Volcano occurred repeatedly in prehistoric times at intervals of several thousands of years. In 1880 a *dacitic* dome was extruded in the middle of the lake. Thus the possibility still exists that major pumice eruptions may be repeated in the future as well. Given the dense population of El Salvador and the close vicinity of the capital city, such outbreaks would have very serious consequences.

In the course of further archaeological investigations, later tephra eruptions from smaller volcanoes were discovered and dated by radiocarbon techniques [14]. However, these eruptions had only local effects on the people living in the vicinity. The volcanic products in question are:

the tephra of the volcano of Laguna Caldera (approx. 15 km NNW of the capital city around 510 ± 135 A.D.),
the San Andrés tuff in the Zapotitán Basin from a flank eruption of the volcano of Boquerón around 800 – 1,100 A.D. [15],
the El Playon tephra from the year 1658.

Honduras, with only a few small extinct volcanoes on the Gulf of Fonseca, is linked with the volcanic chain of the Pacific border. On the other hand, Pleistocene to Holocene volcanoes are located in the interior of the country in the region of the Comayagua fault system, which strikes approximately N/S, and these will be discussed in connection with other volcanoes in the hinterland.

The series of volcanoes in northwestern **Nicaragua** commences with the isolated volcano **Coseguina,** which became famous in 1835 because of its violent Plinian eruption. WILLIAMS (1952 a) has summarized the many descriptions together with his own observations. He esti-

[14] I obtained the data from a still unpublished report by SHEETS: 1978 Research; Preliminary Results.
[15] Recently obtained data from SHEETS (1979).

mated that the volume of extremely finely distributed pyroclastics that were ejected was a maximum of 10 km^3. The crater, approximately 2 km across and almost 200 m deep, which collapsed during the eruption, presents an impressive sight when seen from the air on the usual flying route between Managua and El Salvador.

The volcanic chain of the **Cordillera de los Marabios,** which follows to the southeast, stretches over 72 km and is built up of five separate groups of volcanoes (Fig. 128). It is close to the southwestern boundary of the Nicaragua Depression. Six of its volcanoes were active in historical times and some of them have remained active up to the present. MCBIRNEY & WILLIAMS (1965) and MARTÍNEZ & VIRAMONTE (1973) provided general descriptions of the chain of volcanoes, their eruptive activity and the resulting volcanic products.

They distinguish between the following groups of volcanoes — in each case the largest volcano is given first:

San Cristóbal with El Chonco, El Casita, La Teta, Pelona, Moyotepe;
Telica with Aguero, Santa Clara and San Jacinto;
El Rota with the two domes of Los Bosques;
El Hoyo with Las Pilas, Asososca, Cerro Negro, Las Cacaticas and unnamed cones north of Cerro Negro;
Momotombo with El Montoso, El Colorado and La Guatusa.

They distinguish further between three generations of volcanoes, the oldest stretching back presumably to the Pliocene with highly incised forms and old rocks, an intermediate generation with fresher forms and sometimes recent activity, and a young generation with well-preserved cones, fresh and vegetation-free lava fields. The two older generations are characterized by a sequence of andesites, basalts and dacites, and the youngest generation almost exclusively by olivine basalts and mafic tholeiitic basalts. MCBIRNEY & WILLIAMS (1965, p. 40) also stressed the predominance of basalts in contrast to the more differentiated volcanoes of southeastern Nicaragua.

Good sketches of individual volcanoes are found in the catalogue of active volcanoes of Nicaragua (MOOSER, MEYER-ABICH & MCBIRNEY 1958), in VIRAMONTE (1973) and MARTÍNEZ & VIRAMONTE (1973).

The most active volcano of the Cordillera de los Marabios is the **Cerro Negro,** which erupted for the first time in 1850 and which was producing ash and basaltic to andesitic lavas right up until quite recently (Fig. 129). The most recent eruptions and their products were described by VIRAMONTE (1973) and by ROSE et al. (1974), who concentrated in particular on the ashes. The southeastern corner-pillar of the Marabios Chain is formed by **Momotombo,** which for a long time exhibited Strombolian activity producing scoriae and lavas. A thermal field was drilled at the foot of this volcano to extract energy; in 1977 the project was still in the experimental stage. **Apoyeque** (Fig. 131), the top of which is formed by a massive pit-crater, lies along an imaginary line extending from the Marabios Chain, but it is lower than this and no longer belongs to it. During prehistoric times Apoyeque produced thick layers of pumice, which cover the human footprints found at Acahualina near Managua, the age of which has been determined by ^{14}C dating on a fossil soil to be 5,945 ± 145 years BP (see COLLINS et al. 1976, p. 110).

Apoyeque is the first of a number of small pit craters and scoria cones to which the **Nejapa Pits** described by MCBIRNEY (1955) belong (Fig. 131). The row of craters deviates from the general direction and runs north-south. The boundary between the northwestern and southeastern segment of Nicaragua is placed by STOIBER & CARR (1974, p. 320) in the NNW/SSE-trending family of faults along which the 1972 Managua earthquake occurred.

The series of volcanoes in **southeastern Nicaragua** is offset by 15 km to the southwest. In contrast to the volcanic cones of the Marabios Range, it is characterized by different volcanic forms, in particular by a series of large calderas and by greater differentiation of the products. The series begins with the large calderas of the **Masaya-Nindiri** Volcanoes (MCBIRNEY 1956, UI 1973) (Fig. 132). Masaya is the only volcano in Central America to display lava lake activity, which earned it the name of "Infierno de Masaya" during the time of the Spanish Conquest. The lava lake was active again from 1970 to 1977. The caldera of **Apoyo** is followed by the strato volcanoes **Mombacho, Concepción** and **Madera,** with which the series ends in Lake Nicaragua.

SÁENZ (1971 b), in a review of the volcanoes of **Costa Rica,** notes 68 units of which the majority are concentrated in the Cordillera de Guanacaste in the northwest and the Cordillera Central in the centre of the country. Small extinct volcanoes are scattered over the northern Atlantic Coastal Plain. No volcanic edifices are known to the south of the Valle Central. Pliocene/Pleistocene dacitic volcanic rocks were, however, detected by BALLMANN (1976) at the top of the eastern Cordillera de Talamanca (Fig. 78). On the basis of tephra layers and apparent lavas of recent age, HARRIS (1971) assumed that volcanic activity occurred during the Upper Pleistocene in the western part of the same range, although this assumption has not been confirmed by anyone else. The next large volcano, Barú (=Chiriquí), occurs on the other side of the Panamanian frontier.

The NW/SE-trending range of volcanoes, which separates the Atlantic and Pacific Lowlands from each other in the northwestern part of Costa Rica, is referred to as the **Cordillera de Guanacaste** (Figs. 65 and 66). This range is offset by 40 km southwest of the volcanic range of Nicaragua. Little geological or volcanological research has yet been carried out on this range; however, the morphological data are quite well known thanks to the activity of the Instituto Geográfico de Costa Rica. It is from this Institute that I obtained the aerial photograph of the volcano **Rincón de la Vieja** which is shown in Fig. 133 and in which craters in various states of preservation can be seen. The volcano ejected jets of ash in 1861, 1940 – 1955, 1966 – 67 and 1969. Although only few analyses have been carried out, the lavas of the Cordillera de Guanacaste would appear to be *quartz-bearing andesites* and *latiandesites.*

Ahead of the volcanic chain to the southwest is a plateau built up of **Quaternary ignimbrites** of *rhyodacitic* or *quartz-latitic* composition, in which DENGO (1962 b) distinguished between a Bagaces Formation consisting of grey ignimbrites and uncemented glowing tuffs, and a cover of white pumice tuffs welded at their base, which was designated the Liberia Formation. In the landscape and in aerial photographs the latter stands out particularly clearly because of its light colouration, sparse Chaparal vegetation and steep scarps. More details on the topography are to be found in WEYL (1969/70).

In his general map of Guanacaste, DENGO indicates NW/SE-trending fault lines which probably also determine the alignment of the Guanacaste volcanoes. The ignimbrites may have emerged as the result of fissure eruptions from parallel, but not localized faults; however, they may also have originated from caldera-like depressions which HEALY (1969) believed he could see in an aerial photograph. He claimed that the young strato volcanoes were formed in these depressions.

CARR & STOIBER (1977, p. 152) assume a segment boundary, which had not been postulated in 1974, between the Cordillera de Guanacaste and the Cordillera Central and running close to the volcano **Arenal,** which stands by itself. Arenal was regarded up until 1968 as extinct, although its relatively unincised steep slopes and a lava stream, which has not yet been attacked by erosion and which is still relatively unweathered, all indicate a youthful age. This was recently confirmed by ^{14}C datings on charcoal and pottery remains which were recovered from beneath the lava flow and from the tephra. The carbon was dated at 1,525 $\pm$ 20 years A.D., and the fragments of pottery were dated between 1,200 and 1,400 A.D. (MELSON & SÁENZ 1974, p. 432).

On July 29, 1968, following 10 hours of violent earthquake activity, a series of major explosions occurred. Blocks of rock hurled out with an initial velocity of 600 km/sec turned an area of 12 km^2 into a zone densely covered with impact craters over which glowing clouds spread out, and on the western flank of the volcano three new craters formed (Fig. 134). A total of 78 people were killed by the eruption in this fortunately thinly populated area. Following a repose period from July 31 to August 3, violent ash and steam eruptions occurred up to September 10, and these were followed by a further repose period with fumarolic activity, and from September 19 onwards an eruption of andesitic block lava occurred.

Between September 1971 and August 1972 new eruptions occurred with the production of a second lava flow, and in June 1975 a third eruption phase commenced with ashes and volcanic bombs being ejected; there was also a hot pyroclastic flow and subsequent lava activity, which lasted until 1975.

The eruptive sequence of explosions, glowing clouds, ash ejections and peaceful lava activity point to a gradual decline in the gas pressure, which was originally calculated to be 4,900 bar, and this decline is also apparent in the subsequent repose. Sections through the products of prehistoric eruptions reveal a similar eruption cycle; the amount of material produced during the last eruption around 1,500 A.D., namely 0.17 km^3, is however twice as much as that produced during the eruption of 1968.

The lavas of Arenal are *quartz-andesites,* and analysis shows that there was a slight decrease in SiO_2 , Na_2O and K_2O, while Fe, MgO and TiO_2 increased during the eruption of the first lava (analyses made by MELSON & SÁENZ (1974, p. 433)). Since 1968 Arenal has been monitored by a network of seismological stations (Anonymous 1974 d). The eruptions justifiably attracted a great deal of attention because they differ completely from those known so far from Costa Rica, and furthermore they took place in a region where it is planned to build a large hydro-electric power station. (For the literature see the list of volcanoes.)

Extensive reports exist on the eruptions of **Poás, Irazú** and **Turrialba,** which are volcanoes in the **Cordillera Central.** In addition, WILLIAMS (1952) produced one of the first more modern studies, but despite their easy accessibility, none of these volcanoes has been systematically investigated. It took a series of violent ash eruptions from Irazú between 1963 and 1966, which at times made it impossible to keep cattle on the slopes of the volcano and which caused a great deal of pollution in the capital city, to prompt a number of volcanological, geophysical and petrographical studies (MURATA 1964, DÓNDOLI & SÁENZ 1966, KRUSHENSKY & ESCALANTE 1967). KRUSHENSKY (1972, 1973) mapped the southern and eastern flank of the volcano and was able to distinguish between a number of lavas, ash flow tuffs, lahars, ash layers and epiclastic rocks which he took together as the Irazú Group and placed in the Pleistocene (cf. map in Fig. 135). A large number of chemical rock analyses carried out following the eruptions indicate that the chemistry of the lavas, which may be defined as *quartz-latiandesites,* is very uniform and also that the ash flow tuffs have a high SiO_2 content (60 – 63%).

The ash of Irazú Volcano caused widespread damage to property, crops, and livestock. Much of the damage to property was caused by accelerated erosion and repeated floods of water, mud, and rock debris from the slopes of the volcano. Debris flows and erosion were especially severe in the Río Reventado watershed and valley on the southwest slope of Irazú (WALDRON 1967).

Topographically the volcanoes of the Cordillera Central differ from the other volcanoes of Central America as a result of their gentle lee-side slopes, which drop away to the Valle Central, whereas the weather-side in the north drops steeply to the Atlantic Lowland Plain and is furrowed by countless landslides. Also, the volcanoes have grown together to form one single mountain chain broken only by broad saddles, and this chain borders the Valle Central in the north. Its slopes are inhabited and agriculturally utilized up to great altitudes.

Only two large Pleistocene volcanoes are known in **Panama,** the **Chiriquí (Barú)** on the western border and the **El Valle** in central Panama. Apart from the data provided by SAPPER (1913) and TERRY (1956), I am not aware of any other information on these volcanoes. Between them there are a number of smaller extinct volcanoes which were discovered in particular when DEL GIUDICE & RECCHI carried out their mapping work (1969). TERRY (1956, p. 11) had already drawn attention to Cerro Santiago with its particularly fresh ashes. A crater is shown at this point in the general geological map of Panama. A large number of thermal fields, which MÉRIDA (1973) mentions, may be regarded as the last manifestation of volcanic activity. Like the extinct volcanoes, they are concentrated in western Panama.

In my opinion, the actual problem here is not the volcanic activity but its decline compared with the Tertiary. This corresponds to a much lower seismic activity in Panama compared with the rest of Central America, so that the probable reason for this phenomenon is Panama's different position relative to the colliding plates in Central America. It is no longer located in the area of the northeastwards advancing Cocos Plate but to the north of the Nazca Plate which is moving eastwards towards the South American continent.

Quaternary volcanoes are dotted widely throughout the **hinterland** of the Pacific Volcanic Chain. Their products are, in part, strongly alkaline in character and they therefore play a particular role — which is discussed in the section on the petrology — in the genesis of the magmas.

In **Honduras,** Pleistocene to Holocene volcanoes occur in the interior of the country near the N/S-striking Comayagua Fault System (WILLIAMS & MCBIRNEY 1969, pp. 63 – 72). A number of these volcanoes is grouped around the plateaus of Tegucigalpa; they are described as effusions of olivine-bearing tholeiitic basalts, but according to the RITTMANN/STRECKEISEN definition they are composed of *tholeiite hawaiite* and *tholeiite mugearite.* A group of still very fresh small volcanoes on the northern side of the **Lago de Yojoa,** which according to reinterpretation of the analyses reported by WILLIAMS & MCBIRNEY produced *olivine-bearing hawaiite, latite* and *trachyte,* are particularly important. MERZMAN (1976) detected aegirine augite trachyte, fayalite trachyte, anorthoclase fayalite trachyte, olivine basalt and hortonolite and discussed the origin of this alkaline rock series on the basis of $^{87}Sr/^{86}Sr$ ratios. Further young alkaline rocks (alkaline olivine augite basalt according to WILLIAMS & MCBIRNEY 1969, pp. 97 – 98, *olivine mugearite* according to the RITTMANN/STRECKEISEN definition) are known from the **Utila Island** in the Gulf of Honduras.

In **Nicaragua** MCBIRNEY (1964; also in MCBIRNEY & WILLIAMS 1965, pp. 25 – 29) described three extinct, but from the freshness of their form definitely Upper Quaternary volcanoes on the northeastern flank of the Nicaragua Graben: **Cerro El Ciguatepec, Cerro San**

Jacinto and the **Las Lajas Caldera.** Like the extinct and in part highly eroded Pliocene-Pleistocene volcanoes in the interior of El Salvador, these also attest to a very recent shifting of volcanic activity into the present Pacific border zone.

Further into the interior of Nicaragua, fresh lavas are known from the valley of Esteli, and small volcanic cones have been discovered west and south of Ocotal (MCBIRNEY & WILLIAMS 1965, pp. 28/29, PIÑEIRO & ROMERO 1962, p. 70). In addition, MCBIRNEY describes a small group of volcanic cones from the Atlantic Coastal Plain between the Río Curinhuas and the Laguna de Perlas. A sample chemically analysed and referred to as basaltic scoria is according to our definition an *alkaline olivine mugearite* with the sigma value 9.04 and thus belongs to the alkaline rocks.

A few small extinct volcanic cones made up of pheno-olivine basalts with titanaugites rise above the coastal plain of northeastern Costa Rica (MALAVASSI & CHAVES 1970). As far as is known, their rocks also belong to the alkaline series (ROBIN & TOURNON 1978).

If we consider the chain of Central American volcanoes, we are struck by an at first seemingly trivial fact: within the individual groups of volcanoes, or within the segments as defined by STOIBER & CARR, **the altitude of the volcanoes** is relatively constant both absolutely, i.e. above sea level, and relatively, i.e. the height above the base. Although deviations do occur, they are for the most part restricted to one or two of the larger volcanoes within a group while most of them keep closely to the average height. An attempt was made in Fig. 121 to express this regularity in graphic form. ROSE et al. (1977, Fig. 11) similarly depicted the height of volcanoes as a function of their position along the WNW-trending Quaternary Central American Volcanic Chain.

WEYL (1961 a, p. 123) suggested the reason for this was that each group shared a common magma chamber: "Once a volcano has built up to a height corresponding to the energy of the eruption, it is easier, after a period of repose, for the magma to seek a new and shorter way to the surface by travelling sideways rather than by passing all the way up the previously active vent. A new volcano builds up alongside the extinct one until it, too, attains an altitude corresponding to the energy of its eruption." The volcanoes of Santa Ana and Izalco, which was formed in 1770, were given as examples of this process.

From their study of the Santa Maria/Santiaguito in Guatemala, ROSE et al. (1977, pp. 82 – 83) arrive at an explanation for the constant heights which fits in with modern petrogenetic knowledge; because of the general importance of their findings, they are quoted here verbatim: "The heights (above sea level) of the high composite cones of the Central American chain vary systematically along the length of the chain, being highest at the extreme ends and lowest in the middle where the volcanoes lie within the Nicaraguan depression. In the Guatemalan highlands the composite cones have heights in the range 3,500 – 4,200 m, and the presence of dormant volcanoes here as throughout the entire arc suggests that for each part of the arc there is a critical cone height above which lavas cannot be erupted out of the central vent of a volcano. Such a critical cone height would follow if magma in Central America rose buoyantly, due to density contrasts between it and the crust-mantle section above the magma source. The critical cone height would be reached when the hydrostatic pressure at the magma source was equal to the lithostatic pressure in the wall rocks at the same point. If the magma sources for the entire chain all lie deeper than the depth of isostatic compensation for the volcanic arc, the systematic variation in the height of the composite cones along the chain would reflect systematic variations in the density of the primary magmas, the highest densities occurring where the cones are lowest.

Table 10. Averages of Quaternary volcanic rocks from Central America.

	1	2	3	4	5	6	7	8	9	10	11
Number of analyses	10	9	12	12	8	8	5	6	11	10	21
SiO_2	61.5	51.9	52.4	65.5	65.7	49.5	50.0	58.2	67.3	56.8	55.2
Al_2O_3	18.3	17.9	19.1	15.6	16.6	17.7	16.3	17.6	15.8	17.9	17.4
Fe_2O_3	3.4	3.4	3.9	1.3	2.1	4.2	5.2	2.6	2.6	3.4	2.8
FeO+MnO	1.8	6.0	5.7	2.6	2.4	6.1	7.6	4.5	0.9	4.3	4.7
MgO	2.1	5.6	4.3	1.0	1.7	6.7	5.1	2.6	0.6	3.8	4.8
CaO	5.6	8.7	9.3	2.4	4.3	11.7	10.5	6.0	2.4	7.7	7.7
Na_2O	4.4	3.3	3.1	3.5	3.8	2.2	2.5	4.0	3.2	3.1	3.4
K_2O	1.5	1.2	0.9	3.6	1.5	0.6	1.1	1.9	3.5	1.5	2.1
TiO_2	0.4	1.1	0.9	0.6	0.6	1.1	1.2	0.9	0.5	0.7	1.0
Colour Index	10.0	26.9	24.3	11.9	10.8	37.0	36.6	14.6	13.8	19.3	23.6
Serial Index	1.9	2.3	1.7	1.3	1.3	1.2	1.9	2.3	1.9	1.6	2.5

1. Santa Maria Volcano, Guatemala, corresponds to *quartz-bearing latiandesite.*
2. Obrajuelo Volcano, SE-Guatemala, corresponds to *andesite.*
3. Izalco Volcano, El Salvador, corresponds to *quartz-bearing andesite.*
4. Coatepeque Volcano, El Salvador, corresponds to *rhyodacite.*
5. Ilopango Volcano, El Salvador, corresponds to *dacite.*
6. Cerro Negro Volcano, Nicaragua, corresponds to *olivine-bearing andesite.*
7. Masaya Volcano, Nicaragua, corresponds to *andesite.*
8. Zapatera and Omotepe Volcanoes, Nicaragua, correspond to *quartz-bearing latiandesite.*
9. Bagaces- and Liberia-Formation, NW Costa Rica, corresponds to *rhyodacite.*
10. Cordillera de Guanacaste, NW Costa Rica, corresponds to *quartz-bearing latiandesite.*
11. Cordillera Central, Costa Rica, corresponds to *quartz-bearing latiandesite.*

This model predicts that as a volcano approaches and then reaches its critical height, eruptions from the central vent are first inhibited and then stopped altogether, so leading to the volcano's dormancy. If the repose episodes during the approach to dormancy are sufficiently long to allow significant fractionation to occur at the top of the magma column feeding the volcano, the central vent will erupt increasingly fractionated liquids, and the high density crystal cumulates will be stored in the magma column. Such storage, if accompanied by the eruption of the fractionated liquids, will in itself further inhibit eruption from the central vent by increasing the density of the magma in the feeding column, and so reducing its tendency to rise buoyantly. When a volcano becomes dormant in this way, all further activity must be by way of flank eruptions of various kinds."

The connection between the increasing cone height and the extent of fractionation in the erupted lavas appears in Santa Maria and other Central American volcanoes (ROSE et al. 1977, p. 82); however, there are a number of exceptions to this rule, as is evident from the data for heights and mean SiO_2 contents given in the catalogue of volcanoes.

In 1961 the **petrology of the volcanic products** was discussed in the first edition on the basis of 87 analyses, some of quite early date. Now, 363 published and unpublished analyses have kindly been put at my disposal for evaluation by Prof. MCBIRNEY. Since there are so many, it is not possible to list them here in table form; therefore, I must refer the reader to the literature given in the list of volcanoes. The average values of particular groups of volcanoes are given in Table 10. All the available analyses were incorporated in the following diagrams.

Table 11 gives an overview of the volcanic products; the sub-areas selected in the table are the segments given by STOIBER & CARR (1974). A comparison of the rocks mentioned here with the designations chosen by the various authors reveals that determination on the basis of the content of phenocrysts usually makes the rocks seem much more basic than is indicated by the norms calculated from the chemical analyses. Therefore our table does not, for example, contain the basalts frequently mentioned in the literature, which should for the most part be defined as andesites and olivine andesites according to RITTMANN/STRECKEISEN.

The table furthermore reveals an accumulation of three groups of rock: the *rhyolites, rhyodacites, dacites,* the *quartz-latiandesites,* and the *quartz-andesites, andesites* and *olivine andesites.* Furthermore, it can be seen that sometimes quite different rock groups predominate in the individual segments, and this is also expressed in the diagrams in Figs. 143 to 145. They reveal quite clearly that the mass of the volcanic products belong to the calc-alkaline series and have a strong to medium series character from which only a few occurrences deviate, and these will be examined separately.

The differences in the composition of the volcanic products along the chain of volcanoes were dealt with by MCBIRNEY (1969 a). To start with, they appear to be uniform in collective diagrams, but this uniformity disappears when specific regions are considered separately.

Table 11. Distribution of Quaternary Central American volcanic rocks based on 386 chemical rock analyses.

	Central Guatemala	Eastern Guatemala	El Salvador	Western Nicaragua	Eastern Nicaragua	Costa Rica Cordillera de Guanacaste	Costa Rica Cordillera Central	Other areas	Total
Alkalirhyolite	.	.	.	.	.	2	.	.	2
Rhyolite	.	11	3	1	1	.	.	3	19
Rhyodacite	4	2	10	4	1	8	.	1	30
Dacite	2	5	10	5	6	3	1	2	34
Plagidacite	3	4	.	.	.	.	.	.	7
Quartz-latite	.	.	.	.	.	.	2	2	4
Quartz-latiandesite	8	10	21	7	6	4	18	4	78
Latiandesite	.	3	.	.	2	.	3	1	9
Olivine-bearing latiandesite	.	.	.	.	.	1	.	1	2
Quartz-tholeiite mugearite	.	.	5	.	.	.	.	.	5
Quartz-andesite	28	3	10	10	6	10	1	4	72
Andesite	11	14	22	8	12	.	2	2	71
Olivine andesite	.	25	11	6	6	.	.	.	48
Tholeiite hawaiite	.	1	1	.	.	.	.	.	2
Alkali-olivine hawaiite	.	.	.	.	.	.	.	1	1
Others	.	.	2	.	.	.	.	1	3
Total	56	78	73	41	40	26	27	22	386

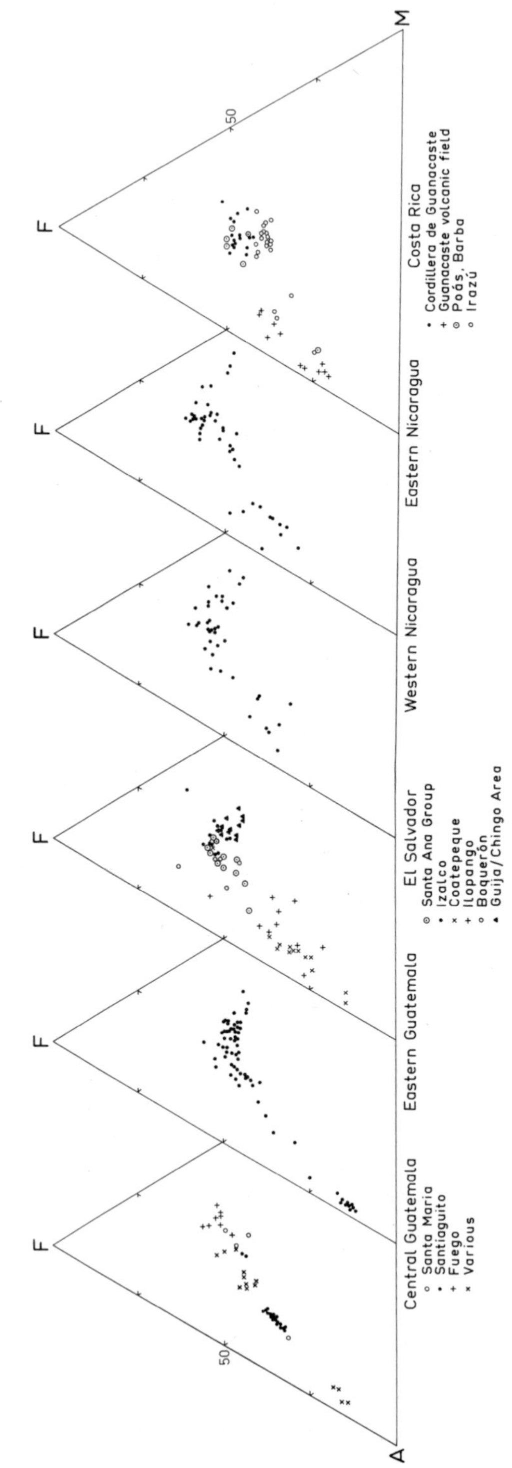

Fig. 143. AFM diagram for geographical groups of Central American volcanoes.

It is then found that the large volcanoes are built up of "andesites" of uniform composition, but that later products exhibit a clear discontinuity and break down on the one hand into SiO_2-rich pumices and acid extrusions and on the other into basic "basalts". A petrographic gap occurs between the two types, and this is particularly apparent in the AFM plots. MCBIRNEY assumes that differentiation processes in the entire volcanic chain, which may vary from volcanic cluster to cluster, are the reason for this.

This bi-modality in the composition manifests itself very clearly when the SiO_2 content, as given in the available analyses, is considered. On the one hand Fig. 146 shows the number of analysed rocks as a function of the SiO_2 content, and on the other it shows the range of scatter of the SiO_2 content in various rock groups. Although this representation tells us nothing about the proportions of the various volcanic products, it does give some idea of the frequency of their occurrence. There is a high percentage of relatively basic to intermediate rocks, "andesites" in the broadest sense, then comes a gap around 57% SiO_2 which is followed by a considerable percentage of acid rocks, "dacites" in the broadest sense. What we have to examine, therefore, is whether these rocks are magmas from two different sources or whether they are differentiation products of a homogeneous magma. According to the geological findings, differentiation is the most likely cause because the acid volcanic masses are mainly the pumice covers of Guatemala and El Salvador and a few small obsidian volcanoes. The "andesitic" rocks, on the other hand, can probably be regarded as the products of a less differentiated magma.

MCBIRNEY (1969 a) raised the question of what caused the change in the chemistry of the volcanic products along the volcanic chain. He observed a continuous increase in basic products from Guatemala to Nicaragua and Costa Rica and he assumed that this might possibly reflect a decrease in the depth of the magma source. On the other hand, there is no evidence in the volcanoes that the different crustal structure — continental crust in the northwest, oceanic crust in the southeast — has any effect.

The paper by PICHLER & WEYL (1973) presented a detailed discussion of the volcanic products; in this paper the volcanic chain was also broken down according to region, but it was not yet divided up into segments. The characterization of the products according to this paper can therefore be summary in form, and is to a large extent apparent from the diagrams.

No data are available from western Guatemala. In central Guatemala, Santa Maria displays a clear division into basic and acid products, whereas Santiaguito very uniformly produces *quartz-andesites.* In eastern Guatemala WILLIAMS et al. (1964) made a division according to volcanoes with basic, *andesitic* to *olivine andesitic* rocks and *rhyolitic* to *dacitic obsidian* domes and interpreted this as the result of differentiation. It is noticeable here, however, that there are no transitions between the two groups, which is what one would expect in the case of differentiation.

The large strato volcanoes of El Salvador consist of "andesites" in the broadest sense while Coatepeque and Ilopango have produced acid pumices. The pumice of Coatepeque is *rhyodacitic* to *rhyolitic,* while that of Ilopango is *dacitic* so that the externally identical pumice covers differ distinctly in petrographic terms. In both cases normative cordierite occurs relatively frequently, presumably as a result of leaching of alkalis.

The volcanoes of northwestern Nicaragua are composed for the most part of very basic rocks, *andesites, olivine andesites* and *mela-andesites,* and in this respect they deviate distinctly from those of neighbouring regions. The volcanoes of southeastern Nicaragua exhibit greater differentiation with *andesites, quartz-andesites* and *quartz-latiandesites.*

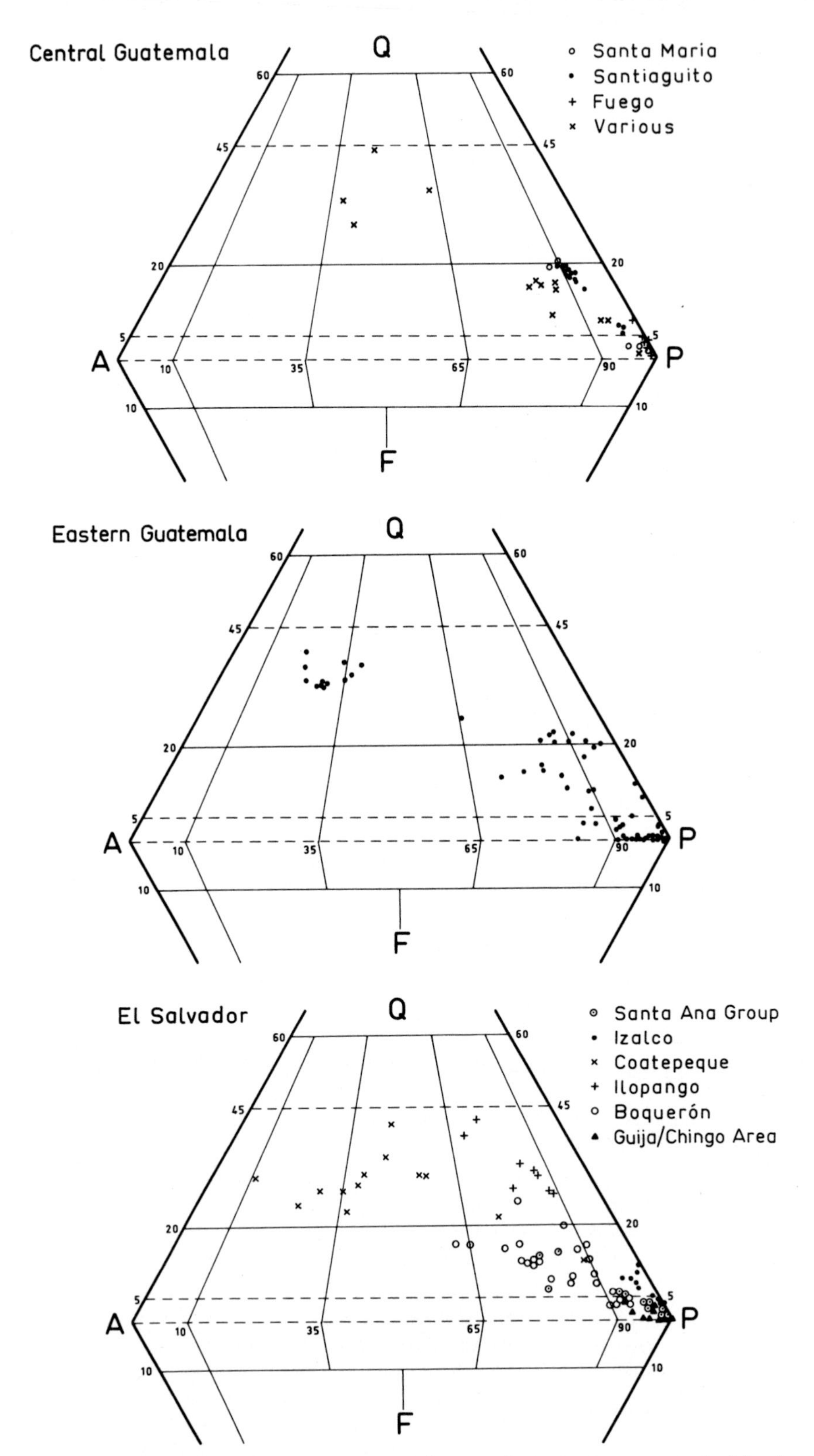

Fig. 144. Distribution of Central American Quaternary volcanic rocks in the STRECKEISEN double-triangle. Separated on a geographical basis.

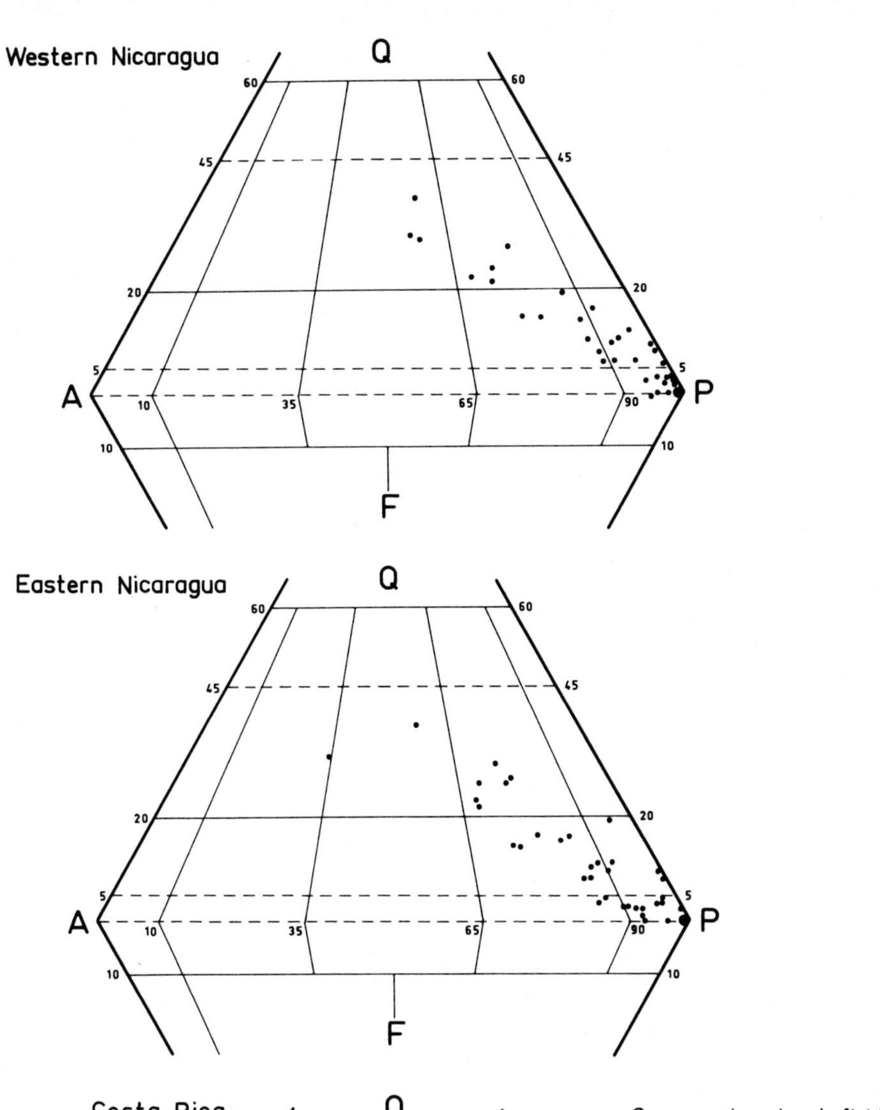
Western Nicaragua
Q
60
45
20
5
A
10
35
65
90
P
10
F
Eastern Nicaragua
Q
60
45
20
5
A
10
35
65
90
P
10
F

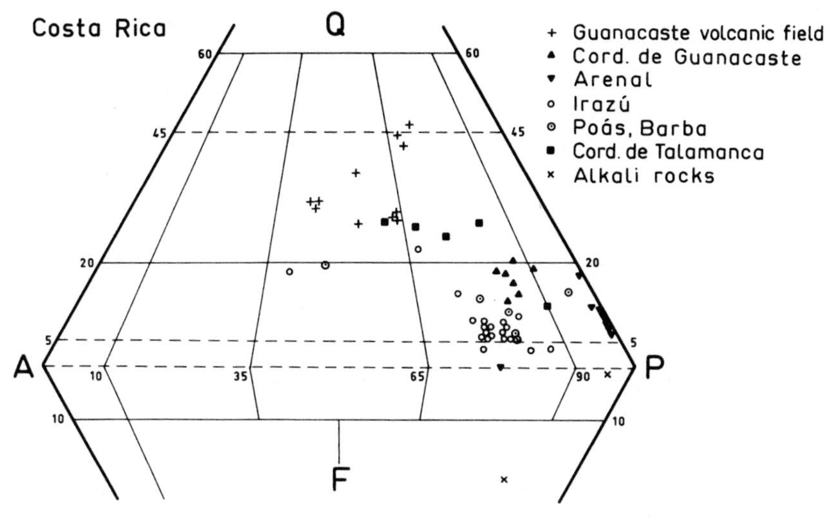
Costa Rica
Q
60
45
20
5
A
10
35
65
90
P
10
F
+ Guanacaste volcanic field
▲ Cord. de Guanacaste
▼ Arenal
○ Irazú
◎ Poás, Barba
■ Cord. de Talamanca
× Alkali rocks

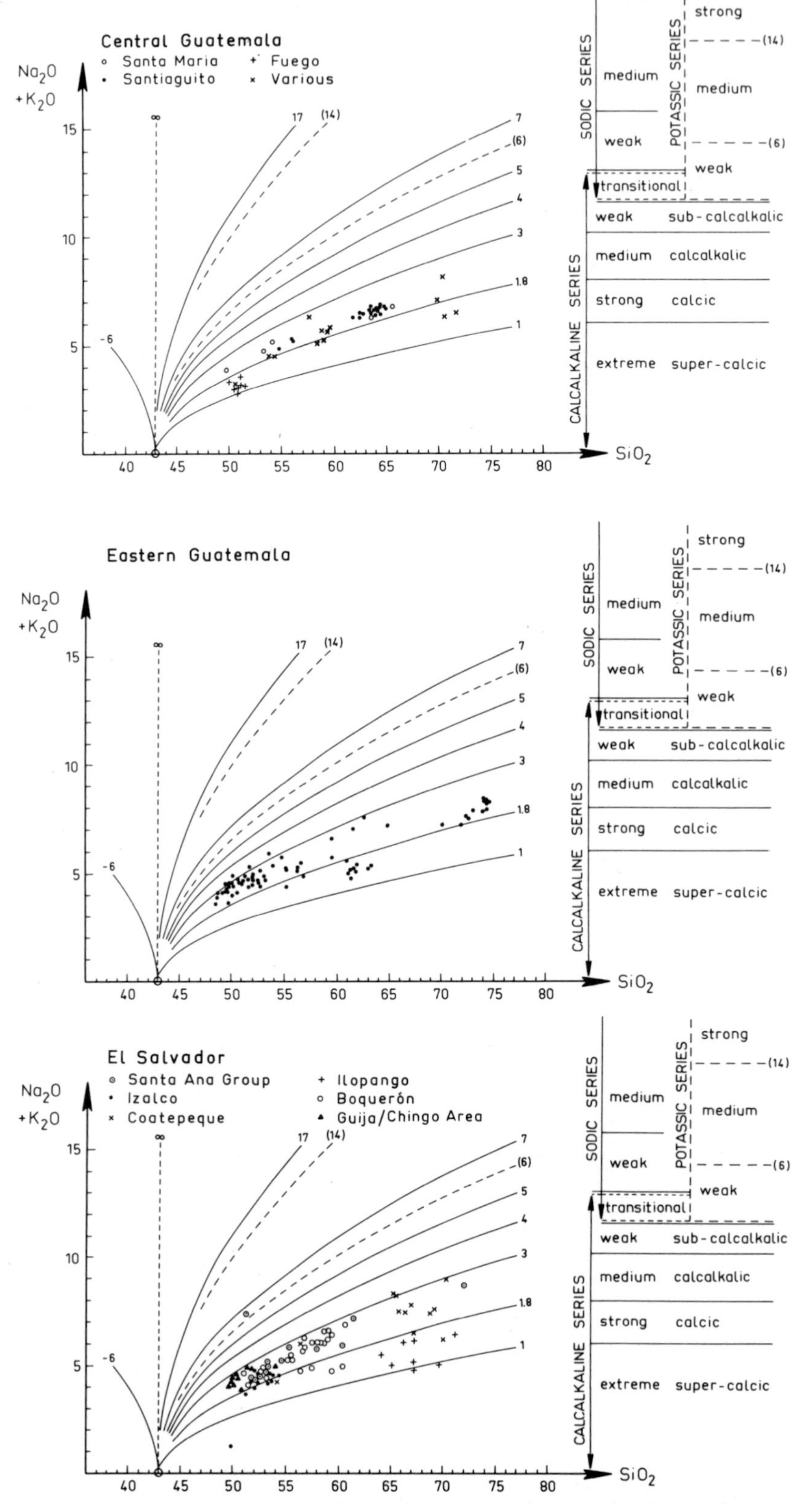

Fig. 145. Serial index (alkali versus SiO_2) of Central American volcanic rocks. Separated on a geographical basis.

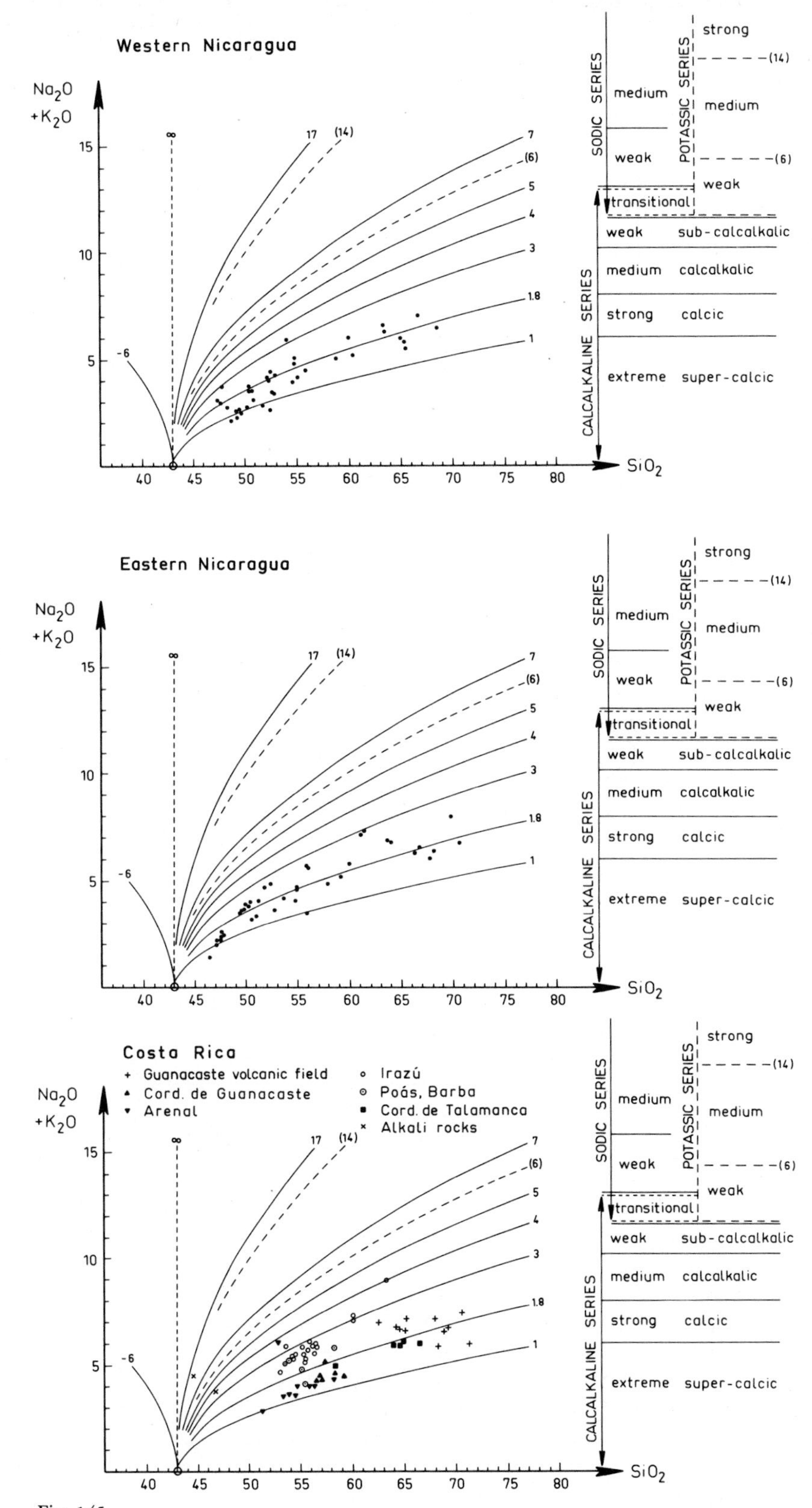
Western Nicaragua
Na2O
+K2O
15
10
5
∞
17
(14)
7
(6)
5
4
3
1.8
1
-6
40
45
50
55
60
65
70
75
80
SiO2
SODIC SERIES
medium
weak
POTASSIC SERIES
strong
(14)
medium
(6)
weak
transitional
CALCALKALINE SERIES
weak
sub-calcalkalic
medium
calcalkalic
strong
calcic
extreme
super-calcic
Eastern Nicaragua
Na2O
+K2O
15
10
5
∞
17
(14)
7
(6)
5
4
3
1.8
1
-6
40
45
50
55
60
65
70
75
80
SiO2
SODIC SERIES
medium
weak
POTASSIC SERIES
strong
(14)
medium
(6)
weak
transitional
CALCALKALINE SERIES
weak
sub-calcalkalic
medium
calcalkalic
strong
calcic
extreme
super-calcic
Costa Rica
+ Guanacaste volcanic field
▲ Cord. de Guanacaste
▼ Arenal
○ Irazú
◎ Poás, Barba
■ Cord. de Talamanca
× Alkali rocks
Na2O
+K2O
15
10
5
∞
17
(14)
7
(6)
5
4
3
1.8
1
-6
40
45
50
55
60
65
70
75
80
SiO2
SODIC SERIES
medium
weak
POTASSIC SERIES
strong
(14)
medium
(6)
weak
transitional
CALCALKALINE SERIES
weak
sub-calcalkalic
medium
calcalkalic
strong
calcic
extreme
super-calcic

Fig. 145.

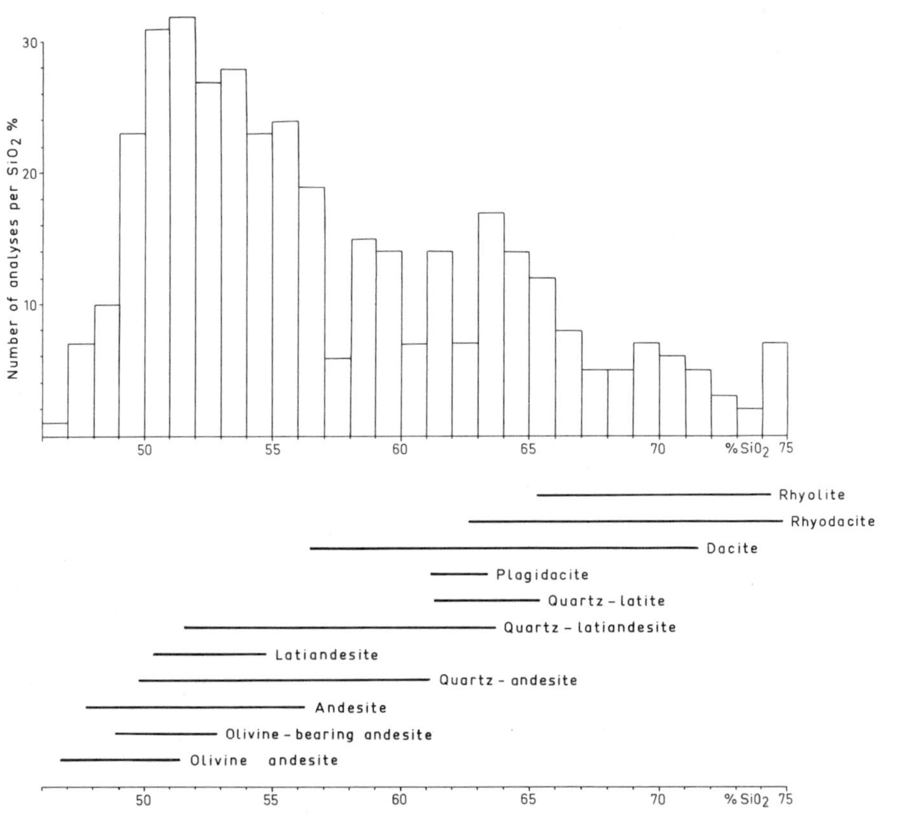

Fig. 146. Distribution of the Quaternary volcanic rocks of Central America according to the SiO_2 content and SiO_2 ranges of the predominant rock classes.

In Costa Rica the ignimbrite plateaus of Guanacaste consist mainly of *rhyodacites* and *dacites.* They, too, often contain a considerable excess of aluminium which probably results, in part, from leaching but which could also be primary in nature, because fresh rocks were analysed. The volcanoes of the Cordillera de Guanacaste are composed of *quartz-latiandesites* and Arenal of *quartz-andesites.* Mainly *quartziferous latiandesites* occur in the Cordillera Central.

This review, in particular the STRECKEISEN and AFM plots, reveals clearly the individuality of the individual segments as regards the composition of their volcanic products. It is at present probably impossible to say whether this is due to the different amount of plunge of the subduction zone in the segments, as was stated by STOIBER & CARR, or to the different composition of the crust, or whether it simply reflects the different development of the magmas as they ascend into the crust, and their development in higher volcanic chambers. In my opinion, however, the studies that have been carried out on individual groups of volcanoes seem most definitely to point in favour of the individual development of the magmas.

The sporadic occurrence of **alkaline rocks** in Costa Rica (TOURNON 1973), Nicaragua (MCBIRNEY & WILLIAMS 1965, p. 28) and Honduras (WILLIAMS & MCBIRNEY 1969, pp. 70 – 71, MERZMAN 1976) deserves special attention. All these occurrences are situated in the hinterland of the active volcanic chain in a manner typical of other circumpacific volcanic regions, e.g. Japan or Indonesia (RITTMANN 1953, KUNO 1966, MIYASHIRO 1972). This is seen as a correlation with the depth of the subduction zone and thus with the depth of the magma source, and it may also apply in Central America. The occurrences of alkaline rocks and their analytical data have been compiled by PICHLER & WEYL (1976, Table 4), and they are shown in the general map in Fig. 120 together with the respective serial index (σ) after RITTMANN; the increase in the alkali content is plotted in the attached diagrams.

Occurrences of particular alkaline rocks from the volcanic chain itself (Cerro de los Alpes, San Miguel in El Salvador and La Vega in Guatemala) have sometimes erroneously been defined as such, but some of them may also be local differentiates.

CARR (1978) sees the relationships slightly differently: "In the narrow volcanic front where large volumes of calc-alkaline rocks erupted. Above the transverse structures and behind the volcanic front olivine normative basalts are common and nepheline normative basalts are present. Volcanoes are usually small and frequently occur in clusters that extend 100 or more km landward from the volcanic front. The third structural setting is the area more than 150 km landward from the volcanic front. Here strongly undersaturated lavas are erupted from vents that have no relation to the structure of the inclined seismic zone. The increase in alkalinity of recent lavas across these two arcs is interpreted to be the result of three different magma-forming processes at three different structural settings rather than a simple increase in alkalinity of partial melts with depth to the seismic zone".

CARR's conclusion is as follows: "In Central America there is no increase in alkalinity of magmas with seismic zone depth for either the volcanic front magmas or the basalt magmas that occur above the transverse breaks in the arc." This assumption was confirmed by the lack of deep-focus earthquakes in the hinterland of the volcanic front as is apparent from the data given by MOLNAR & SYKES (1969, cf. our Figs. 164 and 165) or GRASES (1975, Mapa 6).

ROBIN & TOURNON, in a work which unfortunately I did not find out about until my own book was in the press, describe further finds of alkaline rocks from the Caribbean side of Costa Rica. These two authors believe that "the alkali lavas in southern Central America are set in diverse types of structural units among the folded sediments of the Cordillera de Talamanca, as well as the eastern termination of the Nicaragua graben, and on the coastal Caribbean plain of Nicaragua. Thus, the tectonic control of the alkaline magmatism remains undeciphered, and the origin of the series is still enigmatic. In any case, it does not seem possible to explain the recent Central American alkali magmatic series by a simple tectonic model . . ." (ROBIN & TOURNON 1978, p. 1640).

In addition to the lateral change in the alkaline content (which can be expressed in RITTMANN's serial index), HATHERTON & DICKINSON (1969) state that the potassium content increases greatly towards the descending subduction zone, assuming the same SiO_2 content in the volcanic products. A yardstick for measuring the respective depth of the subduction zone could perhaps be obtained from this. NIELSON & STOIBER (1973), however, have voiced certain doubts because according to their calculations the relationship of the potassium content is not very significant for the depth of the subduction zone. However, they also, with some reservation, give approximate values for the possible depth of the source of the potassium and thus of the magma. A depth of 100 km is assumed for Central America.

Four regions, in which Quaternary volcanism of sufficient extent is encountered transverse to the longitudinal axis of the volcanic zone, are available for checking the question whether it is possible to assume a lateral increase in potassium in Central America. These regions are:

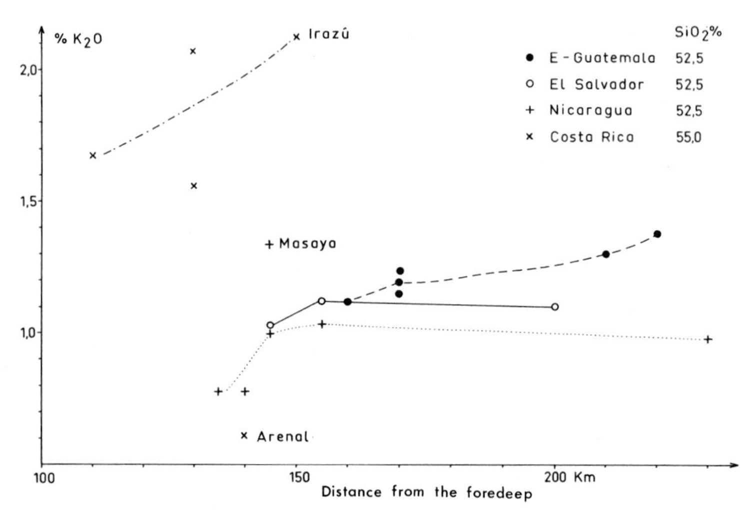

Fig. 147. Potassium contents of the Quaternary volcanic rocks of Central America as a function of the distance from the Middle America Trench.

eastern Guatemala, western El Salvador, Nicaragua and Costa Rica. In these regions rocks with a SiO_2 content between 50 and 50% or 52.5 and 57.5% can be compared — their potassium content was converted to a mean SiO_2 content of 52.5% or 55.0% respectively. Thus volcanic products with the same SiO_2 content can be correlated with each other. Altogether 103 analyses were available for this purpose. However, they were not related to the hypothetical depth of the subduction zone but to the distance from the centre of the foredeep in the Middle America Trench. The result is recorded in Fig. 147.

The result reveals altogether a slight tendency for the potassium content to increase more, relative to the total alkaline content (serial index σ), with increasing distance from the foredeep. However, the individual values are quite widely scattered. In eastern Guatemala they are clearly higher in the northern regions of Ipala and Chiquimula than those in the southern parts. In El Salvador there is no increase in the northern Guija region compared with the Sierra de Apaneca/Izalco region. In Nicaragua the decline in the potassium in the northeastern flank of the trench compared with parts of the main volcanic chain is remarkable even though it is only attested by relatively few values (5). Finally, in Costa Rica the Arenal Volcano with its extremely low alkaline products stands out from all the other volcanoes with their alkali-rich products. If we disregard this fact, then a vague increase in the serial indexes and in the potassium is apparent from west to east in the Cordillera de Guanacaste and the Cordillera Central.

These findings agree with the results of CARR et al. (1979) that "in Central America there is no evidence for an increase in potassium content of lavas with the depth of the seismic zone" (p. 387).

The question of the **formation and development of the magma** must be considered in connection with the general discussion on the formation of magma in the circumpacific belt. The suspicion that close links exist here between volcanism, seismicity and regional tectonics has been confirmed by many examples from prehistoric or recent active zones. These relationships within the region of Central America were discussed from a petrochemical standpoint by

PICHLER & WEYL (1973, 1975, 1976). The question of the formation of the magma was posed by MCBIRNEY in 1969 and answered as follows, mainly with petrological considerations in mind:

"Underthrusting of an oceanic plate beneath the continental margin causes a sharp depression of the isotherms and an inversion of the temperature gradient in rocks immediately above the plate. Water driven from wet oceanic crustal rocks streams up into a hotter horizon and lowers its melting temperature. Magma will be produced where the melting temperature is depressed below the prevailing temperature."

At the same time, MOLNAR & SYKES (1969) defined the relationship between volcanism and seismicity in terms of the plate theory, with specific reference to an underthrusting of the Cocos Plate under the Caribbean-Middle America Plate. STOIBER & CARR (1974) refined these theories by pointing to a clear segmentation within the subduction zone, which influences the form, structure and activity of the volcanoes, although they expressed reservations about the chemistry of the volcanic products. It would seem that everyone agrees that the site of magma formation is a seismic gap located at a depth of 100 to 150 km in the subduction zone above which lies the front of the active volcanoes. The decline in seismic activity at this depth is seen as the commencement of anatexis in the subduction zone or in the neighbouring mantle.

The Sr^{86}/Sr^{87} ratio (PUSHKAR 1967, 1968) is regarded as an important indication of the origin of the magma in the upper mantle. It is uniformly low with values around 0.704 in the volcanic rocks of Central America while with few exceptions crustal rocks from northern Central America have significantly higher values between 0.707 and 0.748. From this PUSHKAR concludes that the volcanic rocks originate at depths with a low Sr^{86}/Sr^{87} ratio and thus come from the upper mantle or maybe from basaltic material in the oceanic crust or in deep parts of the continental crust. If they had assimilated any radiogenic strontium from the upper crust, this must have been leached out. The volcanic rocks of Central America could certainly not have formed from anatexis of the crust.

PUSHKAR's theory is confirmed by trace element analyses on lavas of Santa Maria and Santiaguito (ROSE et al. 1977): "All the lavas have identical Sr^{87}/Sr^{86} values of 0.70397 ± 0.00008. These ratios are permissive of a homogeneous mantle or eclogite derivation, and preclude the incorporation of any radiogenic crustal material. Trace element data suggest that upper mantle peridotite may be a more compatible source rock than eclogite, but if ultramafic material does give rise to the parent liquid, low Ni and Cr abundances in the lavas demand that the liquids be fractionated before approaching the surface" (p. 85).

The **further development of the magmas** has been discussed in a number of examples, mainly on the basis of petrological characteristics and their experimental interpretation, with consideration being given both to assimilation of different crust as well as to differentiation of the ascending magma:

On the basis of a detailed analysis of the petrological and chemical data which he determined in the **Santa Maria/Santiaguito** volcanic group, ROSE (1972) comes to the following conclusion: the primary magma is pyroxene-andesitic, and its point of origin is situated at a depth of 100 to 120 km in the upper mantle. The magma changed into dacite magma in a magma chamber as a result of fractionated crystallization and its H_2O pressure rose in the process by several hundred atmospheres. When Santa Maria erupted in 1902, there was a sudden pressure release in a shallow magma chamber, and this led to the discharge of several cubic kilometres of pumice. The subsequent ascent of the magma took place slowly, resulting in the formation of the dome of Santiaguito, and in the process residual gases were released. This sequence of events is reflected in the crystallization and re-absorption of the phenocrystals.

In a subsequent paper ROSE et al. (1977) confirm and modify this theory: "The basaltic andesites and associated pyroclastics, which make up the volcanic succession and comprise the latter 40% of the volcano's eruptive output, show systematic increases in SiO_2, K_2O, Rb, and Zr, and decreases in MgO and CaO, up the succession. These evolutionary changes anticipate the dacites which first appeared after a period of repose in Santa Maria, first as pyroclastic debris in the catastrophic flank eruption of 1902, and later as the endogenous dacite dome of Santiaguito which appeared in 1922 and has continued to grow to the present day."

This development also indicates a possible way in which the large pumice covers of Guatemala may have formed. WILLIAMS (1960) had not yet commented on this, and KOCH & MCLEAN (1975) only comment indirectly by regarding changes in the mineral content of various tephra layers as an indication of vertical zoning in the magma, with biotite pointing to a higher level and hypersthene providing evidence of high H_2O pressure in a deep layer. In one of their tephra layers (Y) the low SiO_2 content is typical of deep zones in the magma chamber.

In the case of the eruption of **Izalco** in 1966, ROSE & STOIBER (1969) calculate that a shallow, cylindrical magma chamber about 2 km in diameter and 40 km long was involved. The rocks of Izalco, a mafic cone, show a continuous increase in SiO_2, K_2O, Zr and La, and a decrease in CaO, MgO, Co and Cr from the bottom of the rock sequence to the top. These trends are consistent with progressively increasing crystal fractionation (WOODRUFF et al. 1978; see also KILBRIDGE 1979, and p. 226).

The lavas of the **Boquerón Volcano** in El Salvador appear to have evolved from basaltic magma by fractionation of the phenocryst phases (FAIRBROTHERS et al. 1978): "The temporal variation in the chemical composition of the lavas is composed of three components. First, there is a crudely cyclical alternation of basalts and andesites [16]. Second, these cycles are progressively shifted toward higher SiO_2 contents (Fig. 148). Third, approximately in the middle of the stratigraphic section sampled, there is an abrupt change in chemical variation trends from an Al-rich and Fe-poor trend to an Fe-rich and Al-poor trend. This change is interpreted to have been caused by an increased proportion of plagioclase fractionation and a decreased proportion of augite fractionation. The crudely cyclical change in SiO_2 content with time is interpreted as a combination of crystal fractionation that increases SiO_2 content, followed by influxes of basaltic magma that mix with residual magma to decrease SiO_2 content. Successive cycles are shifted toward higher SiO_2 content because there is a significant volume of fractionated magma remaining in the chamber before each influx of basalt" (p. 1).

UI (1973) distinguishes between five groups of rocks with different developmental histories in the very complex **Masaya edifice** of Nicaragua: 1) calc-alkaline basalt and andesite; 2) Fe-rich calc-alkaline basalt; 3) crystal accumulate of 2; 4) Mg-rich and low-K basalt, and 5) calc-alkaline dacite.

In his opinion the magma in the upper mantle formed where the melting temperature was lowered by an influx of water. The differences between rock groups 1 and 2 can be explained by different degrees of crystallization and by the separation of plagioclase and Fe-Ti oxides as

---

[16] A definition of the Boquerón lavas according to the RITTMANN norm gives 10 *quartz-latiandesites,* 6 *quartz-bearing andesites,* 5 *quartz-tholeiite mugearites,* 1 *quartz-andesite,* and 1 *leuco quartz-latiandesite.*

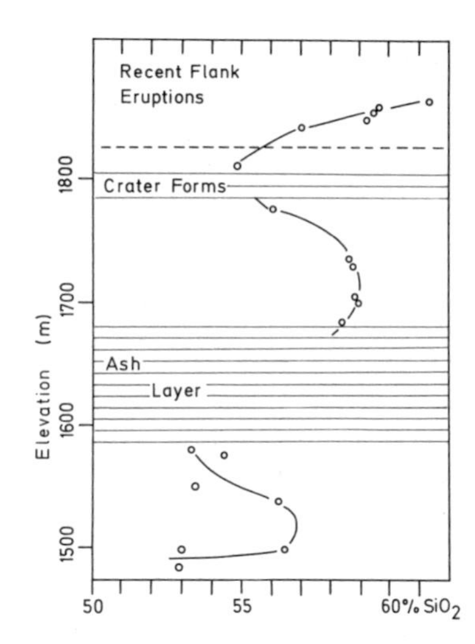

Fig. 148. Change of SiO_2 content with stratigraphic position for Boqueron lavas (Redrawn from FAIRBROTHERS et al. 1978, Fig. 4).

the magma ascended in a magma chamber within the crust. Group 4 can be explained in terms of the crystallization of olivine and augite. The composition of the dacites lies along an extension of the variation curve of basalts and andesites so that they in all likelihood formed as a result of crystallization differentiation.

Further examples of differentiation trends can be derived from the K_2O: SiO_2 ratio (PICHLER & WEYL 1973, p. 392). They are found in Poás, the volcanoes of southeastern Nicaragua and on Santa Maria/Santiaguito, while Telica and Izalco, for example, produced undifferentiated magmas according to this ratio.

The relationships between the Quaternary **volcanism and tectonics** are discussed in Chapter 8. At this point, however, I would like to add a few comments from the earlier literature, which are still valid today, even if they have been variously interpreted.

DOLLFUS & MONT-SERRAT (1868, pp. 292 – 297) recognized that the volcanoes of Central America are arranged in rows and they determined their direction to be SE/NW between Costa Rica and the Gulf of Fonseca, then bending in a WNW direction in El Salvador and Guatemala. They stressed that individual volcanoes could be offset by up to 20 km from these rows and they recognized that the volcanoes of Lake Atitlán, the group from Fuego and Acatenango and in particular those of the Departamento Chiquimula in eastern Guatemala, are arranged at right angles to the main direction of the NS-oriented rows.

Even before the above authors, VON SEEBACH (1865), speaking at a meeting of naturalists in Hannover, said that "the Fuego volcanic group, like almost all composite volcanoes in Central America, is oriented on a transverse fissure almost at right angles to the main volcanic chain of Central America" (VON SEEBACH 1898, p. 224).

SAPPER (1897, 1913, 1937) also stressed the fact that the volcanoes were arranged in rows in which, in his opinion, fissures formed the ascent routes for the magma. He realized that the rows were offset: "The volcanoes of Central America are not arranged along one longitudinal fissure but instead are distributed among a number of shorter individual fissures which are offset from each other" (1897, p. 681). He felt that the volcanism was caused by the transition from the continent to the Pacific Ocean, and the displacement of the individual rows was caused by varying block movement: "In the process of the Pacific Basin subsiding, this did not take place continuously but discontinuously in individual blocks, and volcanic eruptions occurred at the faults or fracture zones which separated the moving block from the stationary continent."

STOIBER & CARR (1974) picked up these ideas again but in a quite different connection, and we will come back to them several times in our discussion of tectonics and earthquake activity.

5.5 Individual areas

5.51 El Salvador

With a territory covering approximately 21,000 km^2, El Salvador is the smallest of the Central American republics and, with the exception of the extreme north of the country, it is located in the region of the Tertiary and Quaternary volcanic formations. Geological research on El Salvador was pursued in the fifties mainly by the "Instituto Tropical de Investigaciones Científicas" (ITIC), as it was then known, and by the "Servicio Geológico Nacional", which was independent up until 1963. The results of this period of research are contained in a series of journals put out by both institutions (cf. Chapter 1.3) as well as in a number of individual publications. They have been summarized by MEYER-ABICH (in HOFFSTETTER 1960, pp. 99 – 128) and in the first (German) edition of this book (WEYL 1961 a). The main results of the research were the stratigraphic classification of the Mesozoic in the northern part of the country (DÜRR & STOBER 1956), the classification of the Tertiary and Quaternary volcanic products on the basis of their mode of occurrence, the morphology and the fossil soils (DÜRR & KLINGE 1960, DÜRR 1960), contributions on the history of Quaternary volcanism (WILLIAMS & MEYER-ABICH 1955, MEYER-ABICH 1956, 1958) and on the petrography of the volcanic products (WEYL 1952 – 1955), insights into the fault-dominated structure of the country (DÜRR 1960, GREBE 1961) and studies on the physical geography of the country, and in particular of its coastal regions (GIERLOFF-EMDEN 1956 – 1959) [17].

A new phase in the research work was initiated by the activity of a Geological Mission of the Bundesanstalt für Bodenforschung (Federal Institute for Geosciences and Natural Resources), Hannover on which WIESEMANN (1973/1977) has reported [18]. Whereas during the fifties only provisional topographical maps and initial series of aerial photographs were available, the Mission was able to prepare geological maps of the entire country (1967 – 1971) on the basis of a topographical map (scale 1 : 100,000) and of a complete series of aerial photographs; it then prepared a 1 : 500,000 scale general map in 1974, which was followed in 1978 by a 1 : 100,000 geological map comprising six sheets.

The mapping activity required a lithological-stratigraphic division of the country which went far beyond that arrived at by DÜRR & KLINGE (1960). Since fossils are lacking, with the exception of the Cretaceous in the north and of the Quaternary floras and faunas, it was only possible to base this division on the mode of occurrence of the various identified forma-

[17] The literature was fully quoted in the first edition of this book, therefore there is no need to repeat it in the list of references contained in the present edition.

[18] I am especially grateful to my colleague Dr. G. WIESEMANN and to the Bundesanstalt für Geowissenschaften und Rohstoffe (Federal Institute for Geosciences and Natural Resources) for letting me have unpublished maps and reports and for providing many stimulating suggestions.

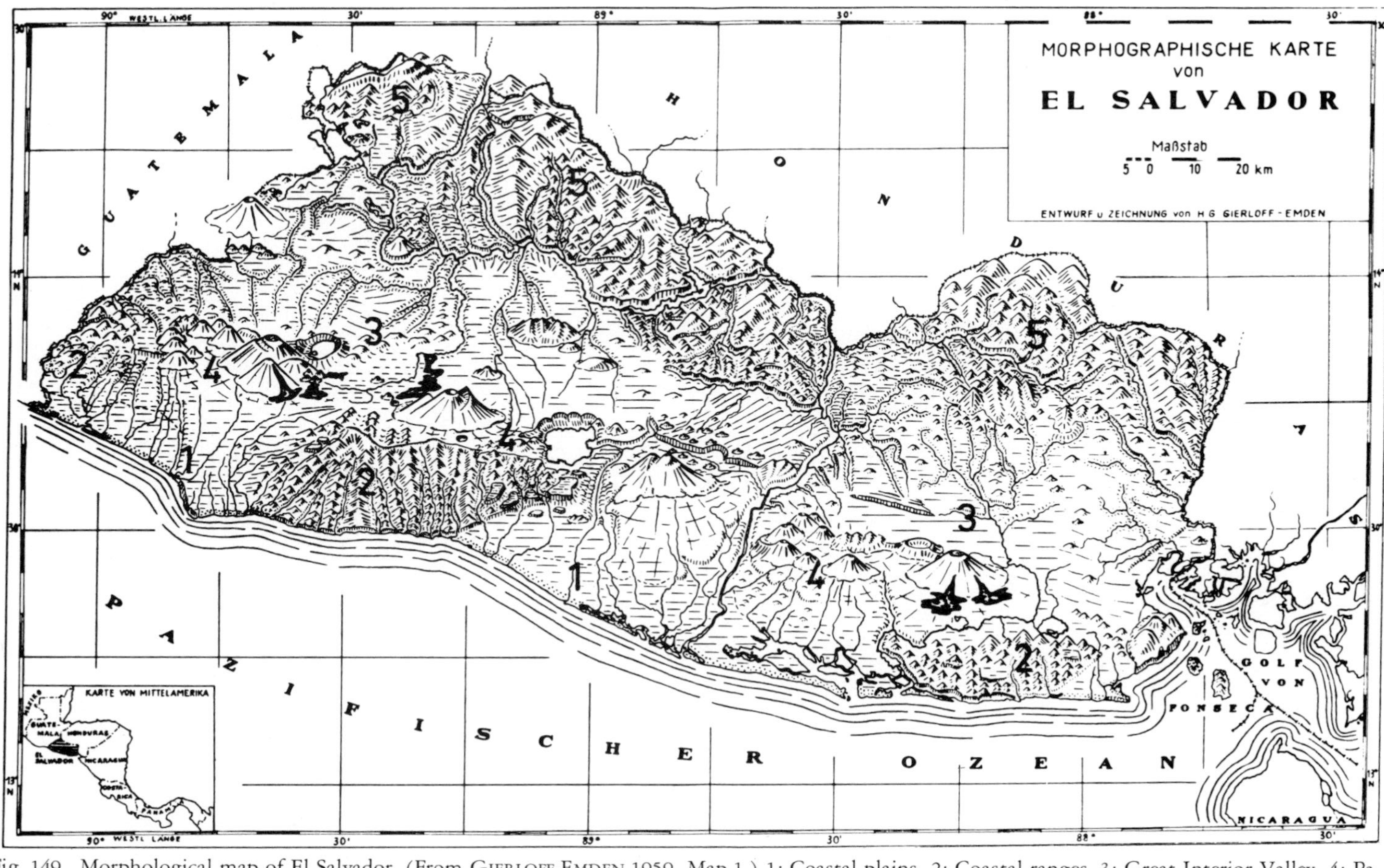

Fig. 149. Morphological map of El Salvador. (From GIERLOFF-EMDEN 1959, Map 1.) 1: Coastal plains, 2: Coastal ranges, 3: Great Interior Valley, 4: Pacific Volcanic Chain, 5: Northern mountains.

tions. At the same time an attempt had to be made to harmonize the stratigraphic sequence in El Salvador with that in the neighbouring countries which had developed over the same period of time. This necessitated making certain adjustments in the course of the work. In addition, since 1976 work had been going on in co-operation with the University of Texas in Austin to arrive at radiometric datings, but the results of this work are not yet available.

WIESEMANN proposes to give a detailed description of the geology of El Salvador. Therefore the following text has been kept brief, and it is based largely on the works by WIESEMANN (in particular the paper published in 1975). Individual questions relating to the geology of El Salvador have been discussed in the chapters on the Tertiary and Quaternary volcanism and the structure of northern Central America.

El Salvador can be divided up into the following morphological-geological units (Fig. 149):

Coastal plains in the west and in the central part of the country with alluvial deposits, spits and mangrove swamps ("Esteros"). Morphological-oceanographic works by GIERLOFF-EMDEN (in particular 1959) and sedimentological studies by WEYL (1953), v. d. BRELIE & TEICHMÜLLER (1953) and by HINTZ (1955) have been published on these areas.

Coastal ranges with the Tacuba, Balsam and Jucuarán Ranges. Built up from volcanic rocks of the Pliocene Bálsamo Formation. Beds and peneplains dipping gently towards the coast in the Balsam Range, which is referred to in the local dialect as "Cumbre". Partial covering of thick red soils. A mostly steep escarpment to the north sloping down to the

Great Interior Valley (STIRTON & GEALEY 1949), a heterogeneous basin and low mountain topography with more or less eroded extinct volcanoes and the intramontane basins of Metapán, Río Lempa, Río Titihuapa and Olomega (cf. map in Fig. 151). In the southern part of this Great Interior Valley comes the

chain of Pleistocene and sometimes still active **volcanoes** which were mentioned in detail in Chapter 5.4.

The **northern mountain ranges** with Monte Cristo, Los Esesmiles and the regions around Chalatenango and the Río Torola. Built up predominantly of Tertiary volcanics of the Chalatenango and Morazán Formations. Plutonic rocks and, in the region of Metapán, Cretaceous-Lower Tertiary sediments. In Cerro Pital the mountain ranges reach a height of 2,730 m, and there is a transition to the Honduran ranges.

GIERLOFF-EMDEN (1959) (Fig. 149) provided a similar, somewhat more detailed morphological division of El Salvador.

The **stratigraphic sequence** of El Salvador commences in the northern border region of Metapán with **Mesozoic** beds which, after DÜRR & STOBER (1956), are referred to as the Metapán Formation and which are subdivided into a lower Todos Santos "series", a middle Cobán "series" and an upper Subinal "series" (see Table 12). Based on the subdivision of Honduras, WIESEMANN (1975) modified the designations as follows: The subjacent (?) Jurassic-Cretaceous red beds correspond to the Todos Santos Formation of northern Central America; above these, the limestones and marls of the Albian-Cenomanian are assigned to the Yojoa Group, and the superjacent (?) Upper Cretaceous-Lower Tertiary conglomerates, sandstones and intercalated volcanics are assigned to the Valle de Angeles Formation (=Group). Unlike in Honduras and Guatemala, here the Upper Mesozoic is not very developed, and it probably represents the marginal facies of the thicker and more complete Mesozoic sequence further to the north. More recent marine beds are unknown; the Tertiary and Quaternary are predominated by terrestrial volcanism, the products and detritus from which are subdivided

Table 12. Volcanostratigraphic scheme of El Salvador. (Adapted from the "Geologische Übersichtskarte der Republik El Salvador 1 : 500 000", Hannover 1974. Revised nomenclature: For Formación de Metapán read Estratos de Metapán, for Subinal-Serie read Valle de Angeles Formation, for Cobán-Serie read Yojoa Group, for Todos Santos-Serie read Todos Santos Formation (see WIESEMANN 1975, Fig. 2).

Esquema volcano-estratigráfico / *Vulkanostratigraphisches Schema*			Extensión verticál de las unidades mapeadas / *Vertikalverbreitung der Kartiereinheiten*			
			Efusivas / *Effusiva* (ácido → básico / *sauer → basisch*)	Piroclastitas / *Pyroklastite* (ácido → básico / *sauer → basisch*)	Epiclastitas volcánicas / *Vulkanische Epiklastite*	Facies de cuencas / *Beckenfazies*
Holoceno / *Holozän*	Formación de SAN SALVADOR / *Folge*	'f: Aluviones, localmente con intercalaciones de piroclastitas; 's: barras costeras; 'm: manglares *'f: Alluvionen, örtlich mit Einschaltungen von Pyroklastiten; 's: Strandwälle, 'm: Mangroven*				
		Efusivas basálticas. 'c: cenizas y tobas de lapilli *Basaltische Effusiva. 'c: Aschen und Lapillituffe*	s5	s5'c	↑	
		Piroclastitas ácidas ("tierra blanca"). 'a: efusivas ácidas *Saure Pyroklastite ("Tierra blanca"). 'a: Saure Effusiva*	s4'a	s4	s4	
		Piroclastitas ácidas, epiclastitas volcánicas ("tobas color café"). 'a: efusivas ácidas *Saure Pyroklastite, vulkanische Epiklastite ("Tobas color café"). 'a: Saure Effusiva*	s3'a, s2	s3, s2	s3	
		Efusivas andesíticas y basálticas; piroclastitas *Andesitische und basaltische Effusiva; Pyroklastite*				
		Piroclastitas ácidas, epiclastitas volcánicas, tobas ardientes y fundidas; efusivas andesíticas *Saure Pyroklastite, vulkanische Epiklastite, mit Glut- und Schmelztuffen; andesitische Effusiva*	s1	s1	s1	↑
Plio–Pleistoceno / *Plio–Pleistozän*	Formación de CUSCATLÁN / *Folge*	Efusivas andesíticas y basálticas *Andesitische und basaltische Effusiva*	c3			
		Efusivas ácidas e intermedias-ácidas (occurrencias aisladas eventualmente = ch2) *Saure und intermediär-saure Effusiva (isolierte Vorkommen evtl. = ch2)*	c2	c1	c1	c1'1
		Piroclastitas ácidas, epiclastitas volcánicas, tobas ardientes y fundidas, edad de ch localmente posible. 'l: sedimentos fluviales y lacustres con intercalaciones piroclásticas *Saure Pyroklastite, vulk. Epiklastite, Glut- und Schmelztuffe. Lokal ch-Alter möglich.* *'l: fluviatil-lakustrine Ablagerungen mit Pyroklastit-Einschaltungen*			↑	
	Formación de BÁLSAMO / *Folge*	Efusivas andesíticas-basálticas *Andesitisch-basaltische Effusiva*	b3	b3		
		Efusivas andesíticas, piroclastitas, epiclastitas volcánicas subordinadas *Andesitische Effusiva, Pyroklastite, untergeordnet vulk. Epiklastite*		b1	b2	
		Epiclastitas volcánicas (en parte fluvial–? lacustre), piroclastitas, corrientes de lava intercaladas *Vulk. Epiklastite (z. T. fluviatil-? lakustrin), Pyroklastite, eingeschaltete Lavaströme*	b1	b1	b1	
?Oligoceno–?Mioceno / *?Oligozän–?Miozän*	Formación de CHALATENANGO / *Folge*	Efusivas ácidas (occurrencias aisladas eventualmente = c2); riolitas *Saure Effusiva (isolierte Vorkommen evtl. = c2); Rhyolithe*	ch2			
		Piroclastitas ácidas, epiclastitas volcánicas con tobas ardientes y fundidas; efusivas ácidas intercaladas. Edad de c localmente posible. / *Saure Pyroklastite, vulk. Epiklastite, mit Glut- und Schmelztuffen; saure Effusiva eingeschaltet. Lokal c-Alter möglich.* Granito, granodiorita / *Granit, Granodiorit*		ch1	ch1	
	Formación de MORAZÁN / *Folge*	Efusivas básicas-intermedias hasta intermedias-ácidas, piroclastitas, epiclastitas volcánicas. Alteración regional por influencia hidrotermal. / *Basisch-intermediäre bis intermediär-saure Effusiva, Pyroklastite, vulk. Epiklastite. Regional hydrothermal zersetzt.*	m2	m2	m2	
		Efusivas intermedias-ácidas, piroclastitas, tobas ardientes, riolitas, epiclastitas volcánicas. Occurrencias aisladas eventualmente más jóvenes. / *Intermediär-saure Effusiva, Pyroklastite, Gluttuffe, Rhyolithe, vulk. Epiklastite. Isolierte Vorkommen evtl. jünger.*	m1	m1	m1	
	Fm. de METAPÁN / *Folge* — Subinal	Conglomerados de cuarzo y caliza rojos, areniscas; intercalaciones de vulcanitas *Rote Quarz- und Kalkkonglomerate, Sandsteine; Vulkanite eingeschaltet*	me3	me3	me3	
?Jurásico–Cretácico / *?Jura–Kreide*	Cobán	Calizas y calizas margosas *Kalksteine u. Mergelkalksteine*				
	Todos Santos	Conglomerados de cuarzo, areniscas, siltitas y lutitas; vulcanitas básicas-intermedias subordinadas. 'a: metasedimentos, metavulcanitas *Quarzkonglomerate, Sandsteine, Schluffe und Tonsteine; untergeordnet basisch-intermediäre Vulkanite.* *'a: Metasedimente, Metavulkanite*	me1'a	me1'a	me1'a	

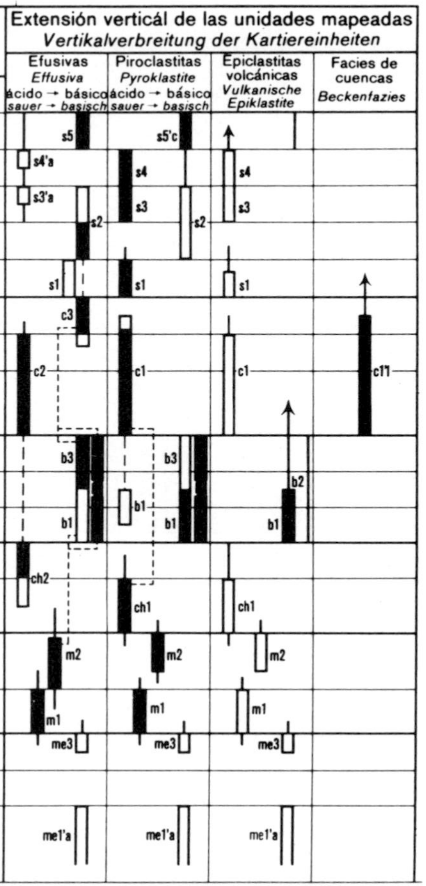

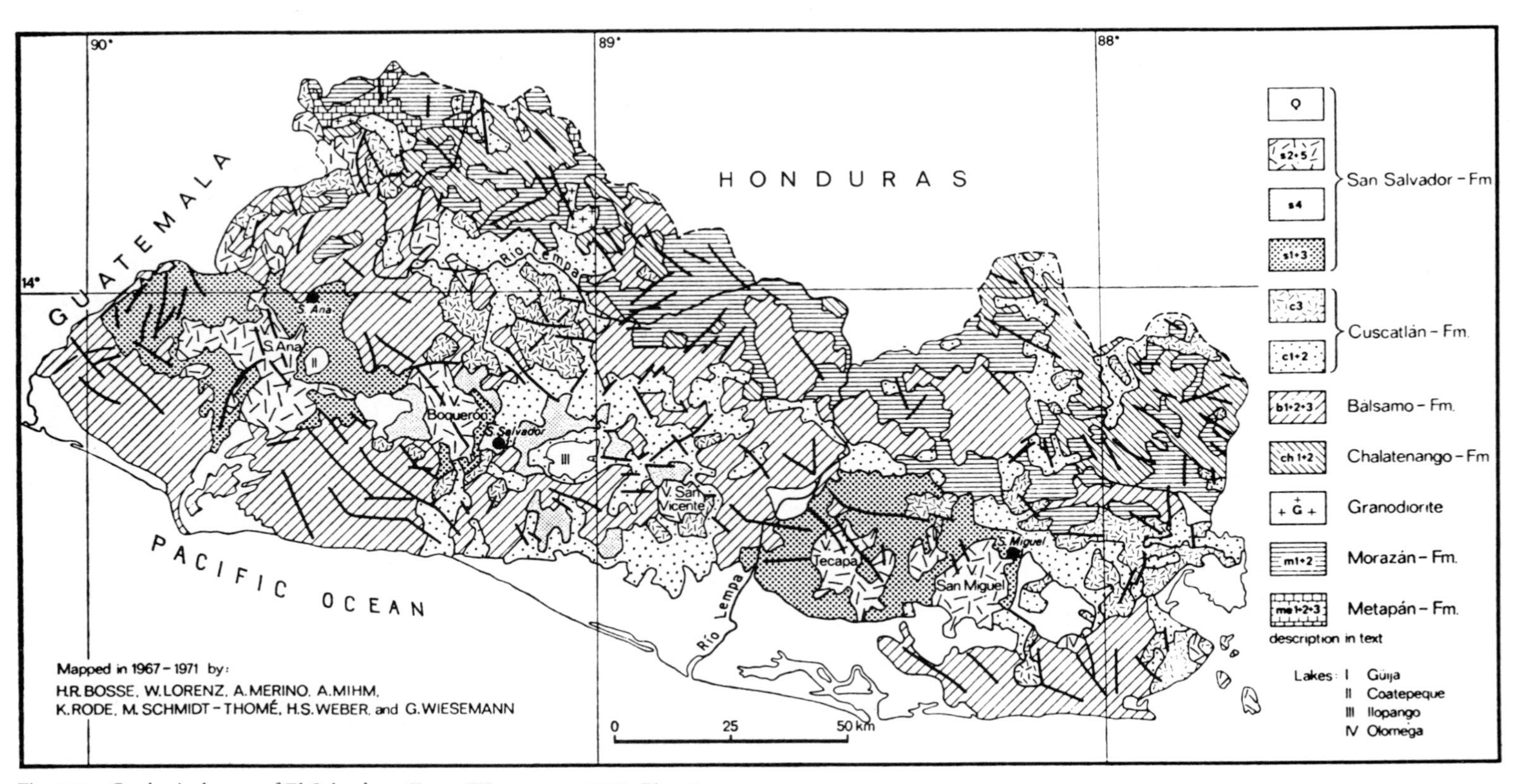

Fig. 150. Geological map of El Salvador. (From WIESEMANN 1975, Fig. 3).

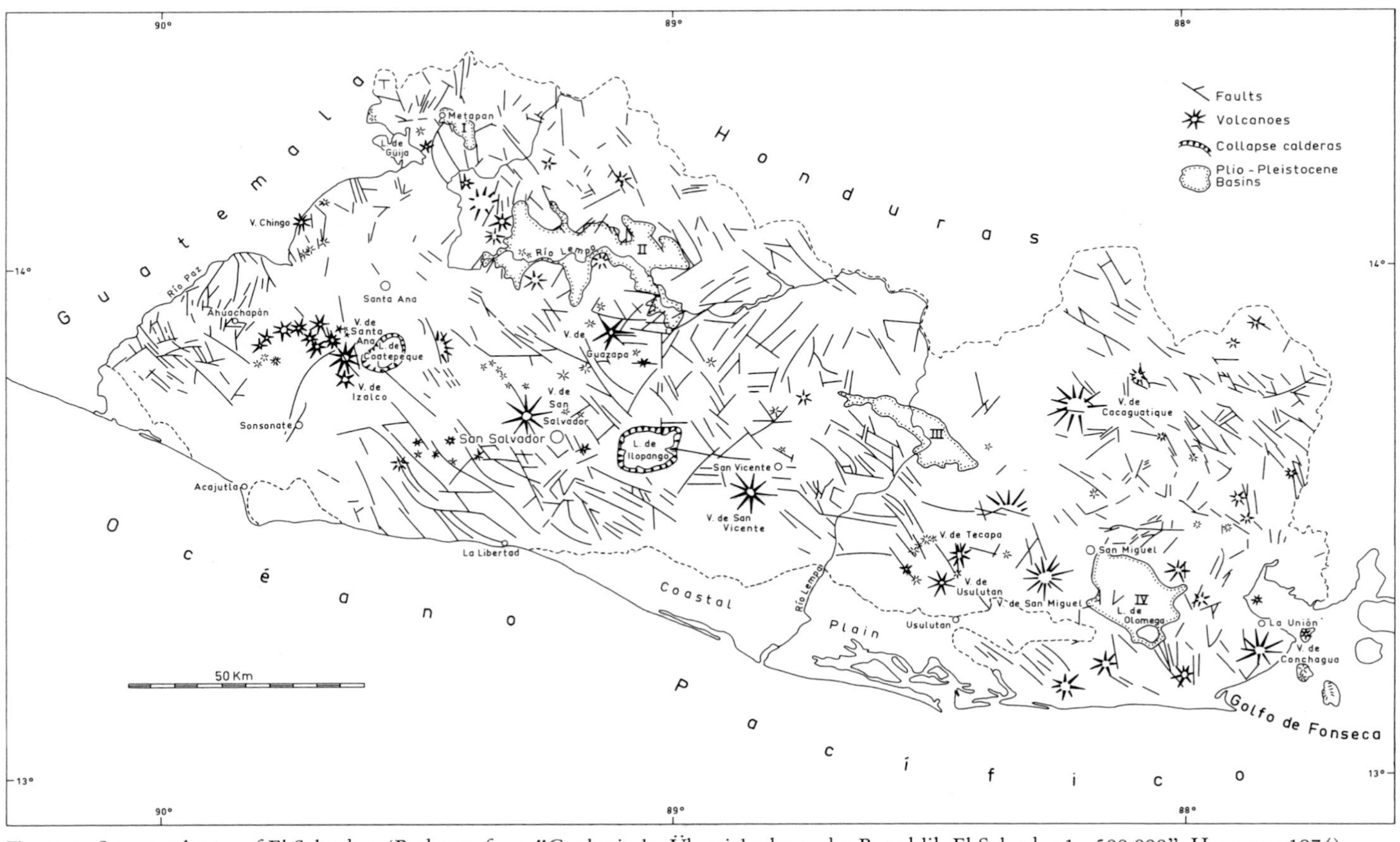

Fig. 151. Structural map of El Salvador. (Redrawn from "Geologische Übersichtskarte der Republik El Salvador 1 : 500,000", Hannover 1974).

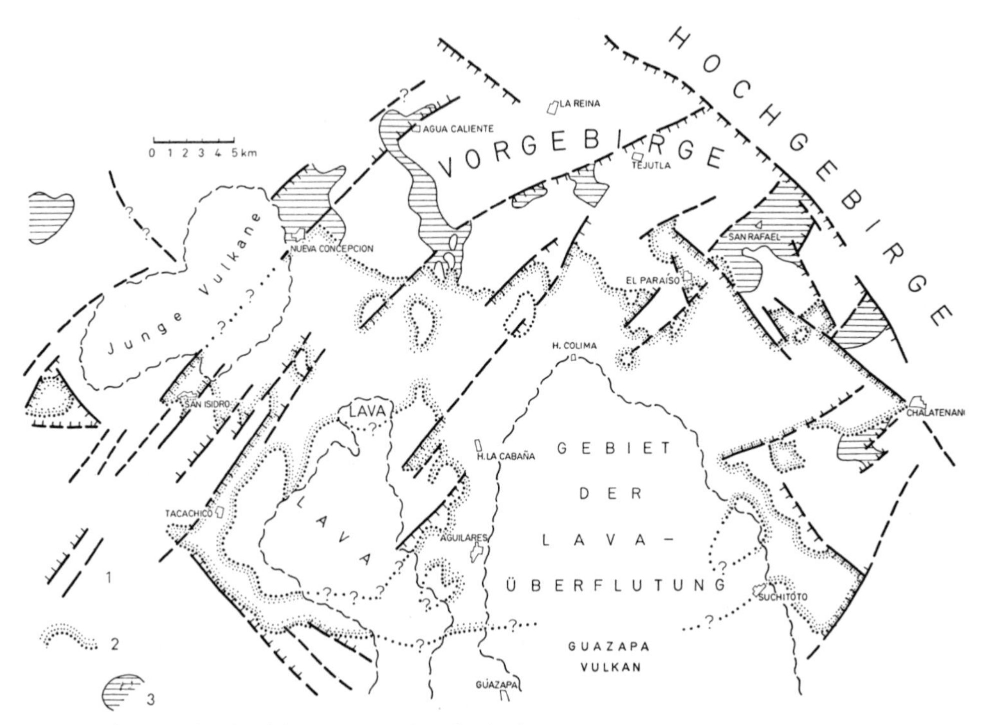

Fig. 152. Tectonic sketch of the Lempa Basin, El Salvador. (From RODE 1975, Fig. 13).

according to the scheme shown in Table 12. Because of the lack of any time marks, the classification according to the stages of the Tertiary had to be based, with reservation, on the situation in neighbouring countries, but it is expected that corroborative evidence will be provided by radiometric datings.

The acidic to basic rocks of the **Morazán Formation** and the acidic rocks of the **Chalatenango Formation** are restricted to the northern part of the country. They interlock with acidic, intermediate and basic plutonic rocks (for modal mineral inventory see WEYL 1961 a, p. 114, chemical analyses idem p. 146).

The more recent **Bálsamo Formation** of intermediate to basic volcanic products, which can be placed with reservation in the Miocene (?) to Pliocene, provides the constituent material for, in particular, the coastal ranges of El Salvador. To the north the Bálsamo Formation extends over the above-mentioned older formations but does not form any mountain ranges corresponding to the coastal ranges (cf. map in Fig. 150). Thick covers of red soils on the volcanic rocks were an important time marker for DÜRR & KLINGE (1960) ("Cumbre Rotlehm").

Acidic to intermediate volcanism, which was concentrated mainly in the central parts of El Salvador, recommenced during the period of the Bálsamo Formation. Ignimbrites were produced, among other places, on the southern slope of the Bálsam Range (WEYL 1954). The rocks were designated the **Cuscatlán Formation** and were placed in the Pliocene to Pleistocene. This formation is also taken to include relatively severely eroded ruins of volcanic edifices, such as Guazapa, which are still covered by a thin layer of red soil. DÜRR & KLINGE

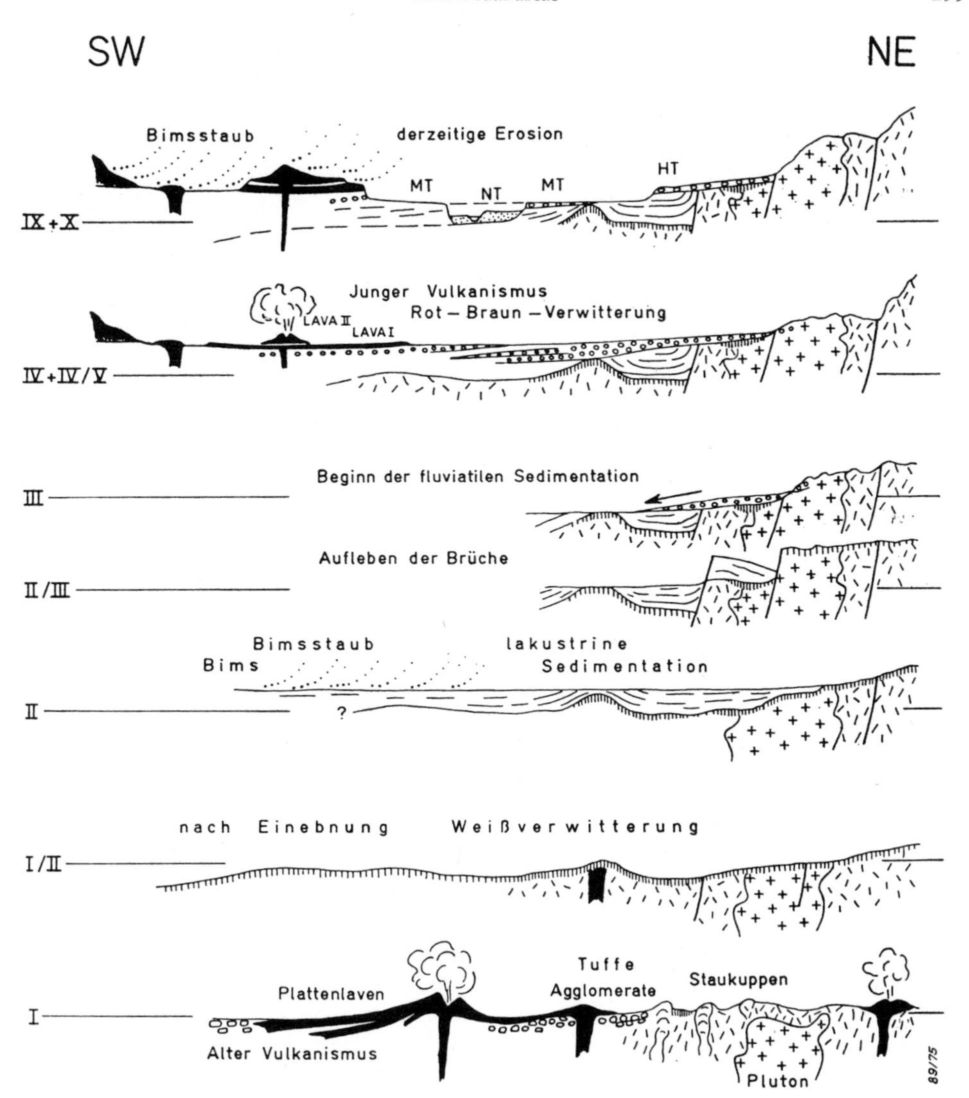

Fig. 153. Evolutionary sketch of the Lempa Basin and surroundings, El Salvador. (From RODE 1975, Fig. 14).

(1960) used this red soil as a time marker and assumed that it formed in the Lower Pleistocene to Upper Pliocene.

The Cuscatlán Formation includes limnic sediments with diatomites, which yielded a rich flora and fauna near Poza Azul in the valley of the Río Sisimico (SCHMIDT-THOMÉ 1975, LÖTSCHERT & MÄDLER 1975, SEIFERT 1977). The flora, the first of any great extent from the tropics of the Americas, was deposited in a freshwater lake. It contains the same families, genera and sometimes even the same species as the present hydrophilic lowland-forest of El Salvador. The delicate substance of many leaves and the frequent occurrence of drip tips indicates that this must have been a rain forest. The extinction of certain species indicates a shift from the fossil to the recent flora. LÖTSCHERT & MÄDLER conclude that the Pleistocene glaciation, evidence of which is found in the glacial scoring of the high mountain ranges of the American tropics, had little if any effect on the lowland and coastal regions, so that the floras remained substantially undamaged.

The most recent formations, which are designated the **San Salvador Formation,** are concentrated in the Pleistocene and Holocene volcanoes, their lavas, pyroclastics and detritus. They include the "tobas color café" and the large pumice covers of the "Tierra blanca" in the vicinity of the volcanotectonic depression of the Lago de Ilopango (WILLIAMS & MEYER-ABICH 1955) and the caldera of the Lago de Coatepeque (MEYER 1964).

While the extensive rock collections of the Geological Mission of the Bundesanstalt für Geowissenschaften und Rohstoffe are still being examined, the reader must be referred to the literature summary in the first edition for a petrographic characterization of the volcanic rocks. Further chemical rock analyses and a brief discussion of the Tertiary volcanic products are found in WEYL (1966 a, pp. 76 – 78).

Altogether the volcanic sequence reveals multiple changes in the chemistry of the rocks and a shift in the main regions of volcanic production:

Morazán Formation	northern border region, acidic, intermediate and basic
Chalatenango Formation	northern border region, acidic with ignimbrites
Bálsamo Formation	southern and central parts of the country, intermediate to basic
Cuscatlán Formation	mainly in the central parts of the country, acidic to basic
San Salvador Formation	between the coastal ranges and the northern border region, main concentration in the chain of Pleistocene volcanoes, acidic to basic.

This alternation suggests that there was a shift in the subduction zone and thus in the zone of magma formation, which is similarly evidenced by the distribution of the Cenozoic volcanics in the rest of northern Central America (cf. Chapter 5.2).

The **structure** of El Salvador is determined by several fault systems, some of which were already noted by earlier observers, but their full significance was not realized until the mapping work carried out by the Geological Mission (see maps. Figs. 150 and 151). WIESEMANN (1975, pp. 564 – 568) reviewed the development of our knowledge on this subject and, on the basis of the mapping activity, arrived at the following conclusions: In El Salvador two or four structural systems of varying importance intersect. The most important fault system strikes WE and extends from the border of Guatemala to the northern edge of the Olomega Basin. It corresponds by and large with the central depression mentioned by earlier authors. In the western part of the country it takes the form of a true graben, the Salvador-Graben sensu stricto, 100 km long and about 20 km wide, ending in an acute angle at the Pacific coast. In the central part of the country the Balsam Range stands out clearly in the terrain as an uplifted southern block, whereas to the east of the Río Lempa it is the uplifted northern block which stands out. This fault system is surmounted by the large volcanoes of the Santa Ana Group as far as San Miguel, and they to some extent mask the faults. Basins like that of the Lago Ilopango, as well as one between Boquerón and Coatepeque, one northwest of Tecapa and one to the east of San Miguel, are located in the region of this fault system.

The fault system also includes a less pronounced N/S-striking fault system.

Another important system strikes NW/SE and determines the position of a number of volcanic eruption centres; it includes four intramontane basins which can be regarded as lying along the line of extension of the Nicaragua Depression. The basins in question are those of

Olomega, Río Titihuapa, Río Lempa and Metapán (see map in Fig. 151), and they are filled with largely fluviatile and limnic sediments from the Cuscutlán Formation. RODE (1964, 1965, 1975) studied the morphology, sediments, soils and mode of occurrence of the Lempa Basin and its surroundings and from this derived the map (Fig. 152) as well as the evolutionary sketch (Fig. 153) of this intramontane basin.

5.52 The Pacific Coastal Plain

Between the volcanoes of Central America and the Pacific coast stretches a coastal plain of varying width which is interrupted by individual groups of mountains. The plain is most extensive in Guatemala; in El Salvador it is broken up by the Balsam and Jucuarán Ranges; in Nicaragua it surrounds the volcanoes of the Marabios Range on both sides; but it is severely constricted again by mountain ranges between Lake Nicaragua and the coast.

TERMER (1936, p. 261) described the **Guatemalan region** as follows:

"From Chiapas the volcanic coastal range is accompanied by a foreland of varying width, which gradually slopes down to the Pacific Ocean. This foreland is between 40 and 60 km wide in the west and it narrows down to 30 km and less in the east. Towards the sea, it is accompanied throughout its entire length by a shallow lagoonal coast with elongated channels cut off from the sea by low beach ridges and containing brackish water whose level fluctuates with the tides. In the eastern section, which we can take as commencing from the line between Escuintla and San José, the lagoons join together in a continuous strip, the Canal de Chiquimulilla, which extends from Iztapa to the mouth of the Rio de los Esclavos. Mangrove swamps border the banks of these waters, and behind them rises the tropical coastal forest. Through this forest run the many rivers coming down from the mountains and discharging into the lagoons or into the open sea.

Behind this comes slowly rising flat country broken up by the river valleys running parallel to each other or merging at acute angles. At an altitude of 100 m above sea level a second zone is reached, namely a mountainous country, intersected by ravines, which in turn leads on into the slope of the coastal cordillera."

Geologically speaking the coastal plain is a uniform, monotonous area covered by pyroclastics and re-deposited sediments from them. In the west the slope of the highland region (TERMER talks of a coastal cordillera) down to the coastal plain is interrupted by two significant terraces at altitudes of 1,400 m and 1,100 m respectively. These terraces can be taken as marking uplift events. Marine sands in recent alluvial deposits, which TERMER found at 75 m above sea level near Caballo Blanco, indicate that the uplift activity continued right into the most recent geological past. Conversely, he discovered that no terraces occur in the southeastern part of the coastal region, while swamping of prehistoric settlements and cultural remains in the mud of coastal lagoons are evidence of subsidence.

These early observations have been supplemented by a study by HORST, KUENZI & MCGEHEE (1976). According to these authors the up to 50 km wide coastal plain rises gently to the foot of the volcanoes with a clear break in the slope coming at the 500 m line, which can be taken as the upper limit. Along the coast the plain is broken up by deltas, spits and behind them mangrove swamps. The sediments on the plain are for the most part fluviatile, but in the upper section lahars occur, while off the coast deltaic, spit and mangrove swamp sediments are found. The material comes almost exclusively from the volcanic province and therefore consists of detritus composed of andesites, basalts, dacites etc. and minerals such as plagioclase, basic silicates and magnetite. The average grain size declines as the coast is approached; the grains are not very rounded and have not undergone very much weathering. The sediments resemble those of the section of coast in El Salvador.

KUENZI et al. (1975, 1979) described the Río Samala as a model for recent sedimentation, and they summarized their results as follows: "The Río Samala fluvial-deltaic system, which heads in the volcanic arc, displays a combination of all of the Holocene features (frequently-shifting braided channel, flanking elongate dendritically branched lakes, and a high-destructive, wave-dominated arcuate delta) that characterize the coastal plain in Guatemala. Historical records suggest that the catastrophic eruption of Santa Maria volcano in 1902 dramatically increased the amount of sediment supplied to the Río Samala in the succeeding few years and produced a rise of 10 – 15 m in the bed of the river, blocking the drainage of its tributaries to produce flanking lakes. Aggradation of sand and gravel produced an elongate fan parallel to flow. Frequently the braided system was diverted laterally into a flanking lake, interrupting deposition of organic-rich muds and locally filling the lake by progradation of a Gilbert-type delta. Contemporaneous with aggradation of the fluvial system, an elongate deltaic platform prograded about 7 km seaward (between 1902 and 1922) in response to the deposition of approximately 7.8 km^3 of deltaic sediments. However, with gradual waning of sediment supplied to the fluvial system, the delta entered a destructive phase and sands were redistributed laterally to prograding shoreface and beach environments to develop the present arcuate delta. A new constructive phase would begin in the event of another great eruption." The detailed report of KUENZI et al. (1979) was published after my own manuscript was completed.

Abstracts are available of further studies on the erosion of volcanoes, in particular of Fuego, and on the sedimentation in the coastal plain of Guatemala (BOOTHBY et al. 1978, VESSEL & DAVIES 1978).

Geophysical surveys revealed considerable thicknesses of the sediments in the coastal plain and shelf region, and as a result, at the end of the sixties, **exploration wells** were drilled by Texaco and Esso; on the mainland these wells reached a depth of approximately 3,820 m, and in the shelf they were sunk to 3,590 m. Although nothing was made known about the stratigraphic sequence through which the wells were sunk, the data indicate that one or more sedimentary basins of considerable depth extend beneath the coastal plain and shelf.

In the fifties, the **Salvadorian section** of the coastal plain was the subject of a large number of investigations which were initiated by the Instituto Tropical de Investigaciones Científicas (ITIC), as it was then called, in San Salvador [19]. The results were summarized by GIERLOFF-EMDEN (1959). He distinguished between a series of natural units, which are separated from each other by rivers and mountain ranges. The most important features are the following:

The coastal plains between the border river Río Paz and the Balsam Range;
The coastal plains between the Balsam Range and the delta of the Río Lempa;
The delta of the Río Lempa;
The coastal plain of Usulután;
The coastal region around the Gulf of Fonseca;
The "Esteros" of Jaltepeque and Jiquilisco.

[19] The works are quoted in the first (German) edition of this book from which the following section has been taken, with a few slight modifications. New data were provided by GIERLOFF-EMDEN (1974) from the four-channel images of the multispectral scanner in the ERTS satellite.

The coastal plains are up to 30 km wide and rise with a gradient of 0.5% to 1% landwards, ending at an altitude of about 100 m without any sharp boundary at the foot of the volcanoes. Towards the sea they are usually delineated from the lagoons and mangrove marshes of the coast by a sharp, low edge. The rivers cut several tens of metres deep into the coastal plain. The sediments of the coastal plain come for the most part from the immediate volcanic hinterland, and accordingly one finds volcanic rocks and minerals, ashes and their decomposition products. The individual volcanoes can also be recognized in the re-deposited material of the rivers and beach sands (HINTZ 1955). The predominant minerals are plagioclase, augite, hornblendes and ore minerals. Some of the sands and silts are already semi-consolidated; in 1956 GREBE & HABERLAND, near the Hacienda La Carrera, found animal and human footprints beneath a 1.5 m thick cover of sediments. The footprints were dated to a time approximately 1500 B.C., thus providing a clue to the rate at which sedimentation and consolidation of the deposits occurs. The rivers and flows of debris and mud during the rainy season also are involved in the process of sedimentation. A considerable amount of material is transported annually; GIERLOFF-EMDEN estimated that the sediment load in the Río Lempa alone was 7.7 million tons per annum. ZOPPIS & DEL GIUDICE (1958) gave the maximum thickness of the exclusively Quaternary alluvial masses in Nicaragua as 30 m.

Towards the sea the beach region proper, with large and small lagoons as well as tidal flats, the "Esteros", lies ahead of the coastal plains. These Esteros give the Pacific coast of Central America its particular character; in El Salvador the largest are the Estero de Jiquilisco and the Estero de Jaltepeque. Geological and sedimentological studies were published by v. d. BRELIE & TEICHMÜLLER (1953), HINTZ (1955), and WEYL (1953 a and b, 1954). The flora and fauna, which have a definite influence on the sedimentation process in the Esteros, were studied by HARTMANN (1956, 1957), LÖTSCHERT (1954, 1955), PETERS (1955), SCHUSTER-DIETERICHS (1957), and ZILCH (1954), to mention only the most important works. GIERLOFF-EMDEN (1959) summarized the geographical results, adding some observations of his own.

Spits up to 40 km long run along the coast. They are built up from beach ridges, sometimes running parallel to the shore and sometimes arcuate, which have prograded seawards through recurved spits as well as frontally (Fig. 154). From the course followed by the beach ridges and from the shapes of the mouths of rivers, GIERLOFF-EMDEN was able to deduce that the process of progradation took place in stages. According to him, the coastal margin has been increasing without interruption for about 3,000 years and has now reached the final stage of a straight shoreline, which at present seems to be stable.

The sands of the spits are made up of plagioclase, volcanic glass, very little quartz, titaniferous iron ores as well as hypersthene, augite, hornblende and, to a secondary extent, epidote, zirconium, garnet, and olivine (HINTZ 1955). They are a remarkable example of an accumulation of large, almost quartz-free sand masses. They probably originated from the groups of volcanoes in the hinterland, and to a lesser extent from the middle and upper reaches of the Río Lempa. Concentrates of heavy minerals, in particular of iron ores, are frequently found. The expected mineable deposits of titaniferous iron ore have not been discovered (cf. Chapter 9.15).

A branched system of canals is protected by the spits, and their water flows in and out with a mean tidal range of 1.85 m. Tidal sand and mud deposits and, above all, extensive mangrove stands spread out between the canals, and sediment is trapped in the tangled roots of the mangroves. Sandy and silty mud and occasionally extremely humous, peaty sands make up the sediment. The mineral content is the same as in the sands of the spits, and two peaks

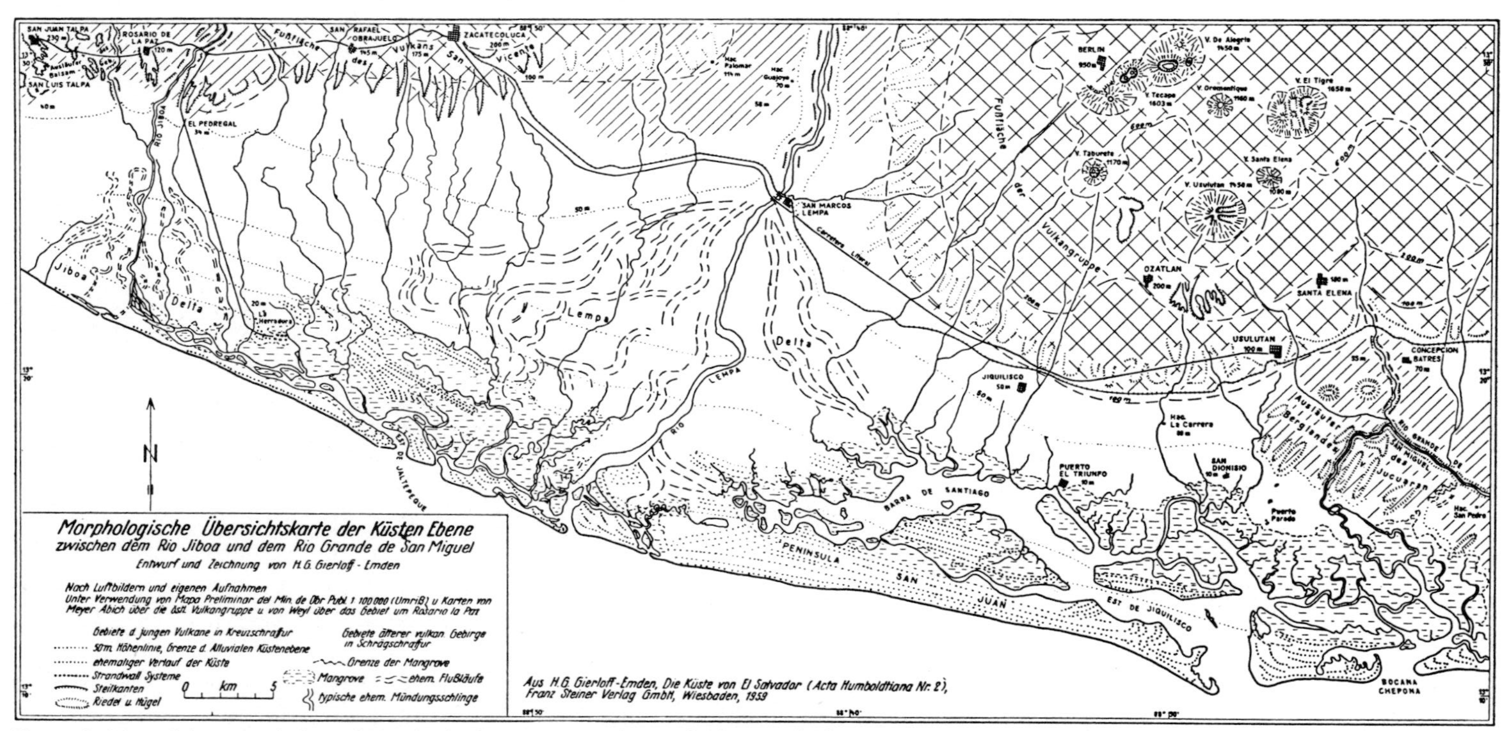

Fig. 154. Map of the coastal plain of El Salvador between Río Jiboa and Río Grande de San Miguel. (From GIERLOFF-EMDEN 1959, Map 8).

in the grain size distribution occur frequently in the fine sand and silt fraction. The structure of the mangrove sediments is determined by the extreme degree of rooting of the mangroves and also by the burrowing activity, in particular of crabs (cf. v. d. BRELIE & TEICHMÜLLER 1953, 1954, WEYL 1953 e, 1954 c).

The supply of large amounts of clastic material, which the rivers bring down in particular from the unconsolidated rocks and deposits in the volcanic region, and the constant re-deposition of this material by the action of surf and littoral currents means that coral reefs can flourish only at a few points on the Pacific coast. The only point on the Salvadorian coast where coral reefs are to be found is situated at Punta Remedios near Acajutla (GIERLOFF-EMDEN 1959). The shell banks which have been thrown up here on the beach have, for several years, provided the raw materials for a cement works.

The almost straight line formed by the Pacific coast of El Salvador and Nicaragua is strikingly interrupted by the Gulf of Fonseca, which, with the adjoining mangrove swamps, extends over 50 km deep into the continent and is surrounded by extensive alluvial plains. GIERLOFF-EMDEN (1959) drafted an accurate map of the Gulf and its surroundings. Since no geological studies have been made of the region of the Gulf of Fonseca, we have to assume on the basis of the topography that it formed as the result of the intersection of two zones of subsidence: on the one hand the large trough which extends from the Caribbean coast of Costa Rica via the lakes of Nicaragua in a northwesterly direction right up to the Gulf of Fonseca, and on the other hand the trough and depression zone of Comayagua, which runs through Central America in a north-south direction. GIERLOFF-EMDEN (1959, p. 141) further points out that the fault lines and volcanic ranges of the Pacific coast bend in the Gulf of Fonseca from an orientation of 105° in El Salvador to an orientation of 130° in Nicaragua. These findings have been taken up again by CARR and STOIBER, in particular, in connection with the modern concept of plate tectonics; this was discussed in greater detail in Chapter 3.22.

6. Seismic activity

Central America lies in the circumpacific seismic belt. Many descriptions of the earthquakes exist — some of them dating from quite early on — and they have been summarized by, among others, MONTESSUS DE BALLORE (1888), SIEBERG (1932) and SAPPER (1937) (Fig. 155). These observations provided information on the geographical distribution, intensity and consequences of seismic activity, and they indicated certain relationships with the geological structure, where the latter was known: The main earthquake zone of Central America is situated near the Pacific coast and coincides with the zone of young volcanoes and frequent faulting. This made it difficult to distinguish between tectonic and volcanic quakes, which was regarded as a particularly important distinction to make, above all in the older literature.

However, since the appearance of the first world-wide review of earthquakes by GUTENBERG & RICHTER (1954), and in particular since the establishment of a World-Wide Standard Seismograph Network (WWSSN), our understanding of the seismic activity in Central America has increased fundamentally. Taking the seismic records from four seismographs, SCHULZ (1964) in El Salvador was able to distinguish between earthquakes in the Pacific shelf region and deep-sea trenches with focal depths dipping towards the continent at between 10 and 120 km below the surface, and a continental earthquake zone with shallow-focus earthquakes. LOMITZ & SCHULZ (1966) were able to confirm this in the course of evaluating the Salvadorian earthquake of May 3, 1965, and they depicted this situation in the map shown in Fig. 156, which separates the two types of earthquake and earthquake region. In subsequent years our knowledge of the seismicity has been expanded and differentiated throughout the entire region of Central America as a result of the numerous earthquakes which have been recorded and evaluated by the WWSSN (MOLNAR & SYKES 1969, DEWEY & ALGERMISSEN 1974, STOIBER & CARR 1974, GRASES 1975, CARR 1976, BOWIN 1976, SPENCE & PERSON in ESPINOSA 1976, CARR & STOIBER 1977 et al.). In particular, the severe earthquakes in Managua on December 23, 1972 and in Guatemala on February 4, 1976 and the evaluation of these events have contributed a great deal to our knowledge of the causes and tectonic setting of earthquake activity (BROWN, WARD & PLAFKER 1973, ESPINOSA (ed.) 1976).

According to these authors, there are three basic types of earthquake that occur in the region of Central America:

1. Earthquakes in the seismic zone dipping towards the continent from the Middle America Trench with focal depths of about 30 to 200 km. They are nowadays without exception

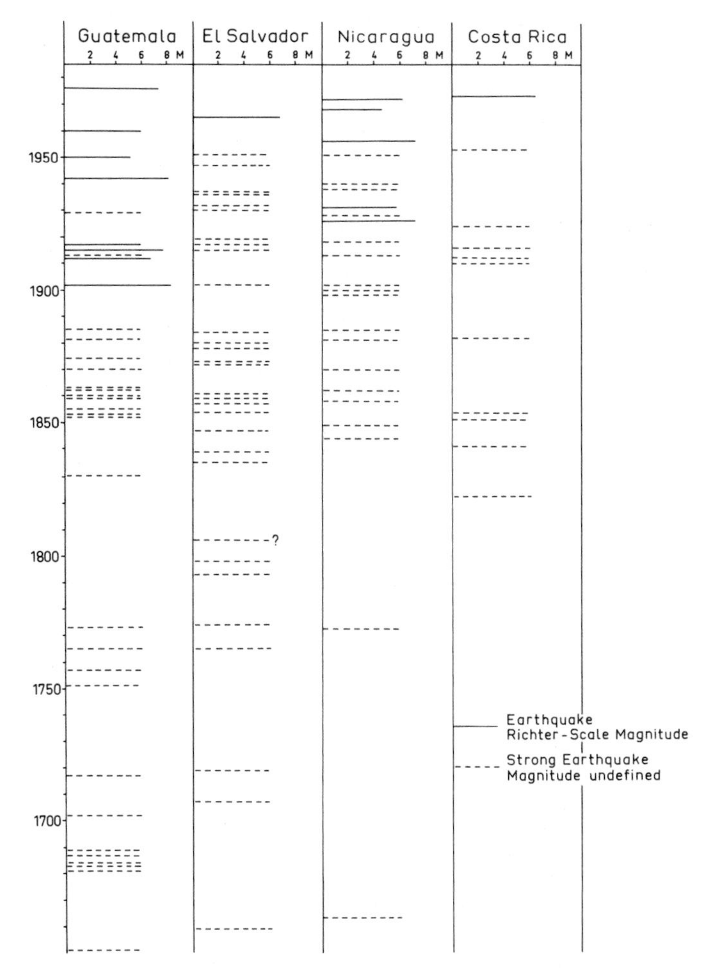

Fig. 155. Central American earthquakes during the last 300 years.

interpreted as the direct expression of subduction of the Cocos Plate beneath the Caribbean Plate.

2. Shallow-focus earthquakes in the region of the Pacific Volcanic Chain and Fracture Zone, which is interpreted, in accordance with various models, as a reaction of the crust to events in the subduction zone.

3. Earthquakes occurring along the major fracture systems in northern Guatemala, which form the boundary between the Caribbean and North American plates or which, in the form of a transform fault, separate the Cocos Plate and Nazca Plate from each other in the Panama Fracture Zone.

CARR & STOIBER (1977) have studied the historically recorded earthquakes that resulted in considerable damage to see how they could be classified according to the above types, and they were able to assign 26 earthquakes with shallow and intermediate focal depths to the

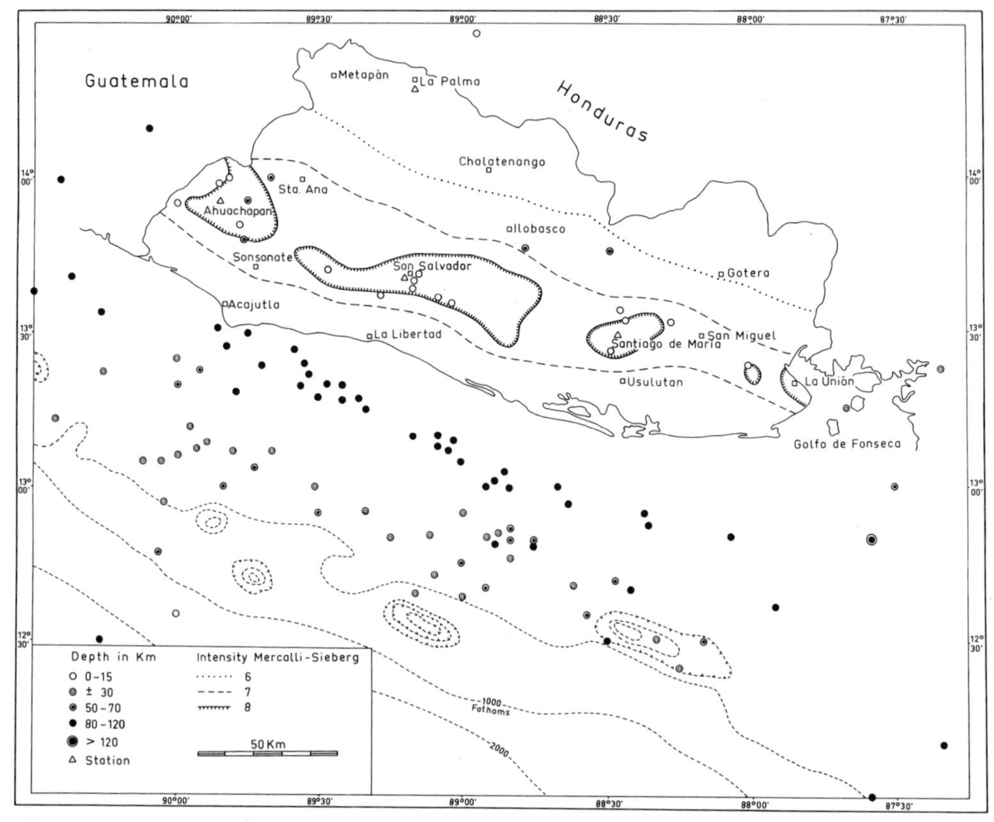

Fig. 156. Seismicity map of El Salvador. (Redrawn from SCHULZ 1965).

first group. The second type comprises 61 earthquakes for which DEWEY & ALGERMISSEN (1974) introduced the concept "shallow-focus volcanic terrane earthquake". Only 4 historical earthquakes could be assigned to the fault zone in northern Guatemala, an area which was, however, struck by an earthquake in 1976. According to these data, the largest number of earthquakes accompanied by severe damage are shallow-focus earthquakes in the volcanic zone. The earthquakes in El Salvador (1965), Managua (1972) and Arenal (1973), which are described below, are typical examples of such earthquakes.

The following table provides an overview of the most recent severe earthquakes in Central America:

Year	Place	Magnitude (RICHTER Scale)	Intensity (MERCALLI Scale)	Number of dead	Property damage in millions of US $
1976	Guatemala	7.5	IX	22,868	1,100
1973	Arenal, C.R.	6.5	VIII – X	23	?
1972	Managua, Ni.	6.2	X	11,000	800
1965	San Salvador	6.2	VIII	?	?
1951	Jucuapa, E.S.	?	VIII	400 – 500	?
1931	Managua, Ni.	5.8	?	1,000	15

The San Salvador earthquake of May 3, 1965 (LOMNITZ & SCHULZ 1966). The earthquake announced itself from February 2, 1965 onwards in a seismic swarm in which up to 300 shocks were recorded daily. The main shock took place on May 3 at 04.01 local time. The magnitude was 6.0 – 6.25 and the epicentre was located between the city of San Salvador and the village of Sto. Tomás on the southern edge of the fault zone, which runs in an E/W direction through El Salvador — its position was thus similar to that of the epicentres of countless historical earthquakes in that country. The intensity achieved the maximum of grade VIII on the modified MERCALLI Scale. The main damage was limited to a range of 15 km around the epicentres, and the characteristics of the subsoil were found to exert a decisive influence; the main damage occurred over the loose pumices of the "Tierra blanca", while far less damage was suffered in the areas over consolidated volcanic rocks. This interrelationship, which was recognized by LOMNITZ & SCHULZ (1966, p. 574), was confirmed by a later exploration of the subsoil by SCHMIDT-THOMÉ (1975 b) in the region of the capital.

Judging by the small extent of the damage zone, the earthquake must have had a shallow focal depth. In fact, the earthquake can be regarded as typical of the zone of shallow-focus earthquakes in the vicinity of active volcanoes and young faults. In El Salvador these faults were designated the "Median Trough" by WILLIAMS & MEYER-ABICH (1955) and by LOMNITZ & SCHULZ (1966); this trough was held to be the foundered crest of a broad geanticline, while later surveys (WIESEMANN 1975) revealed a much more complicated pattern of structures in which, however, an E/W-striking fault system plays a major role (cf. Chapter 5.5).

The earthquake of April 13, 1973 near Arenal, Costa Rica (PLAFKER 1973). The earthquake occurred on April 13, 1973 at 03.34 local time without any foreshocks. The magnitude M_S was 6.5 and the intensity on the modified MERCALLI Scale was VIII to X, depending on the criteria selected. The centre of most severe damage was located south of the Laguna de Arenal, while the epicentre, which was detected by instruments, was recorded approximately 18 km to the north (Fig. 157). Only an approximate estimate could be made of the focal depth, namely 29 $\pm$ 10 km, but at all events this was a shallow-focus earthquake. According to geological maps prepared by the Instituto Costarricense de Electricidad (ICE) the damage area was located in a zone of detected or presumed faults with NW and SW strikes (Fig. 158).

While the immediate damage inflicted on buildings was held within certain limits by the method of construction and by the open type of settlement, ground effects and landslides caused severe property damage and the loss of 23 lives. The largest landslide was estimated to involve a mass of 75,000 to 125,000 m^3. In particular, the steep roadcuts in the lateritic soils and unconsolidated volcanic ashes proved to be particularly prone to slides.

In contrast to the earthquakes at Managua and in Guatemala no faults were detected at the surface of the earth, although this may be due to the thick cover of poorly consolidated material. Nevertheless, on the basis of its geological situation and the characteristic decline of the aftershocks, the earthquake can be regarded as of tectonic and not volcanic origin. The earthquake is of considerable practical importance because of the construction of a major hydroelectric facility on the Río Arenal, and this is why the ICE, as the responsible organization, carried out detailed seismological investigations (Anonymous 1974 d).

The Managua, Nicaragua, earthquake of December 23, 1972 (Anonymous undated, ARCE VELASCO 1973, MATUMOTO & LATHAM 1973, BROGAN et al. 1977, BROWN et al. 1973, CHAVEZ et al. 1973, CLUFF et al. 1976, DEWEY & ALGERMISSEN 1974, SCHWARTZ et al. 1975, WARD et al. 1974, CARTER & RINKER 1976).

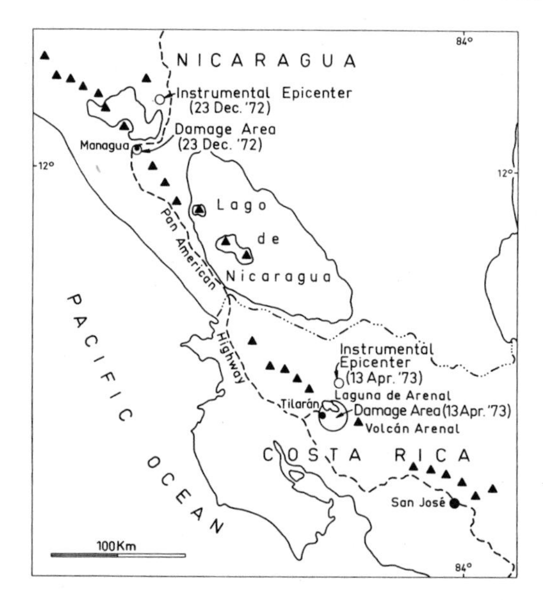

Fig. 157. Index map of the 1972 earthquake of Managua and the 1973 earthquake of Arenal. (Redrawn from PLAFKER 1973a, Fig. 1).

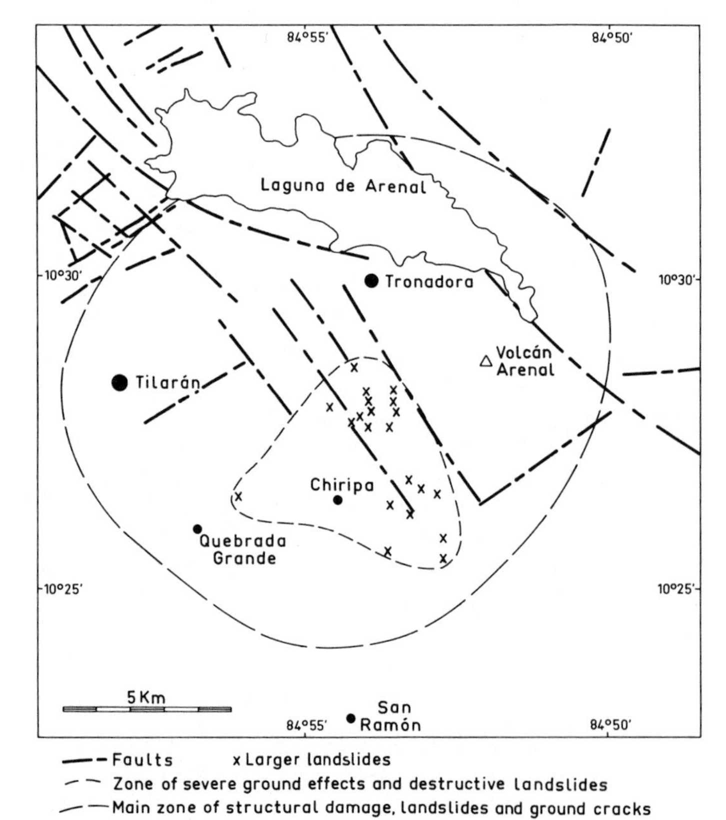

Fig. 158. Damage area of the 1973 earthquake of Arenal. (Redrawn from PLAFKER 1973a, Fig. 2).

The earthquake happened on December 23, 1972 at 12.30 local time and, together with two aftershocks, it destroyed 75% of the city of Managua causing 11,000 (according to other sources 8,000) deaths and left behind 20,000 injured and property damage amounting to approximately 500 million US $. Because of the severity of the losses and damage the earthquake was the subject of many investigations into its cause and type, into the main reasons for the damage and also into various possible ways of rebuilding the city.

Despite the relatively low magnitude of M_s 6.2 and m_b 5.6 of the main shock and m_b 5.0 or 5.2 respectively of the aftershocks, the intensity of the earthquake attained values up to IX on the modified MERCALLI Scale. The reasons why the destruction was so great are as follows:

1. the focus of the earthquake was located only 5 km below the city;
2. five faults opened up with left-lateral strike-slip between 20 and 38 cm and with only slight vertical displacement;
3. the city was built on loose subsoil and
4. most buildings were inadequately constructed to stand up to earthquakes.

The most striking geological effect was the horizontal movement in five approximately parallel faults (Fig. 159), which could be detected over distances up to 5.9 km. Their strike was between 38° and 40° E. The zone of the greatest number of aftershocks, which were recorded between January 4 and January 17, 1973, using portable seismographs, extends in the same direction (Fig. 160). They indicate a fault 10 – 15 km long and 8 – 10 km deep with a strike of 30° to 35°.

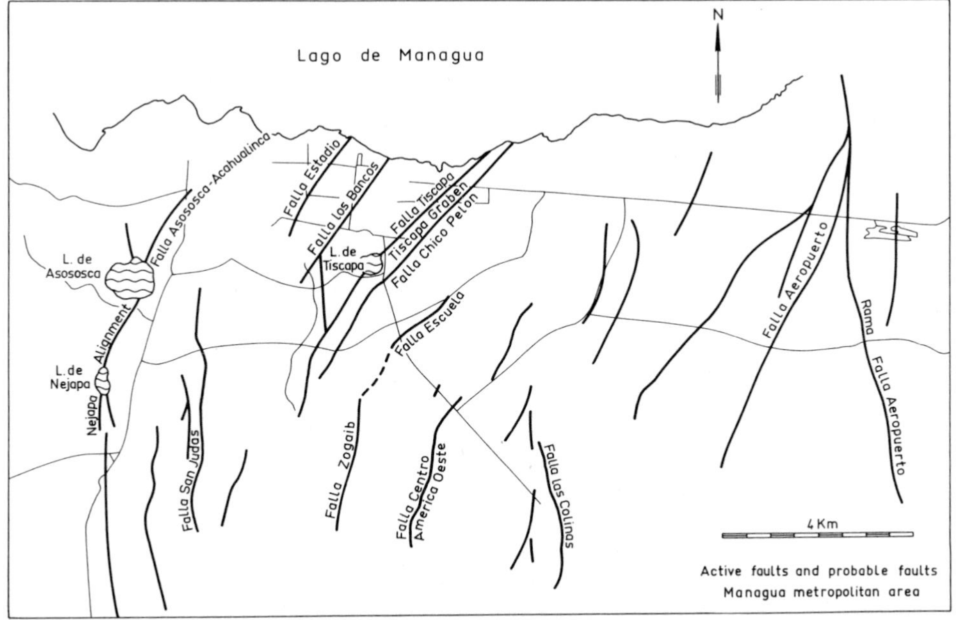

Fig. 159. Fault map of the Managua metropolitan area, Nicaragua. (Redrawn from CLUFF et al. 1976).

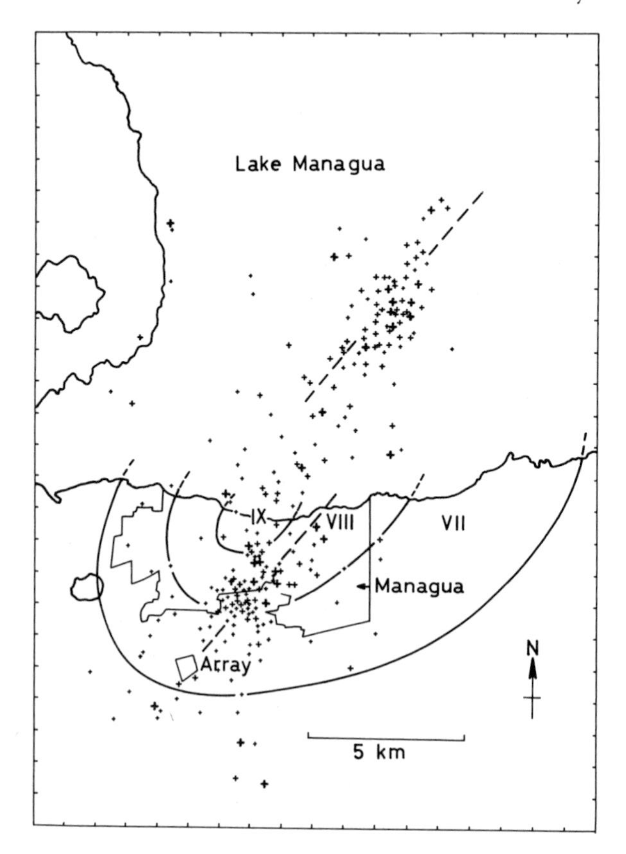

Fig. 160. Intensity map for the main earthquake of Managua of 1972 and epicentres of 300 of the aftershocks based on data from a five-station seismic array. (From MATUMOTO & LATHAM 1973, Fig. 2).

The earthquake cannot be related to the subduction zone, which is already 200 km deep below Managua, but instead belongs to the group of shallow-focus volcanic terrane earthquakes in the region of the NW/SE-striking fault zone and line of active volcanoes in Central America, which in Nicaragua is developed as a large depression zone, the Nicaragua Trough. According to recent surveys (SCHWARTZ et al. 1977) this zone is intersected in the area of Managua by a N/S-trending depression, the "Managua Graben" which is documented in fault scarplets and downthrow faults. NNE/SSE-striking lateral shifts, which were also the direct cause of the 1972 earthquake, occur within the graben. Thus Managua is located exactly in the zone of intersection of several fault systems which have given rise to repeated destructive earthquakes. This finding is confirmed by evaluation of ERTS-1 images (CARTER & RINKER 1976).

Transform faults at a segment boundary in the Central American subduction zone, as defined by STOIBER & CARR (1977), or extension in the relatively thin lithosphere below the chain of volcanoes, are being discussed as higher reasons for the earthquakes.

The Guatemalan earthquake of February 4, 1976 (Anonymous 1976, CLUFF et al. 1976, ESPINOZA (ed.) 1976, PLAFKER 1976, WEICHERT 1976, PAGE (ed.) 1976, FIEDLER 1976, 1977, SCHWARTZ 1977, BUCKNAM, PLAFKER & SHARP 1978, KANAMORI & STEWART 1978, VIVO ESCOTO 1978).

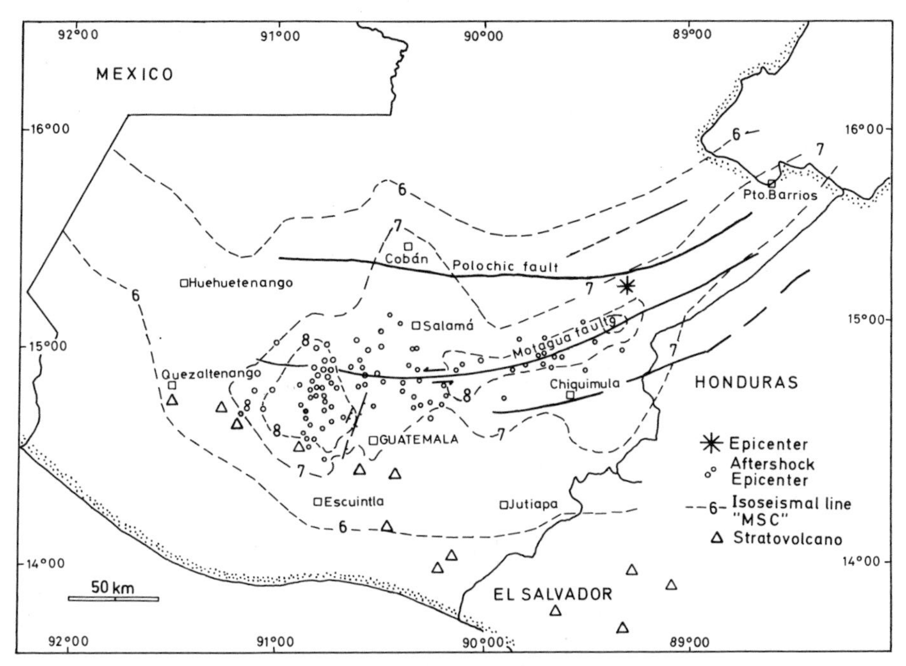

Fig. 161. Map of the Guatemalan earthquake of February 4, 1976. (From ESPINOZA 1976 and PLAFKER 1976).

The Guatemalan earthquake of February 4, 1976, which happened at 03.02 local time, is the most severe earthquake which has struck this country in historical times. It cost 22,868 people their lives and left behind 77,190 injured, 254,750 damaged houses and caused property damage in the order of 1.1 billion US $. From the geological point of view it was one of the most remarkable earthquakes ever to have occurred because it happened on the Motagua Fault Zone, which is taken to be the boundary between the Caribbean and North American plates (cf. the relief map in Fig. 44). It is therefore understandable that extensive investigations were launched immediately after the earthquake, and the above reports are the result.

The earthquake was triggered by left lateral horizontal movement at the Motagua Fault and it affected an area of at least 100,000 km^2. The centre was situated near Los Amates, about 157 km NE of Guatemala City between the Motagua and the San Agustin Fault at a depth of approximately 5 km (Fig. 161). The magnitude M_S was 7.5 and the maximum observed intensity on the modified MERCALLI Scale was IX. A map of the distribution of intensity traces the Motagua Fault in the eastern part of the seismic zone, while the isoseismals spread towards the west over a wide area. Two aftershocks with a magnitude of m_b 5.8 followed on March 7 close to Guatemala City, probably on the N/S-striking Mixco Fault.

The earthquake was characterized by unusually strong left lateral horizontal movement along the Motagua Fault for a distance of approximately 300 km. On average the horizontal shift amounted to 100 cm and attained a maximum of 325 cm about 25 km north of Guatemala City. The direction of movement confirmed the hitherto uncertain view that the North American Plate is moving westwards relative to the Caribbean Plate.

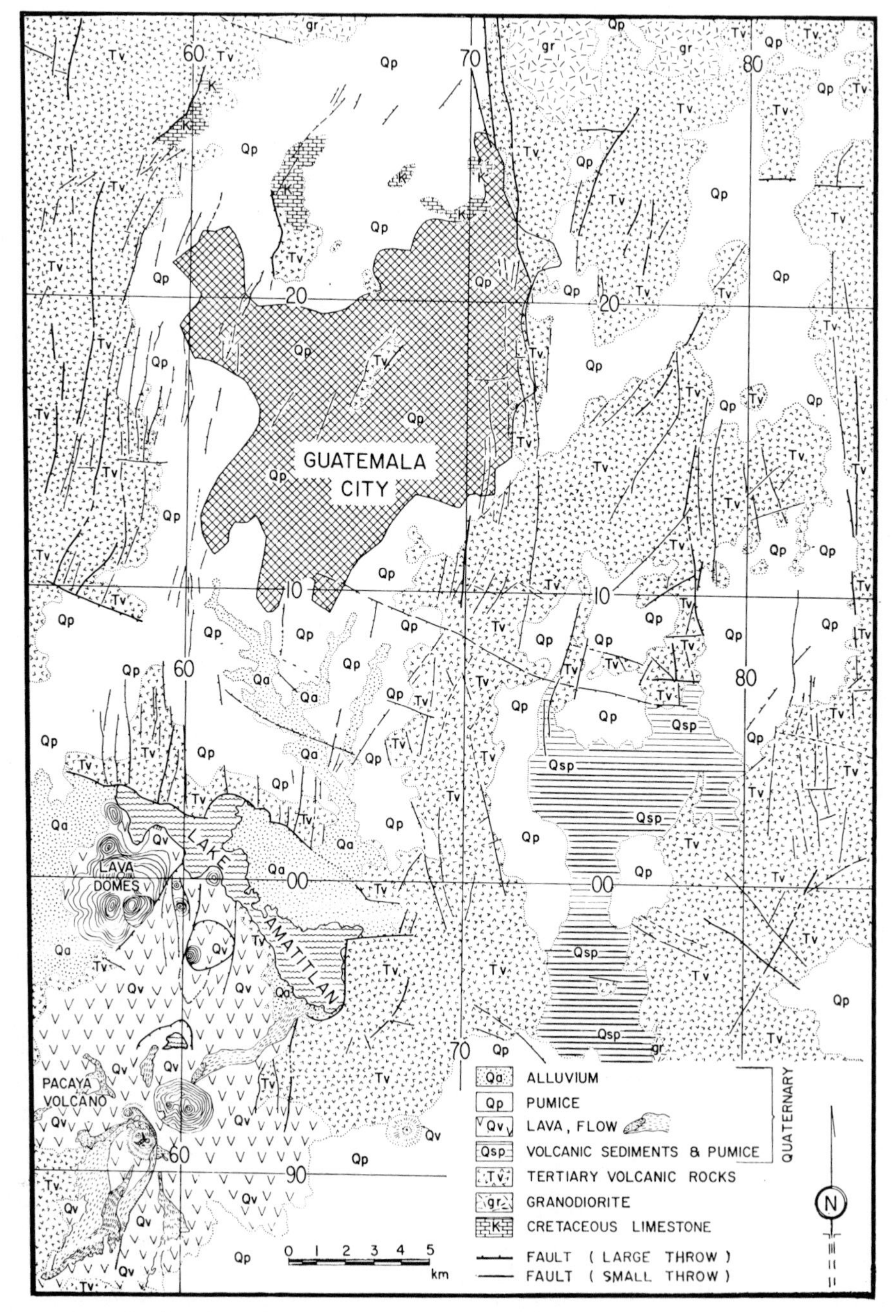

Fig. 162. Generalized geological map of the Guatemala City Graben. (From DENGO et al. 1970, Fig. 4).

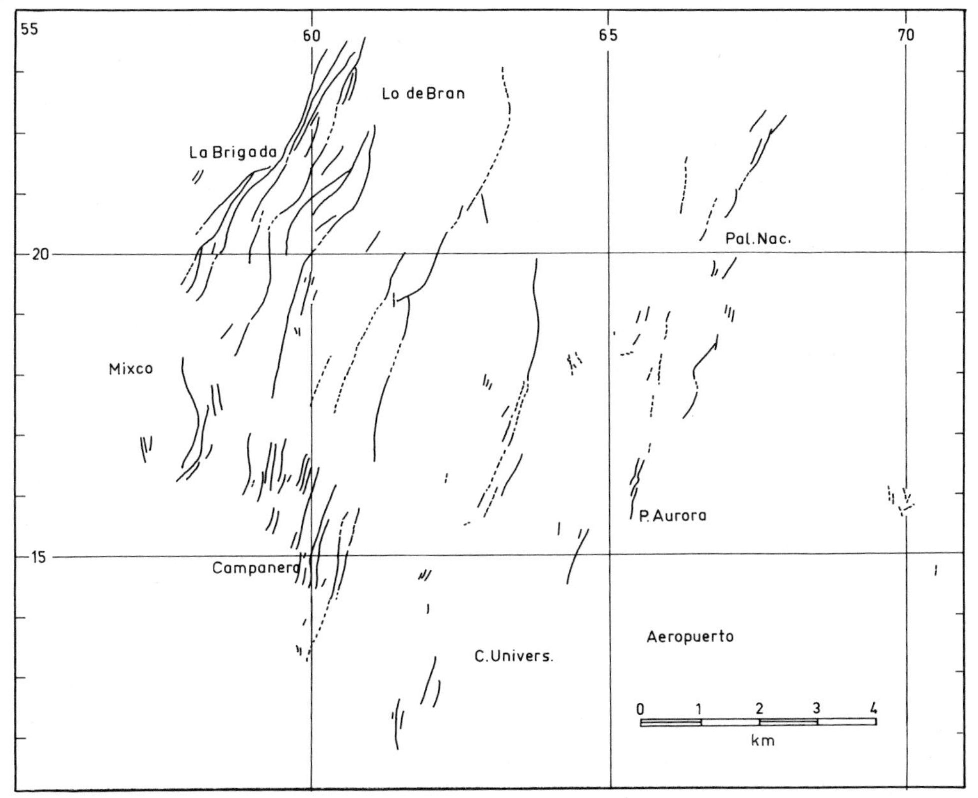

Fig. 163. Faults activated by the 1976 earthquake in the metropolitan area of Guatemala City. (Redrawn from a map of the Instituto Geográfico Nacional, Guatemala 1976).

Afterslips with up to 31 cm of horizontal displacement were detected along the Motagua Fault up to October 1977 (BUCKNAM et al. 1978). The amount of horizontal shift is inversely proportional to the amount of shift first occurring at the time of the earthquake, and it decreases with the square of time following the main shock.This obeyance of regular laws and the occurrence of secondary movement over a range of 50 km exclude local or near-surface factors, such as a thick alluvial cover, and indicate that the cause is to be sought in creep movements in a crust which is about 4 km thick.

The earthquake triggered movement and faults in various fracture zones in Guatemala where countless aftershocks were recorded. The strongest were situated near Tecpan along an NE/SW-striking fault, in the area of the N/S-striking Guatemala Graben, in particular at its western boundary, the Mixco Fault, and at Agua Caliente. In the region of the Guatemala Graben countless NNE/SSW-striking faults opened up with mainly vertical displacement in the decimetres range. Mapping of these faults revealed remarkable agreement with the fault systems discovered by earlier geological mapping activity (Figs. 162 and 163).

The most severe damage occurred in the direct vicinity of the activated faults. In addition, the thick cover of Pleistocene pumice deposits in the seismic region intensified the effect of the shocks. A large number of landslides, some of which contained several million cubic metres of material, occurred in these deposits (HARP et al. 1976, HOOSE & WILSON 1976).

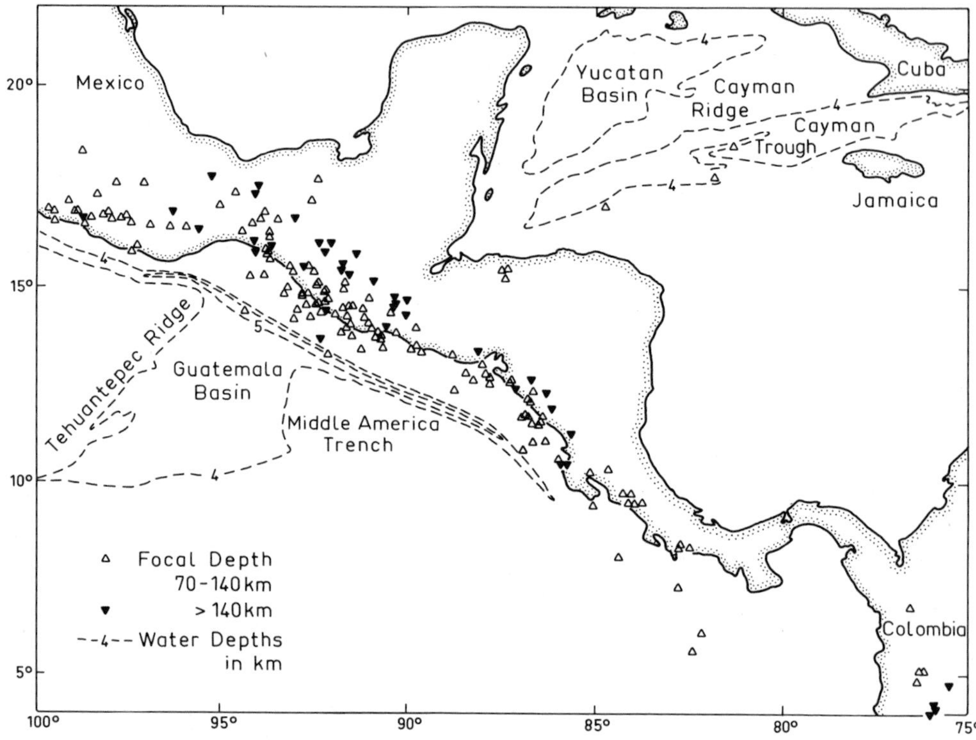

Fig. 164. Shallow-focus earthquakes in Central America, southeastern Mexico and adjacent regions. (Redrawn from MOLNAR & SYKES 1969, Fig. 6).

Earthquakes are of decisive importance when it comes to assessing present-day crustal movements and thus for the group of geotectonic theories gathered under the heading of "plate tectonics". This holds true in particular for Central America where no less than five plates are linked with each other and react with each other in various ways; the plates are the North American, Caribbean, South American, Cocos and Nazca plates. MOLNAR & SYKES (1969) collected some basic information on this situation, and since that time detailed aspects of the problem have been enlarged upon. It was already possible to refer to these data in some cases when describing the last severe earthquakes.

The relocated epicentres in Central America are plotted after MOLNAR & SYKES (1969, Fig. 6 and 7) in Figs. 164 and 165 for the period from 1954 to 1962. Distribution maps for other periods of time (BOWIN 1976, Fig. 13, GRASES 1975, Mapa 6, SPENCE & PERSON 1976, Fig. 2 and 3) yield an essentially similar pattern. The main seismic zone both of shallow-focus earthquakes as well as of intermediate-depth earthquakes runs parallel to the Middle America Trench and to the chain of active volcanoes, and it extends from Mexico to Costa Rica [20]. The focal depth increases from the Middle America Trench towards the continent. This dip is clearly evident in vertical sections perpendicular to the extent of the earthquake zone — these sections were also calculated by various authors — but it must be pointed out that uncertain-

[20] The concept of a Central American Arc is used in analogy to other circumpacific island arcs.

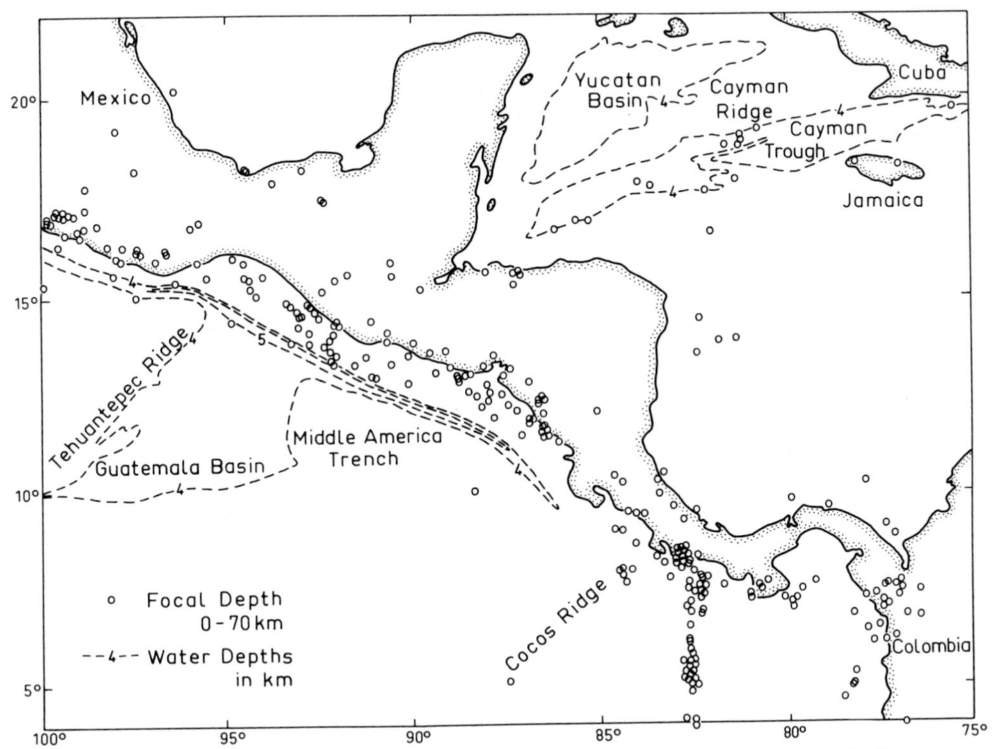

Fig. 165. Intermediate-depth earthquakes in Central America, southeastern Mexico and adjacent regions. (Redrawn from MOLNAR & SYKES 1969, Fig. 7).

ties in the pinpointing of the hypocentres still results in a certain degree of scatter. Based on the evaluation made by CARR (1976) and DEWEY & ALGERMISSEN (1974) five such profiles through Guatemala, El Salvador and Nicaragua are shown in Fig. 166. Since MOLNAR & SYKES it has been generally accepted that this can be interpreted as evidence of a Benioff zone or of a subduction zone between the Cocos Plate and the Caribbean Plate [21]. This zone continues in the northwest as the subduction of the Cocos Plate under the North American Plate where annual rates of 8 cm/a are given for the underthrusting (LARSON & CHASE 1970).

STOIBER & CARR (1974) have drawn attention to the segmentation of the Central American Arc which is detectable in the subduction zone in the form of differences in the angle of dip (Fig. 166), superficially in the delineation of the volcanic chains, in faulting transverse to the main seismic zone and in the distribution of shallow-focus earthquakes within the volcanic zone. The individual segments are shown, according to the two authors, in the map of the volcanoes in Fig. 120. CARR & STOIBER (1974) also recognized that there was an accumulation of intermediate-depth earthquakes about 10 km offshore from the active volcanoes, while immediately beneath the volcanoes only slight seismic activity is recorded. They see this as an expression of partial melting along the subduction zone immediately below the volcanoes.

[21] The rate of movement is given as 2 – 9 cm/a depending on the method of calculation used (MOLNAR & SYKES 1969, Table 2).

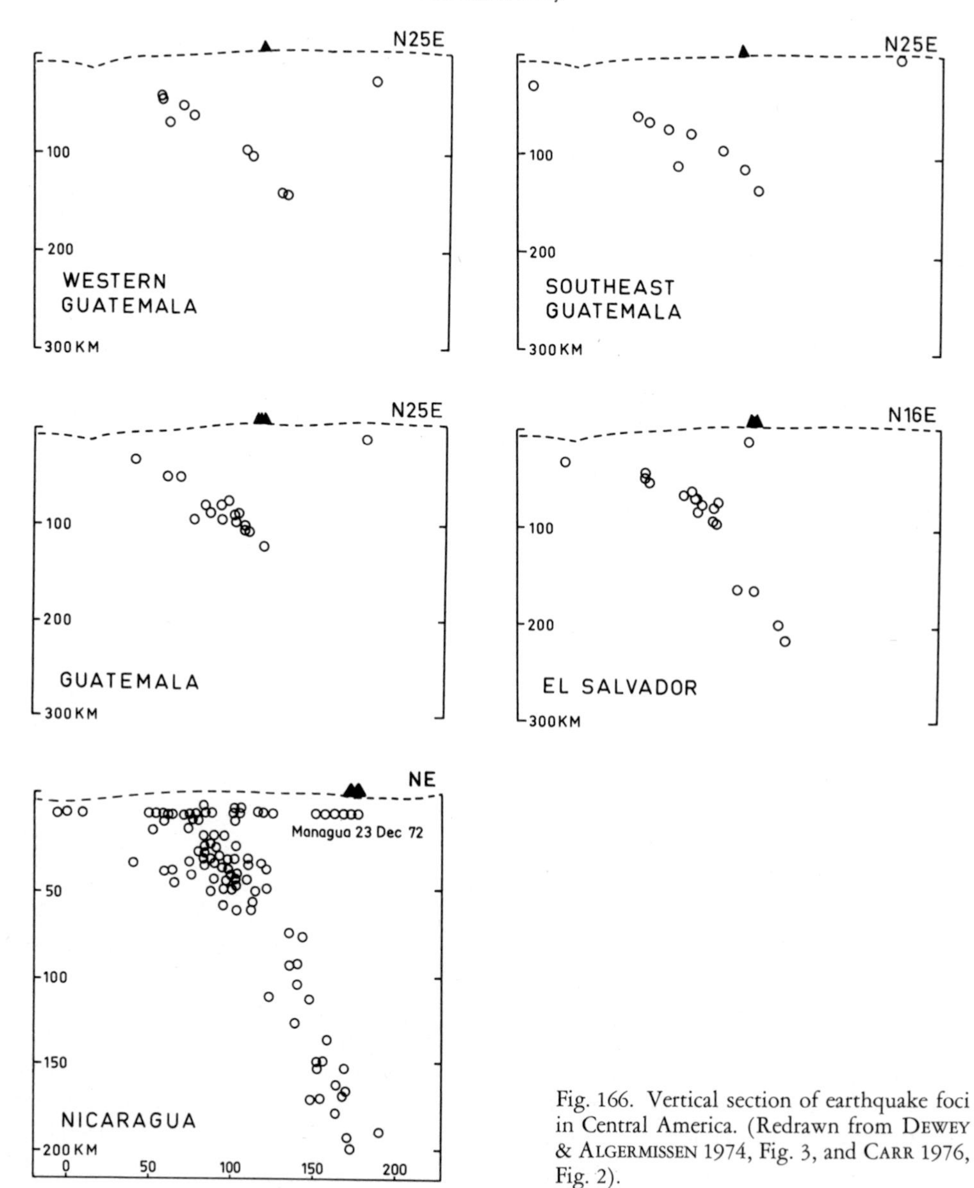

Fig. 166. Vertical section of earthquake foci in Central America. (Redrawn from DEWEY & ALGERMISSEN 1974, Fig. 3, and CARR 1976, Fig. 2).

The profiles prepared by MOLNAR & SYKES (1969, Figs. 10 and 11) confirm the decrease in deep-focus earthquakes beneath the volcanic zone.

The earthquakes with very shallow focal depths, which follow the Central American Arc, have to be clearly separated from the foci in the subduction zone. This zone, which is responsible for most of the destructive earthquakes, is closely related with the young fault tectonics of Central America which are discussed in detail in Chapter 3.2.

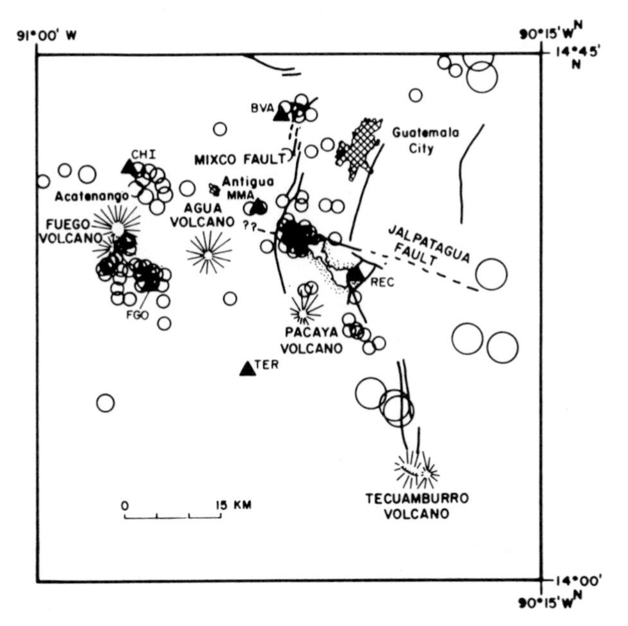

Fig. 167. Volcanoes, epicentres and faults in the region of Guatemala City. (From ESPINOZA 1976, Fig. 10).

In addition there are the earthquakes that occur in the immediate vicinity of active volcanoes, and which precede or accompany the volcanic eruptions. An example of this are the earthquakes occurring in the vicinity of the very active Fuego Volcano in Guatemala (Fig. 167).

The importance of earthquakes for the theory of plate tectonics is examined in Chapter 8.

7. Gravity field

The widely scattered and sometimes inaccessible gravity data on the Caribbean region and thus also on Central America were summarized by BOWIN (1976) and interpreted in line with the theory of plate tectonics. Extracts from his impressive free-air gravity anomaly and Bouguer gravity anomaly maps are given in Figs. 168 and 169. They deviate in some respects from the gravity maps of Costa Rica (Fig. 83) and of Panama (Fig. 103). The main reason for this is probably that the isogales in these maps are spaced at closer intervals than in BOWIN's maps and they therefore are able to cover more details

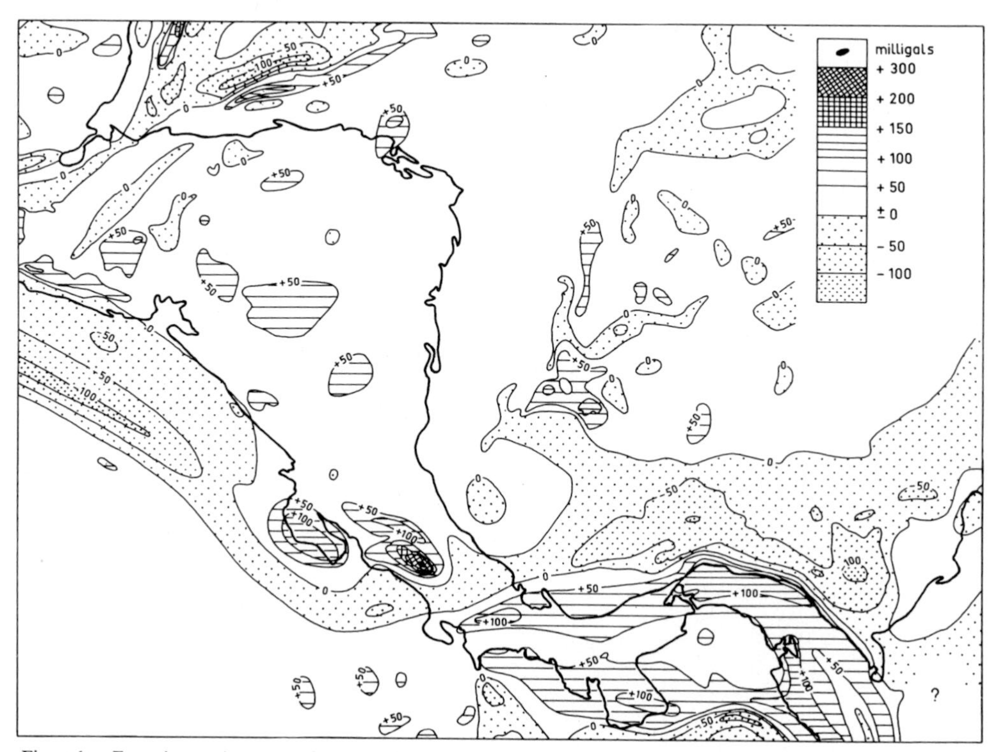

Fig. 168. Free-air gravity anomaly map of Central America. (Redrawn from BOWIN 1976, Fig. 3).

In the **free-air anomaly map** the following regions stand out as exhibiting major gravity anomalies:

The Middle America Trench has a maximum of –120 mgal, i. e. a much smaller anomaly than in the Puerto Rico Trench which is about 2 km deeper. As the Middle America Trench flattens out off Costa Rica so the gravity anomaly also declines. It is replaced over the Nicoya Peninsula by a strong gravity maximum which is probably caused by the high-lying position of the basaltic Nicoya Complex. To the east of this comes a further gravity maximum in the region of the young volcanic Cordillera Central and the northwestern section of the Cordillera de Talamanca of Costa Rica. The interpretation of this gravity maximum alone out of the entire excess of mass in the range is rendered difficult by the fact that a broad zone of low subgravity extends across the eastern Cordillera de Talamanca from the Pacific to a point off the northern coast of Panama. Furthermore, the assumption of heavy crustal rocks below the gravity maximum is disproved by negative Bouguer anomalies in the region of central Costa Rica.

A strong positive free-air anomaly is found both in the northern and southern part of Panama. It is emphasized by positive Bouguer anomalies. Here it is primarily the uplifted regions of oceanic crust of the Basic Igneous Complex which act as disturbing bodies; these are not in isostatic equilibrium and indicate recent uplifting of Panama. The regions of balanced or negative gravity occurring between the gravity maxima in Panama coincide with basins of

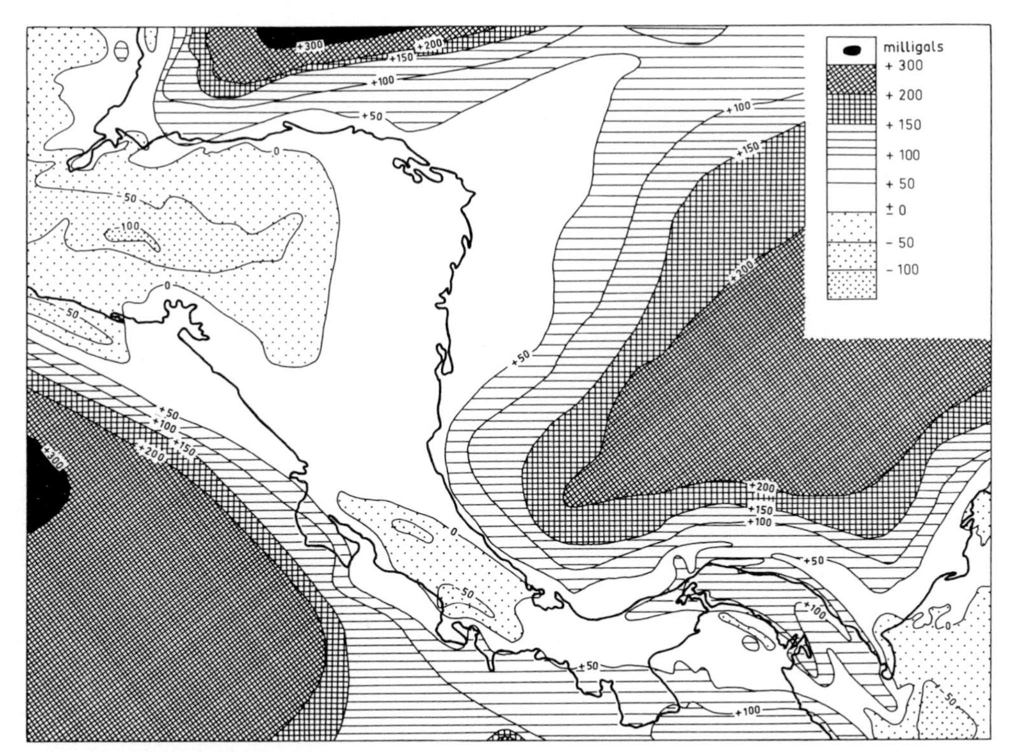

Fig. 169. Simple Bouguer gravity anomaly map of Central America. (Redrawn from BOWIN 1976, Fig. 4).

thick Tertiary sediments, such as those between the Azuero Peninsula and the Cordillera Central or in the basin of the Río Chucunaque and the Río Tuira. They also feature in the distribution of the Bouguer anomalies, in particular according to the map compiled by CASE (Fig. 103).

A trench to the south off the shelf of the Gulf of Panama produces a powerful negative free-air anomaly. The trench continues in arc-form to the south to a point off the western coast of Colombia. A further zone of negative gravity is situated in an arc off the northern coast of Panama. At a few points its relatively narrow centre attains values of up to –100 mgal. Parallel to this, also in a northward-oriented arch, runs a bend in the +50 mgal Bouguer anomaly. The mass deficiency which this indicates is explained by BOWIN (1976, p. 33) who points out that at this point the Caribbean crust is overthrust by the crust of Panama and forced into the depths. CASE (1974) had already developed a similar theory (see p. 173/174).

In the map of the **Bouguer anomalies** the Central American Isthmus, with its shelf margin and the Nicaragua Rise, stands out clearly from the oceanic regions of the Pacific, the Colombia Basin and the Cayman Trench. Stronger negative anomalies are bound to the mountain ranges of northern Central America, which are built up from sialic rock, to the volcanic Cordillera Central and to the Cordillera de Talamanca of Costa Rica, which is formed from granitic rocks and sediments. Reference has already been made to the gravity minimum over the Tertiary basins of Panama, and it continues in the Atrato Basin in Colombia. In Costa Rica the Tertiary basin of the Valle del General is a region of strong negative Bouguer anomalies (see map Fig. 83), which merges with steep gravity gradients into the gravity maximum of the Golfo Dulce where the Nicoya Complex crops out. The gravity gradient is just as steep between the volcanic chain of northwestern Costa Rica and the Nicoya Peninsula, which is built up of oceanic rock. In the shelf off El Salvador a moderate gravitational anomaly provides evidence of a sedimentary basin.

8. Plate tectonics

The present tectonic activity of Central America is expressed in the strong relief and recent uplift, in deep-sea trenches, volcanic chains, strong seismicity and gravity anomalies. These phenomena, which were described in the preceding chapters, can nowadays be regarded as causally related according to the theory of plate tectonics. The Central American-Caribbean area is a key region for understanding plate tectonics.

No less than five lithospheric plates collide here and react with each other. Our knowledge of their present arrangement is derived above all from the evaluation of seismic data by MOLNAR & SYKES (1969), which was supplemented in detail and on the whole confirmed by later works. Of the recent descriptions of the plate boundaries in Central America, one by PLAFKER (1976) and one by BOWIN (1976) are summarized in Fig. 170. Theories on the position and direction of movement of the plates in the geological past have also been discussed in a number of works, and a wide range of related facts have been interpreted in the process, but the problems grow more difficult the further back in time we go.

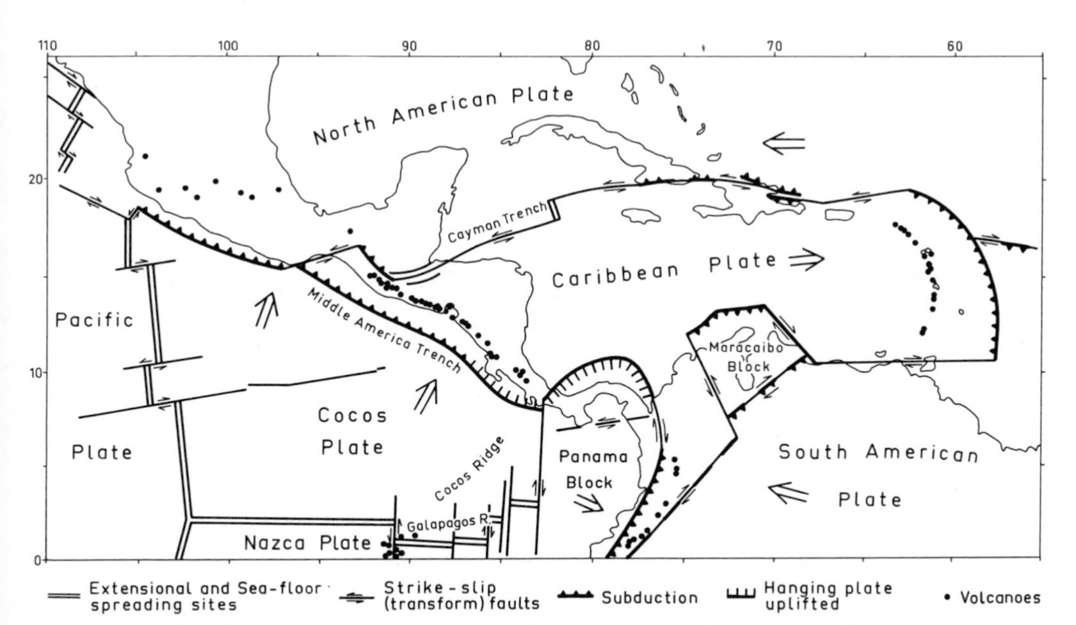

Fig. 170. Inferred present plates and plate boundaries in the Central America-Caribbean region. (Redrawn from PLAFKER 1976, Fig. 6 and BOWIN 1976, Fig. 16).

The present pattern of plate tectonics in the region of Central America can be briefly outlined as follows:

Most of Central America between Guatemala and Costa Rica forms part of the **Caribbean Plate**; Panama is a part of the **Nazca Plate**, or according to BOWIN's theory (1976) it is an independent unit which he refers to as the **Panama Block**. The **North American Plate** extends to the areas of Guatemala located north of the Motagua Fault Zone.

In the southwest the **Cocos Plate** underthrusts the Caribbean Plate, and the result is that a perfect example of a subduction zone has developed at the edge of the Pacific. Its characteristics are as follows: hypocentres dipping towards the continent (Fig. 166), slip vectors pointing in the same direction, fracture tectonics in the upper crust (Fig. 38), the chain of active volcanoes between Guatemala and Costa Rica (Fig. 120), the Middle America Trench (Fig. 5) and a free-air gravity anomaly above it (Fig. 168). It was deduced from the shifting of the individual rows of volcanoes, from transverse discontinuities and from variations in the dipping of the hypocentres that the subduction zone was segmented (STOIBER & CARR 1974). A section through the subduction zone in the area of Guatemala, slightly modified after SEELY et al., is reproduced in Fig. 171. This section is based on seismic data and data obtained from seismic shooting, magnetics and from the surface geology, and it thus summarizes all the available criteria indicating subduction.

In the north the Caribbean Plate, together with the Motagua Fault Zone and the Cayman Trench, abuts the North American Plate. The course of this fracture zone and a discussion of it and its importance with regard to the Guatemalan earthquake of 1976 have been dealt with in detail in Chapters 3.2 and 6, to which the reader's attention is drawn. PLAFKER (1976) discussed various possibilities regarding the movement in the triangle formed by the Cocos, Ca-

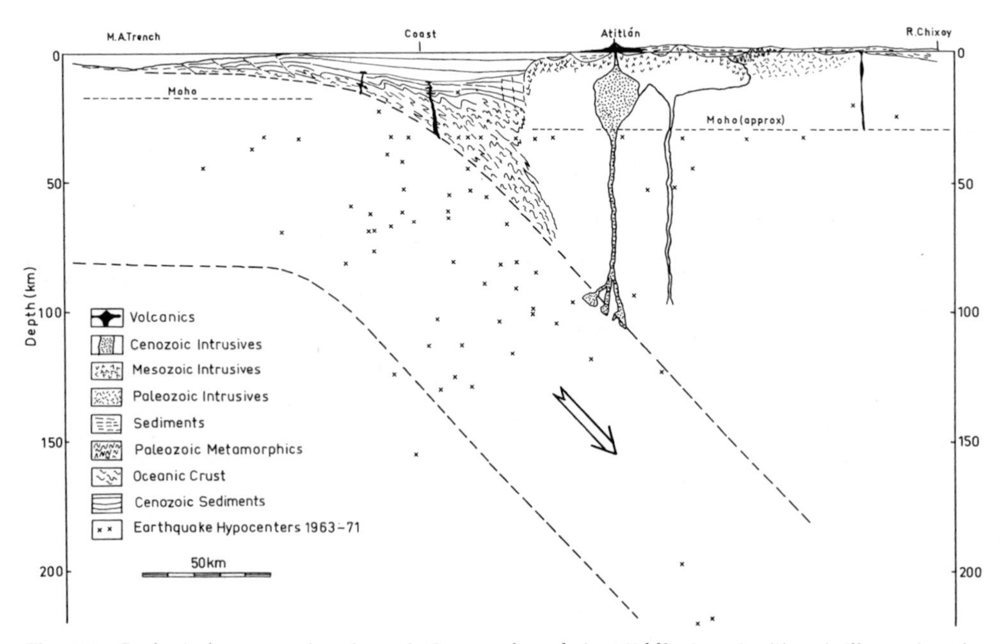

Fig. 171. Geological cross-section through Guatemala and the Middle America Trench illustrating the subduction of the Cocos Plate under the Caribbean Plate. (Redrawn from SEELY et al. 1974, Fig. 12).

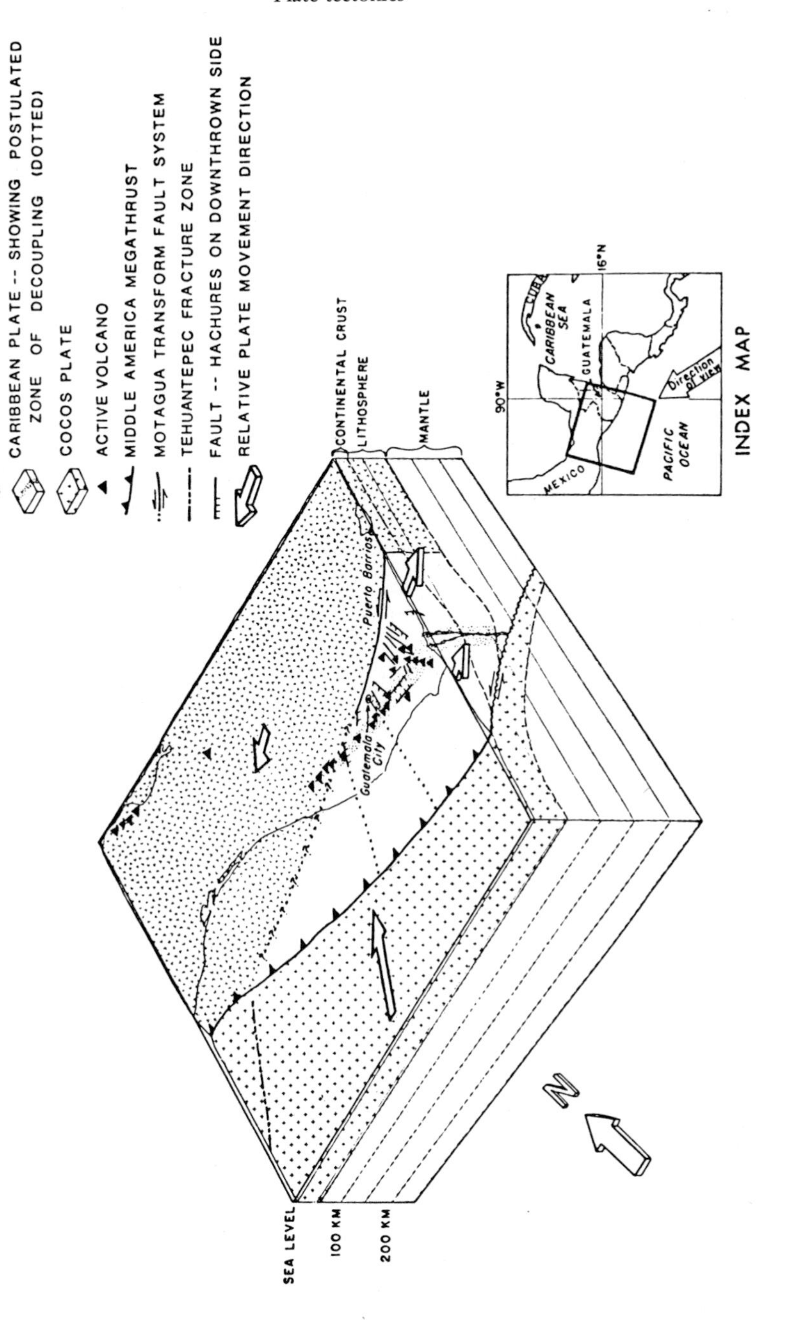

Fig. 172. Block diagram showing the relationships between the lithospheric plates in the Guatemalan region. (From PLAFKER 1976, Fig. 8).

ribbean and North American plates, and he illustrated what he thought was the most likely configuration of the plates and their boundaries in the block diagram reproduced in Fig. 172. According to BURKART (1978) the plate boundary is a shear zone about 200 km wide, which was activated, now at the Polochic Fault, now at the Motagua Fault, or in a shear zone between the two. The horizontal shift along the Polochic Fault is given as 132 ± 5 km.

In the southeast, the phenomena of the subduction of the Cocos Plate underneath the Caribbean Plate end in the area of Costa Rica. The uninterrupted chain of volcanoes ends in the Cordillera Central of Costa Rica, the Middle America Trench and its gravity anomaly flatten out off the Nicoya Peninsula and the seismic activity decreases distinctly in a southeastern direction. The relatively unbroken straight coast with its broad shelf off northern Central America is replaced by the very broken Pacific coastline of Costa Rica and Panama, and the deep-sea floor off this part of the coast is characterized by strong relief with the Cocos Ridge, the Panama Fracture Zone and countless seamounts. Obviously this region is dominated by a fundamentally different structural plan from the region of northern Central America.

The Cocos Plate ends in the east at the Panama Fracture Zone, which extends southwards from the continent at 82° 30′ and which exhibits vigorous shallow-focus seismicity. According to MOLNAR & SYKES (1969) the plate displays right-lateral strike-slip motion; the focal mechanism and the narrowness of the seismic belt are reminiscent of the transform faults of the mid-oceanic ridge system, and the strong submarine relief with its longitudinal depressions fits in with this view.

To the east of the Panama Fracture Zone a broad wedge of the Nazca Plate thrusts northwards and forms the mountain arc of eastern Panama, the parallel gravity anomaly and sedimentation basins off the northern coast of Panama. The sialic magmatism, which commenced in the Upper Cretaceous, and the formation of the elongated Tertiary troughs in eastern Panama are evidence of considerable activity during the Tertiary. The main movement of the Nazca Plate is, however, directed against the South American continent and is expressed, among other things, in the chain of volcanoes in Colombia. BOWIN (1976) regards this northern part of the Nazca Plate as a small independent lithospheric unit which he called the Panama Block (Fig. 170).

So much for the present pattern of plate tectonics. The problems become more complex when we try to look back into the geological past. Here the only criteria that we have to go by are the findings regarding prehistoric magmatism, the structures of the uppermost part of the crust and paleomagnetic data, while the seismic and gravimetric data which are so important in the present are lacking. Thus any conclusions that we draw from the development of volcanism and plutonism and then apply to the development of the subduction of the Cocos and Caribbean plates must be treated with caution:

To the north in El Salvador and Nicaragua a chain of extinct, with all probability Pliocene, volcanoes lies ahead of the chain of active volcanoes in Central America. These Pliocene volcanoes may indicate a zone of magma formation and thus of subduction which was located several tens of kilometres further north than is the case today. Still further north comes the area of the extensive ignimbrite covers of the Miocene and Pliocene and thereafter the region of Middle to Lower Tertiary volcanism of the Matagalpa Group. This, too, could indicate a shift in subduction. If the Upper Cretaceous to Lower Tertiary plutonic rocks of northern Honduras and Nicaragua are also included in this theory, it would mean that the magmatic activity shifted in a southerly/southwesterly direction from the Cretaceous onwards, with the last stages of this activity being restricted to the present margin of the Pacific. An attempt has been made to portray this development in the general map of magmatic activity in northern Central America (Fig. 104). If we also see in this map a shift in the subduction zones according to the theory of plate tectonics, then the Nicaragua Trough and its northwestern continuation below the shelf could be a possibly associated fore arc basin filled with large amounts of sediment. Similar theories regarding a shift in the subduction zone in the recent history of the planet, say from the Cretaceous onwards, have been developed for southern Central America by STIBANE et al. (1977) — in the region of Nicoya — and by CASE et al. (1971) — in the region of Panama.

Closely connected with this question is that of the previous position of the continental nucleus of Honduras and Nicaragua which overlaps with the northern part of the South American Plate in the reconstruction of Pangaea. Here, too, widely different models are offered, to the extent that this question is considered at all. From the geology of the continental margin of British Honduras DILLON & VEDDER (1973) deduced a wedge-shaped opening and southwards-directed rotation in the Yucatán Basin in the Jurassic and the Cretaceous and they assumed left-lateral strike-slip motion along the Cayman Trench (Fig. 173); along with KESLER (1971) they put the amount of slip at probably 150 km. From the reconstruction of a homogeneous salt basin in the Jurassic of the Gulf zone and off the northern coast of Honduras PINET (1972) had expected a left-lateral strike-slip of 1,000 km. MACDONALD (1976 a

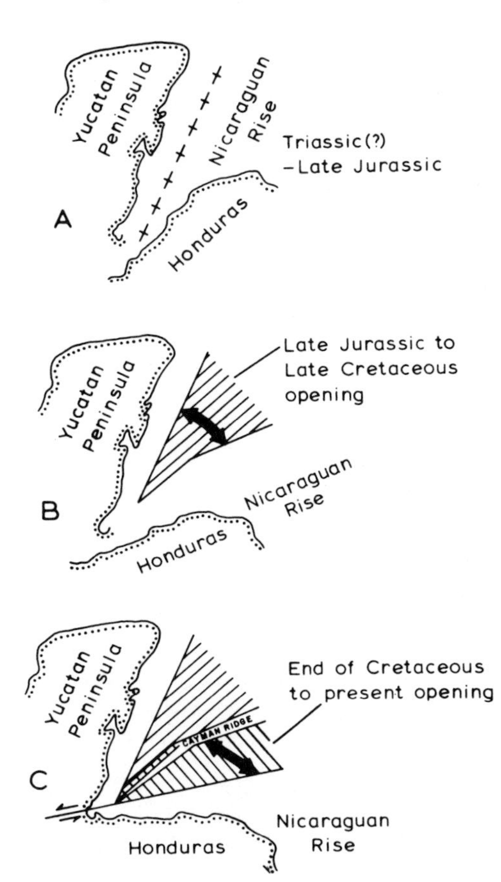

Fig. 173. Diagrams of assumed stages in the development of the northwestern Caribbean. The present coastline is shown for reference. (From DILLON & VEDDER 1973, Fig. 11).

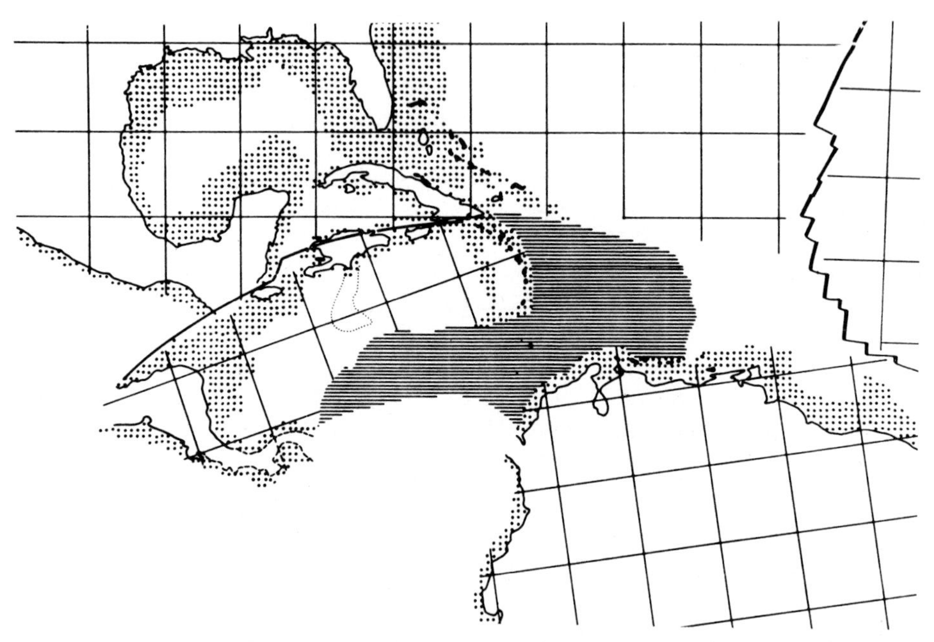

Fig. 174. Eocene reassembly of the Caribbean area, based on geometric principles of plate rotations. (From MacDONALD 1976, Fig. 2). Dotted pattern indicates continental margin or other depths less than 1,000 fm. Horizontal striped pattern indicates oceanic lithosphere which has been subducted.

and b) deduced an even greater displacement from the reconstruction of polar wandering and he calculated that the plates had become displaced by 1,400 km since the Eocene alone. As a result, for the time preceding the Eocene he places the region of Honduras and Nicaragua far out into the Pacific (Fig. 174).

The picture presented by PERFIT & HEEZEN (1978), which explains above all the development of the Cayman Trench and its boundary regions, is different yet again. They have summarized their theory in three paleogeographic sketches which are reproduced in Fig. 175. PERFIT & HEEZEN comment on these sketches as follows:

"A. Late Cretaceous, prior to the collision of Yucatán and Honduras-Nicaragua. NP-CR is the combined mass of the Nicaraguan Plateau and Cayman Ridge. J, C and H represent the approximate locations of Jamaica, Cuba and Hispaniola prior to left-lateral movement along the Cayman Trench.

B. Pre-Oligocene, after subduction to the south ceases; left-lateral motion begins along the Cayman Trench Fault, accompanied by the formation of numerous tensional grabens. The Cayman Ridge (CR) and Oriente Province (C) are split from the emergent arc to the south. Spreading is initiated in the proto-Cayman Trench. Volcanism wanes along the Greater Antil-

Fig. 175. Schematic drawings demonstrating the evolution of the northern Caribbean area and Cayman Trench since the Late Cretaceous. (From PERFIT & HEEZEN 1978, Fig. 10). Large arrows indicate relative motion between the North American and South American plates. Solid lines represent the present land boundaries; broken lines are approximate regions of emergent land and shallow banks in the past. Explanation in the text.

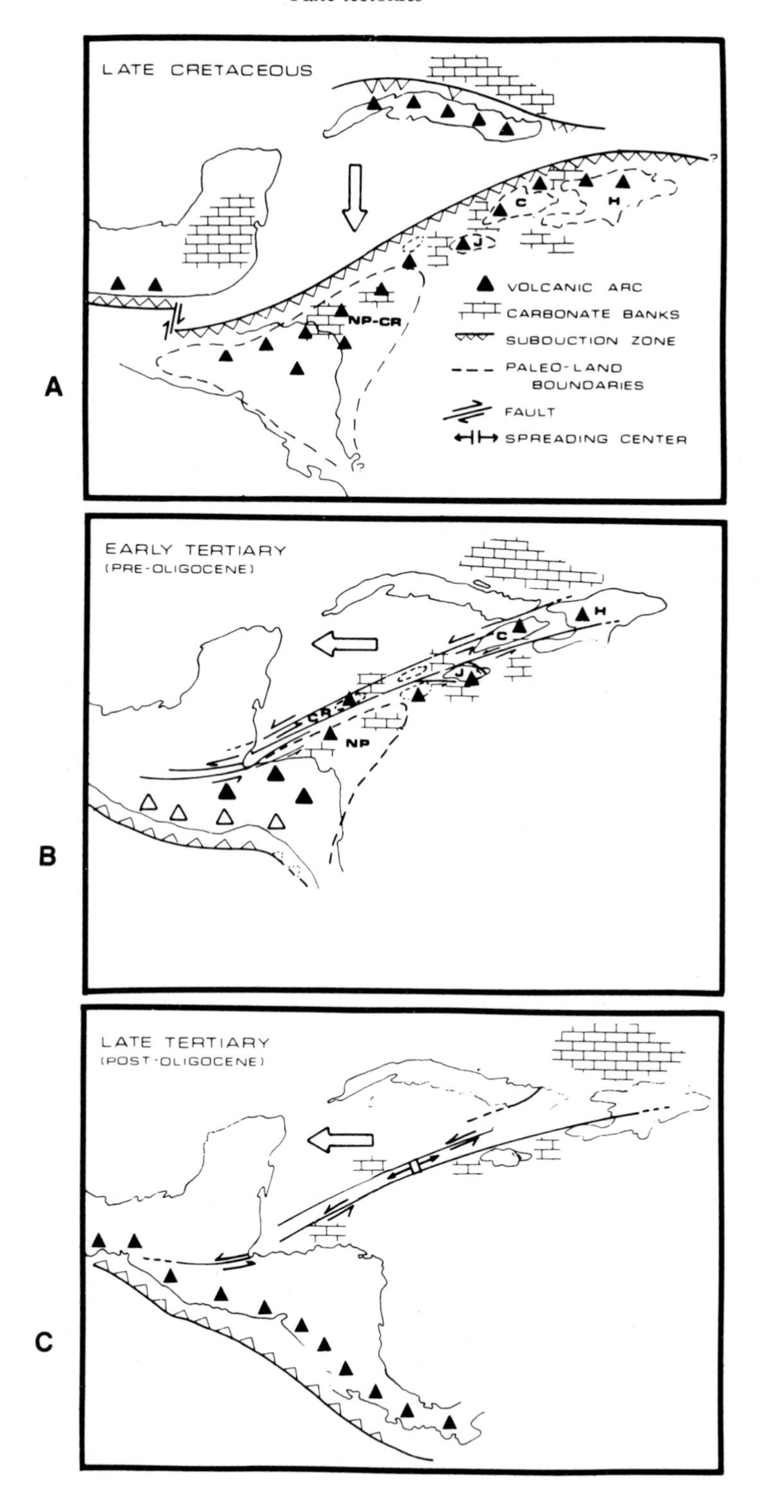
LATE CRETACEOUS
C
H
J
NP-CR
VOLCANIC ARC
CARBONATE BANKS
SUBDUCTION ZONE
PALEO-LAND BOUNDARIES
FAULT
SPREADING CENTER
A
EARLY TERTIARY
(PRE-OLIGOCENE)
H
C
J
CR
NP
B
LATE TERTIARY
(POST-OLIGOCENE)
C

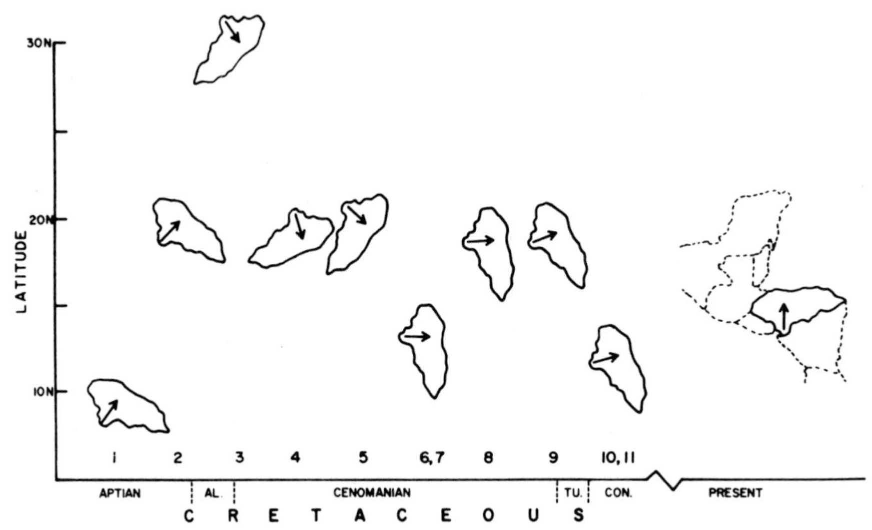

Fig. 176. Schematic presentation of the position of Honduras during part of the Cretaceous relative to its present position. (From GOSE & SWARTZ 1977a, Fig. 1).

les arc and begins along the western coast of Central America (open triangles) due to subduction along the Middle America Trench.

C. Late Tertiary, continued relative left-lateral motion between North America and Caribbean plates; localization of left-lateral shearing and normal faulting along the walls of the Cayman Trench. Mafic floor of the trench is formed as spreading continues from the mid-Cayman spreading centre. Extensive subsidence causes disappearance of Cayman Ridge and Nicaraguan Rise land masses; progressive restriction of carbonate banks. Oriente Province is sutured onto Cuba. Volcanism ceases in the Greater Antilles but increases in western Central America as Middle America Trench extends southeastward (p. 1171)."

Of all the plate tectonic models advanced for the northern Caribbean region and for northern Central America, that put forward by PERFIT & HEEZEN seems to me to fit the geological facts best.

From paleomagnetic measurements carried out on Cretaceous sediments from Honduras GOSE & SWARTZ (1977 a and b) deduced north-south movements and rotations of more than 100° (Fig. 176). They do not think this finding should be restricted to the area of Honduras alone but they also relate it to movements of the Central American continental core south of the Motagua Fault Zone. This core was originally not a part of the Caribbean Plate but was located in the Pacific Ocean.

The **structure and evolution of southern Central America** can largely be understood in terms of plate tectonic models, even though some questions still remain unanswered or contradict the models. The fundamental problems are as follows:

1. The lack of a pre-Cretaceous basement, be it constructed of Paleozoic or older crystalline rock or merely from pre-Cretaceous sediments.
2. The wide distribution of Upper Cretaceous oceanic tholeiitic basalts and intercalated or directly superjacent deep-sea sediments, which can be regarded as oceanic crust.

3. The conversion of this oceanic crust into continental "tectonitic" crust during the Cenozoic.

The lack of a pre-Cretaceous basement, or in other words the commencement of the geological record no earlier than in the Jurassic, is explained by the theory of the super-continent of Pangaea extending into the Triassic and joining together within itself the region of present-day North and South America. However, the overlapping of continental crust from North America and from northern Central America contradicts this model. FREELAND & DIETZ (1971, 1972) had solved this problem by rotating Oaxaca, Yucatán and northern Central America and placing them in the region of the later Gulf of Mexico. From this Gulf these fragments of crust reached their present position by rotation during the Jurassic. However, there is still no direct geological proof of such wide-ranging movement, and it will probably also be difficult to find any such evidence.

Regardless of this unsolved question, we can however assume that the two American plates drifted apart from the start of the Jurassic onwards. LADD (1976 a and b) has traced this drifting by evaluating magnetic data from the northern and southern Atlantic from the beginning up to the Late Tertiary, and for the Jurassic to the Valanginian he discovered a southeastwards-directed relative movement of South America, assuming a stable location for North America (Fig. 177).

There is little direct evidence for the opening up of an oceanic fissure during the Jurassic in the region of southern Central America and the Caribbean. Marine sediments are found only in boundary regions such as the Lower to Middle Jurassic San Cayetano Formation of Cuba, in Trinidad and on the Goajira Peninsula (GEYER 1968).

Violent fracture tectonics have been detected in Honduras where the Upper Triassic-Lower Jurassic El Plan Formation occurs as a 2,000 m thick graben filling in which marine strata are sporadically interbedded. This can be regarded as the result of extension.

SCHMIDT-EFFING (1976 b, p. 214) regards the great differences in the Liassic ammonite faunas, which could only display close relationships if they shared a common shelf, as indirect evidence of an oceanic region between Mexico and Colombia. On the other hand GEYER regards precisely the close relationships between the ammonite faunas of Mexico and Colombia as positive evidence of a direct sea connection, no matter how uncertain the extent of this sea might be (1968, p. 81).

During the Cretaceous (Valanginian to Coniacian), South America moved mainly eastwards (LADD, see Fig. 177), assuming that the position of North America remained stable. Eupelagic sediments and submarine basalts are proof of oceanic conditions in Costa Rica, Panama, large areas of the Antilles and in the floor of the Caribbean, where they were detected by drilling carried out during Legs 4 and 15. From the interbedded and superjacent sediments containing foraminiferal and radiolarian faunas, the basalts can be dated from the Turonian to the Campanian. Certain occurrences may also be older (GALLI 1977) and in Costa Rica the last emissions of basaltic magmas extend into the Earliest Eocene (SCHMIDT-EFFING 1979). The main flow, however, probably occurred within a relatively short period of time and covers an area of more than 1 million km^2 (DONNELLY et al. 1973).

The question arises here whether the oceanic crust situated between the continental nucleus of northern Central America and the South American continent was inserted as a pre-existing protuberance of the Eastern Pacific Plate or whether it was newly formed by sea-floor

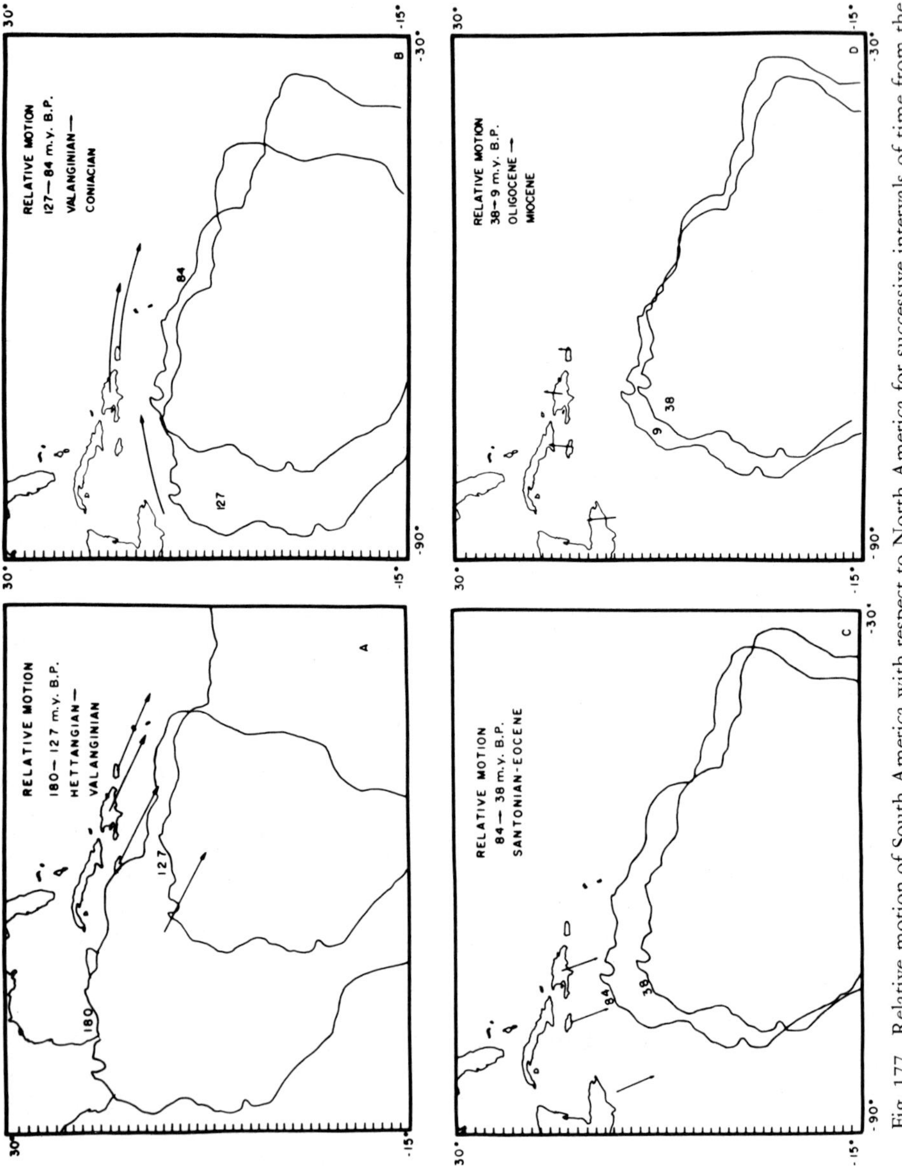

Fig. 177. Relative motion of South America with respect to North America for successive intervals of time from the Early Jurassic to the Late Tertiary. North America is in its present geographic location in all diagrams. (From LADD 1976a, Fig. 2).

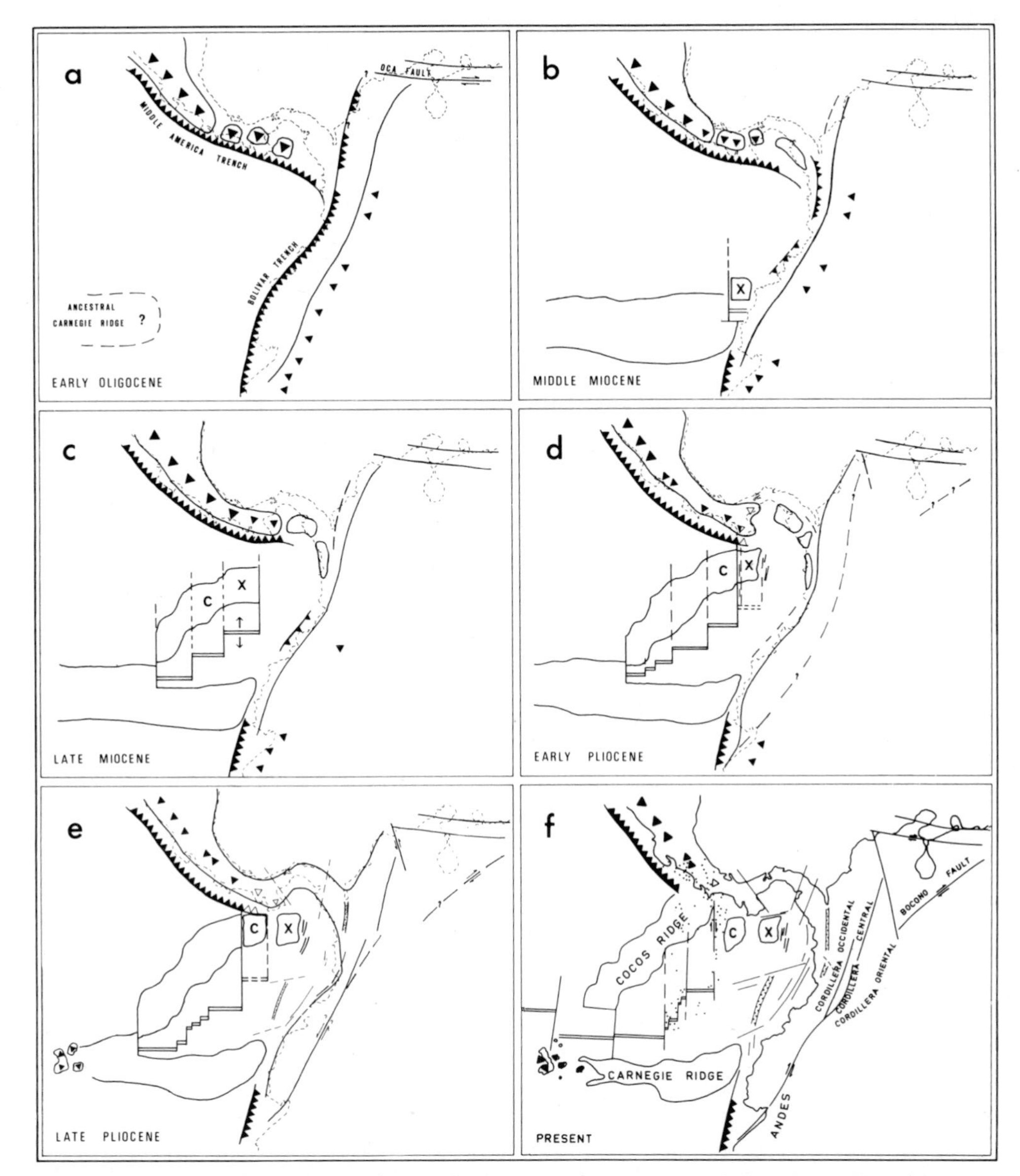

Fig. 178. Structural development of Central America and the Panama Basin. (From VAN ANDEL & HEATH 1973, Fig. 7). Sawtooth lines: Subduction zones; triangles: Volcanoes (solid-active, open-extinct); single lines: Faults or fracture zone; double lines: Spreading centres; dots (in f only): Earthquake epicentres. C and X are fragments broken from the end of the Cocos Ridge after they plugged and deactivated the subduction zone of the eastern Middle America Trench.

spreading. In the Colombia Basin the E/W-trending magnetic anomalies and their classification into index anomalies 27 – 34 are probably evidence of the formation of new oceanic crust in the period between 85 and 66 m.y. BP (CHRISTOFFERSON 1973, 1976). The crust must have formed to the south of the present basin because the anomalies increase in age from S to N. This coincides with the predominantly N/S relative movement of the American plates, which was determined by LADD, and with the assumption of an Upper Cretaceous subduction zone in the area of the Greater Antilles.

As regards the further development of southern Central America, we have to ask when, where and how the oceanic crust was converted into the present-day continental crust, thus forming the land bridge between the American subcontinents.

STIBANE et al. (1977) were able to establish a model of this process in the Nicoya Peninsula of Costa Rica. Here, as early as in the Campanian, they discovered a sharp differentiation into ridges with reef formation and erosion alongside basins with pelagic sediments and thick clastic deposits. This finding can be interpreted as the start of the separation of the Cocos and Caribbean plates and as the consequence of incipient subduction of the Cocos Plate (on this, see p. 120). Another sign that formation of continental crust had commenced is the production of quartz-dioritic intrusive rocks in the region of Panama during the Maestrichtian and Paleocene. And finally, the migration of vertebrate faunas from North and South America at the end of the Cretaceous required at least island chains as the bridge.

It is therefore probable that subduction, and the sialic magmatism triggered by it, commenced towards the end of the Cretaceous and gave rise to the formation of an isthmic island chain, which has been preserved along with still bathyal marine regions. The paleogeographic maps of Panama (Fig. 91) may serve as a model of this process.

Further development of the isthmus during the Cenozoic is characterized by the great mobility of the crust along with the formation of deeply subsided troughs and their thick fillings of sediment, by the uplifting of islands and ridges, by the onset of violent volcanism of increasingly sialic character, and with quartz-dioritic to granodioritic and granitic plutonism, and finally by the uplifting of the isthmus and the rise of mountain chains to over 3,000 m. This development is closely connected with the development of the Cocos Plate and of the Nazca Plate in the region of the Panama Basin, a process which has been documented above all by the works of VAN ANDEL et al. (1971), HEATH & VAN ANDEL (1973), HEY (1977) and HEY et al. (1977).

According to a theory developed by VAN ANDEL, HEATH et al. (1973), which is explained by a series of sketches in Fig. 178, there was a subduction zone off the Pacific margin of Central America as far back as in the Early Tertiary and it extended as far as Panama (a and b). The presence of this zone is indicated in Costa Rica and Panama by the formation of deep troughs and above all by violent sialic magmatism. At the same time, an ancestral Carnegie Ridge was moving towards South America, with which it collided about 25 m.y. ago (b). As a result of this, it split open from east to west and formed the Cocos Ridge and the smaller Malpelo and Coiba Ridges within the Panama Basin (b, c). As a result of drifting apart in the Galapagos Rift Zone these ridges moved northwards until they reached the subduction zone of the Middle America Trench, and they progressively deactivated this zone from east to west (d, e). At the same time, the transform fault at the eastern edge of the Cocos Plate was displaced to its present position in the Panama Fracture Zone, and this enlarged the northern parts of the Nazca Plate (d, e). The (in themselves) deactivated ridges reached the area off Costa Rica at the latest by the time of the Upper Pliocene and here, too, they plugged the previously exist-

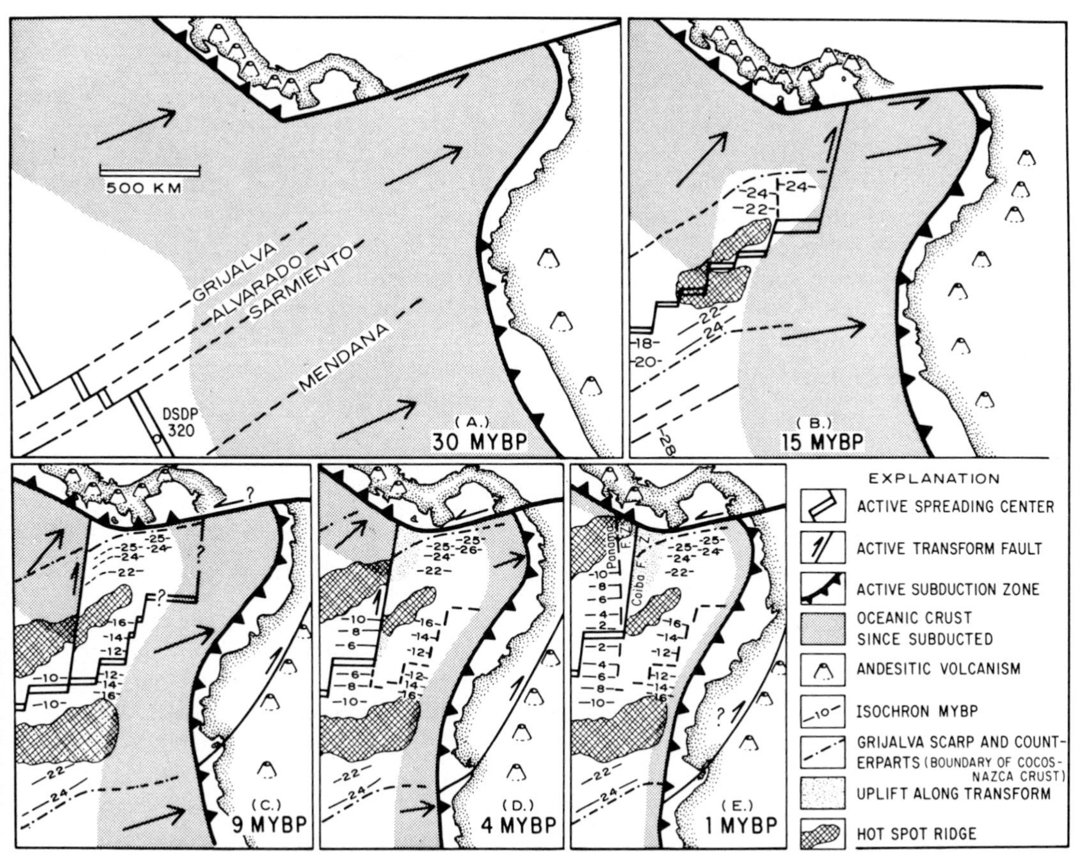

Fig. 179. Tectonic reconstruction tracing the inferred history of the eastern Panama Basin and southern Central America. (From LONSDALE & KLITGORD 1978, Fig. 11). Explanation in the text.

ing subduction zone. The present situation is outlined in Fig. 178 f, in which it is claimed that a new westward shift of the Panama Fracture Zone from 83° W to 85° W is evident.

This model was developed above all on the basis of newly surveyed paleomagnetic patterns from the eastern part of the Panama Basin, and this led to the theory of Cenozoic plate development which was put forward by LONSDALE & KLITGORD (1978) and which is illustrated here in Fig. 179: In the Early Cenozoic the Basic Igneous Complex of Panama, Colombia and (not explicitly mentioned) of Costa Rica was obducted as fragments of the original Farallon Plate and joined with the Caribbean Plate. It can be assumed that from about the Middle Oligocene onwards the Farallon Plate moved northeastwards and was subducted under the Caribbean and South American plates (Fig. 179 A). The formation of troughs such as the Terraba Trough in Costa Rica, the violent vertical movements in Panama, calc-alkaline volcanism (the material evidence of this in Panama is still not very good) and granodioritic plutonism can all be regarded as consequences of this subduction.

During the Middle Miocene (Fig. 179 B) the Farallon Plate began to split up into the northeastwards-moving Cocos Plate and the eastwards-moving Nazca Plate, while the Malpelo portion and the eastern portion of the later Carnegie Ridge built up over hot spots.

The continuation of the magmatism in Panama and vigorous vertical movements throughout all of southern Central America are further manifestations of the reaction of the Caribbean Plate. Because of increased spreading and thus the formation of new oceanic crust between the plates, the Malpelo and Carnegie Ridges are separated from each other (Fig. 179 C). Towards the end of the Miocene the subduction in eastern Panama slows down, the Cocos Plate is separated off by an N/S-trending transform fault and is further subducted. As a result, the Malpelo and Cocos Ridges are separated. The consequences of this are that magmatism becomes extinct in eastern Panama, trough formation dies out and the inter-ocean link becomes more shallow. In western Panama and Costa Rica, on the other hand, plutonism and volcanism last through to the Pliocene, and renewed trough formation with over 2,000 m thick sediments occurs in the Mio/Pliocene Charco Azul Formation.

The subduction of the Cocos Plate off Costa Rica and off the most westerly part of Panama continued in the Pliocene (Fig. 179 D), thus resulting in the volcanism lasting into the Pleistocene (Volcán de Chiriquí) and the Late Pliocene plutonism of the Cerro Colorado. The eastern splinter of the Nazca Plate moved towards the South American continent. The present situation was reached by the transform fault between the Cocos and Nazca Plates being shifted from the Coiba Fracture Zone into the more westerly situated Panama Fracture Zone (Fig. 179 E).

This highly simplified outline of the development explains, on the one hand, the unusually pronounced relief of the Panama Basin and, on the other, the development of the continental crust with its magmatism and vigorous movements. In broad terms the plate movements deduced from paleomagnetic patterns seem to be temporally coordinated with the reaction in the area of the Caribbean Plate and of its present continental portions. In particular, this explains the different development of western and eastern Panama.

On the other hand, the violent horizontal motion and the disappearance of large sections of crust as a result of plate tectonic activity make it difficult to reconstruct the paleogeographic conditions and to describe early distributions of facies. As a rule, such reconstructions are based on the present position in the system of geographical coordinates (e.g. Fig. 75). However, given the variations in the required plate movements, it is not yet reasonable to make any shift into a prehistoric system of coordinates. Thus, such paleogeographic reconstructions probably at present apply only within an inherently stable plate, but they do retain their significance there. The maps by WILSON (Fig. 27) or RECCHI (Fig. 91) can be taken as models of this. They in particular retain their value as a means of illustrating the development of limited regions.

9. Deposits

9.1 Ore deposits

9.11 Introduction

Due to the strong magmatic activity that has taken place there from the Paleozoic to the present, Central America possesses a large variety of ore deposits. However, the hope that deposits of worldwide economic significance would be found was not fulfilled until quite recently with the discovery of large porphyry copper mineralizations in Panama and nickel-rich deposits formed by weathering on the serpentinites of Guatemala.

ROBERTS & IRVING (1957), on the basis of the wartime search for strategic mineral deposits, provided an initial overview of the ore deposits of Central America. A report by the Bundesanstalt für Bodenforschung, Hannover, remained unpublished (GREBE et al. 1959). Of more recent date is the metallogenetic map of Central America which was published by ICAITI (Instituto Centroamericano de Investigacion y Tecnología Industrial) (1979) together with a detailed explanatory volume by DENGO & LEVY (Publ. Geol. ICAITI, No. 3, Guatemala 1970). This volume deals with the wealth of literature and unpublished reports to which it is difficult or impossible to gain access. KESLER's (1978) brief review of the metallogenesis of the Caribbean region is based on countless individual studies conducted by the author and his students. A review of the mining activity in Central America was published in the "Engineering and Mining Journal" of November 1977.

Recent and older reports on the various countries are as follows:

For Guatemala:	Anonymous 1969 a.
For El Salvador:	GREBE 1955 a and b.
For Honduras:	ELVIER 1969, 1970, 1974.
For Nicaragua:	Boletin del Servicio Geológico Nacional No. 7, Managua 1963.
For Costa Rica:	Anonymous 1978
For Panama	FERENČIĆ 1971.

A review of the ore production of Central America up to 1976 is given in Table 13.

Along with LEVY (1970) we can assign the ore deposits of Central America to three provinces corresponding essentially to the large morphotectonic units as defined by DENGO (1968/1973) and also to the basic sub-division of the material in this book (Fig. 180):

Table 13. Production statistics for Central America (based on various deviating sources).

	1970	1971	1972	1973	1974	1975	1976
Gold (kg)							
Honduras	168	85	64	85	96	71	60
El Salvador		76					
Nicaragua	3 200			2 400	1 920	1 810	1 700
Costa Rica					520		
Silver (kg)							
Guatemala				71			
Honduras	111 000 [1]	113 000	108 000	64 500	64 500	65 100	59 000
El Salvador		5 560					
Nicaragua	5 750			9 800	9 800	7 000	1 200
Costa Rica					22		
Lead (t)							
Guatemala				71	75	89	83
Honduras [1]	18 000	20 200	23 500	22 600	23 500	25 000	23 300
Nicaragua				3 224	3 746	2 000	1 200
Zinc (t)							
Guatemala [2]				112			
Honduras	23 500	26 000	26 500	28 500	26 500	28 000	27 600
Nicaragua				20 600	18 500	14 000	15 500
Cadmium (t)							
Honduras	210	110	280	245	225	230	235
Copper (t)							
Guatemala [2]					4 085	11 082	10 318
Nicaragua	4 590			1 936	1 957	500	650
Antimony (t) [2]							
Guatemala				1 200	800	1 533	2 627
Honduras		18	36	213 [1]	30	40	25
Tungsten (t) [2]							
Guatemala				578	150	200	

[1] partly exports
[2] concentrate

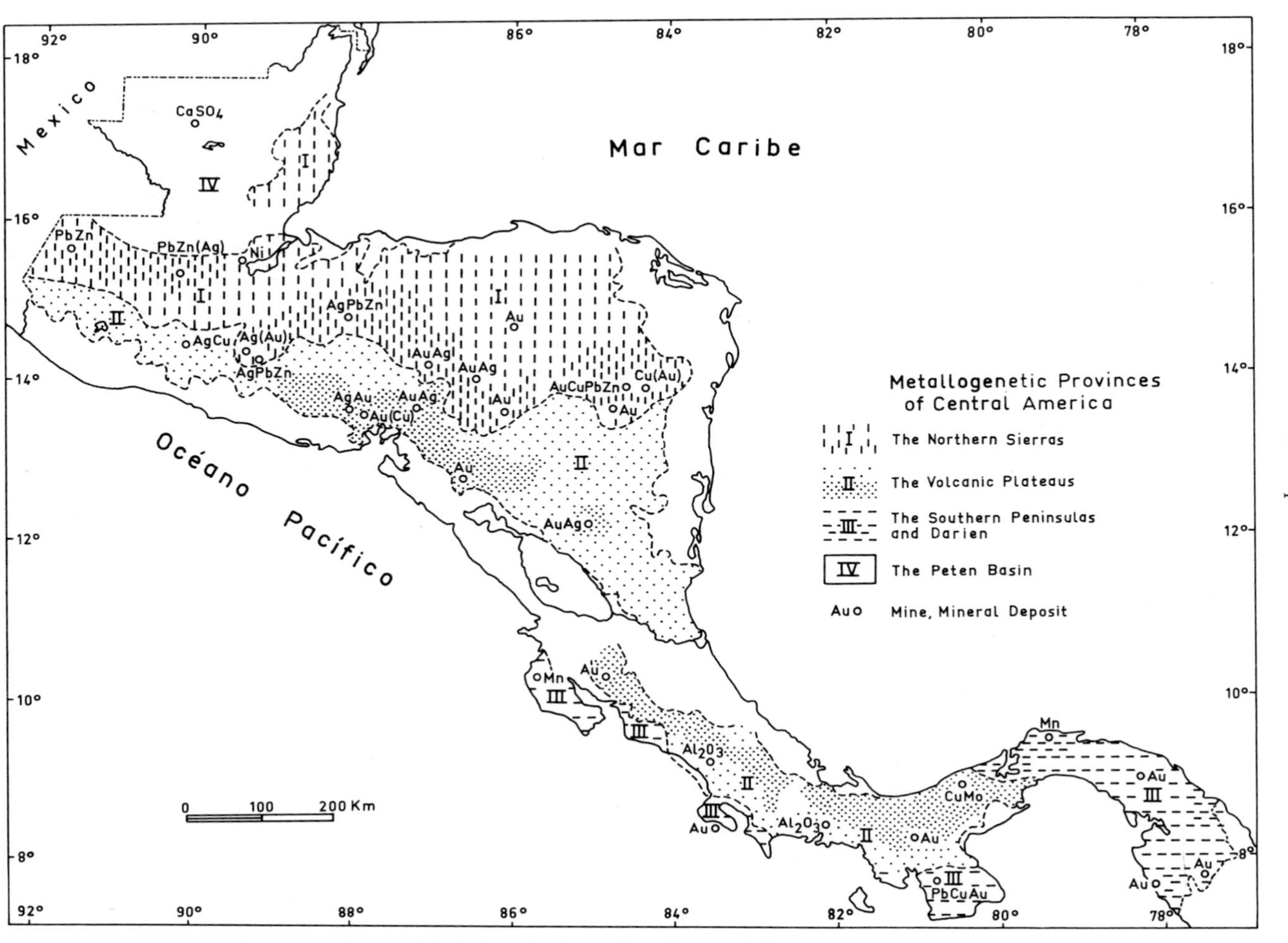

Fig. 180. Metallogenetic provinces of Central America. (Redrawn from LEVY 1970).

the province of the northern Sierras,
the province of the Pacific peninsulas and of Darién in southern Central America,
the province of the volcanic plateaus.

LEVY distinguishes between five phases of mineralization within these provinces:

a Late Paleozoic mineralization,
a Cretaceous mineralization,
a Laramide mineralization,
a post-Laramide-Tertiary mineralization,
a Quaternary mineralization.

According to LEVY most of the deposits within these provinces can be assigned with certainty, others only with reservation, to one or the other of the mineralizations. It should, however, be noted that as the area is becoming better known, an increasing number of magmatic activity phases are being revealed which are verified by numerous radiometric datings (Fig. 116). According to these studies an almost continuous intrusion activity involving acid "granitic" magmas predominated in northern Central America from the Cretaceous to the Tertiary so that it might cause problems to make an excessively precise distinction between the associated mineralization phases.

KESLER (1978) distinguishes similar regions which he characterizes according to their "metallogenetic maturity". The most mature region is northern Central America with its cratonic structure. It contains a broad distribution of precious, non-ferrous and ferrous mineralizations which are linked with terrestrial volcanics. In addition, it contains small deposits of tungsten, antimony and mercury. Southern Central America is younger; it is the site of submarine manganese production and of large porphyry copper mineralizations while only modest amounts of precious metals are represented and tungsten, antimony and mercury are missing. In his concise description of individual deposits KESLER distinguishes between the following types: "porphyry disseminated, massive sulphide deposits, limestone replacement deposits, vein deposits, late Cenozoic precious metal and base metal deposits, older precious metal and base metal deposits, manganese deposits, ophiolite chromite deposits, laterite deposits" (Fig. 181).

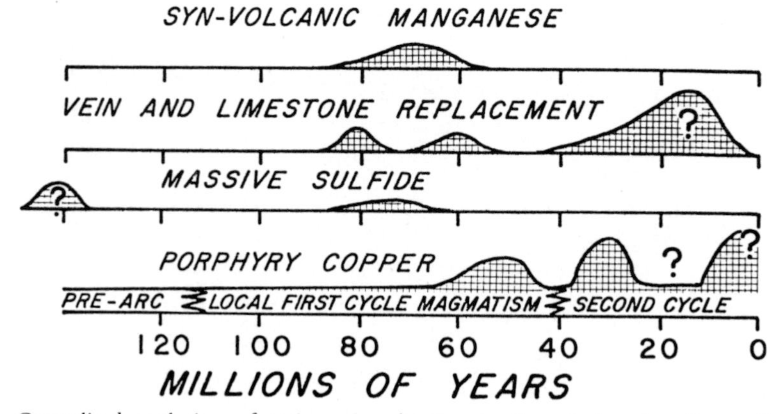

Fig. 181. Generalized evolution of major mineralization in Central America. (From KESLER 1978, Fig. 6).

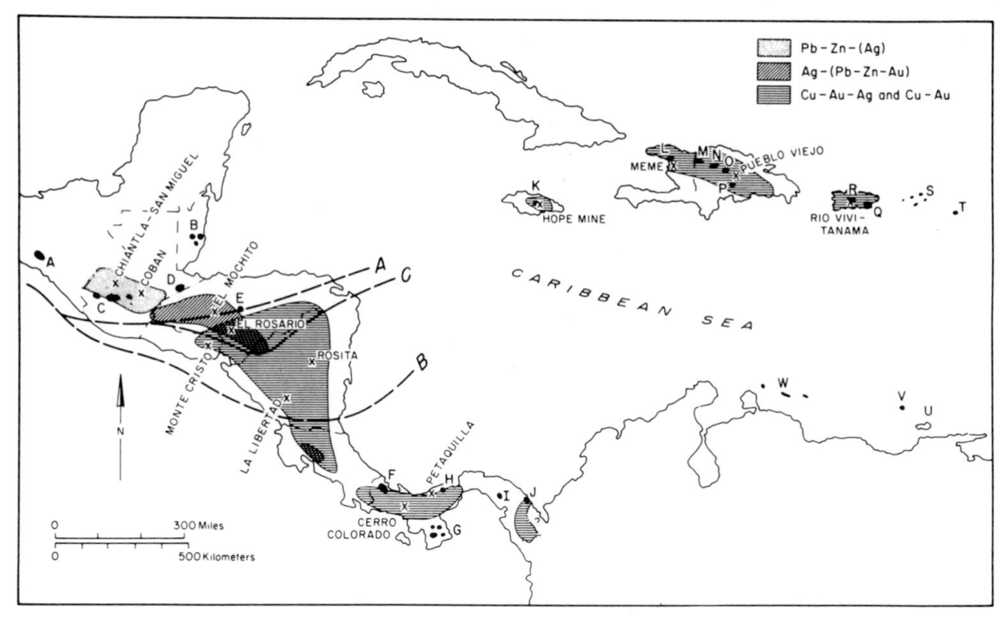

Fig. 182. Distribution of ore deposit zones in Central America and the Caribbean. Lines A, B and C show alternative southern boundaries for the Central American craton. (From CUMMING & KESLER 1976, Fig. 1).

CUMMING & KESLER (1976) have pointed out the relationship that exists between the contents of the deposit and the structure of the crust: In the northwest are situated the important lead-zinc deposits (Chiantlá, San Miguel, Cobán). These have a low precious metal content, but the latter increases in a SE direction so that a region with lead-zinc-silver-gold mineralization exists (Montecristo, La Libertad, Cerro Colorado, Petaquilla and Río Pito) (Fig. 182). This sequence is paralleled by a change in the non-ferrous metal content of intrusive rocks which was detected by means of taking a large number of samples (CUMMING & KESLER 1976, Table 1 — an extract is given in the following table).

Non-ferrous metal contents of Central American intrusive rocks (in ppm) according to CUMMING & KESLER (1976, Table 1). The letters correspond to the entries in Fig. 182.

	Cu	Pb	Zn
A. Chiapas, Mexico	15	11	52
B. Belize	9	30	52
C. Western Guatemala	9		33
D. Quiquimula, Guatemala	18	14	50
E. Minas de Oro, Honduras	36	12	59
F. Bocas de Toro, Panama	83	7	77
G. Azuero, Panama	102	4	91
H. Petaquilla, Panama	112	4	57
I. Río Guayaba, Panama	194	6	35
J. Río Pito, Panama	132	4	57

Lead isotope values from 15 samples exhibited a parallel tendency: The highest values for radiogenic lead are found in the lead-zinc zone in the north while the lowest values are found in the copper zone in the south.

The zoning of the ore deposits is similar to that on the western coast of North America (SILLITOE 1972) where the lead-rich deposits are located in the vicinity of the old continental core and the Cu-Au-rich deposits overlie the younger and much less developed continental margin. The same also applies in the case of the relationship of the northern "continental" to the southern "oceanic" crustal structure of Central America and its mineral deposits. In addition, it can be expected that more radiogenic lead was mobilized from the older continental crust and incorporated into the deposits than was the case further south.

In the following chapters I have adopted LEVY's classification, taking into account more recent literature, as far as it was available to me.

9.12 The province of the northern Sierras

Corresponding to the great age of the province of the northern Sierras, the multiple deformation phases and the accompanying magmatism, this is where the greatest variety of ore deposits is to be found, even if most of them are not worth working or can be exploited only for short periods of time. According to LEVY (1970) the deposits can be assigned to four mineralization phases: an Upper Paleozoic, a Cretaceous, a Laramide and a post-Laramide phase.

The **Upper Paleozoic mineralization** includes many gold quartz veins in the plutonic and metamorphic rocks. The "Anderson & Rey" mine in Olancho, Honduras, is an example of this. The gold quartz veins themselves have not been mined to any great extent but they are regarded as the parent rock for the placer deposits of the Río Guayapa, Patuca and Río Frío in Olancho, Honduras, which have been exploited by small mining operations ever since the 16th century.

Non-mineable sporadic occurrences of copper, lead and zinc ores are linked with the granites of the Maya Mountains. Similarly, non-mineable deposits of alluvial gold and cassiterite are derived from the same granites. A new exploration carried out using geochemical methods came to the conclusion "that the area as a whole presents little hope of major economic mineral deposits being discovered" (BATESON & HALL 1977).

Ilmenite veins in the valley of the Río Cuilco, Guatemala, which were investigated in a UNDP study project, are linked with a basic intrusion occurring in gneiss in the form of biotite amphibolite.

In the period from 1967 to 1970 a mission from the Bundesanstalt für Bodenforschung, which was working in Guatemala, discovered geochemical enrichment of copper, zinc and lead in river sediments of the Sierra de Chuacús; this phenomenon can be regarded as indicative of a weak epigenetic mineralization of the metamorphic rocks. In addition, small occurrences of barite and disthene were found. The muscovite and quartz deposits of the Sierra de Chuacús, which are associated with pegmatite and aplite veins, are very well known and they have been described in detail by ROBERTS & IRVING (1957). The muscovite was exploited during the years 1926–1942 and the quartz in the period 1941–1944.

Compared with the more recent mineralization, that dating from the Paleozoic is remarkably limited in extent. Presumably this is a result of long-term erosion right down to deep-lying zones which occurred during and after the Paleozoic and which carried away the more highly mineralized upper stockworks.

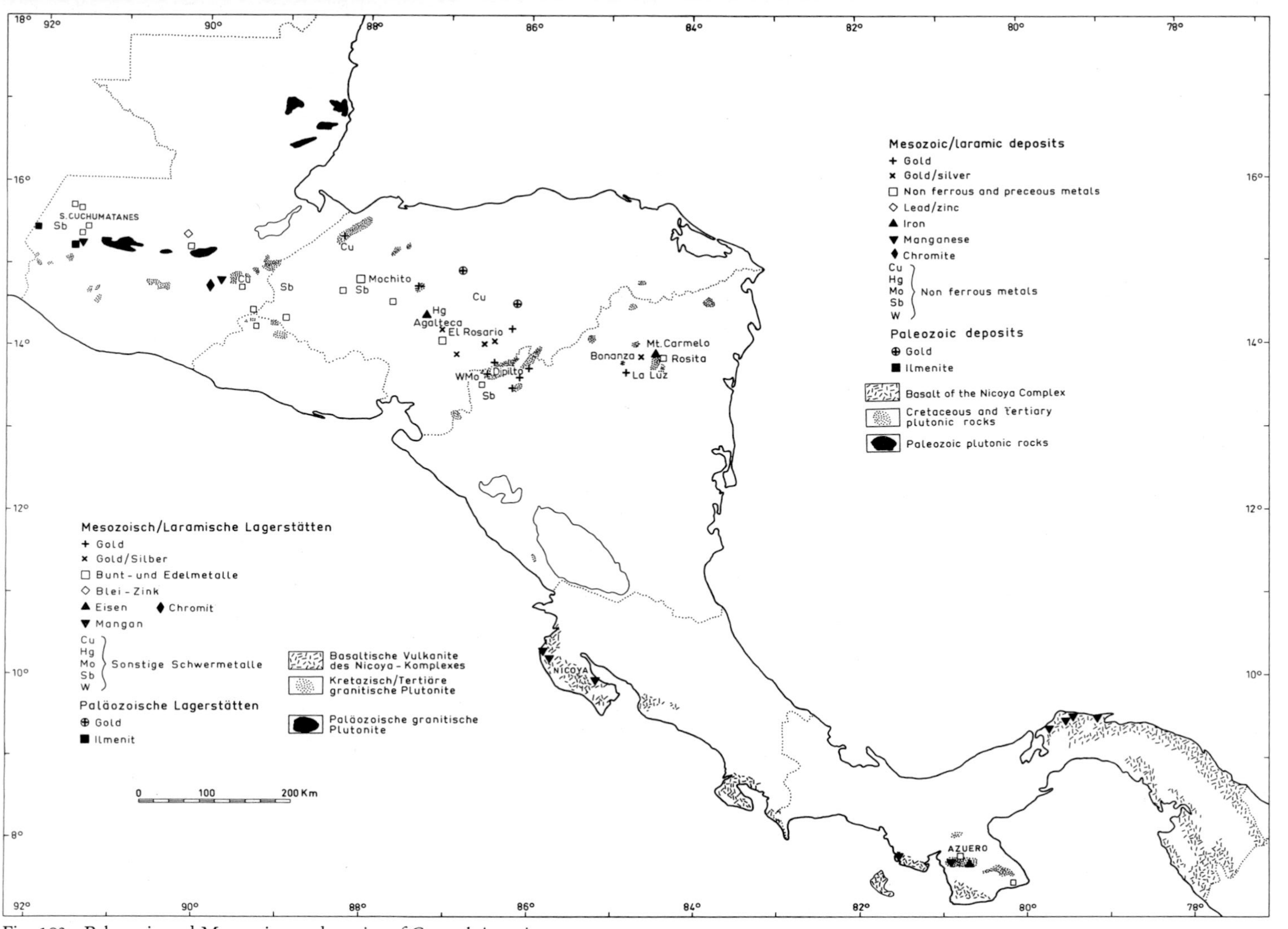

Fig. 183. Paleozoic and Mesozoic ore deposits of Central America.

In the **post-Paleozoic metallogenesis** of northern Central America LEVY distinguishes between a Cretaceous and a Laramide mineralization but he stresses that as more and more radiometric dating is carried out on magmatic source rocks there will be changes in the allocation of deposits to the various groups and that the boundaries are unclear, particularly since the start and end of the Laramide orogenesis in Central America is poorly defined. In fact, the radiometric dates which have become available since his classification was published (1970) reveal that intrusion activity probably occurred almost continuously from the Lower Cretaceous right through to the Tertiary (Fig. 116) so that a true break did not occur until the onset of the Tertiary continental volcanism in the northern part of Central America, but even this was still accompanied by intrusion activity. No attempt will be made therefore to distinguish between a Cretaceous and a Laramide metallogenesis; however, the reader's attention is drawn to the time differentiation indicated in the metallogenetic map prepared by ICAITI.

The spectrum of the deposits which are grouped together here as Late Mesozoic formations ranges from pegmatic-pneumatolytic wolframite/molybdenum veins through contact-metasomatic iron ores and copper deposits to gold-bearing quartz veins and hydrothermal replacement deposits of lead-zinc sulphide ores. Most of the rather modest output of ore in Central America is accounted for by these deposits. A few examples will now be given, although no claim is made that the list is or indeed ever could be exhaustive. Most of the examples are identified by name in the general map (Fig. 183).

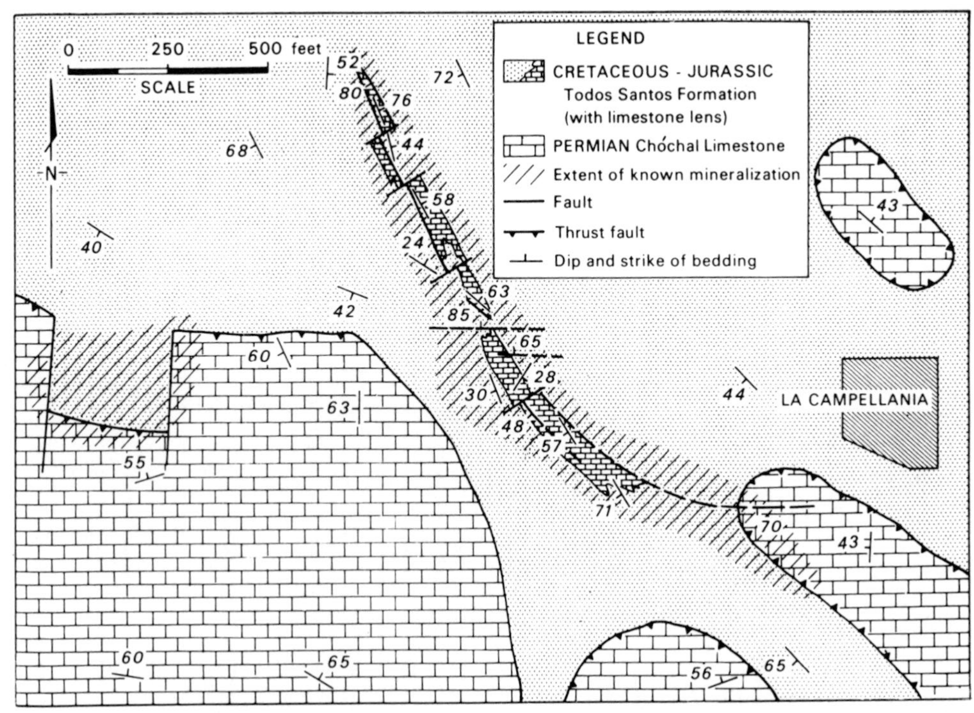

Fig. 184. Geological map of the area of the Santo Domingo mine showing mineralization in a thin lense of Todos Santos limestone at and near its contact with overthrust Chóchal limestone. (From KESLER & ASCARRUNZ 1973, Fig. 6).

Replacement deposits of lead-zinc ores in Paleozoic and Mesozoic limestones occur in Guatemala in a narrow belt between the Mexican border and the region of Cobán/Alta Verapaz. The main districts are the region of Chiantlá-San Sebastián on the southern slope of the Altos Cuchumatanes, the San Miguel region in the Altos Cuchumatanes and the Cobán region in Alta Verapaz.

A new survey conducted in the region of Chiantlá (Fig. 184) (ASCARRUNZ & KESLER 1971; KESLER, ASCARRUNZ & ANDERSON 1972, 1973) revealed that the mineralizations are preferentially associated with the Paleozoic carbonates of the Chóchal and Esperanza Formation, but also occur in conjunction with Cretaceous limestones. The main zones of mineralization are the NE-striking faults, but taken overall their distribution follows the NW-striking structure of the Altos Cuchumatanes. The age is given as Cretaceous or younger. The main minerals are pyrite, sphalerite and galena as well as small quantities of pyrargyrite, chalcopyrite, malachite and azurite. Barite occurs as gangue. According to ROBERTS & IRVING the primary ores have been converted into smithsonite, anglesite, cerusite and limonite down to depths of 25 m.

In the region of San Miguel the same ores are bound to limestones and dolomites of the Cretaceous Ixcoy Formation, and they are also found in association with faults and strata. Mining commenced during the Second World War and continued in subsequent years.

In the Cobán region both the Permian Chóchal Limestone as well as the Cretaceous Cobán Limestones are mineralized. Mining commenced at about the same time as in the other districts, and in 1970 the last remaining mine, the Suquinay, was still in operation. In recent years, however, many attempts have been made to reactivate mining for lead-zinc ores.

The total annual output of lead and zinc in Guatemala in the years of maximum production amounted to 7,000–8,000 tons of lead and 6,000 to 10,000 tons of zinc.

KESLER & ASCARRUNZ (1973) discussed the origin of the ores compared with that of the lead-zinc ore deposits of North America. The Guatemalan ores cannot positively be assigned to either of the two North American types, but come somewhere between the two. S-isotopes and inclusions of liquid in the ores point to temperatures of formation below 250 °C. The distribution of S-isotopes is similar to that found in the sulphate rocks of the Jurassic and the Cretaceous while, on the other hand, the occurrence of mineralization along tectonic fault zones would seem to indicate that the ore solutions originated from great depths. It is assumed, for example, that the interplay between deep-reaching fracture zones, sulphur in the basin sediments and the existence of a hitherto undetected deep-lying magma body all led to the formation of the deposits.

It is just as difficult as in the foregoing case of the lead-zinc ore deposits to discover any direct link with a body of magma in the case of the **antimonite-scheelite mineralization** which was found by COLLINS & KESLER (1969) in the upper part of the Permo-Carboniferous Tactic Schists near Ixtahuacan in NW Guatemala. In-the-field observations point to a possible connection with a young volcanic centre located about 4 km south of the deposit. However, the Cuilco-Polochic Fault runs between the two. The occurrence is interpreted as a telescoping xenothermal deposit formed at temperatures in the range of 400 – 500 °C as the result of a rapid chemical reaction of hydrothermal alkali solution with Chóchal Limestones or calcareous strata in the upper Tactic Schists. Strontium isotope measurements indicate that the ore solutions come from the upper mantle zone or from the lower basaltic layer. The deposit was worked on an experimental basis at the end of the sixties.

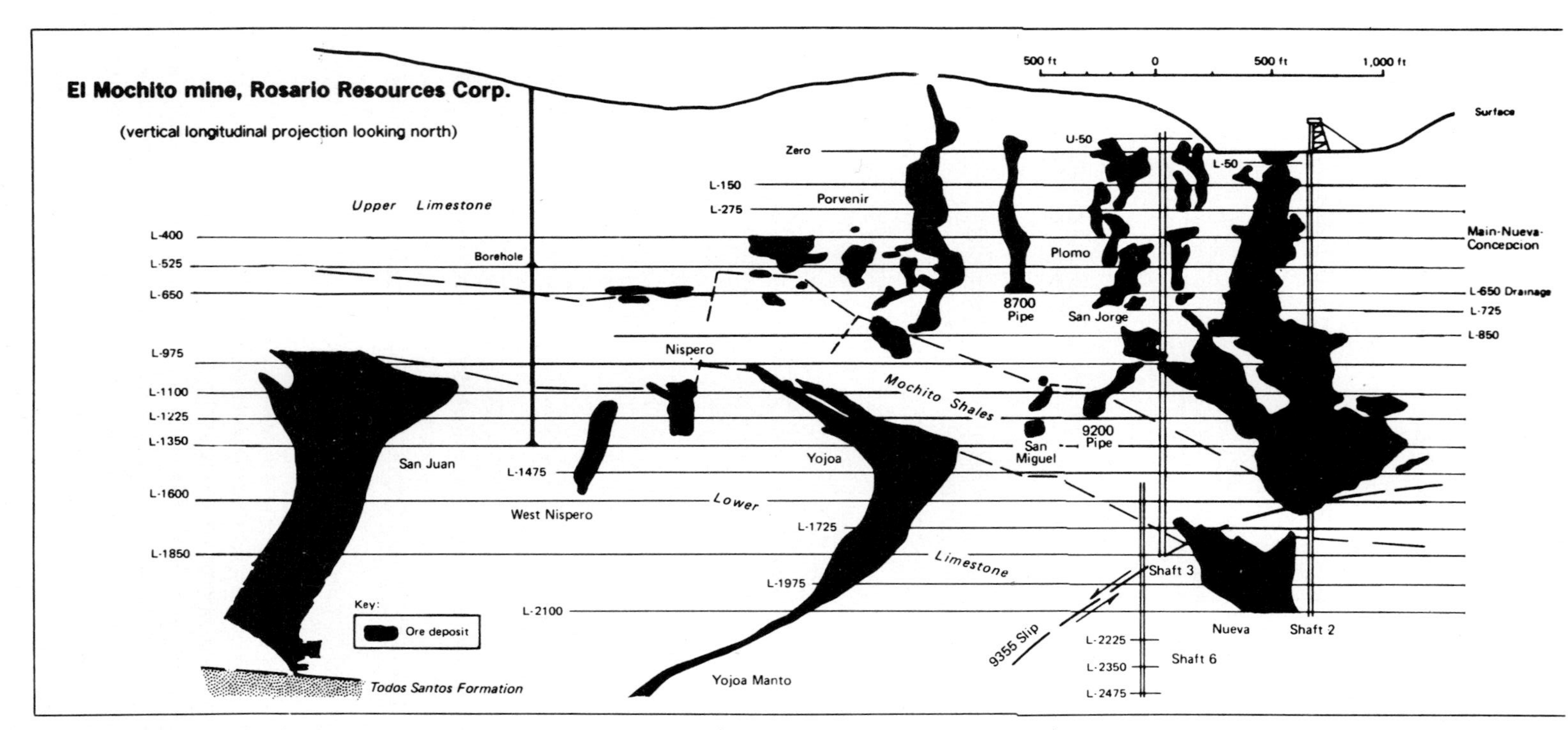

Fig. 185. Vertical longitudinal section through the Mochito mine, Honduras. (From WHITE 1977, pp. 168/169).

Mining of a **chalcopyrite deposit** near Oxec in Alta Verapaz has been going on for the past five years. Lenticular ore bodies may contain up to 20% Cu but the average content of the mined ore is 1.6%, which makes it only just barely profitable to extract. The deposit is located in the vicinity of a diorite mass and it is affected by epidotized andesite veins. I was not able to obtain any more detailed information.

Antimony deposits are located in Paleozoic phyllites in Honduras and Nicaragua. The main ore type is antimonite which occurs in the form of small lenses and veins. The only occurrence which has been worked is the El Quetzal mine in the Departamento Copán, Honduras, which had a total output of 575 tons of ore (62% Sb) during the First World War. Here the antimonite is accompanied by some pyrite, arsenopyrite and sphalerite, and close to the surface it is weathered into antimony oxides. The ores occur in lenses in metamorphic rocks and in the superincumbent "diorites" and "rhyolites" (ROBERTS & IRVING 1957, pp. 39–44). A further deposit near Santa Cruz Cuchilla in Honduras was studied as part of a UN minerals project. The antimony mineralization is bound to silicified limestones, shales or sandstones of the Upper Cretaceous/Late Tertiary Valle de Angeles Group (RIECK 1975).

A hydrothermal replacement deposit is being mined at the **Mochito mine** west of Lake Yojoa in the northern part of Honduras (Departamento Santa Bárbara). This deposit is located in limestones and sandstones of the Atima Formation (Fig. 185). The main ores are galena, sphalerite, argentite, pyrite, chalcopyrite and free silver with calcite and quartz gangue. The ores follow faults and bedding planes in an irregular manner. According to ELVIER (1974, p. 36) the mine has reserves of more than 4.5 million tons of ore containing economically extractable amounts of silver (350 g/ton), gold (0.26 g/ton), lead (8.98%), zinc (9.13%) and cadmium. The mine can produce and process into concentrates or noble metals up to 1,000 tons of ore per day. In 1968 the production figures were:

gold	145 kg
silver	120 t
lead	13,869 t
zinc	14,985 t
cadmium	240 t.

Because of its continuous output of gold and silver between the years 1882 and 1954 the **Rosario mine,** which is situated about 28 km to the north of Tegucigalpa, was one of the world's economically most important mines. Quartz veins were worked which are associated with a dacitic intrusion in the shaly-sandy sediments of the Jurassic El Plan Formation. The ores found there are galena, sphalerite, chalcopyrite, stephanite, argentite, pyrargirite, native silver and electrum. Mining was stopped in 1954 when the high-grade ores were exhausted, but since 1971 exploratory work has been going on in the deeper stockworks in the search for exploitable material (ELVIER 1974, p. 36).

A contact-metasomatic iron ore deposit named after the village of **Agalteca** is located in Honduras on the northern slope of the Montes de Comayagua (ROBERTS & IRVING 1957, pp. 64–71 and Pl.7). It consists of a large number of irregular-shaped ore bodies of magnetite which has locally been transformed into hematite (Fig. 186). The ores are situated in the contact zone between Cretaceous Atima Limestone and an underlying diorite intrusion probably of "Laramide" age. Associated rocks are skarns with garnet, epidote, hornblende and plagioclase. At the request of the Banco Central de Honduras the reserves were surveyed in the 60's

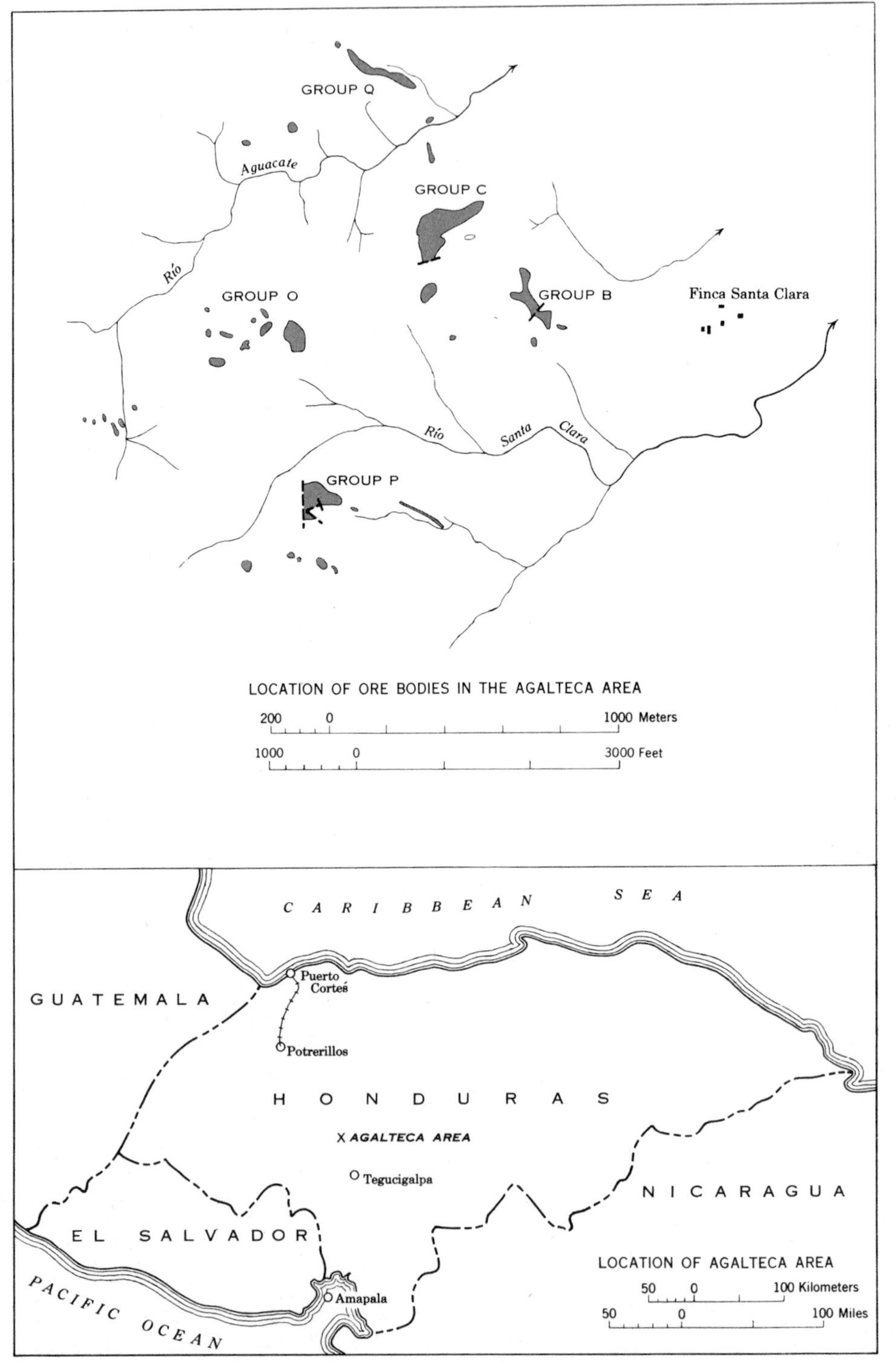

Fig. 186. Location of iron ore bodies in the Agalteca area, Honduras. (From ROBERTS & IRVING 1957, pl. 7).

and found to amount to about 8 million tons of ore with an average Fe content of 60%. It is not known whether or to what extent these reserves are being mined.

The **Monte Carmelo** near Rosita in northeastern Nicaragua is another deposit of the same type (ZOPPIS DE SENA 1957, BENGOECHÉA 1963, pp. 39–41). The ores, which are distributed among 10 separate occurrences, are located in the contact zone between granite intrusions and Cretaceous sediments. The ore consists of magnetite and hematite with an average Fe content of 63.1%. A total of 3.8 million tons of ore have been proven and the probable reserves are put at around 8.3 million tons.

A very variegated pegmatitic-pneumatolytic and hydrothermal mineralization is associated with the large **granite pluton of Dipilto** in the region of the frontier between Honduras and Nicaragua. The pluton, which is lying on top of (?) Paleozoic schists and phyllites, is dated at 83 ± 3 million years (ZOPPIS BRACCHI 1961, p. 43). The Servicio Geológico Nacional of Nicaragua has carried out intensive exploration of the associated ore deposits but the studies have not detected any economically workable deposits (ZOPPIS BRACCHI 1961, ECHÁVARRY & RUEDA 1962). Pegmatitic-pneumatolytic dikes within the pluton contain small quantities of wolframite, scheelite and molybdenite; quartz dikes contain up to 30 g/t gold. The alluvial gold deposits of the Río Patuca in Honduras and of the Río Jícaro and Coco in Nicaragua are derived from these dikes (BENGOECHÉA 1961). In the extreme southeast of the pluton, veins of galena and sphalerite, which contain silver in the oxidation zone, occur in the country rock. The mineralization is bound to the fracture zones and it is finely distributed in the schists.

In the northeastern part of Nicaragua there is a concentration of ore bodies which have been mined on and off since 1899, sometimes even achieving global economic significance. BENGOECHÉA (1963) has given a general description of the district. The most important metals are gold, silver, copper and iron. The deposits are of various types but intrusive rocks ranging from granites via syenites to diorites are regarded as the source rocks (Fig. 187).

Cretaceous sediments in the "**La Luz**" gold ore deposit exhibit a high-temperature (pneumatolytic) mineralization with magnetite, hematite, chalcopyrite, pyrite and gold, while a low-temperature (hydrothermal) mineralization yielded hematite, pyrite, galena and sphalerite. The gold is preferentially bound to the hematite. The country rocks have been transformed into garnet-epidote-pyroxene skarns and have undergone hydrothermal silification and calcitization. The deposit is situated in an inlier of the Cretaceous sediments below a cover of Tertiary volcanics. An andesite stock can probably be regarded as the feeder for the deposit; however, the mineralization did not originate from this stock, but instead from presumed intrusions at a deeper level.

The deposit has been mined on and off since 1906. In 1965, for example, it yielded about 2,200 kg of gold and about 6,600 kg of silver. In 1968 about 2 million tons of ore with a gold content of 2.9 g/ton and 1.3 million tons with a gold content of 2.5 g/ton were discovered.

The "**Bonanza**" gold deposit in the Pis Pis district consists of two families of veins of brecciated quartz with chlorite, epidote, sphalerite, galena, pyrite, chalcopyrite and hematite. Gold is bound to the sulphides. Mining has been carried on since 1889 in a large number of separate mines, and in 1962 production stood at approximately 2,500 kg of gold and 3,500 kg of silver. In 1970, about 11 km west of the Bonanza, mining of lead-zinc ores was commenced in the "Vesubia" mine. The ores, galena and sphalerite, are associated with quartz veins. Reserves of 1.5 million tons (1976) justify the construction of a cable railway to take the ores to the dressing plant at the Bonanza mine (BURN 1969).

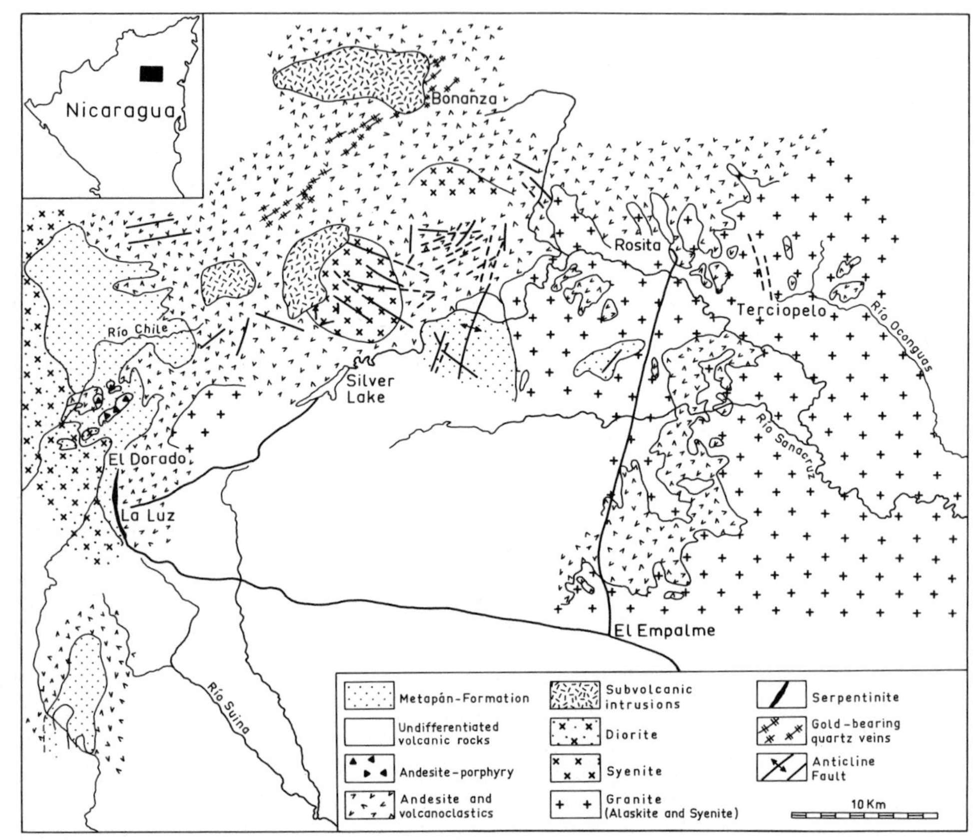

Fig. 187. Geological sketch of the mining district of northeastern Nicaragua. (Redrawn from BENGOECHÉA 1963).

The "**Rosita**" copper deposit is located in the contact zone between an acid intrusive body (alaskite) and siliceous limestone. The skarns formed here are permeated by andesitic veins to which the ore bodies are bound. The Tertiary mineralization consists of oxides, sulphides and carbonates. Mining for precious metals took place between 1906 and 1912; copper and precious metals were extracted between 1959 and 1975, and since 1975 mining for precious metals has again been commenced in the general area of the deposit because the copper was no longer economical. According to BEVAN (1973), the production figures for the Rosita open pit were 475 kg gold, 3,090 kg silver and 3,666 t copper in the report year 1970/71.

The ore concentrates of the Rosita district are transported 120 km on the Río Prinzapolca to the port of Puerto Isabel on the Caribbean coast (BEVAN 1973).

Chromite and nickel deposits occur in the peridotite/serpentinite massifs of Guatemala. The chromite occurs in nests, lenses and irregular masses (ROBERTS & IRVING 1957, pp. 48–55). It contains 40–61% Cr_2O_3 and 10–16% Fe_2O_3 and is of metallurgical quality. Mining took place between 1918 and 1945 in the order of magnitude of several thousand tons. The lateritic nickel ores are discussed in connection with other ores formed by weathering.

9.13 The province of southern Central America

The crustal structure, magmatic development and thus the metallogenesis of southern Central America are fundamentally different from those of the northern part. These differences have been described in detail in the chapters on regional geology. Our knowledge of the ore deposits in the southern part of Central America and their classification according to type have been enhanced above all by the UNDP programmes conducted over periods of several years in Panama and Costa Rica; only very few of the results of these programmes have ever been published (FERENČIĆ 1971, FERENČIĆ, DEL GIUDICE & RECCHI 1971). We therefore have to fall back on unpublished reports where available (Anonymous 1969 a, 1972 c, 1972 d, 1974 a, 1977). In addition to the general maps of the ore deposits, two maps of the mineral and fuel reserves of Panama and Costa Rica are given together with the sources on which they are based, in Figs. 188 and 189. These maps, however, show a large number of unmineable deposits.

The metallogenesis commences with the Nicoya Complex in Costa Rica or the Basic Igneous Complex in Panama. A large number of small occurrences of **manganese ores**, which were mined during the First World War, are linked with this Complex in the area of the Nicoya Peninsula, in the region of the Gulf of San Blas and in the Bahia de Montijo. From 1915 to 1920 Costa Rica's output amounted to about 32,000 tons while that of Panama between 1916 and 1919 was about 60,000 tons. Under the present circumstances they are of no economic significance. A detailed description of the occurrences, with references to earlier literature, can be found in ROBERTS & IRVING (1957, pp. 104–134).

According to these authors the manganese ores occur in the form of irregular nests and lenses of various sizes. The primary ores are replacements of the basic volcanics and sediments along fault zones and strata, and in addition to jasper they contain the manganese silicates braunite, bemenite and rhodonite. In many places the silicate ores have been converted into the oxides pyrolusite, psilomelane and wad. The deposits accordingly belong to the type of manganese deposit found in the Schist-Hornstone Formation.

A completely different genesis was assumed, on the other hand, by SCHMIDT-EFFING (1979) who discovered manganese nodules in the oldest sub-complex of Brasilito (Nicoya); these nodules fit in very well genetically with the abyssal milieu of formation of this sub-complex. They are associated with radiolarites which, modified by contact-metamorphism, were formerly incorrectly regarded as hydrothermal replacements of the basalts. SCHMIDT-EFFING therefore identifies the manganese ores as manganese nodules of the "Olympic peninsula type" which have been modified by contact-metamorphism.

KUYPERS & DENYER (1979), in a paper published after my own manuscript was completed, present a new study of the manganese deposits. They point out that "the manganese deposits in basalt and radiolarite were both formed by volcanic exhalations. This conclusion is based upon the following: the restriction of hydrothermal alteration to rocks that underlie the manganese deposits; the coprecipitation of manganese minerals and colloidal silica; and the presence of organic borings in the top of mineralized zones." Considering the associated ophiolitic rock assemblage, they conclude "that the volcanic exhalations occurred on the ocean floor below the calcium carbonate compensation depth". (See also DENYER & KUIJPERS 1979).

GOOSSENS (1971) supposed that the Basic Igneous Complex was the primary location of many heavy metals which were mobilized by subsequent magmatic processes and which as-

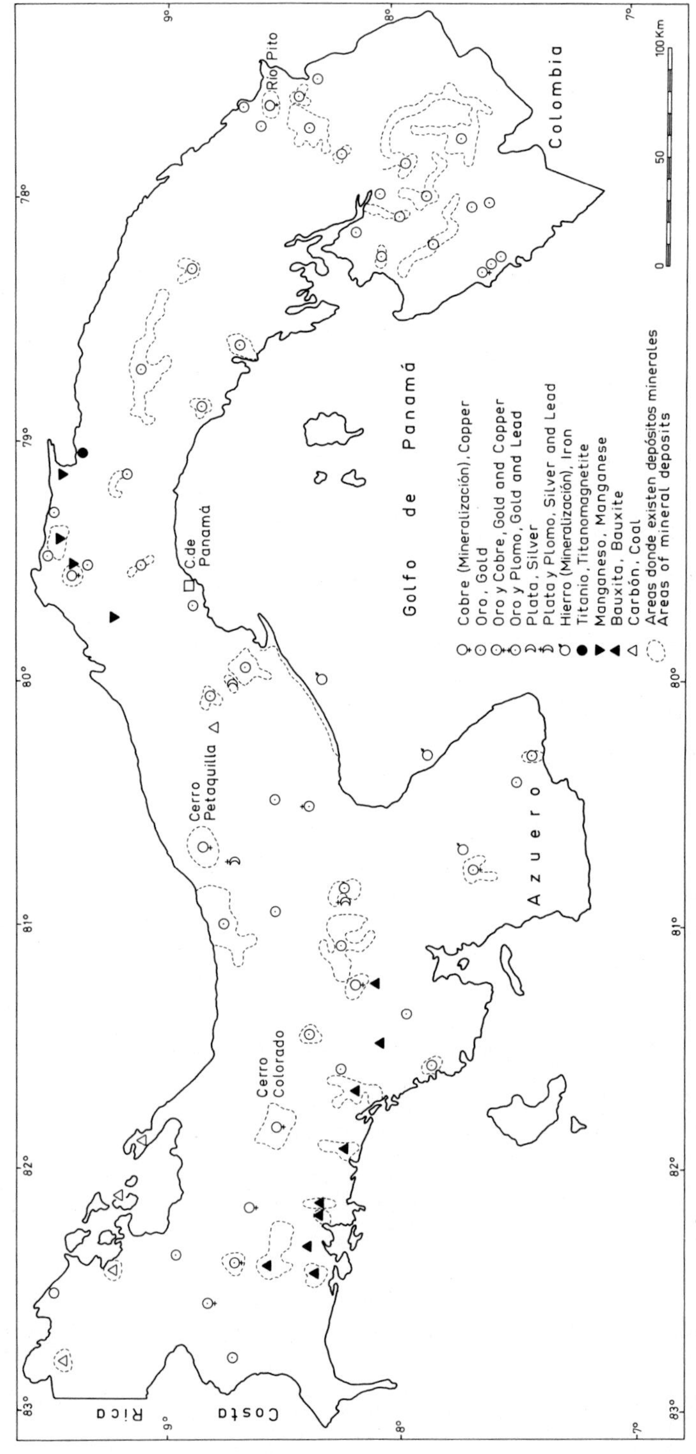

Fig. 188. The mineral deposits of Panama. (Redrawn from "Atlas Nacional de Panamá", Panamá 1975).

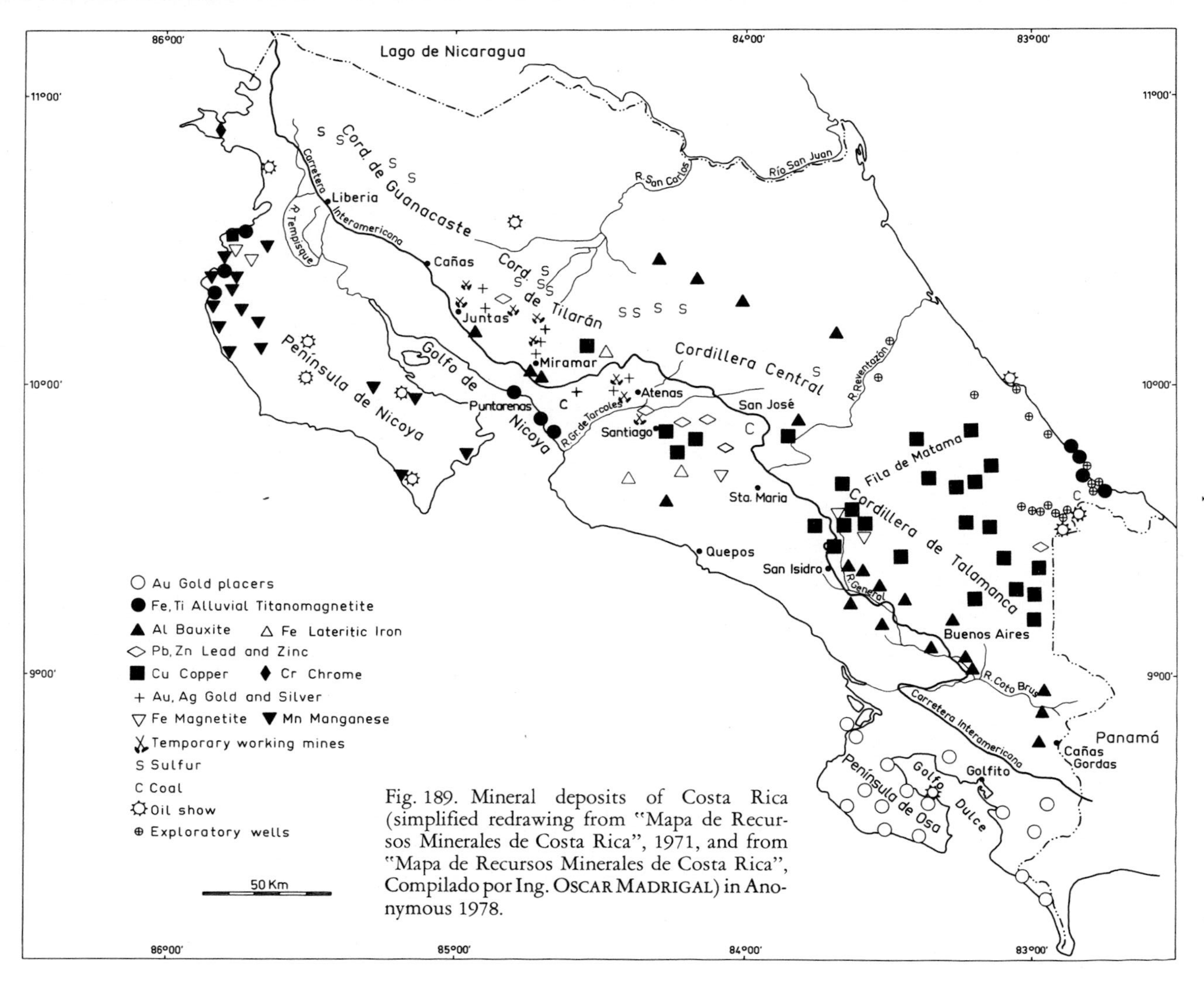

Fig. 189. Mineral deposits of Costa Rica (simplified redrawing from "Mapa de Recursos Minerales de Costa Rica", 1971, and from "Mapa de Recursos Minerales de Costa Rica", Compilado por Ing. OSCAR MADRIGAL) in Anonymous 1978.

cended via deep-reaching fault zones. This assumption has recently been confirmed by the discovery of sulphidic Cu, Zn, Co and Fe ores in the Nicoya Complex of Punta Gorda, Nicoya (FLORES 1976) and of chromite lenses in the peridotite/serpentinite of the Santa Elena Peninsula (JAGER 1977).

The other mineralizations are closely linked with the plutonic activity of southern Central America, which commenced as early as in the Upper Cretaceous, and also with the Tertiary volcanism in this region. It is probably not possible to draw a distinct line between the two, because according to recent data many of the sub-volcanic deposits can be assigned to intrusion activity. They are dealt with in the following section along with the deposits of the volcanic plateaus.

As shown in Chapter 4.23 and Table 7, as well as in Fig. 97, the plutonic activity in Panama can be divided up into several phases in which first quartz-dioritic, then quartz-dioritic-granodioritic and finally granodioritic material was produced while the potassium content simultaneously increased from phase to phase. The large porphyry copper mineralizations discovered in the last decade are associated with this plutonic activity. A similar situation seems to exist in Costa Rica where, however, according to present knowledge only one (Upper Miocene) intrusion phase exists.

A contact **hydrothermal copper mineralization** is linked with an Upper Cretaceous quartz diorite on the Azuero Peninsula (Iguana, Barro). In addition, small lenses and veins of hematite/magnetite were found (Huacas). FERENČIĆ (1971) identified the region of the quartz-diorite pluton as a metallogenetic copper province (Fig. 190).

A promising copper mineralization was discovered on the **Río Pito**, Comarca de San Blas, close to the boundary between Panama and Colombia (Anonymous 1972c). The occurrence is located on the north side of the Serranía del Darién, north of the San Blas Fault. Here, on the southwest flank of a pluton of presumably Mid-Eocene age, which is approximately 200 km^2 in extent, the mineralization intersects roof or boundary zones together with remains of roof materials consisting of upper Cretaceous volcanics. The mineralized zone contains tonalites, granodiorites, porphyric dacites and other volcanics. It is permeated by faults, fissures and fracture zones with quartzose sections containing pyrite, auriferous and argentiferous chalcopyrite and a little molybdenite. Five potentially ore-bearing zones were identified geochemically by detecting high heavy metal anomalies in soil samples and in river sediments (Fig. 191).

A porphyry copper impregnation at **Cerro Petaquilla** on the northern slope of the Cordillera Central is probably of greater economic significance. Here a complex of andesitic-basaltic lavas is intruded by an Oligocene pluton of intermediate composition. In the centre it consists of quartz-monzonite which toward the edge changes into granodiorite, porphyric granodiorite and finally into dacite, but which could also be a contact product of the volcanic rocks. The contact zone has undergone severe hydrothermal alteration and it is impregnated with copper ores, mainly chalcopyrite, molybdenite and pyrite. The impregnation zone is about 8 – 10 km^2 in extent, but it is likely that it is even larger than that (FERENČIĆ 1971, p. 85, FERENČIĆ et al. 1971, pp. 191–192). The age of the mineralization is Middle Oligocene, based on radiometric dating of the intrusions (KESLER et al. 1977, p. 1148).

The youngest and most significant porphyry copper mineralization was discovered at **Cerro Colorado** in Chiriquí Province. It is associated with a series of shallow intrusions which permeate a thick andesite cover (Fig. 192). The intrusion sequence is as follows:

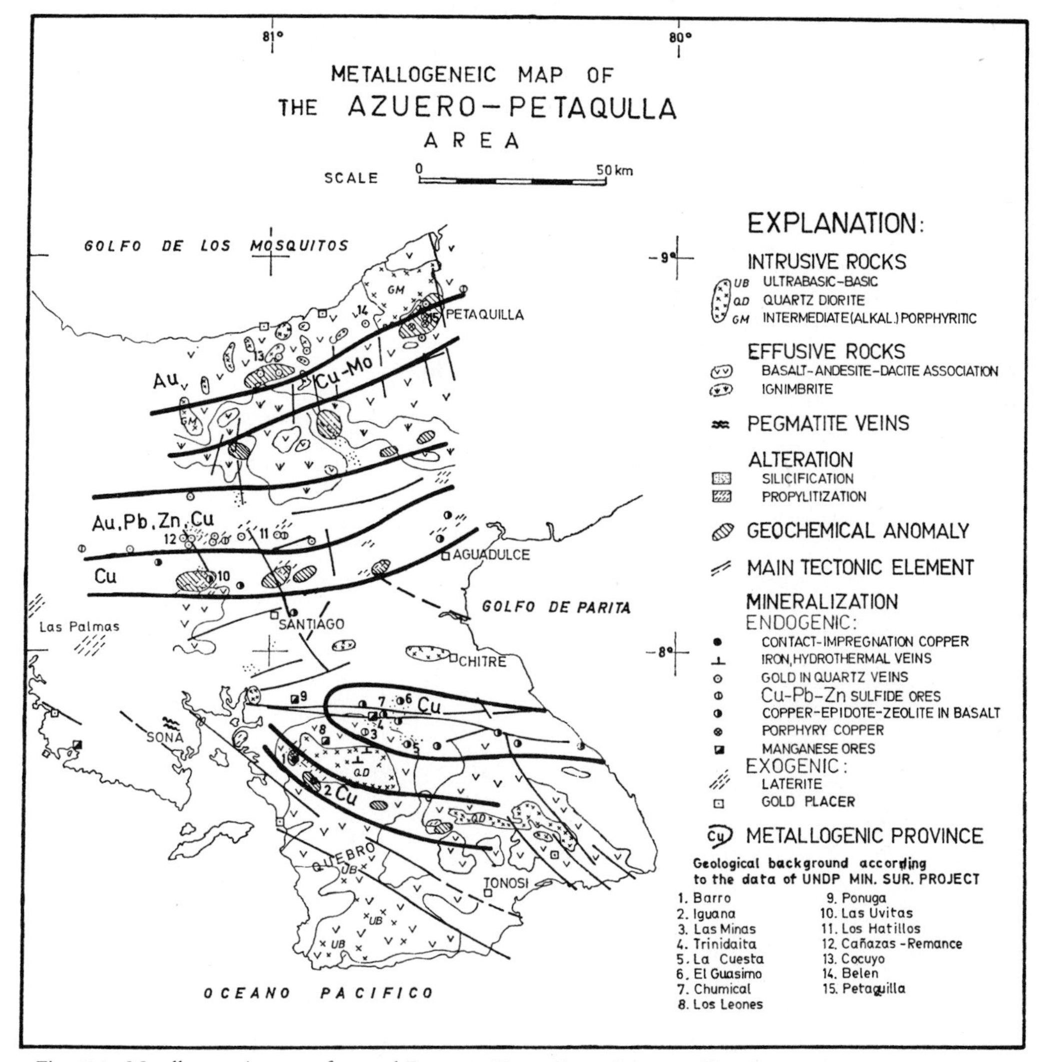

Fig. 190. Metallogenetic map of central Panama. (From FERENČIĆ 1971, Fig. 3).

1. Intrusion of fine-grained rhyolitic "ore porphyries"
2. Intrusion of a quartz-monzonite series
3. Intrusion of a quartz-porphyry.

All three intrusions contain notable primary copper contents between 0.05 and 0.7%. The age of the intrusions is given as 7.2 to 5.9 million years at the start and 4.2 million years at the end (these dates have been rounded off according to CLARK et al. 1977). A secondary enrichment of the ore occurred after intrusions were eroded between 4.2 and 3.0 million years ago. A more recent effusion phase produced rhyodacitic veins and trachyandesitic tuffs, mainly in the form of ignimbrites.

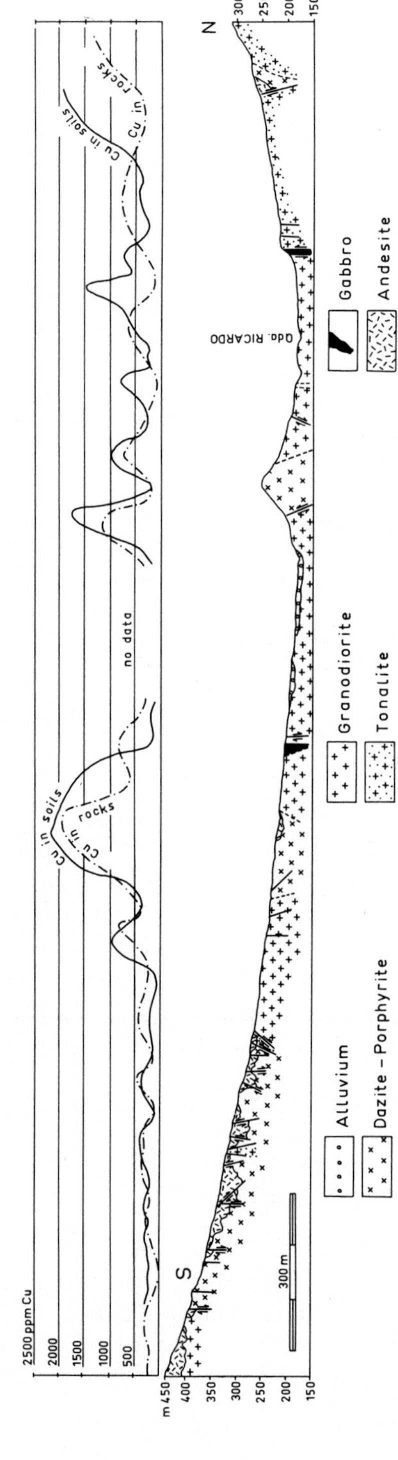

Fig. 191. Geological and geochemical section through the Río Pito area. (Redrawn from Proy. Min. II, Inf. Técn. No. 4).

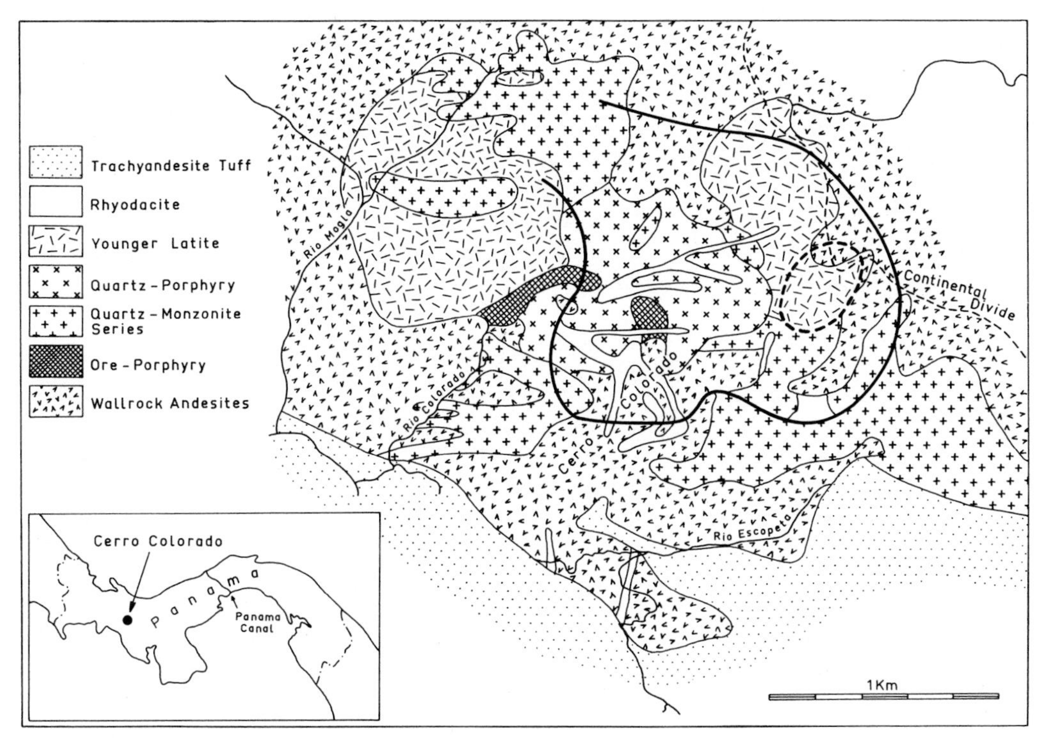

Fig. 192. Geological map of the Cerro Colorado area, Panama. (Redrawn from CLARK et al. 1977, Fig. 1).

The reserves at Cerro Colorado are put at a maximum of 3 billion tons of 0.6% ore. Large-scale mining has been planned and developed by the company Empresa de Cobre Colorado SA, and actual mining may possibly commence in 1978 if both Empresa's partners (80% the Government of Panama and 20% Texas Gulf) give their approval.

The links between magmatic development, mineralization and plate tectonics were pointed out by KESLER et al. (1977). A report on this is given in Chapter 4.23. The already very extensive literature on the copper province of Panama has been summarized in the above-mentioned works by CLARK et al. (1977) and KESLER et al. (1977).

Following the discovery of the Panamanian deposits, the exploration of occurrences of the same type, which was commenced in 1970 in the Cordillera de Talamanca of Costa Rica, achieved promising results [22]. In particular, a deposit was found in the **Fila de Matama,** a northeastern branch of the Cordillera. Stock-shaped intrusive bodies of mostly tonalitic composition occur here in a sequence of basic and intermediate volcanics and volcanogenic sediments of Oligocene to Miocene age. The mineralizations are linked with porphyric tonalites of which one, the "Nari tonalite", was examined in greater detail.

Chalcopyrite, chalcocite, bornite, malachite, molybdenite and pyrite occur in the intrusive body. In addition, galena and sphalerite occur as infilling in the volcanic-sedimentary encasing rock. A characteristic feature is a strong hydrothermal alteration of the tonalite and its coun-

---

[22] I am grateful to Dr. HORST SCHNEIDER for the following hitherto unpublished data.

try rocks in conjunction with sericitization, propylitization and silicification. Initial drilling results indicated a mean Cu content of 0.5 %.

Geological and geochemical survey work carried out by the UN in the Eastern Cordillera of Talamanca between Buenos Aires and Amubri (Anonymous 1974 a) and studies conducted by private companies provide further evidence of the presence of non-ferrous metals, in particular copper (see Fig. 189).

The discovery of these deposits in Panama and Costa Rica has demonstrated that the circumpacific copper belt is present in Central America as well, and that copper exploration in this region should be seen in a new light.

9.14 The province of the volcanic plateaus (Fig. 193)

From a metallogenetic standpoint it appears sensible to group together in this province the regions of Tertiary and Quaternary volcanism of Central America, although the magmatic development followed a different course in the northern and southern parts. This development has been described in detail in the chapters on magmatism.

In the northern part of Central America the volcanic activity commenced in the Oligocene when there was widespread production of mainly basic to intermediate magmas, although some acid magmas were also present. Together with their detritus they are described under various names usually as the Matagalpa Formation. Acid magmas, which cover large areas of northern Central America mainly in the form of ingnimbrites, erupted chiefly during the Miocene but also continued through to the Pliocene.

In the southern part of Central America, in the region of Costa Rica, there is a continuous rise in the SiO_2 content from the Upper Cretaceous tholeiitic submarine volcanism of the Nicoya Complex via intermediate Lower to Middle Tertiary flows to the acid Quaternary volcanism. In Panama, volcanic activity with the eruption of phenobasaltic to phenoandesitic lavas occurred practically throughout the entire Tertiary. The activity climaxed in the Middle and also in the Upper Miocene.

The intrusion of quartz-dioritic to granitic magmas is temporally and spatially linked with the volcanism. Radiometric datings have been summarized in Table 9 and Fig. 116. It has been found more and more that this plutonic activity, which extends right into the sub-volcanic stockwork, must be the source of the ores while the volcanic rocks form the country rock of the deposits. ELVIER (1974, p. 21), for example, pointed out that the plutonic rocks in Honduras are important indicators of ore deposits because the two are closely linked with each other. LEVY (1970) still emphatically denied any link between the ore deposits of the province and the intrusive rocks.

The metallogenesis of the volcanic province is described a little differently by LEVY (1970, pp. 42–47) compared with FERENČIĆ (1971). According to LEVY the main mass of the deposits is of the type of the epithermal subvolcanic gold-silver formation, which is linked with quartz veins. Gold occurs in native form, in pyrite or in combination with silver as electrum. Silver occurs in the form of argentite or pyrargirite. In the region of the Gulf of Fonseca the precious metal content declines in favour of copper, lead and zinc sulphides. The mineralization is linked with silicification and pyritization of the country rock.

LEVY divided the province into four sub-provinces:

a) Sub-province of the Gulf of Fonseca with the gold-silver deposits in the eastern part of El Salvador, in the southern part of Honduras and in the adjacent regions of Nicaragua. The

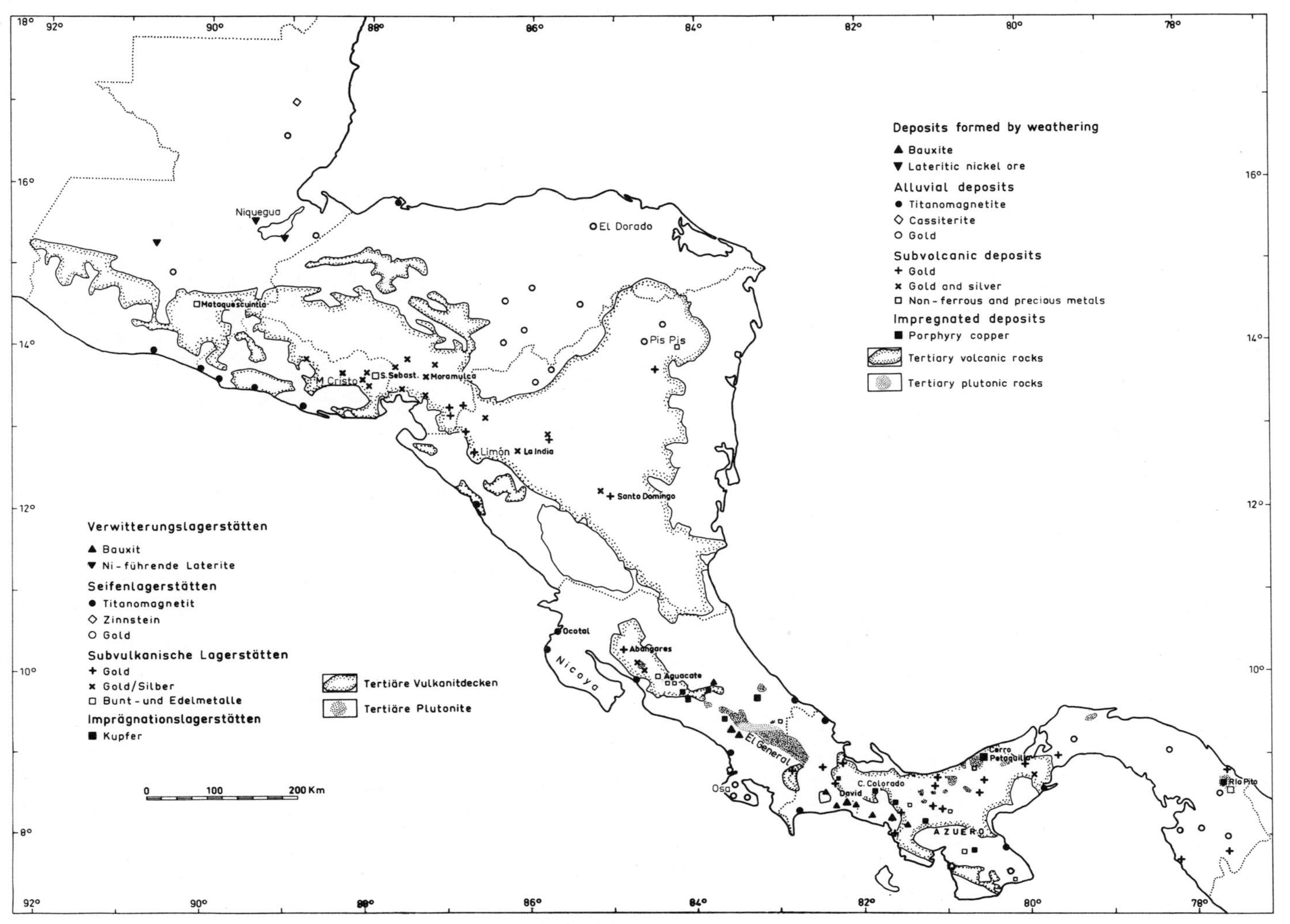

Fig. 193. Tertiary subvolcanic ore deposits and deposits formed by weathering.

most important mines in El Salvador were or are: Potosi, El Dorado, Montecristo (= San Cristobal), San Sebastián, Los Encuentros, Hormiguero and Divisadero. In Honduras: El Tránsito, Corpus, San Martín, Moramulca, Victorina, Sábana Grande and El Porvenir. In Nicaragua: Mina El Limón, La India, El Chorreadero and La Grecia.

b) Sub-province La Libertad and Santo Domingo north of Lake Nicaragua.

c) Sub-province of Abangares and Aguacate in Costa Rica.

d) Sub-province of Serranía de Tabasara in Panama.

The majority of the deposits located in these sub-provinces were described by ROBERTS & IRVING (1957). Only a few are at present being mined.

FERENČIĆ distinguished within the Cenozoic volcanic regions between three metallogenetic provinces:

a) A copper-lead-zinc province in central Panama, which is connected with the contact zone between quartz diorite and dacites.

b) A province extending from Panama via Costa Rica to El Salvador and containing gold-silver and copper-lead-zinc mineralization in various mineral parageneses.

c) A porphyry copper province in the northern part of central Panama with mineralization around the Cerro Petaquilla; the Cerro Colorado probably also belongs in this province.

Provinces a) and c) were discussed by FERENČIĆ in connection with the deposits of southern Central America. The contents of his province b) correspond to LEVY's province of the volcanic plateaus. This province was the main source of supply for gold and silver in Central America while copper, lead and zinc production was less important here. The following examples can be given:

In **Costa Rica,** in the northwestern continuation of the Cordillera de Talamanca, mining for gold, and to a lesser extent also for nonferrous metals, commenced at the start of the 19th century but then came to a standstill in 1950. It has been reported that mining recommenced in two mines in 1972, and the statistics for 1974 show that 3,000 ounces of gold were produced.

The main districts are as follows:

Aguacate district:	which contains the Sacra Familia series of veins, which in turn contain pyrite, arsenopyrite, realgar as well as rhodonite, sphalerite, chalcopyrite and galena.
Abangares district:	Quartz veins in andesitic volcanic rocks with pyrite, chalcopyrite, sphalerite, galena and in places antimonite.
Miramar and Esparata district:	Andesitic lavas and pyroclastic rocks permeated by basalts, having a similar ore paragenesis to the Abangares district. In addition, realgar, rhodonite and secondary manganese oxides.

An initial geological survey conducted by CHAVES & SÁENZ (1974) confirmed that the region is built up mainly of basic volcanics which were described as the Aguacate and Monteverde Formation (see map in Fig. 77). It was a surprise to discover a pluton covering an area of approximately 200 km^2 and containing predominately monzonite and adamellite. The pluton intruded into the Aguacate Formation of the (?) Middle to Upper Miocene and is thus provisionally dated in the Middle Miocene.

According to CHAVES & SÁENZ the mineralization occurs in definite zones. In the centre, right up against the intrusive body, come complex sulphides together with gold, silver, zinc and lead. Around this comes a second zone containing zinc, lead and silver sulphides; and the best defined zone, on the western edge of the mineralization in the sub-volcanic region, contains the gold and silver-bearing quartz veins. This thus confirms the author's assumption (1961 a, p. 179) that plutons are to be regarded as source rocks for ore.

The gold and silver deposits of **Nicaragua** are described as low temperature quartz fillings of pre-existing fissures. The filling occurred preferentially in a NE-striking system. The main country rock is made up of andesites of the Matagalpa Formation (=Group) and of the lower part of the Miocene Coyol Group. The La India and Limón mines achieved quite considerable economic significance.

The La India mine (about 80 km north of Managua), which was closed down around 1954, worked two parallel-striking quartz veins containing free gold. The annual production amounted to 900 kg of gold and 1,200 kg of silver.

The Limón mine, which was discovered in 1940 about 40 km north of the town of León, is working 5–8 m thick quartz veins which are superjacent on andesites of the Coyol Group. Mining is limited by a very low geothermal gradient. In the forties output amounted to around 900 kg of gold and 300 kg of silver, and between 1965 and 1971 production was about 1,800 kg gold and approximately 2,000 kg silver per annum. According to an unpublished report by ICAITI (Anonymous 1972 a) the reserves are almost exhausted, although further deposits are being worked in the vicinity.

The **Honduran** section of the metallogenetic province of the volcanic plateaus is described as follows by ELVIER (1974, p. 35): The deposits are associated with epigenetic quartz veins containing free gold and hydrothermally formed sulphides. The individual veins are relatively narrow but occur in large numbers with a constant strike and at constant depth. With increasing depth the precious metal content of the mineralization gives way to lead, zinc and copper ores.

The Moramulca mine was operating in 1970 and it possessed reserves in excess of 200,000 tons of ore with a gold content of 5 g per ton and a silver content of 13 g per ton.

In **El Salvador** gold and silver are bound to pyrite and chalcopyrite on quartz veins in the Tertiary volcanic rocks. The richest districts are located in the northeastern part of the country where mining was carried out on a large scale by several operations between 1890 and 1918 and then again between 1937 and 1953. In 1950 the production of 916 kg of gold and 15.5 tons of silver was the country's third most important export (3%). In 1953 most of the mines were closed, and at the beginning of the seventies mining was commenced again at the San Sebastián and Montecristo mines.

The San Sebastian mine, which is situated about 5 km north of the town of Santa Rosa de Loma, works gold-bearing quartz veins which are superincumbent on an andesite breccia in the vicinity of the contact with a monzonite vein. The breccia is silicified and sericitized. The monzonite vein is regarded as the apophysis of a deeper intrusive stock which is said to crop out about 4 km from the mine (LEVY 1970, p. 47), although it is not recorded in the General Geological Map of El Salvador. The following ore minerals occur here: pyrite, chalcopyrite, chalcocite, bornite, fahlore, molybdenite and pyrrhotite. In addition to the noble metals, the mine briefly produced up to 1.25 t of copper per annum.

Because of its ore paragenesis and gangue, and also because it is bound to the monzonite vein, the deposit differs from the neighbouring Montecristo deposit, which MÜLLER-KAHLE (1962) states is definitely not linked with intrusions. The following statements are taken from MÜLLER-KAHLE's work:

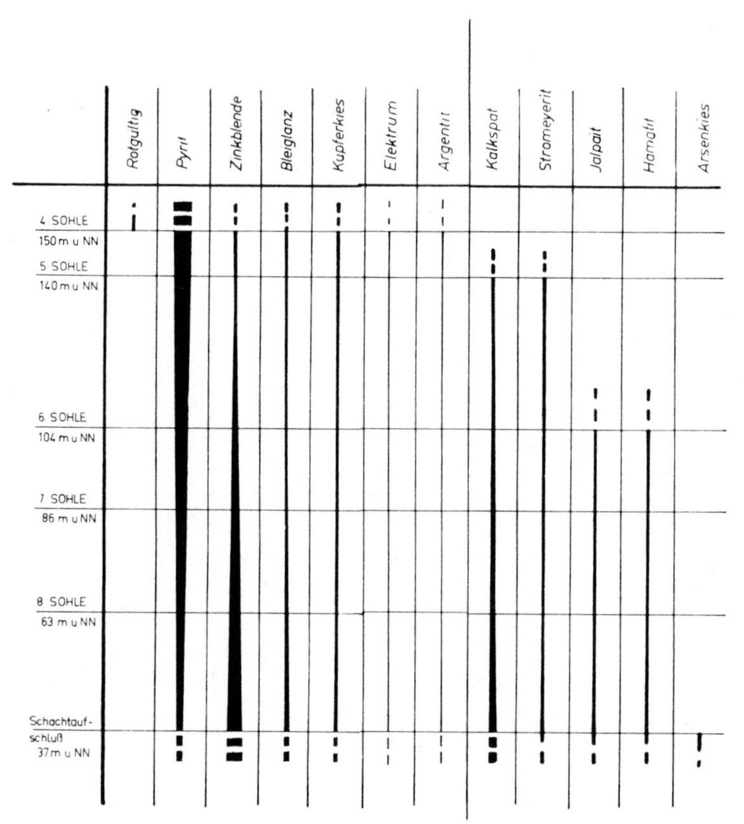

Fig. 194. Ore parageneses and depth levels in the Montecristo mine, El Salvador. (From MÜLLER-KAHLE 1962, Fig. 2).

The deposit is bound to a Tertiary andesite which has undergone extensive propylitization. The deposit is formed from residual hydrothermal solutions of the Tertiary volcanism and does not extend beyond the propylitized parts of the andesite cover. A presumably Pleistocene oliviniferous andesite in the upper strata is not mineralized. The content of ores and gangue material and their primary differences in depth are shown after MÜLLER-KAHLE in Fig. 194. This is a low-temperature mineralization whose temperatures of formation are indicated by the presence of argentite with conversion lamellae (179°) and of stromeyerite (93°).

In the years 1952 to 1955 the Montecristo mine produced about 180 kg of gold and 7,800 kg of silver. In the following years production declined, and in 1960 the mine was closed down. It was started up again in 1971 and produced about 80 kg of gold and approximately 5,780 kg of silver.

In **Guatemala** the copper-silver deposit of Mataquescuintla is regarded as belonging to the province of the volcanic plateaus. It is located about 70 km to the southeast of the capital. Mining on a modest scale has been pursued here since 1694, and it is claimed that in 1886/87 a total of 1,200 kg of silver were extracted. There is no news of any more recent mining activity. The deposit is bound to volcanic tuffs with intercalated rhyolites and andesites, and it occurs at the intersection of two or more fault systems. It is a body conical in shape with a diameter of 120 m at the top and 150 m at the bottom. The mineralization consists of a large number of small veins of iron glance and pyrite with associated copper and silver sulphides. The silver is enriched up to amounts of 3,000 g/t in the upper part of the ore body, but below 107 m the content declines to 130 g/t.

A review of the deposits of the volcanic plateaus reveals two clearly different types: On the one hand there are the purely volcanogenic gold-silver deposits associated with intermediate volcanism (Honduras, Nicaragua, Montecristo in El Salvador), and on the other there are deposits which are clearly related to the Tertiary plutonic activity (San Sebastián in El Salvador, Sierra de Tilarán in Costa Rica). In addition there are the porphyry copper mineralizations, which as far as is known are limited to the southern part of Central America.

9.15 Alluvial deposits and deposits formed by weathering

The **lateritic nickel ores** formed by weathering at Lago Izabal in Guatemala are very important economically. The ores occur in terraces on top of deeply weathered serpentinite and they contain about 30% Fe and 1.6 to 2.0% Ni. The ore is about 7 m thick, but the ore-waste contact is undulating and requires close checking. The reserves are estimated to be > 100 million tons. Prospecting work and preparations for mining and smelting have been going on since 1960 at El Estor on Lago Izabal. According to WHITE (1977) production commenced in 1977 and the aim is to attain an output of 28 million lb. of nickel per annum. The concession is owned by the Compañia Eximbal (Exploraciones y Explotaciones Minerales Izabal S.A.), and it extends over an area of 429 square kilometres.

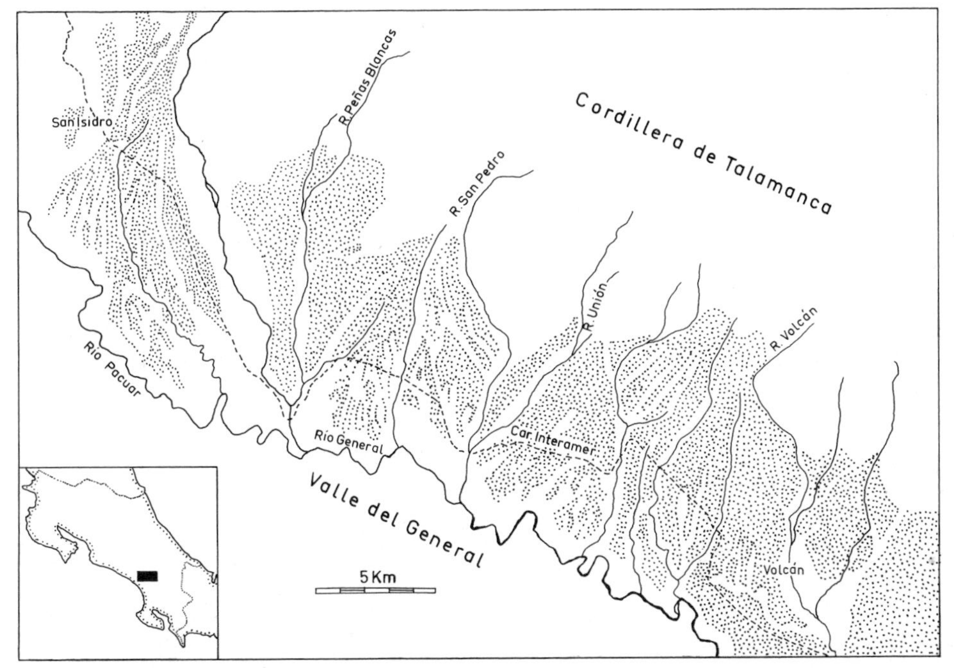

Fig. 195. Bauxite-bearing terraces (dotted) in the western Valle del General, Costa Rica. (Redrawn from CASTILLO et al. 1970).

Bauxite deposits occur in Costa Rica and Panama (Fig. 193). The most important of these deposits are located in the Valle del General in Costa Rica where they cover extensive terraced areas of Plio-Pleistocene age which are incised by waters flowing from the Cordillera de Talamanca (KOJAN 1958, FERNÁNDEZ 1962, CASTILLO 1966, CASTILLO et al. 1970). The terraced areas are quoted as being 248 km^2 in extent, although only a small portion contains exploitable amounts of bauxite. In 1968 Alcoa acquired the mining rights in the region of San Isidro del General but gave them up again in 1976 for economic reasons (Fig. 195).

DÓNDOLI (1970, 1973) describes a further bauxite occurrence in the region of Paraiso de Cartago in the eastern Valle Central of Costa Rica. In addition, some laterite deposits, which are very rich in SiO_2, have been discovered south of the Río San Juan near the towns of Venecia de San Carlos and El Guacimo.

The Panamanian occurrences are concentrated in the western Province, Chiriquí. They are superjacent sometimes on gravels and sometimes on basic volcanic rocks (QUIRÓS 1966). The reserves are put at 36.6 million tons of ore with an Al_2O_3 content in excess of 30% and 31.7 million tons with contents of 25–35%.

The following average contents are given:

	Al_2O_3	Fe_2O_3	SiO_2
Valle del General	43.1	13.6	14.9%
Paraiso (Costa Rica)	39.4 – 49.1	2.93 – 15.8	13.7 – 25.4%
David (Panama)	45.9	16.5	15.5%
Tolé (Panama)	45.5	28.1	7.5%

Heavy mineral sands with concentrations of **titanomagnetite** are widespread on the coasts of Central America and there was hope that they would prove mineable. Reserves of over 10 million tons were estimated to exist on the coast of El Salvador alone. A provisional study carried out by the Bundesanstalt für Bodenforschung (Reporte Final 1967–1971, Hannover 1974) has shown, however, that the main occurrences NW of Acajutla, west of La Libertad and in the regions of the estuaries of the Río Lempa and the Río Jiboa contain reserves of 16,000 to 70,000 tons respectively and are therefore not worth investigating in detail. Altogether the reserves of titanomagnetite in El Salvador probably do not exceed 150,000 tons of Fe and 20,000 tons of Ti. According to HINTZ (1955) the heavy minerals derive from the volcanic rocks produced during the Quaternary.

In Guatemala LLUNGGREEN (1958 b) investigated heavy mineral sands on the Pacific coast near Iztapa.

Large deposits with reserves in the order of 3.8 million tons were discovered along the Pacific coast of Costa Rica (FERNÁNDEZ & DONDOLI 1965, Anonymous in Informe Tecn. y Notas Geol., 42, undated). The reserves are, however, distributed over many beaches and they also contain sands with ore contents of less than 10%. Further placer deposits have been found on the Caribbean coast between the mouth of the Río Estrella and that of the Río Sixaola. The reserves are put at 540,000 tons. (See Fig. 189 and WEYL 1969 a, b.)

Gold placers are found along the rivers of the province of Darién, Panama, and on the Osa Peninsula of Costa Rica. The Panamanian occurrences were economically important during the colonial period, and those on the Osa Peninsula are still being worked today on a very small scale.

9.2 Energy raw materials

The lack of adequate energy sources is a barrier to economic development in Central America, and the need to import oil and gasoline imposes a considerable burden on the economies of the various countries.

At the present time the only really important domestic source of energy is **hydro power**, which is being vigorously exploited. The following are some of the dam installations and power plants which have been completed or are still under construction (the list makes no claim to be complete):

The "Río Chixoy" hydroelectric project in Guatemala,
Hydroelectric installations on the Río Lempa and Lago de Guija in El Salvador,
Hydroelectric plant on the Río Lindo (Lago de Yojoa) in Honduras,
Hydroelectric plant on the Río Viejo in Nicaragua,
The "Garita", "Cachi" and "Arenal" hydroelectric plants and the Boruca project in the valley of Río Térraba, in Costa Rica,
The "Río Banano" project in Panama.

Altogether, however, only a fraction of the available hydroelectric power in Central America has been tapped.

The basins containing extremely thick Cretaceous and Tertiary sediments were, or indeed still are, regarded as **potential petroleum-bearing regions**. Intensive explorations have been carried out here in recent decades but, as far as can be judged by an outsider, they do not seem to have met with much success. The boreholes plotted in Fig. 196 were taken from the annual reports in the Bull. Amer. Ass. Petrol. Geologists. They are concentrated in the following regions: The Petén Basin in Guatemala and Belize, the Mosquitia Basin, in particular on the shelf of the Nicaragua Rise, the Limón Basin in Costa Rica and northwest Panama and the basins in eastern Panama. A smaller number of exploratory wells were sunk on the Pacific coast and in the shelf region between Nicaragua and Guatemala.

Only in the provinces of Alta Verapaz and Petén in northwestern Guatemala adjacent to the Mexican border have major oil accumulations been discovered in Lower and Middle Cretaceous carbonates. The oil accumulations have been found in two distinct structures. The "Las Tortugas" structure is a circular, piercement salt dome with a steeper flank and salt overhang. The "Rubelsanto" structure is a northwesterly-trending anticline with a steeper northeastern limb and with a possible faulting (MURRAY 1976).

The "Rubelsanto" oilfield has been producing since 1976. Most of the crude oil is used for bitumen road surfaces in the near vicinity of the oil field, and the remainder is used by local industry. Since large reserves are anticipated, consideration is being given to the construction of a pipeline to the Caribbean.

A new discovery 9.3 miles northeast of Rubelsanto seems to be far more attractive, having greater formation porosity and apparently larger areal extent as well as a higher oil and gas potential (International Petroleum Encyclopedia 1978, p. 110).

CASTILLO (undated) has reported on the history of petroleum exploration in Costa Rica. A brief summary of the offshore petroleum potential of Central America was given by MORRIS (1979).

According to present knowledge, there are **no coal** deposits worth mining in Central America. Small occurrences of coal do, it is true, occur in the El Plan Formation of Honduras and

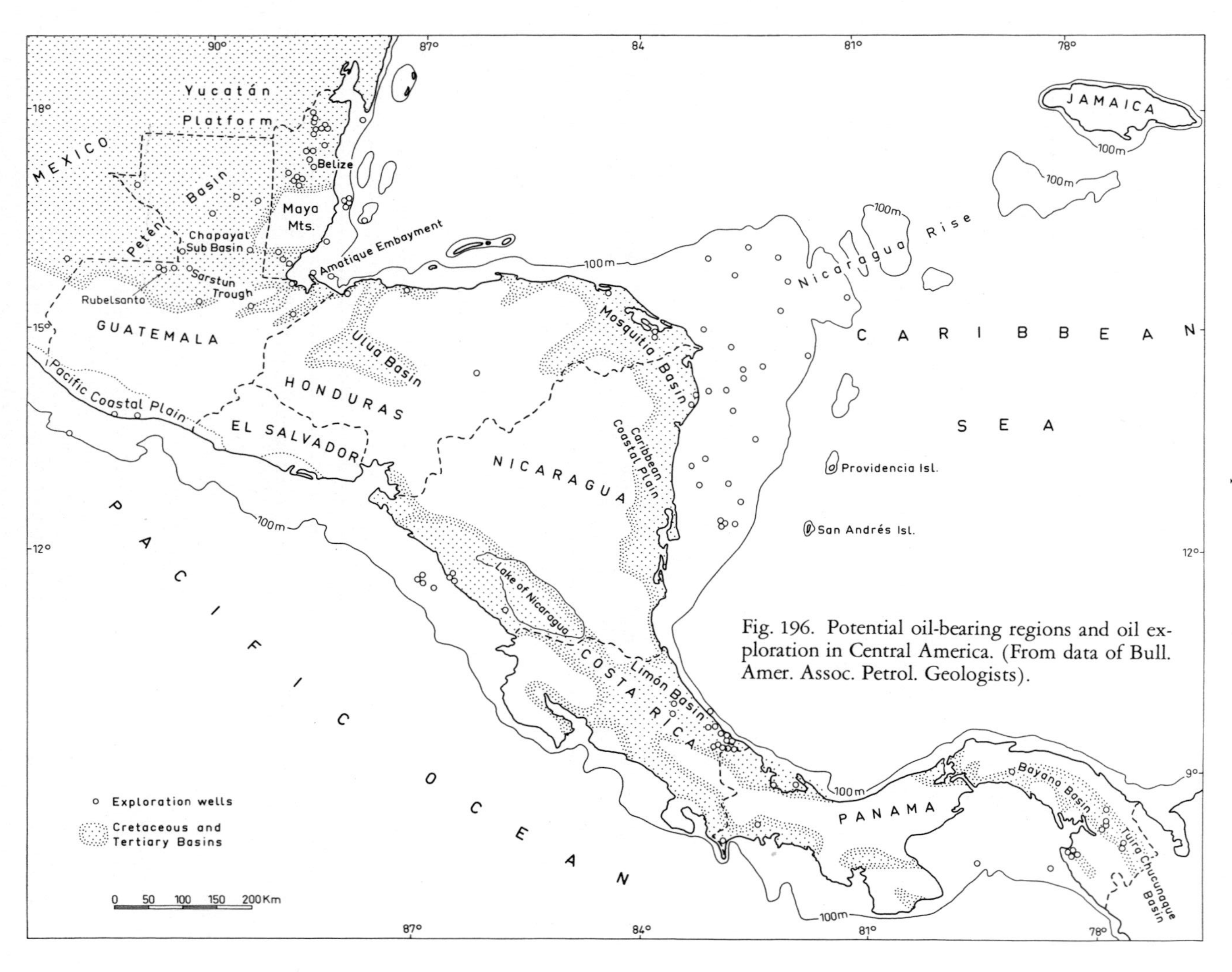

Fig. 196. Potential oil-bearing regions and oil exploration in Central America. (From data of Bull. Amer. Assoc. Petrol. Geologists).

in strata dating presumably from the same period in the Departamento San Marcos of Guatemala; however, it does not seem feasible to extract them. Lignites occur in the Late Tertiary limnic-brackish deposits of southern Central America (Figs. 188, 189) and within intramontane basins in the north, but they are not worth mining either. BOHNENBERGER & DENGO (1968) were commissioned by ICAITI to compile all the available data on coal deposits in Central America, and they came to the following conclusion, "A systematic inventory of the very young lignites should eliminate all but a very few localities, which could be further evaluated".

Since it is a region of active volcanism, Central America should offer good conditions for the extraction of **geothermal energy**. But, the development of this type of energy requires long-term studies and considerable investments, and consequently El Salvador is so far the only country to have built a geothermal power plant with an output of 60 MW. It is located in a well-known fumarolic area near Ahuachapan in the western part of the country (Fig. 126). Preliminary studies on the development of this area go back as far as 1953; the first exploratory boreholes were drilled in the early sixties (DÜRR 1960, ALONSO 1969, BAILEY 1969); the plant has been in operation now for several years. The steam is taken from a reservoir in Pleistocene lavas. (More details and a list of unpublished reports can be found in ROMAGNOLI et al. 1976).

In Nicaragua, 20 boreholes have been sunk at the foot of the Momotombo Volcano, and from the results obtained it seems realistic to plan a geothermal plant with an initial capacity of 30 MW increasing later to 100 MW. In Guatemala, studies are being carried out in the region of the Lago de Atitlán as well as in a thermal field near Zuñil in the western highlands. FERNÁNDEZ-RIVAS (1970), whose publication also contains a contribution from HOSTAS, gave a review of the energy sources on Guatemala. In Costa Rica, preliminary studies have been carried out in the Cordillera de Guanacaste and in Panama, mainly in the region of the El Barú Volcano near the western border of the country (MÉRIDA 1973).

10. Maps

The topographical survey (scale 1 : 50,000) of the countries of Central America is almost completed. In addition, a large number of general maps (scale 1 : 250,000) have appeared. The mapmaking and the sale of the maps are regulated by the geographical institutes (Instituto Geográfico de . . . .) in the individual countries.

Since the appearance of the first (German) edition of this book in 1961, the status of the geological surveys has undergone a fundamental change. At that time only the general map (scale 1 : 1,000,000) prepared by ROBERTS & IRVING (1957) and based to a large extent on the maps of SAPPER (in particular 1937) and TERRY (1956) existed; in the meantime, however, new general maps of the entire region have appeared or are in preparation. In addition, general geological maps of admittedly varying factual accuracy, and drawn on different scales, exist for all the countries in Central America; also a large number of specialized geological maps are available, and a list of those known to the present author is given below. The printing and distribution of these maps is controlled, for the most part, by the respective geographical institutes.

Central America in general

Geological Map of Central America 1 : 1,000,000. — (Enclosure in ROBERTS & IRVING: Mineral Deposits of Central America). — U.S. Geol. Surv. Bull., **1034**, Washington 1957.

Mapa Metallogenético de América Central 1 : 2,000,000 — Metallogenetic Map of Central America 1 : 2,000,000. — Instituto Centroamericano de Investigación y Tecnología Industrial (ICAITI), Guatemala 1969.

Preliminary geologic-tectonic and bathymetric maps of the Caribbean Region. — U.S. Geol. Surv. open file mape 75 – 146, 1975 (by J. E. CASE & T. L. HOLOCOMBE).

Guatemala (Fig. 197):

Mapa Geológico de la República de Guatemala 1 : 500,000. Primera Edición. — Instituto Geográfico Nacional, Guatemala, C.A. 1970.

ND 15 – 7G. Quezaltenango 1 : 250,000 (with explanations by BONIS, S.: Geología del Area de Quezaltenango, República de Guatemala). — Instituto Geográfico Nacional de Guatemala 1965.

ND 16 – 5G. Chiquimula, Instituto Geográfico National 1969.

Geologische Übersichtskarte 1 : 125,000, Baja Verapaz und Süden der Alta Verapaz (Guatemala). — Bundesanstalt für Bodenforschung, Hannover 1971 and Instituto Geográfico Nacional 1967 – 1969. — Mapa Geológico General 1 : 125,000, Baja Verapaz y parte Sur de la Alta Verapaz (Guatemala). — Bundesanstalt für Bodenforschung, Hannover 1971 e Instituto Geográfico Nacional de Guatemala 1967 – 1969.

Mapa de Reconocimiento Geológico del Cinturón plegado de Alta Verapaz, Guatemala 1 : 125,000. — Instituto Geográfico Nacional, Guatemala C.A. 1967. (Enclosure in S. B. BONIS: Reconocimiento

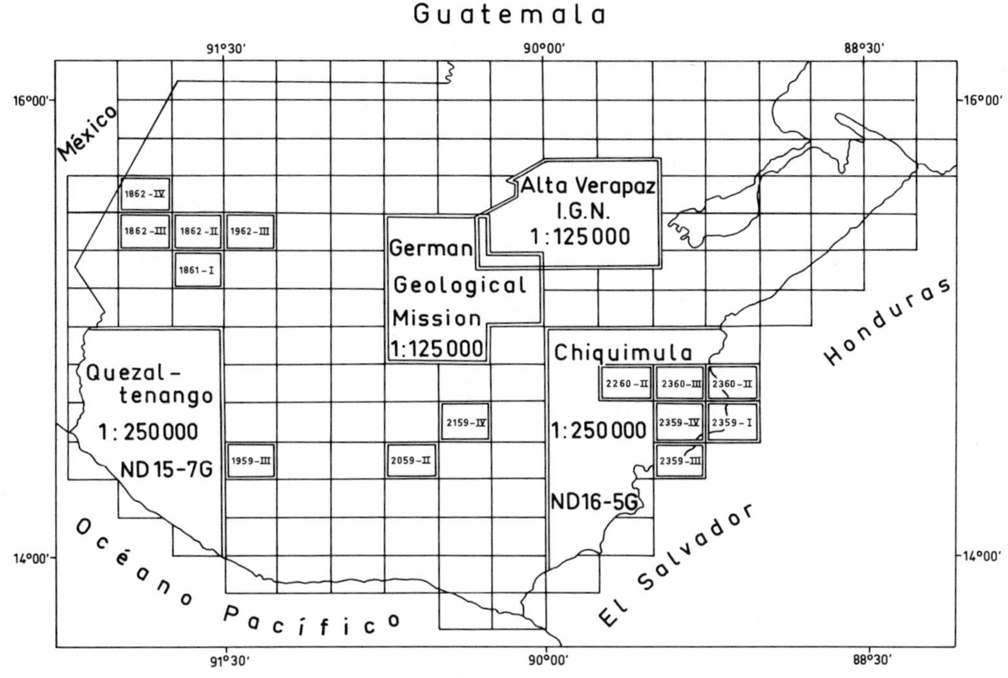

Fig. 197. Index of geological maps of Guatemala.

Geológico del Cinturón plegado de Alta Verapaz, Guatemala). — Bol. Geol. **5**, Instituto Geográfico Nacional, Guatemala, C.A. 1967.

Mapa Geológico de Guatemala 1 : 50,000. — Instituto Geográfico Nacional, Guatemala C.A. with the following sheets:

Hoja 1861 I.G. Santa Bárbara, 1972
Hoja 1862 II.G. San Sebastián Huehuetenango, 1972
Hoja 1862 III.G. Cuilco, 1968
Hoja 1862 IV.G. La Democracia, 1967
Hoja 1959 III.G. Río Bravo, 1970
Hoja 1962 III.G. Chiantlá, 1968
Hoja 2059 II.G. Amatitlán, 1975
Hoja 2159 IV.G. San José Pinula, 1975
Hoja 2260 II.G. Chiquimula, 1966
Hoja 2359 I.G. Chanmagua, 1966
Hoja 2359 III.G. Cerro Montecristo, 1966
Hoja 2359 IV.G. Esquipulas, 1966
Hoja 2360 II.G. Timushán, 1966
Hoja 2360 III.G. Jocotán, 1966
Hoja 2062 II.G. Tiritibol, 1976
Hoja 2162 III.G. Cobán, 1976

Honduras (Fig. 198):

Mapa Geológico de Honduras 1 : 500,000, Primera Edición. — Instituto Geográfico Nacional, Tegucigalpa 1974.

Mapa Geológico de Honduras 1 : 50,000. — Instituto Geográfico Nacional, Tegucigalpa with the following sheets:

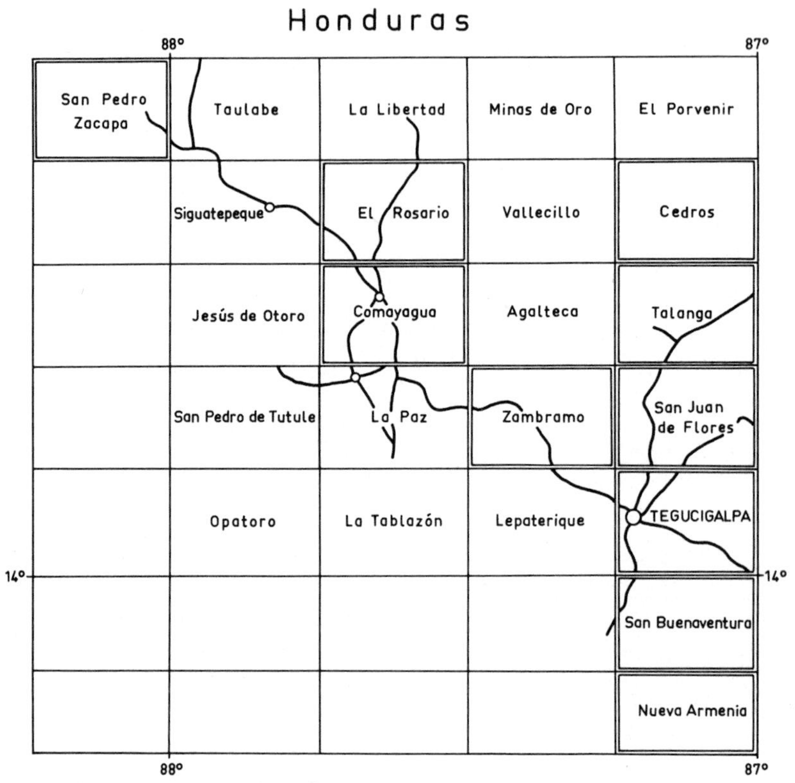

Fig. 198. Index of geological maps of Honduras.

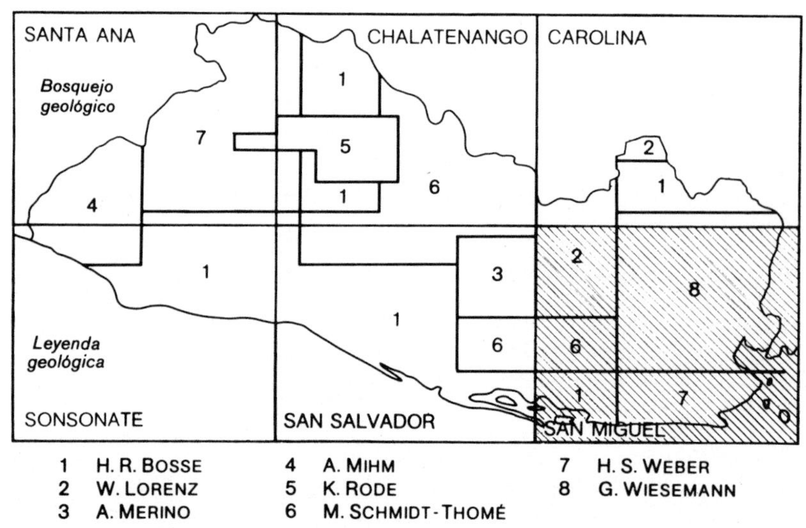

Fig. 199. Index of geological maps of El Salvador. (From Geologische Karte der Republik El Salvador 1 : 100 000, Hannover 1978).

Hoja 2559	IV.G.	San Pedro Zacapa
Hoja 2659	I.G.	El Rosario
Hoja 2659	II.G.	Comayagua
Hoja 2758	I.G.	San Juan de Flores
Hoja 2758	II.G.	Tegucigalpa
Hoja 2758	IV.G.	Zambrano
Hoja 2757	I.G.	San Buenaventura
Hoja 2757	II.G.	Nueva Armenia
Hoja 2759	I.G.	Cedros
Hoja 2759	II.G.	Talanga

El Salvador (Fig. 199):

Geologische Übersichtskarte der Republik El Salvador 1 : 500,000. — Herausgegeben von der Bundesanstalt für Bodenforschung, Hannover 1974. — Mapa Geológico General de la República de El Salvador 1 : 500,000. — Publicado por la Bundesanstalt für Bodenforschung, Hannover 1974.

Geologische Karte der Republik El Salvador/Mittelamerika 1 : 100,000 herausgegeben von der Bundesanstalt für Geowissenschaften und Rohstoffe, Hannover 1976, 1977, 1978. — Mapa Geológico de la República de El Salvador/América Central 1 : 100,000. — Publicado por Bundesanstalt für Geowissenschaften und Rohstoffe, Hannover 1976, 1977, 1978 with the following sheets:
Santa Ana
Chalatenango
Carolina
Sonsonate
San Salvador
San Miguel.

Nicaragua (Fig. 200):

Mapa Geológico Preliminar 1 : 1,000,000. — Instituto Geográfico Nacional, Managua 1973.

Mapa Geológico de Nicaragua Occidental 1 : 250,000. — Catastro e Inventario de Recursos Naturales, Primera Edición, Managua 1972 with the following sheets:
ND – 16 – 11 Esteli
ND – 16 – 14 Chinandega
ND – 16 – 15 Managua
NC – 16 – 3 Granada
NC – 16 – 4 San Carlos with part of ND – 16 – 16 Juigalpa

Mapa Geológico 1 : 50,000. — Catastro e Inventario de Recursos Naturales. Many sheets of the region covered by the 1 : 250,000 map, in black and white, showing boundaries of formations. Index map not yet available.

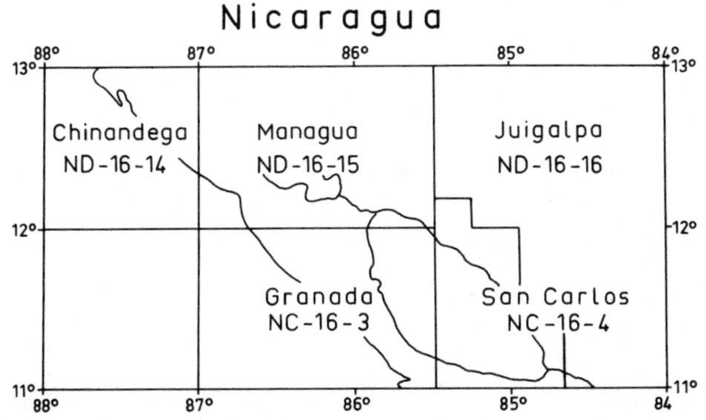

Fig. 200. Index of geological maps of Nicaragua.

Costa Rica (Fig. 201):

Mapa Geológico de Costa Rica 1 : 700,000. Edición Preliminar. Dirección de Geología, Minas y Petróleo, San José 1968. — Cartography and sales: Instituto Geográfico Nacional.

Mapa de Recursos Minerales de Costa Rica 1 : 750,000. — Dirección de Geología, Minas y Petróleo, San José 1971.

Mapa de Recursos Minerales de Costa Rica 1 : 750,000. — (Compilado por O. MADRIGAL). – In Anonymous 1978, Mapa 1 – 2.

Mapa Geomorfológico de Costa Rica 1 : 1,000,000, elaborado por J. P. BERGOEING & L. G. BRENES Q. — Instituto Geográfico Nacional, San José 1978.

Mapa geofísico preliminar de Costa Rica 1 : 500,000. Instituto Geogr. Nac, San José 1974. (Elaborado por J. DE BOER).

Costa Rica, Zona Norte, Mapa 6, Geología 1 : 200,000 (by E. MALAVASSI V.) in G. SANDNER, H. NUHN et al.: Estudio Geográfico de la Zona Norte de Costa Rica, San Jose 1966. — Also in MALAVASSI & MADRIGAL 1970.

Mapa Geológico Generalizado de la Provincia de Guanacaste y Zonas Adyacentes 1 : 300,000. — Instituto Geográfico de Costa Rica, San José 1962. (Enclosure in G. DENGO: Estudio Geológico de la Región de Guanacaste, Costa Rica). — Instituto Geográfico de Costa Rica, San José, C.R. 1962.

Mapa Geológico del Valle Central 1 : 150,000. — Dirección de Geología, Minas y Petróleo, San José 1968.

Reconnaissance Geologic Map and Cross Sections of Central Costa Rica 1 : 100,000. — U.S. Geol. Surv. Misc. Invest. Ser. Map I-899, Washington 1976. (R. D. KRUSHENSKY, V. E. MALAVASSI & M. R. CASTILLO).

Mapa Geológico 1 : 50,000, Hoja Barranca. — Dirección de Geología, Minas y Petróleo, San José 1967.

Mapa Gelógico del Mapa Básico Río Grande 1 : 50,000. — Dirección de Geología, Minas y Petróleo, San José 1970. (R. CASTILLO).

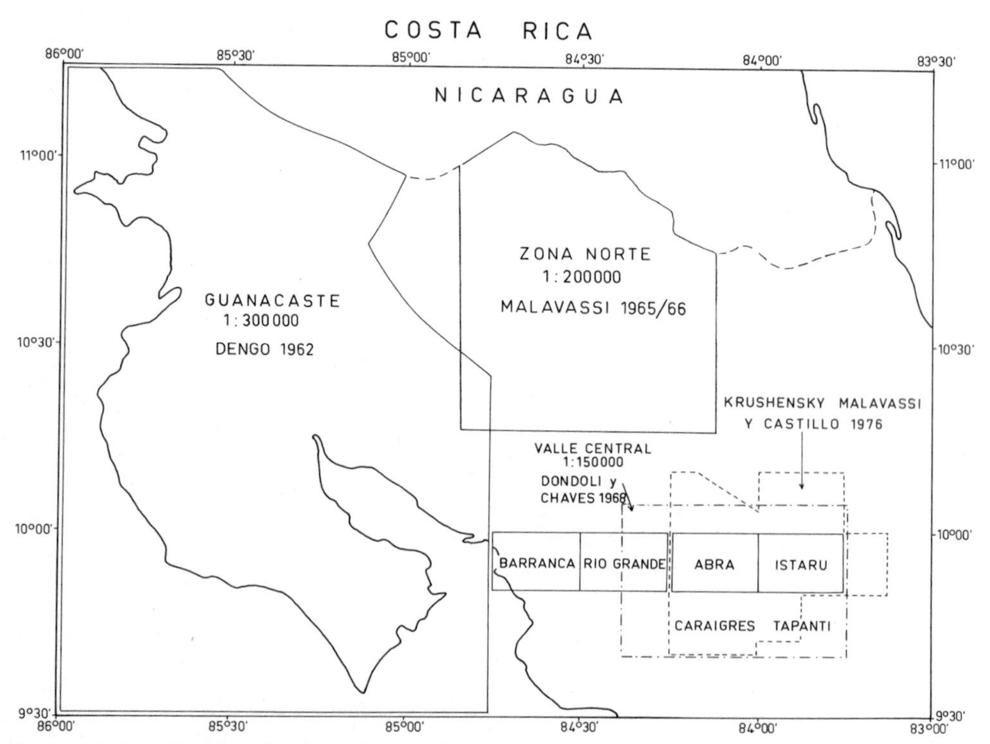

Fig. 201. Index of published geological maps of Costa Rica.

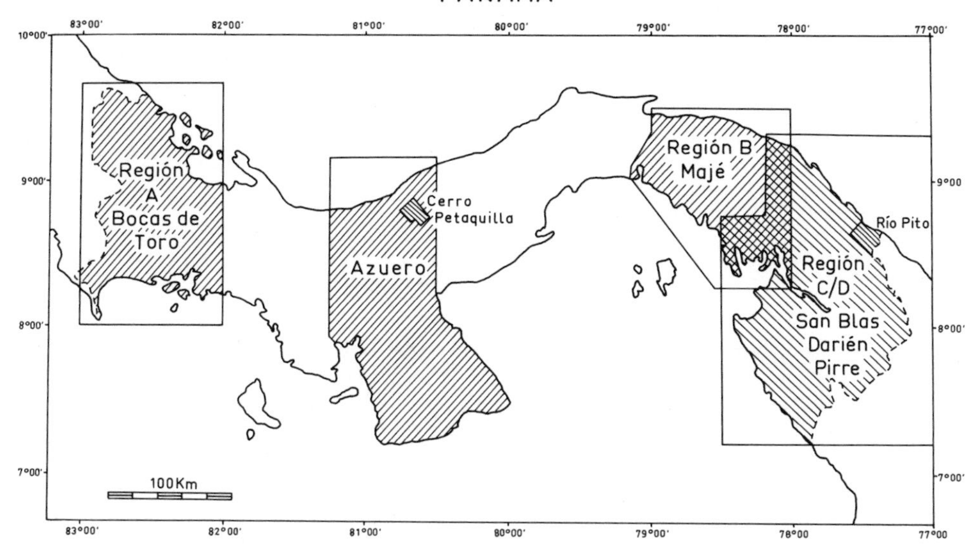

Fig. 202. Index of geological maps of Panama.

Mapa Geológico de los Mapas Básicos Abra y una parte de Río Grande, Costa Rica 1 : 50,000. — Dirección de Geología, Minas y Petróleo, San José 1969. (R. CASTILLO M.).

Geologic Map and Cross Section of the Abra Quadrangle, Costa Rica 1 : 50,000. — U.S. Geol. Surv., Misc. Invest. Ser., Map I-992, Washington 1977. (R. CASTILLO M. & R. D. KRUSHENSKY).

Geologic Map and Cross Sections of the Istarú Quadrangle, Costa Rica 1 : 50,000. (Enclosure in R. D. KRUSHENSKY: Geology of the Istarú Quadrangle, Costa Rica). — Geol. Surv. Bull., 1358, Washington 1972.

For unpublished maps see Anonymous 1978, Mapa 1 – 1.

Panama (Fig. 202):

República de Panamá. Mapa Geológico 1 : 1,000,000. — Dirección de Recursos Minerales, Panamá 1976. — Also in: Atlas Nacional de Panamá, Panamá 1975.

Mapa Geológico Región "A" Bocas del Toro 1 : 250,000. — República de Panamá, Administración de Recursos Minerales. Naciones Unidas, Programa para el Desarrollo, Proyecto Minero Fase II. Panamá 1971.

Mapa Geológico del Area del Proyecto Minero de Azuero 1 : 250,000. República de Panamá, Administración de Recursos Minerales. Naciones Unidas, Programa para el Desarrollo, Panamá 1969. (2 sheets).

Mapa Geológico Región "B" Majé 1 : 250,000. — República de Panamá, Administración de Recursos Minerales. Naciones Unidas, Programa para el Desarrollo, Panamá 1971.

Mapa Geológico Región "C" y "D" San Blas, Darién-Pirre 1 : 250,000. — República de Panamá, Administración de Recursos Minerales. Naciones Unidas, Programa para el Desarrollo, Panamá 1971.

Panamá, Mapa Geológico 1 : 15,000, Área 65 — Cerro Petaquilla, Mapa 5, Proyecto Minero de Azuero. — República de Panamá, Administración de Recursos Minerales, Naciones Unidas, Programa para el Desarrollo, Panamá 1969.

Panamá, Mapa Geológico 1 : 40,000, Área General — Río Pito. — República de Panamá, Dirección de Recursos Minerales, Panamá 1971.

Panamá, Mapa Geológico 1 : 8,000, Área 101 — Río Pito. — Dirección General de Recursos Minerales, Panamá 1971.

References

The following bibliography is a continuation of that given in the first (German) edition (WEYL 1961 a). Works which were listed there are mentioned here only if they are explicitly referred to in the text. On the other hand, the present bibliography also lists works which are not mentioned in the text but were nevertheless important for the writing of this edition of the book.

The bibliography includes theses submitted to US universities, some of which were kindly made available to me. Others exist in published form, but the majority could only be consulted through the intermediary of "Dissertation Abstracts Intern. (USA)".

As an initial guide, bibliographies, comprehensive works and periodicals appearing in Central America have been listed in Chapter 1.

The spelling of the works published in English and Spanish varies with regard to the use of capital initial letters. Where possible I have followed the spelling used in the original title, unless it was given entirely in capitals. Readers who are unfamiliar with German might like to note the following abbreviations:

N. Jb. = Neues Jahrbuch (New Annals)
Mh. = Monatshefte (Monthly Journal)
Abh. = Abhandlungen (Proceedings)
Geol. Jb. = Geologisches Jahrbuch (Geological Annals)
Geol. Rdsch. = Geologische Rundschau (Geological Review)

Finally, I would like to thank sincerely all those colleagues who assisted me by letting me have copies of their works.

ACEITUNO, M. A., BOHNENBERGER, O. H., BLOUNT, D. N., GODOY, J. et al. (1966): Libreto Guia de las Excursiones. — Segunda Reunión de Geólogos de América Central, Guatemala, C.A.

AKERS, W. H. (1972): Planktonic Foraminifera and Biostratigraphy of some Neogene Formations, Northern Florida and Atlantic Coastal Plain. — Tulane Stud. in Geol. and Paleont., **9**, 1 – 139, New Orleans.

ALLEGRE, C. J. & CONDOMINES, M. (1976): Fine Chronology of Volcanic Processes Using ^{238}U-^{230}Th Systematics. — Earth and Planetary Sci. Letters, **28**, 395 – 406, Amsterdam.

ALONSO, E. H. (1969): La Energía Geotérmica en El Salvador, América Central y en Cerro Prieto, México. — Publ. Geol. del ICAITI, **2**, 31 – 39, Guatemala C.A.

ANDEL, T. H. VAN, HEATH, G. R., MALFAIT, B. T., HEINRICHS, D. F. & EWING, J. I. (1971): Tectonics of the Panama Basin, Eastern Equatorial Pacific. — Geol. Soc. Amer., Bull., **82**, 1489 – 1508, Boulder/Col.

ANDEL, T. H. VAN, HEATH, G. R. et al. (1973): 2. Site 155; 3. Site 156; 4. Site 157; 5. Site 158. — Initial Rep. of the Deep Sea Drilling Project, **16**, 19 – 229, Washington (U.S. Governm. Print. Office).

ANDERSON, T. H. (1968): First evidence for glaciation in Sierra los Cuchumatanes Range, Northwestern Guatemala. — Geol. Soc. Amer. Abstr. for 1968, Spec. Pap. **121**, 387, Boulder/Col.

— (1969): Geology of the San Sebastián Huehuetenango quadrangle, Guatemala, Central America. — Ph. Diss. Texas Univ., 218 p., Austin, Texas.

ANDERSON, T. H., BURKART, B., CLEMONS, R. E., BOHNENBERGER, O. H. & BLOUNT, D. N. (1973): Geology of the Western Altos Cuchumatanes, Northwestern Guatemala. — Geol. Soc. Amer., Bull., **84**, 805 – 826, Boulder/Col.

ANDERSON, T. A. & SCHMIDT, V. A. (1978): Mesozoic crustal evolution of Middle America and the Caribbean: Geologic considerations. — Eos, Transact. Amer. Geophys. Union, **59**, 404, Washington, D.C.

Anonymous (1962): Red gravimétrica de la República de Guatemala. — Inst. Geogr. Nac., Guatemala, C.A.

Anonymous (1964): Inventario de Estudios Básicos sobre Recursos Naturales de Centroamérica. — 127 p., ICAITI, Guatemala.

Anonymous (1969 a): Bergwirtschaftlicher Jahresbericht Guatemala 1968. — Bundesanstalt f. Bodenforschung. Deutsche Geol. Mission in Guatemala, Ber. Nr. 6, Guatemala.

Anonymous (1969 b): Regional Reconnaissance, Geochemical Stream Sediment Survey. — Tecn. Rep. for the Government of El Salvador by the United Nations as Participating and Executing Agency for the UNDP. United Nations, New York.

Anonymous (1969 c): Porphyry Copper Mineralization at Cerro Petaquilla, Province of Colon, Panama. — UN Development Program, Mineral Survey of the Azuero Area. Techn. Rep. Publ. No. **3**, Panama.

Anonymous (1972 a): Geología y Mineralogía del Estudio de Cuencas Multinacionales, Golfo de Honduras, Golfo de Fonseca y Río San Juan. — 117 p. ICAITI, Guatemala, C.A.

Anonymous (1972 b): The Geology of Western Nicaragua. — Final Technical Report Vol. IV, Tax Improvement and Natural Resources Inventory Project, Nicaragua. — 221 p. Managua, Nicaragua.

Anonymous (1972 c): Reconocimiento de las mineralizaciones de cobre y otro en la región del Río Pito, Comarca de San Blas. — Programa de las Naciones Unidas para el Desarrollo, Proyecto Minero (Fase II), Informe técn., **4**, 91 p., Nueva York.

Anonymous (1972 d): Geología general de las regiones oriental y occidental. — (Panamá) Programa de las Naciones Unidas para el Desarrollo, Proyecto Minero, (Fase II), Informe técn., **1**, 53 p., Naciones Unidas, Nueva York.

Anonymous (1972 e): Geología de la región noroeste de Honduras. — Programa de las Naciones Unidas, Investigación de los Recursos Minerales en Áreas seleccionadas, Informe técn., 28 p., Nueva York.

Anonymous (1973): Bergwirtschaftlicher Bericht El Salvador. 1970/71. — Bundesanstalt f. Bodenforschung, Deutsche Geol. Beratergruppe in El Salvador, Ber. **22**, Hannover.

Anonymous (1974 a): Talamanca Proyecto de Investigación Minera, NU — DGMP Costa Rica COS/72/004, Informe técn. **2**, 94 p., San José, C.R.

Anonymous (1974 b): El Mochito Field Trip. — IV. Reunión de Geólogos de América Central, Tegucigalpa.

Anonymous (1974 c): Excursión Post-Congreso. — IV. Reunión de Geólogos de América Central, Tegucigalpa.

Anonymous (1974 d): Descripción y propósito de la red sísmica instalada para auscultación continua de Proyecto de Arenal. — Inst. Costarricense de Electricidad, 16 p., San José, C.R.

Anonymous (1976): Guatemala's 76 Erdbeben der Karibischen Platte, 48 p. — Münchener Rückversicherungsgesellschaft, München.

Anonymous (1977): The Caribbean: New Faces in a mixed mining scene. — Engineering and Mining J., 55 – 59, Nov. 1977, New York.

Anonymous (1978): Diagnóstico del Sector Minero, 91 p. — Organización de los Estados Americanos, República de Costa Rica, San José.

Anonymous (undated): Arenas Magnetíticas de Costa Rica. — Direcc. de Geol., Minas y Petróleo, Inform. Técn. y Notas Geol., **42**, 1 – 27, San José, C.R.

Anonymous (undated): Managua. Un Estudio sobre el Terremoto del Año 1972. — A study of the 1972 earthquake. — 35 p. Münchener Rückversicherungs-Gesellschaft, München.

Anonymous (undated): Nombres Geográficos de Costa Rica, Vol. I, Litonomía e Insunimía; Vol. II, Oronomía. — Inst. Geogr. Nac., San José, C.R.

ARCE VELASCO, J. (1973): Región de Managua, Tectónica y Sismicidad. — Ministerio de Economía, Industria y Comercio, 54 p., Managua, Nicaragua.

ARDEN Jr., D. D. (1969): Geologic History of the Nicaraguan Rise. — Transact. Gulf Coast Assoc. Geol. Soc., **19**, 295 – 309, Tallahassee, Florida.

— (1975): Geology of Jamaica and the Nicaragua Rise. — In: The Oceans Basins and Margins, **3**, 617 – 661, New York and London (Plenum Press).

ARGALL Jr., G. O. (1973): Panama Porphyry: drilling. — World Mining, February, 32 – 37.

ASCARRUNZ, K. R. & KESLER, S. T. (1971): Distribution of mineralization of the Altos Cuchumatanes and Cobán area, Guatemala. — Unpubl. Rep. Inst. Geogr. Nac., Guatemala, C.A.

ATWOOD, M. G. (1972): Geology of the Minas de Oro quadrangle, Honduras, Central America. — M. A. Thesis, 88 p., Wesleyan Univ.

AZAMBRE, B. & TOURNON, J. (1977): Les intrusions basiques du Río Reventazon (Costa Rica). — C.R. somm. Soc. géol. Fr., **2**, 104 – 107, Paris.

AZÉMA, J., GLAÇON, G., TOURNON, J. & VILA, J.-M. (1979): Precisiones acerca del Paleoceno de Puerto Quepos y sus alrededores, provincia de Puntarenas, Costa Rica. — Inst. Geogr. Nac., Informe Semestral Julio a Diciembre 1978, 77 – 87, San José.

AZÉMA, J., SORNAY, J. & TOURNON, J. (1979): Découverte d'Albien supérieur à Ammonites dans le matériel volcano-sedimentaire de "complexe de Nicoya" (province de Guanacaste, Costa Rica). — C. R. somm. Soc. géol. Fr., 1979, fasc. **3**, 129 – 131, Paris.

AZÉMA, J., TOURNON, J. & SORNAY, J. (1979): Presencia de amonites del Albiano Superior en las Formaciones del "Complejo de Nicoya". El yacimiento de la loma Chumico, provincia de Guanacaste, Costa Rica. — Inst. Geogr. Nac., Informe Semestral Julio a Diciembre 1978, 71 – 88, San José.

AZÉMA, J. & TOURNON, J. (1979): La péninsule de Santa Elena: Un massif ultrabasique charrie en marge pacifique de L'Amérique centrale. — Note transmise à la séance du 19/11/1979 de l'Acad. Sc., Paris.

BAILEY, D. G. F. (1969): Exploration for Geothermal Energy in El Salvador. — Publ. Geol. del ICAITI, **2**, 25 – 31, Guatemala, C. A.

BALLMANN, P. (1974): Informe sobre una travesía de Talamanca. — Naciones Unidas, Proyecto de Investigación Mineral NU-DGMP, 1 – 37, San José.

— (1976): Eine geologische Traverse des Ostteils der Cordillera de Talamanca, Costa Rica (Mittelamerika). — N. Jb. Geol. Paläont. Mh., 1976, H. 8, 502 – 512, Stuttgart.

BANDY, O. L. (1970): Upper Cretaceous-Cenozoic Paleobathymetric cycles, eastern Panama and northern Colombia. — Transact. Gulf Coast Assoc. Geol Soc., **20**, 181 – 193, Tallahassee, Florida.

BANDY, O. L. & CASEY, R. E. (1973): Reflector Horizons and Paleobathymetric History, Eastern Panama. — Geol. Soc. Amer., Bull., **84**, 3081 – 3086, Boulder/Col.

BANKS, N. G. & RICHARDS, M. L. (1969): Structures and bathymetry of western Bartlett trough, Caribbean Sea. — Amer. Assoc. Petrol. Geol., Mem., **11**, 221 – 228, Tulsa, Oklahoma.

BARR, K. W. & ESCALANTE, G. (1969): Contribución al Esclarecimiento de la Edad del Complejo de Nicoya, Costa Rica. — Publ. Geol. del ICAITI, **2**, 43 – 47, Guatemala, C.A.

BASS, M. N. & ZARTMANN, R. E. (1969): The basement of Yucatán Peninsula (Abs.). — Eos, Transact. Amer. Geophys. Union, **50**, 313, Richmond, Virginia.

BATESON, J. H. (1972): New Interpretation of Geology of Maya Mountains, British Honduras. — Amer. Assoc. Petrol. Geol., Bull., **56**, 956 – 963, Tulsa, Oklahoma.

BATESON, J. H. & HALL, I. H. S. (1971): Revised Geologic Nomenclature for Pre-Cretaceous Rocks of British Honduras. — Amer. Assoc. Petrol. Geol., Bull., **55**, 529 – 530, Tulsa, Oklahoma.

— — (1977): The Geology of the Maya Mountains, Belize. — Inst. Geol. Sci., Overseas Mem. (G.B.) 1977, Nr. **3**, 1 – 43, London.

BELLON, H. & TOURNON, J. (1978): Contribution de la géochronométrie K-Ar à l'étude du magmatisme de Costa Rica, Amérique Centrale. — Bull. Soc. géol. Fr., 1978 (7), **20**, 6, 955 – 959, Paris.

BENETT, F. D. & RACCICHINI, S. M. (1978): Subaqueous sulphur lake in Volcan Poás. — Nature, **271**, 342 – 344, London.

BENGOECHÉA, A. J. (1961): Los Placeres Auríferos de Area de Quilalí, Río Jícaro, Departamento de Nueva Segovia. — Bol. Serv. Geol. Nac. de Nicaragua, **5**, 59 – 99, Managua.

— (1963): Distrito Minero de Noreste. — Bol. Serv. Geol. Nac. de Nicaragua, **7**, 13 – 50, Managua.

BENGSTON, N. A. (1926): Notes on the Physiography of Honduras. — Geogr. Review, **16**, 403 – 413, New York.

BERGGREN, W. A. (1972): A Cenozoic time-scale — Some implications for regional geology and Paleobiography. — Lethaia, **5**, 195 – 215, Oslo.

BERGOEING, G., J. P. (1978): Modelado glaciar en la Cordillera de Talamanca, Costa Rica. — Inst. Geogr. Nac., Informe Semestral Julio a Diciembre 1977, 33 – 44, San José, C.R.
— (1978): Aspectos geomorfológicos de Isla de Chira. — Inst. Geogr. Nac., Informe Semestral Julio a Diciembre 1977, 45 – 52, San José, C.R.
— (1979): Península de Santa Elena, Costa Rica. — Estudio geomorfológico preliminar. — Inst. Geogr. Nac., Informe Semestral Julio a Diciembre 1978, 19 – 30, San José.
BERGOEING, G., J. P. & BRENES, O. L. G. (1978): Regiones morfográficas de Costa Rica. — Inst. Geogr. Nac., Informe Semestral Julio a Diciembre 1977, 53 – 58, San José, C.R.
— — (1978): Laguna de Hule, una caldera volcánica. — Inst. Geogr. Nac., Informe Semestral Julio a Diciembre 1977, 59 – 63, San José, C.R.
BERGOEING, G., J. P., MALAVASSI, V., E. & PROTTI, Q., R.(1979): Tres posibles edificios volcánicos del sector cerros del Aguacate. — Inst. Geogr. Nac., Informe Semestral Julio a Diciembre 1978, 35 – 45, San José, C. R.
BERGOEING, G., J. P., MORA, C., S. & JIMÉNEZ, R., R. (1979): Evidencias de vulcanismo Plio-Cuaternario en la fila Costẽna. — Inst. Geogr. Nac., Informe Semestral Julio a Diciembre 1978, 47 – 65, San José, C. R.
BERTRAND, J. & VUAGNAT, M. (1975): Sur la présence de basalts en coussins dans la zone ophiolitique méridionale de la Cordillère centrale du Guatemala. — Schweiz. Miner. Petrogr. Mitt., **55**, 136 – 142, Zürich.
BEVAN, P. A. (1973): Rosita Mine — A Brief History and Geological Description. — Canada Mining Metallurg. (CIM) Bull., **66**, No. 763, 80 – 84.
BIRNIE, R. W. (1973): Infrared Radiation Thermometry of Guatemalan Volcanoes. — Bull. volcanol., **37**, 1 – 36, Napoli 1973.
BLOUNT, D. N. (1967): Geology of the Chiantla quadrangle, Guatemala. — Ph. D. Diss., 135 p., Louisiana State Univ.
— (1969): Stratigraphy and Petrology of the Chiantla Area, Guatemala. — Publ. Geol. del ICAITI, **2**, 67 – 72, Guatemala, C.A.
BLOUNT, D. N. & MOORE Jr., C. H. (1969): Depositional and Non-Depositional Carbonate Breccias, Chiantla quadrangle, Guatemala. — Geol. Soc. Amer., Bull., **80**, 429 – 442, Boulder/Col.
BOER, J. de (1979): The Outer Arc of the Costa Rican Orogen (Oceanic Basement Complexes of the Nicoya and Santa Elena Peninsulas). — Tectonophysics, **56**, 221 – 259, Amsterdam.
BOHNENBERGER, O. H. (1966): Nomenclatura de las Carpas Santa Rosa en Guatemala. — Publ. Geol. del ICAITI, **1**, 47 – 51, Guatemala, C.A.
— (1967 a): Two hypothetical sections across Pacaya Volcanic Complex. — Excursion Guide Book for Guatemala, Pl. 10, Geol. Bull. 4, Inst. Geogr. Nac., Guatemala, C.A.
— (1967 b): Geologic reconnaissance of the Soloma and Barillas Quadrangles, Huehuetenango, Guatemala. — ICAITI, Guatemala, C.A.
— (1969): Los Focos Eruptivos de Guatemala. — Publ. Geol. del ICAITI, **2**, 23 – 24, Guatemala, C.A.
— (1978): Data Sheets of the Post-Miocene Volcanoes of the World, Sheet No. 3. Region Guatemala, El Salvador, Honduras, Nicaragua, Costa Rica, Panama. — Ed. by Working Group on the World Volcanological Map IAVCEI, Rome, Italy.
BOHNENBERGER, O. H., BENGOECHÉA, A., DÓNDOLI, C. & MARROQUIN, A. (1966): Report on active volcanoes in Central America during 1957 – 1965. — Segunda Reunión de Geólogos de América Central, Guatemala, C.A. Bull. Volc. Erupt., **9**,**1**, 2 – 19, Tokyo 1969.
BOHNENBERGER, O. H. & DENGO, G. (1968): Coal Resources in Central America. — Coal Geol. Symp., Geol. Soc. Amer. Meet., 1 – 13, Mexico City.
BOLD, W. A. VAN DEN (1963): Ostracods and Tertiary stratigraphy of Guatemala. — Amer. Assoc. Petrol. Geol., Bull., **47**, 696 – 698, Tulsa, Oklahoma.
— (1966): Ostracodes zones in Caribbean Miocene. — Amer. Assoc. Petrol. Geol., Bull., **50**, 1029 – 1031. Tulsa, Oklahoma.
— (1967): Miocene Ostracoda from Costa Rica. — Micropaleontology, **13**, 75 – 86, New York.
— (1969): Estado de nuestro conocimiento de los Ostrácodos Fósiles de América Central. — Publ. Geol. del ICAITI, **2**, 51 – 57, Guatemala, C.A.
— (1971) Ostracodes Association, Salinity and Depth of Deposition in the Neogene of the Caribbean Region. — Bull. Centre Rech. Pau-SNPA, **5**, Suppl., 449 – 460.

— (1972 a): Contribution of Ostracoda to the Correlation of Neogene Formations of the Caribbean Region. — Transact. VI. Caribb. Geol. Conf., 485 – 490, Caracas.
— (1972 b): Ostracodes of the La Boca Formation, Panama, Canal Zone. — Micropaleontology, **18**, 410 – 442, New York.
— (1973): La posición estratigráfica de la Formación La Boca, Panamá, Canal Zone. — Publ. Geol. del ICAITI, **4**, 167 – 170, Guatemala, C.A.
— (1974): Ostracodes Associations in the Caribbean Neogene. — Verh. Naturf. Ges. Basel, **84**, 214 – 221, Basel.
— Distribution of the *Radimella confragus* group (Ostracoda, Hemicytherinae) in the Late Neogene of the Caribbean. — J. Paleont., **49**, 692 – 701, Menasha, Wisc.
BONIS, S. B. (1965): Geología del Área de Quezaltenango, República de Guatemala. 84 p. — Inst. Geogr. Nac., Guatemala, C.A.
— (1967 a): Geologic reconnaissance of the Alta Verapaz fold belt, Guatemala. — Ph. D. Diss., 104 p., Louisiana State Univ.
— (1967 b): Reconocimiento geológico del cinturón plegado de Alta Verapaz, Guatemala. — Bol. Geol., **5**, Inst. Geogr. Nac., Guatemala, C.A.
— (1967 c): Excursion guide book for Guatemala (Ed.). — Inst. Geogr. Nac., Geol. Bull., **4**, 1 – 71, Guatemala, C.A.
— (1968): Age of Guatemalan serpentinite. — Geol. Soc. Amer., Abstr. for 1967, Spec. Pap. **115**, 18, Boulder/Col.
— (1969): A Summary of the Geology of Guatemala. — Publ. Geol. del ICAITI, **2**, 76 – 80, Guatemala, C.A.
— (1975): Guatemala. — In: The Encyclopedia of World Regional Geology I: Western Hemisphere. 305 – 309, Stroudsburg, Pennsylvania.
— (1976): The 1974 eruption of Volcan de Fuego, Guatemala (Abstr.). — Eos (Amer. Geophys. Union Transact.) **57**, 345, Washington, D.C.
BONIS, S. B., BOHNENBERGER, O., STOIBER, R. E., & DECKER, R. W. (1966): Age of pumice deposits in Guatemala. — Geol. Soc. Amer., Bull., **77**, 211 – 212, Boulder/Col.
BONIS, S. & SALAZAR, O. (1974): The 1971 and 1973 Eruptions of Volcán Fuego, Guatemala, and some socio-economic Considerations for the Volcanologists. — Bull. volcanol., **37**, 394 – 400, Napoli.
BOOM, G. VAN DEN (1972): Petrofazielle Gliederung des metamorphen Grundgebirges in der Sierra de Chuacús, Guatemala. — Beih. geol. Jb., **122**, 5 – 49, Hannover.
BOOM, G. VAN DEN & REYES GIRÓN, J. (1969): Mitteilung über Disthen-Vorkommen in der Sierra de Chuacús (Zentralguatemala). — Bundesanstalt f. Bodenforschung, Deutsche Geol. Mission in Guatemala, Guatemala.
BOOTHBY, D. R., GREER, E. W., DAVIES, D. K. & BONIS, S. (1978): Coastal sedimentation of volcanogenic sands, Guatemala. — Geol. Soc. Amer., Abstr. and Programs, **10**, 1, 2, Boulder/Col.
BOSC, E. A. (1971): Geology of the Agustín Acasaguastlan Quadrangle and Northeastern Part of El Progreso Quadrangle. — Ph. D. Thesis, Rice Univ., Houston, Texas.
BOWIN, C. O. (1968): Geophysical study of the Cayman Trough. — J. Geophys. Res., **73**, 5159 – 5173, Richmond, Virginia.
— (1976): Caribbean Gravity Field and Plate Tectonics. — Geol. Soc. Amer. Spec. Pap., **169**, 1 – 79, Boulder/Col.
BOWLES, F. A., JACK, R. N. & CARMICHAEL, I. S. E. (1973): Investigation of Deep-Sea Volcanic Ash Layers in Equatorial Pacific Cores. — Geol. Soc. Amer. Bull., **84**, 2371 – 2388, Boulder/Col.
BROGAN, G. E., MARVIN, R. N., WEAWER, K. D., CLINE, K. M., LLOYD, S. C. & SCHWARTZ, D. P. (1977): Active fault map of Managua, Nicaragua. — 8th Caribb. Geol. Conf. Abstr. 27 – 28, Amsterdam.
BROWN Jr., R. D., WARD, P. L. & PLAFKER, G. (1973): Geologic and Seismologic Aspects of the Managua, Nicaragua, Earthquakes of December 23, 1972. — Geol. Surv. Prof. Pap. **838**, Washington, D.C.
BROWNE, H. L. (1961): Paleontological and stratigraphic report of Limón Province, Costa Rica and northeastern Panama. — GSCR-8, 16 p., unpublished.
— (1965): Notes on Correlation and Paleogeography in Limón Province, Costa Rica. — Elf Petroleros de Costa Rica S.A. GSCR **28**, 11 p., unpublished.
BUCKNAM, R. C., PLAFKER, G. & SHARP, R. V. (1978): Fault movement (afterslip) following the Guatemala Earthquake of February 4, 1976. — Geology, **6**, 170 – 173, Boulder/Col.

BURK, C. A. & DRAKE, C. L. (eds.) (1974): The Geology of Continental Margins. — 1009 p. Berlin-Heidelberg-New York (Springer).

BURKART, B. (1965): Geology of the Esquipulas, Chanmagua, and Cerro Montecristo Quadrangles, Southeastern Guatemala. — Ph. D. Diss., Rice Univ. 121 p., Houston, Texas.

— (1978): Offset across the Polochic fault of Guatemala and Chiapas, Mexico. — Geology, **6**, 328 – 332, Boulder/Col.

BURKART, B. & CLEMONS, R. E. (1972): Late Paleozoic Orogeny in Northwestern Guatemala. — Transact. VI. Caribb. Geol. Conf., 210 – 213, Caracas.

BURKART, B., CLEMONS, R. E. & CRANE, D. C. (1973): Mesozoic and Cenozoic Stratigraphy of Southeastern Guatemala. — Amer. Assoc. Petrol. Geol., Bull., **57**, 63 – 73, Tulsa, Oklahoma.

BURN, R. G. (1969): The Pis-Pis gold mining district of NE Nicaragua. — Mining Mag., **120**, 169 – 175, London.

BURRI, C. & SONDER, R. A. (1936): Beiträge zur Geologie und Petrographie des jungtertiären und rezenten Vulkanismus in Nicaragua. — Z. Vulkanol., **17**, (1936/37), 34 – 92, Berlin.

BUTTERLIN, J. (1977): Géologie Structurale de la Région des Caraïbes (Mexique — Amérique Centrale — Antilles — Cordillère Caraïbe). — Paris, New York, Barcelone, Milan (Masson). 259 p.

BUTTERLIN, J. & BONET, F. (1960): Découverte d'une série éocène dans la presqu'île du Yucatan (Mexique). — Transact. II. Caribb. Geol. Conf., 33 – 39, Mayaguez, P.R.

— — (1963 a): La Paleogeografía de la Margen meridional del Golfo de México en el Paleoceno. — Ingenería Hidráulica en México, Mexico, D. F.

— — (1963 b): Mapas geológicas de la Península de Yucatán. — Ingenería Hidráulica en México, 19 p., México, D.F.

— — (1966): Les Formations Cénozoiques de la partie Mexicaine de la Presqu'île du Yucatan. — Transact. III. Caribb. Geol. Conf., 75 – 90, Kingston, Jamaiaca.

CARBALLO H., M. A. & FISCHER, R. (1979): La Formación San Miguel (Mioceno, Costa Rica). — Inst. Geogr. Nac., Informe Semestral Enero a Junio 1978, 45 – 144, San José, C.R.

CARPENTER, R. H. (1954): Geology and ore deposits of the Rosario Mining district and the San Juancito Mountains, Honduras, Central America. — Geol. Soc. Amer. Bull., **65**, 23 – 38, Boulder/Col.

CARR, M. J. (1974): Tectonics of the Pacific Margin of Northern Central America. — Ph. D. Thesis, Dartmouth College, Hanover, N.H.

— (1976): Underthrusting and Quaternary faulting in northern Central America. — Geol. Soc. Amer., Bull., **87**, 825 – 829, Boulder/Col.

— (1977): Volcanic Activity and Great Earthquakes at Convergent Plate Margins. — Science, **197**, 655 – 657, Lancaster.

— (1978): Magma type and structure of the inclined seismic zone in the Central American Arc. — Geol. Soc. Amer., Abstr. and Programs, **10**, 3, 98 – 99, Boulder/Col.

CARR, M. J., ROSE, W. I. & MAYFIELD, D. G. (1979): Potassium content of lavas and depth to the seismic zone in Central America. — J. Volcanol. Geotherm. Res., **5**, 387 – 401, Amsterdam.

CARR, M. J. & STOIBER, R. E. (1974): Intermediate Depth Earthquakes and Volcanic Eruptions in Central America, 1961 – 1972. — Bull. volcanol., **37**, 326 – 337, Napoli.

— — (1977): Geologic setting of some destructive earthquakes in Central America. — Geol. Soc. Amer. Bull., **88**, 151 – 156, Boulder/Col.

CARTER, W. D. & RINKER, J. N. (1976): Structural Features related to Earthquakes in Managua, Nicaragua, and Córboda, México, — Geol. Surv. Prof. Pap., **939**, 123 – 128, Washington, D.C.

CASE, J. E. (1974): Oceanic Crust Forms Basement of Eastern Panama. — Geol. Soc. Amer., Bull., **85**, 645 – 652, Boulder/Col.

— (1975 a): Geophysical Studies in the Caribbean Sea. — In: NAIRN, A. E. M. & STEHLI, F. G. (eds.): The Ocean Basins and Margins, **3**, 107 – 180, New York and London (Plenum Press).

— (1975 b): Geologic framework of the Caribbean Region. — In: Geology, Geophysics and Resources of the Caribbean. Rep. of the IDOE Workshop on Geol. and Marine Geophys. of the Caribb. Region and its Resources, 3 – 26, Kingston, Jamaica (Mayaguez, P.R.).

CASE, J. E., DURÁN, L. G., LÓPEZ, A. & MOORE, W. R. (1969): Investigación gravimétrica y magnética a traves del Istmo de Panamá (Noroeste Colombiano). — Geol. Colombiana, **6**, 5 – 16, Bogotá.

— — — — (1971): Tectonic Investigations in Western Colombia and Eastern Panama. — Geol. Soc. Amer., Bull., **82**, 2685 – 2712, Boulder/Col.

CASE, J. E. & HOLCOMBE, T. L. (1975): Preliminary geologic tectonic and bathymetric maps of the Caribbean region. — U.S. Geol. Surv. open-file maps 75 – 146.

CASE, J. E., MOORE, W. R., DURAN, L. G. & LÓPEZ, R., A (1971): Junction of the Andes and Panamanian Chains in Northern Colombia. — Transact. 5th Carib. Geol. Conf., Geol. Bull. **5.**, 11 – 15, Queens College, New York.

CASTILLO M., R. (1966): Aspectos Geológicos de los Yacimientos de Arcilla y Laterita de Costa Rica. — Publ. Geol. del ICAITI, **1**, 5 – 21, Guatemala, C.A.

— (1969): Geología de los mapas básicos Abra y partes de Río Grande, Costa Rica. — Direcc. de Geol., Minas y Petróleo, Inform. Técn. y Notas Geol., **33**, 1 – 40, San Pedro de Montes de Oca, C.R.

— (undated): Sinopsis Histórica sobre las Exploraciones Petroleras en Costa Rica. — Rev. de la Univ. de Costa Rica, **41**, 47 – 61, San Pedro de Montes de Oca, C.R.

— (1977): Geoquímica ambiental de la Península de Nicoya, Costa Rica. — Rev. Biol. Trop., **25** (2), 219 – 255, San José.

CASTILLO M., R., MADRIGAL G., R. & SÁNDOVAL M., F. (1970): Nota geotécnica sobre el yacimiento de laterita bauxítica del Valle de el General. — Direcc. de Geol., Minas y Petróleo, Inform. Técn. y Notas Geol., **34**, 1 – 22, Ciudad Universitaria, C.R.

ČEPEC, P. (1975): Die Kreide-Coccolithen aus der Referenz-Lokalität der Sepur-Folge bei Lanquin in Guatemala. — Geol. Jb., **B 14**, 87 – 109, Hannover.

CHAVES, R. (1969): Características Físicas, Químicas y Mineralógicas de los Materiales Eruptados por el Volcán Arenal. — Inst. Geogr. Nac., Informe Semestral, 1969 I, 49 – 67, San José, C.R.

CHAVES, R. & SÁENZ, R. (1973): Efectos del Terremoto de Managua (23 de diciembre de 1972). — Informe Semestral Enero a Junio 1973, 45 – 67, Inst. Geogr. Nac., San José, C.R.

— — (1974): Geología de la Cordillera de Tilarán. (Proyecto Aguacate 2da Fase). — Direcc. de Geol., Minas y Petróleo, Inform. Técn. y Notas Geol., **53**, 1 – 49, San Francisco de Dos Ríos, San José, C.R.

CHAVES, V. M., HANSEN, F. & QUESADA, D. (1973): Isosistas de Managua Terremoto del 23 de Diciembre. — Ministerio de Economía, Industria y Comercio, Catastro e Inventario de Recursos Naturales, 21 p., Managua.

CHRISTOFFERSON, E. (1973): Linear Magnetic Anomalies in the Colombia Basin. — Geol. Soc. Amer., Bull., **84**, 3217 – 3230, Boulder/Col.

— (1976): Colombian Basin magnetism and Caribbean plate tectonics. — Geol. Soc. Amer., Bull., **87**, 1255 – 1258, Boulder/Col.

CLARK, A. H., FARRAR, E. & KENTS, P. (1977): Potassium-Argon Age of the Cerro Colorado Porphyry Copper Deposit, Panama. — Economic Geol., **72**, 1154 – 1158, Lancaster.

CLEMONS, R. E. (1966): Geology of the Chiquimula quadrangle, Guatemala, Central America. — Ph. D. Diss., Texas Univ., 124 p., Austin, Texas.

— (1969): Geologic History of the Chiquimula Region, Guatemala. — Publ. Geol. del ICAITI, **2**, 72 – 75, Guatemala, C.A.

CLEMONS, R. E., ANDERSON, T. H., BOHNENBERGER, O. H. & BURKART, B. (1974): Stratigraphic Nomenclature and Recognized Paleozoic and Mesozoic Rocks of Western Guatemala. — Amer. Assoc. Petrol. Geol., Bull., **58**, 313 – 320, Tulsa, Oklahoma.

CLEMONS, R. E. & BURKART, B. (1971 a): Stratigraphy of Northwestern Guatemala. — Bol. Soc. Geol. Mexicana, **32**, no. 2, 143 – 158, (1969), México, D.F.

— — (1971 b): Geology of the Western Sierra de los Cuchumatanes, Guatemala: A Preliminary Report. — Transact. V. Caribb. Geol. Conf. (Queens College Geol. Bull. **5**), 117 – 118, New York.

CLEMONS, R. E. & LONG L. E. (1971): Petrologic and Rb-Sr Isotopic Study of the Chiquimula Pluton, Southeastern Guatemala. — Geol. Soc. Amer., Bull., **82**, 2729 – 2740, Boulder/Col.

CLUFF, L. S., NICCUM, M. R., BROGAN, G. E. & CLINE, K. M. (1976): Surface Fault Hazard zoning: Methods for regional planning and for specific Investigations. — 15th Ann. Symp. on Engineering Geology and Soils Engieering, Geol. Dept. Boise State Univ., 15 p., Boise, Idaho.

COLLINS, E. M. & KESLER, S. E. (1969): High temperature telescoped Tungsten-Antimony Mineralization, Guatemala. — Mineralium Deposita, **4**, 65 – 71, Berlin.

COLLINS, D. E., NICCUM, M. R. & BICE, D. C. (1976): Preliminary Summary of Late Pleistocene and Holocene Volcanic and Sedimentary Stratigraphy of the Managua Area, Nicaragua. — Publ. Geol. del ICAITI, **5**, 105 – 113, Guatemala, C.A.

CRAFFORT, T. C. (1975): SO_2 Emission of the 1974 Eruption of Volcán Fuego, Guatemala. — Bull. volcanol, **39**, 4, 1 – 21, Napoli.
CRANE, D. C. (1965): Geology of the Jocotán and Timushan Quadrangles, Southeastern Guatemala. — Ph. D. Thesis, Rice Univ., Houston, Texas.
CSERNA, Z. DE (1960): Orogenesis in time and space in Mexico. — Geol. Rdsch., **50**, 595 – 605, Stuttgart.
CUMMING, G. L. & KESLER, S. E. (1976): Source of Lead in Central American and Caribbean Mineralization. — Earth and Planetary Sci. Letters, **31**, 262 – 268, Amsterdam.
CURRAN, D. W. & MACDONALD, W. D. (1977): Stratigraphy and Paleomagnetism of Tertiary Volcanic Rocks of the Siguatepeque Area, Honduras. — V. Reunión de Geólogos de América Central, 13 p., Managua.
DAVIES, D. K., QUEARRY, M. W. & BONIS, S. B. (1978): Glowing avalanches from the 1974 eruption of the volcano Fuego, Guatemala. — Geol. Soc. Amer. Bull., **89**, 369 – 384, Boulder/Col.
DENGO, G. (1960): Notas sobre la geología de la parte central del litoral pacífico de Costa Rica. — Inst. Geogr. de Costa Rica, Informe Semestral, Julio a Diciembre, 43 – 58, San José, C.R.
— (1962 a): Tectonic-Igneous Sequence in Costa Rica. — Petrologic Studies: A Volume to Honor A. F. BUDDINGTON, 133 – 161, Geol. Soc. Amer., Boulder/Col.
— (1962 b): Estudio Geológico de la Región de Guanacaste. — 112 p., Inst. Geogr. de Costa Rica, San José, C.R.
— (1965): Geological Structure of Central America. — 23 p., Internat. Conf. Trop. Oceanography, Miami, Nov. 18 – 24.
— (1966): Estructura Geológica e Historia Tectónica de América Central. — Contrib. II. Reunión de Geólogos de América Central, Noviembre 17 – 21, Guatemala, C.A.
— (1967): Geological Structure of Central America. — Studies in Tropical Oceanography, 5, Proc. Internat. Conf. Trop. Oceanography, Univ. Miami, 56 – 73, Miami.
— (1968/1973): Estructura geológica, historia tectónica y morfología de América Central. — Inst. Centroamericano de Investigación y Tecnología Industrial (ICAITI). — 52 p., México, Buenos Aires (Centro Regional de Ayuda Técnica) 1. y 2. edición.
— (1969 a): Relación de las Serpentinitas con la Tectónica de América Central. — Simpos. Panamericano de Manto Superior, **II**, 23 – 28, México, D.F.
— (1969 b): Problems of tectonic relations between Central America and the Caribbean. — Transact. Gulf Coast Assoc. Geol. Soc., **19**, 311 – 320, Tallahassee, Florida.
— (1972): Review of Caribbean Serpentinites and Their Tectonic Implications. — Geol. Soc. Amer. Mem., **132**, 303 – 312, Boulder/Col.
— (1975 a): Paleozoic and Mesozoic Tectonic Belts in Mexico and Central America. — In: NAIRN, A. E. M. & STEHLI, F. G. (eds.): The Ocean Basins and Margins, **3**, The Gulf of Mexico and the Caribbean, 283 – 323, New York and London (Plenum Press).
— (1975 b): Belize. — In FAIRBRIDGE, R. W. (ed.): The Encyclopedia of Worlds Regional Geology I: Western Hemisphere, 116, Stroudsburg, Pennsylvania (Dowden, Hutchinson & Ross, Inc.).
DENGO, G. & BOHNENBERGER, O. (1969): Structural development of northern Central America. — Amer. Assoc. Petrol. Geol., Mem., **11**, 203 – 220, Tulsa, Oklahoma.
DENGO, G., BOHNENBERGER, O. & BONIS, S. (1970): Tectonics and volcanism along the Pacific Marginal Zone of Central America. — Geol. Rdsch., **59**, 1215 – 1232, Stuttgart.
DENGO, G. & LEVY, E. (1970): Anotaciones al Mapa Metalogenético de América Central. — Publ. Geol. del ICAITI, **3**, 1 – 15, Guatemala, C.A.
DENYER, P. & KUIJPERS, E. P. (1979): Mineralizasiones de manganeso intercaladas en basaltos del Complejo de Nicoya, Guanacaste, Costa Rica. — Inst. Geogr. Nac. Informe Semestral Julio a Diciembre 1978, San José.
DEUX, D. & COURBON (eds.) (1976): Spelunca, Spécial No. 1, (Guatemala) (suppl. au No. 3).
DEWEY, J. W. & ALGERMISSEN, S. T. (1974): Seismicity of the Middle America arc-trench system near Managua, Nicaragua. — Seismol. Soc. Amer., Bull., **64**, 1033 – 1048.
DILLON, W. P. & VEDDER, J. G. (1973): Structure and development of the continental margin of British Honduras. — Geol. Soc. Amer., Bull., **84**, 2713 – 2732, Boulder/Col.
DILLON, W. P., VEDDER, J. G. & GRAF, R. J. (1972): Structural profile of the northwestern Caribbean. — Earth and Planetary Sci. Letters, **17**, 175 – 180, Amsterdam.

DIXON, C. G. (1955): Geology of Southern British Honduras with Notes on Adjacent Areas. — 85 p., Georgetown.
DÓNDOLI, C. (1970): Localización de un horizonte laterítico bauxítico en la zona de Paraíso de Cartago. — Direc. del Geol., Minas y Petróleo, **36**, 1 – 12, Ciudad Universitaria, C.R.
— (1973): Depósitos laterítico bauxítico en la zona de Paraíso de Cartago, Costa Rica. — Publ. Geol. del ICAITI, **4**, 31 – 36, Guatemala, C.A.
DONLEY, M. W. (1971): The Pumice Landscape of Guatemala: A Study on Basin Morphology and Landforms in the Upland Tropics. — Ph. D. Thesis, Univ. of Oregon.
DONNELLEY, T. W. (1975): The Geological Evolution of the Caribbean and Gulf of Mexico — Some Critical Problems and Areas. — In: NAIRN, A. E. M. & STEHLI, F. G. (eds.): The Ocean Basins and Magins, **3**, New York and London (Plenum Press).
— (1977): Metamorphic rocks and structural history of the Motagua Suture Zone, eastern Guatemala. — 8th Caribb. Geol. Conf., Abstr., 40 – 41, Amsterdam.
DONNELLY, T. W., CRANE, D. & BURKART, B. (1968): Geologic History of the Landward Extension of the Bartlett Trough. — Transact. 4th Caribb. Geol. Conf., 225 – 228, Trinidad and Tobago.
DONNELLY, T. W., KAY, R., ROGERS, J. J. W., MACDONALD, W., MACGILLAVRY, H. J., BEETS, B. J. & SCHUBERT, C. (1976): The Caribbean Basaltic "Crust" — a phanerozoic flood basalt of unparalleled size, its tectonic implications, and its significance in future DSDP operations. — Eos (Transact. Amer. Geophys. Union), **57**, 409 – 410, Washington, D.C.
DONNELLY, T. W., MELSON, M. & KAY, R. (1973): Basalts and dolerites of Late Cretaceous age from the Central Caribbean. — Initial Rep. Deep Sea Drilling Project, **15**, 989 – 1011, Washington, D.C.
ECHÁVARRI P., A. F. & RUEDA G., J. (1962): Estudio geológico económico de los yacimientos de Tungsteno y Molibdeno de Macuelizo, Nueva Segovia. — Bul. Serv. Geol. Nac. de Nicaragua, **6**, 23 – 43, Managua.
EDGAR, N. T., EWING, J. I. & HENNION, J. (1971): Seismic Refraction and Reflection in the Caribbean Sea. — Amer. Assoc. Petrol. Geol., Bull., **55**, 833 – 870, Tulsa, Oklahoma.
EGGERS, A. A. (1972): Geology of the Amatitlán Quadrangle, Guatemala. — Ph. D. Diss., Dartmouth College, Hanover, N. H.
— (1978): Baseline gravity and magnetic survey of Pacaya volcano, Guatemala. — Geol. Soc. Amer. Abstr. and Programs, **10**, 3, 104, Boulder/Col.
EGGERS, A. A., KRAUSSE, J., RUSH, H. & WARD, J. (1976): Gravity changes accompanying volcanic activity at Pacaya Volcano, Guatemala. — J. Volcanol. Geotherm. Res., **1**, 229 – 236, Amsterdam.
ELVIER, A. R. (1969): Historia Minera de Honduras. — Publ. Geol. del ICAITI, **2**, 1 – 7, Guatemala, C.A.
— (1970): Evolución y Perspectivas de Desarrollo de la Minería en Honduras. — Direcc. General de Minas e Hidrocarburos, 11 p., Tegucigalpa.
— (1974): Geología de Honduras. — 44 p., Tegucigalpa.
— (1976): Síntesis de la Geología de Honduras. — Publ. Geol. del ICAITI, **5**, 1 – 4, Guatemala, C.A.
ENGELS, B. (1962): Microtectónica del sureste de Nicaragua. — Bol. Serv. Geol. Nac., **6**, 44 – 49, Managua.
— (1964 a): Resumen del estudio geológico sobre la Tectónica interna de la región esquistosa de Nueva Segovia, Nicaragua. — Bol. Serv. Geol. Nac., **8**, 11 – 52, Managua.
— (1964 b): Geologische Problematik und Strukturanalyse Nicaraguas. — Geol. Rdsch., **54**, 758 – 795, Stuttgart.
ESCALANTE, G. (1966): Geología de la Cuenca Superior del Río Reventazón, Costa Rica. — Publ. Geol. del ICAITI, **1**, 59 – 70, Guatemala, C.A.
ESKER III, G. C. (1969): Upper Cretaceous Planktonic Foraminifera from Guatemala. — Publ. Geol. del ICAITI, **2**, 56 – 57, Guatemala, C.A.
ESPINOZA, A. F. (ed.) (1976): The Guatemalan Earthquake of February 4, 1976, a preliminary report. — U.S. Geol. Surv. Prof. Pap., **1002**, 90 p., Washington, D.C.
EVERETT J. R. (1970): Geology of the Comayagua quadrangle, Honduras, Central America. — Ph. D. Diss., 152 p., Texas Univ., Austin, Texas.
EVERETT, J. R. & FAKUNDINY, R. H. (1976): Structural Geology of El Rosario and Comayagua Quadrangles, Honduras, Central America. — Publ. Geol. del ICAITI, **5**, 31 – 42, Guatemala, C.A.
FAIRBRIDGE, R. W. (1975): Central America — Regional Review, 235 – 236; Costa Rica, 251 – 252; El Salvador, 270 – 272; Honduras, 327 – 330; Nicaragua, 400 – 401. — In: The Encyclopedia of World

Regional Geology, I: Western Hemisphere, Stroudsburg, Pennsylvania (Dowden, Hutchinson & Ross).

FAIRBROTHERS, G. E., CARR, M. J. & MAYFIELD, D. G. (1978): Temporal Magmatic Variation at Boqueron Volcano, El Salvador. — Contrib. Mineral. Petrol., **67**, 1 – 9, Berlin, Heidelberg, New York.

FAKUNDINY, R. H. (1970): Geology of the El Rosario quadrangle, Honduras, Central America. — Ph. D. Thesis, Univ. of Texas, Austin.

FAKUNDINY, R. H. & EVERETT, J. R. (1976 a): Re-Examination of Mesozoic Stratigraphy of the El Rosario and Comayagua Quadrangles, Central Honduras. — Publ. Geol. del ICAITI, **5**, 5 – 17, Guatemala, C.A.

— — (1976 b): Metamorphic and Intrusive Rocks of the El Rosario and Comayagua Quadrangles, Central Honduras. — Publ. Geol. del ICAITI, **5**, 71 – 77, Guatemala, C.A.

FERENČIĆ, A. (1970): Porphyry Copper Mineralization in Panama. — Mineralium Deposita **5**, 383 – 389, Berlin.

— (1971): Metallogenetic Provinces and Epochs in Southern Central America. — Min. Deposita, **6**, 77 – 88, Berlin.

FERENČIĆ, A., GIUDICE, D. DEL & RECCHI, G. (1971): Tectomagmatic and Metallogenetic Relationships of the Central Panama — Costa Rica Region. — Transact. 5th Caribb. Geol. Conf. (Queens College Geol. Bull., **5**,), 189 – 197, New York.

FERNÁNDEZ C., M. (1962): Notas geológicas sobre los depósitos de laterita y bauxita localizadas en Costa Rica. — Dept. de Geol. Min. y Petrol., Informe Técn. **10**, 31 p., San José, C.R.

FERNÁNDEZ C., M. & DÓNDOLI B., C. (1965): Sobre los depósitos costeros de arenas magnetíticas de las Playas de Costa Rica. — Direcc. de Geol. Minas y Petróleo, Informes Técn. y Notas Geol., **18**, 66 p., Ciudad Universitaria, San Pedro de Montes de Oca, C.R.

FERNÁNDEZ P., R. (1966, 1967, 1968): Estado de volcanes de Costa Rica. — Inst. Geogr. Nac., Informe semestral, 1966, II, 23 – 24, 1967, II, 19 – 20, 1968 II, 11 – 12, San José, C.R.

— (1968): La actividad del Poás en el año 1953 y su transformación de seudo geiser en volcán humeante. — Inst. Geogr. Nac., Informe Semestral 1968, I, 31 – 38, San José, C.R.

FERNÁNDEZ-RIVAS, R. (1970): Geothermal Resources of Guatemala, Central America. — U.N. Symposium on the Development and Utilization of Geothermal Resources, Pisa 1970. — Geothermics, Spec. Iss., **2**, 1015 – 1025.

FERRARI, B. & VIRAMONTE, J. (1973): Contribución al conocimiento de la geomorfología regional de Nicaragua. — Publ. Geol. del ICAITI, **4**, 95 – 104, Guatemala, C.A.

FIEDLER, G. (1976): The Motagua Fault Earthquake. — Preliminary Rep. of the Unesco-Ceresis Mission on Geology-Seismology, Caracas, Venezuela.

— (1977): Das Erdbeben von Guatemala vom 4. Februar 1976. — Geol. Rdsch., **66**, 309 – 335, Stuttgart.

FIGGE, K. (1966): Die stratigraphische Stellung der metamorphen Gesteine NW-Nicaraguas. — N. Jb. Geol. Paläont., Mh., **1966**, 234 – 247, Stuttgart.

FINCH, R. C. (1972): Geology of the San Pedro Zacapa quadrangle, Honduras. — Ph. D. Diss., Texas Univ., Austin, Texas.

FINCH, R. C. & CURRAN, D. W. (1977): Mesozoic Stratigraphy of Central Honduras: A Review. — V. Reunion de Geólogos de América Central, 35 p., preprint. Managua.

FISHER, R. L. (1961): Middle America Trench: Topography and Structure. — Geol. Soc. Amer., Bull., **72**, 703 – 720, Boulder/Col.

FISHER, S. P. & PESSAGNO, E. A. (1965): Upper Cretaceous strata of Northwestern Panama. — Amer. Assoc. Petrol. Geol., Bull., **49**, 433 – 444, Tulsa, Oklahoma.

FORTH, D. (1971): Geology of the Sacapulas quadrangle, Guatemala. 113 p. — M. S. Thesis, Louisiana State Univ., Baton Rouge.

FREELAND, G. L. & DIETZ, R. S. (1971): Plate Tectonic Evolution of Caribbean — Gulf of Mexico Region. — Nature, **232**, July 2, 20 – 23, London.

— — (1972): Plate Tectonic Evolution of the Caribbean — Gulf of Mexico Region. — Transact. 6th Caribb. Geol. Conf., 259 – 264, Caracas.

FUDALI, R. F. & MELSON, W. G. (1972): Ejecta velocities, Magma Chamber Pressure and Kinetic Energy associated with the 1968 Eruption of Arenal Volcano. — Bull. volcanol., **35**, 383 – 401, Napoli.

GALL, F. (1966): Cerro Quemado, Volcán de Quezaltenango. — 115 p., Guatemala (José de Pineda Ibarra).

GALLI-OLIVIER, C. (1977): Edad de emplazamiento y período de acumulación de la ofiolita de Costa Rica. — Rev. Cienc. Tec., Univ. de Costa Rica, **1**, 81 – 86, San José, C.R.

— (1979): Ophiolite and island-arc volcanism in Costa Rica. — Geol. Soc. Amer. Bull., **90**, 444 – 452, Boulder/Col.

GALLI-OLIVIER, C. & SCHMIDT-EFFING, R. (1977): Estratigrafía de la cubierta sedimentaria supra-ofiolítica cretácea de Costa Rica. — Rev. Cienc. Tec., Univ. de Costa Rica, **1**, 87 – 96, San José, C.R.

GARAYAR, S., J. (1971): Geología y depósitos de minerales de una parte de las Mesas de Esteli, Cordillera Norte y Montañas de Dipilto. — Catastro e Inventario de Recursos Naturales, Div. Geol., Arch. Accesible Informe **10**, 1 – 89, Managua, Nicaragua, C.A.

— (1977): Generalidades Tectónica de Nicaragua. — V. Reunión de Geólogos Centroamericanos, Managua, Nicaragua, C.A.

GARAYAR S., J. & VIRAMONTE, J. G. (1971): Sobre el hallazgo de peridotitas en Nicaragua. — Catastro e Inventario de Recursos Naturales, Div. Geol., Arch. Accesible Informe, **17**, 1 – 24, Managua, Nicaragua, C.A.

— — (1973): Hallazgo de peridotitas en Nicaragua. — Publ. Geol. del ICAITI, **4**, 105 – 114, Guatemala, C.A.

GEISTER, J. (1975): Riffbau und geologische Entwicklungsgeschichte der Insel San Andrés (westliches Karibisches Meer, Kolumbien). — Stuttgarter Beitr. Naturk., Ser. B, Nr. **15**, 203 p., Stuttgart.

GEYER, O. (1968): Über den Jura der Halbinsel Goajira (Kolumbien). — Mitt. Inst. Colombo-Alemán Invest. Cient., **2**, 67 – 83, Santa Marta, Colombia.

GIERLOFF-EMDEN, H. G. (1959): Die Küste von El Salvador. — Acta Humboldtiana, Ser. geogr. ethnogr., **2**, 183 p., Wiesbaden (Steiner).

— (1974): Anwendung von Multispektralaufnahmen des ERTS-Satelliten zur kleinmaßstäblichen Kartierung amphibischer Küstenräume am Beispiel der Küste von El Salvador. — Kartogr. Nachr., **24**, 54 – 76, Bonn-Bad Godesberg.

GIUDICE, D. DEL (1960): Apuntes sobre la geología del Departamento Nueva Segovia. — Bol. Serv. Geol. Nac. Nicaragua, **4**, 17 – 37, Managua.

— (1973): Características geológicas de la República de Panamá. — Conf. Reunión Continental sobre la Ciéncia y el Hombre, México, D.F.

GIUDICE, D. DEL & RECCHI, G. (1969): Geología del Area del Proyecto Minero de Azuero. — Informe Técnico preparado para el Gobierno de la República de Panamá, Programa para el Desarrollo de las Naciones Unidas, Proyecto Minero de Azuero. 48 p., Panamá.

GOLOMBEK, M. P. & CARR, M. J. (1978): Tidal triggering of seismic and volcanic phenomena during the 1879 –1880 eruption of Islas Quemadas Volcano in El Salvador, Central America. — J. Volcanol. Geotherm. Res., **3**, 299 – 307, Amsterdam.

GOMBERG, D. N., BANKS, P. O. & MCBIRNEY, A. R. (1968): Preliminary zirkon ages from the Central Cordillera. — Science, **161**, 121 – 122, Washington, D.C.

GÓMEZ P., L. D. (1973): Criptonemiales calcáreas fósiles en las calizas terciarias de Patarrá, Costa Rica. — Rev. Biol. Trop., **21**, (1), 107 – 110.

— (1974): *Ficus padifolia* HBK en la diatomita Pliocena/Pleistocénica de la Formación Bagaces, Gte., Costa Rica. — Veröff. Überseemus. Bremen, R. A. **4**, Nr. **15**, 141 – 148, Bremen.

GÓMEZ P., L. D. & VALERIO, C. E. (1973): Lista preliminar de los Moluscos fósiles de la Formación Río Banano (Mioceno), Limón, Costa Rica II. — Inst. Geogr. Nac., Informe Semestral Enero a Junio, 11 – 22, San José, C.R.

GOOSSENS, P. J. & FERENČIĆ, A. (1971): Discussions. About the Paper "Metallogenetic Provinces and Epochs in Southern Central America" FERENČIĆ, Mineralium Deposita **6**, 77 – 89 (1971). — Mineralium Deposita, **6**, 258 – 260 (Berlin).

GOOSSENS, P. J., ROSE Jr., W. I. & FLORES, D. (1977): Geochemistry of tholeiites of the Basic Igneous Complex of northwestern South America. — Geol. Soc. Amer., Bull., **88**, 1711 – 1720, Boulder/Col.

GOSE, W. A. & SWARTZ, D. K. (1977 a): Paleomagnetic results from Cretaceous sediments in Honduras: Tectonic implications. — Geology, **5**, 505 – 508, Boulder/Col.

— — (1977 b): Paleomagnetic results from Cretaceous sedimentary rocks in Honduras. — 8th Caribb. Geol. Conf. Abstr., 63 – 64, Amsterdam.

GRASES, J. (1975): Sismicidad de la Región asociada a la Cadena Volcánica Centroamericana del Cuaternario. — 106 p. UVC-OEA, Caracas.

GREBE, W.-H. (1955 a): Die Lagerstätten der zentralamerikanischen Republik El Salvador. — Mitt. Geol. Staatsinst. Hamburg, **24**, 40 – 45, Hamburg.
— (1955 b): La Minería en El Salvador, Centro América. — Anal. Serv. Geol. Nac. de El Salvador, Bol. **1**, 1 – 62, San Salvador.
— — (1961): Zur Geologie der altvulkanischen Gebirge in El Salvador (Mittelamerika). — Beih. geol. Jb., **50**, Hannover.
GREBE, W.-H., KÜRSTEN, M., PUTZER, H. & SCHNEIDER, H. (1959): Bericht der Bundesanstalt für Bodenforschung über Lagerstätten in Mexiko und Mittelamerika. — 195 p., Hannover (Unpublished).
GROSS, W. H. (1975): In Central America large expansion in mineral output yet to come. — Northern Miner (Canada), **61**, 33 – 51, Toronto.
GUNN, B. M. & MOOSER, F. (1971): Geochemistry of the Volcanics of Central Mexico. — Bull. volcanol., **34**, 577 – 616, Napoli.
GUZMÁN, E. J. & CSERNA, Z. DE (1963): Tectonic history of Mexico. — Amer. Assoc. Petrol. Geol., Mem., **2**, 113 – 129, Tulsa, Oklahoma.
HALL, I. H. S. & BATESON, J. H. (1972): Late Paleozoic Lavas in Maya Mountains, British Honduras, and Their Possible Regional Significance. — Amer. Assoc. Petrol. Geol., Bull., **56**, 950 – 963, Tulsa, Oklahoma.
HARP, E. L., WIECZOREK, G. F. & WILSON, R. C. (1976): Seismic-induced landslides from the 4 February 1976 Guatemala Earthquake. — Geol. Soc. Amer., Abstr. Programs (USA), **8**, No. 6, 905.
HARRIS, S. A. (1971): Quaternary Vulcanicity in the Talamanca Range of Costa Rica. — Canadian Geographer, **15**, 141 – 145.
HARROUCH, M. (1966): Relación del tecto-volcanismo con la actividad sísmica en el vulcanismo Santa Ana-Ahuachapán. — Centro de Estudios y Invest. Geotécn., 19 p., San Savador.
HASTENRATH, S. (1973): On the Pleistocene Glaciation of the Cordillera de Talamanca, Costa Rica. — Z. Gletscherk. Glazialgeol., **9**, 105 – 121, Innsbruck.
— (1974): Spuren pleistozäner Vereisung in den Altos de Cuchumatanes, Guatemala. — Eiszeitalter u. Gegenwart, **25**, 25 – 34, Öhringen/Württ.
HATHERTON, T. & DICKINSON, W. R. (1969): The Relationship between Andesitic Volcanism and Seismicity in Indonesia, the Lesser Antilles, and Other Island Arcs. — J. Geophys. Res., **74**, 5301 – 5310, Richmond, Virginia.
HEALEY, J. (1969): Notas sobre los volcanes de la sierra volcánica de Guanacaste, Costa Rica. — Inst. Geogr. Nac., Informe Semestral, Enero — Junio, 37 – 47, San José, C.R.
HEATH, G. R. & ANDEL, T. H. VAN (1973): Tectonic and Sedimentation the the Panama Basin: Geologic Results of Leg 16, Deep Sea Drilling Project. — Initial Rep. of Deep Sea Drill. Proj., **16**, 899 – 913, Washington, D.C.
HEEZEN, B. C. & RAWSON, M. (1977 a): Visual observations of contemporary current erosion and tectonic deformation on the Cocos Ridge. — Marine Geol., **23**, 173 – 196, Amsterdam.
— — (1977 b): Visual observations of the sea floor subduction line in the Middle-America Trench. — Science, **196**, 423 – 426, Lancaster.
HELBIG, K. M. (1959): Die Landschaften von Nordost-Honduras. Aufgrund einer geographischen Studienreise im Jahre 1953. — Petermanns Mitt., Erg.-H., **268**, 270 p. Gotha.
HENNINGSEN, D. (1966 a): Die pazifische Küstenkordillere (Cordillera Costeña) Costa Ricas und ihre Stellung innerhalb des süd-zentralamerikanischen Gebirges. — Geotekt. Forsch., **23**, 3 – 66, Stuttgart.
— (1966 b): Notes on stratigraphy and paleontology of Upper Cretaceous and Tertiary sediments in Southern Costa Rica. — Amer. Assoc. Petrol. Geol., Bull., **50**, 562 – 566, Tulsa, Oklahoma.
— (1966 c): Estratigrafía y Paleogeografía de los Sedimentos del Cretácico Superior y del Terciario en el Sector Sureste de Costa Rica. — Publ. Geol. del ICAITI, **1**, 53 – 57, Guatemala, C.A.
HENNINGSEN, D. & WEYL, R. (1967): Ozeanische Kruste im Nicoya-Komplex von Costa Rica (Mittelamerika). — Geol. Rdsch., **57**, 33 – 47, Stuttgart.
HESS, H. H. (1938): Gravity anomalies and island arc structure in particular reference to the West Indies. — Amer. Philosoph Soc. Proc., **79**, 71 – 96, Philadelphia.
HESS, H. H. & MAXWELL, J. C. (1953): Caribbean research project. — Geol. Soc. Amer., Bull., **64**, 1 – 6, Boulder/Col.
HEY, R. (1977): Tectonic evolution of the Cocos-Nazca spreading center. — Geol. Soc. Amer., Bull., **88**, 1404 – 1420, Boulder/Col.

HEY, R., JOHNSON, G. L. & LOWRIE, A. (1977): Recent plate motions in the Galapagos area. — Geol. Soc. Amer., Bull., **88**, 1385 – 1403, Boulder/Col.

HINTZ, R. A. (1955): Beiträge zur Geologie El Salvadors VII. Stoffbestand und regionale Verteilung vulkanogener Sedimente. — N. Jb. Geol. Paläont., Abh., **102**, 37 – 76, Stuttgart.

HIRSCHMANN, T. S. (1963): Reconnaissance Geology and Stratigraphy of the Subinal Formation (Tertiary) of the El Progreso Area, Guatemala, C.A. — M. A. Thesis, Indiana Univ.

HODGSON, G. (1971): Geología y anotaciones mineralógicas de la Planicie Noroeste y de la Precordillera Occidental. — Catastro e Inventario de Recursos Naturales, Div. Geol., Arch. Accesible Informe, **13**, 112 p., Managua, Nicaragua, C.A.

HODGSON, G. & FERREY, C. J. (1971): Geología y anotaciones de la Planicie Sureste del Lago de Nicaragua. — Catastro e Inventario de Recursos Naturales, Div. Geol., Arch. Accesible Informe, **13**, Managua, Nicaragua, C.A.

HOLCOMBE, T. L. (1975): Caribbean Bathymetry and Sediments. — In: Geology, Geophysics and Resources of the Caribbean. Report of the IDOE Workshop on Geol. and Marine Geophys. of the Caribbean Region and its Resources, 27 – 62, Kingston, Jamaica (Mayaguez, P.R.).

HOLCOMBE, T. L., VOGT, P. R., MATTHEWS, J. E., MURCHISON, R. R. (1973): Evidence for sea-floor spreading in the Cayman trough. — Earth and Planetary Sci. Letters, **20**, 357 – 371, Amsterdam.

HOOSE, S. N. & WILSON R. C. (1976): Liquefaction-caused ground failure during the Guatemala Earthquake of February 4, 1976. — Geol. Soc. Amer., Abstr. Programs (USA), **8**, No. 6, 925, Boulder/ Col.

HORNE, G. S., ATWOOD, M. G. & KING, P. A. (1974): Stratigraphy, Sedimentology, and Paleoenvironment of Esquias Formation of Honduras. — Amer. Assoc. Petrol. Geol., Bull., **58**, 176 – 188, Tulsa, Oklahoma.

HORNE, G. S., CLARK, G. S. & PUSHKAR, P. (1974): Geochronology of the Basement Rocks in the Sierra de Omoa, Northwestern Honduras. — IV. Reunión de Geólogos de América Central, Preprint, Tegucigalpa.

— — — (1976): Pre-Cretaceous Rocks of Northwestern Honduras: Basement Terrane in Sierra de Omoa. — Amer. Assoc. Petrol. Geol., Bull., **60**, 566 – 583, Tulsa, Oklahoma.

HORNE, G. S., PUSHKAR, P. & SHAFIQULLAH, M. (1976 a): Laramide Plutons on the Landward continuation of the Bonacca Ridge, Northern Honduras. — Publ. Geol. Del ICAITI, **5**, 84 – 90, Guatemala, C.A.

— — — (1976 b): Preliminary K-Ar age data from the Laramide Sierras of Central Honduras. — Publ. Geol. del ICAITI, **5**, 91 – 98, Guatemala, C.A.

— — — (1976 c): Laramide Plutons on the Landward continuation of the Bonacca Ridge, Northern Honduras. — Transact. VII. Conf. Géol. des Caraïbes, 583 – 588, Saint François (Guadeloupe).

HORST, O. H., KUENZI, W. D. & MCGEHEE, R. V. (1976): Sedimentación Reciente en la Planicie Costera del Suroeste de Guatemala y su relación con la Actividad Volcánica. — Publ. Geol. del ICAITI, **5**, 113 – 131, Guatemala, C.A.

ISSIGONIS, M. J. (1973): The geology and geochemistry of the porphyry-type copper and molybdenum mineralization at Cerro Colorado, Panama. — M. Sc. Thesis, Univ. of Toronto.

ISSIGONIS, M. J., KESLER, S. E. & KENTS, P. (1973): Geology of the Cerro Colorado porphyry copper deposits and its relation to other intrusive systems in Panama. — (Abstr.) Mining. Eng., **25**, (12), 57.

— — — (1974): Geology of the Cerro Colorado porphyry copper and its relation to other intrusive systems in Panama. — (Abstr.) Economic Geol., **69**, 150 – 151, New Haven.

JACOBS, C., BÜRGL, H. & CONLEY, D. L. (1963): Backbone of Colombia. — Amer. Ass. Petrol. Geol., Mem., **2**, 62 – 72, Tulsa, Oklahoma.

JAGER C., G. (1977): Geología de las Mineralizaciones de Cromita al Este de la Península de Santa Elena, Provincia Guanacaste, Costa Rica. — Unpubl. Thesis, Escuela Centroamer. Geol., Univ. de Costa Rica, San Pedro, C. R.

JORDAN, T. H. (1975): The Present-Day Motions of the Caribbean Plate. — J. Geophys. Res., **80**, 4433 – 4439, Richmond, Virginia.

JOSEY, W. L. (1970): Metamorphic petrology and structural geology of the Santa Barbara Quadrangle, Guatemala. — M. S. Thesis, Louisiana State Univ., 96 p., Baton Rouge.

KANAMORI, H. & STEWART, G. S. (1978): Seismological Aspects of the Guatemala Earthquake February 4, 1976. — J. Geophys. Res., **83**, 3427 – 3434, Richmond, Virginia.

KARIM, M., CHILANGAR, G. V. & HOYLMAN, H. W. (1966): Northeast Nicaragua oil and gas indications. — World Oil, 84 – 96, March 1966.
KEIGWIN Jr., L. D. (1976): Late Cenozoic planktonic foraminiferal biostratigraphy and paleooceanography of the Panama Basin. — Micropaleontology, **22**, 419 – 442, New York.
— (1978): Pliocene closing of the Isthmus of Panama, based on biostratigraphic evidence from nearby Pacific and Caribbean Sea cores. — Geology, **6**, 630 – 634, Boulder/Col.
KESEL, R. H. (1973): Notes on the lahar landforms of Costa Rica. — Z. Geomorph. N.F., Suppl. **18**, 78 – 91, Berlin-Stuttgart.
KESLER, S. E. (1971): Nature of Ancestral Orogenic Zone in Nuclear Central America. — Amer. Assoc. Petrol. Geol., Bull., **55**, 2116 – 2129, Tulsa, Oklahoma.
— (1972): Western Extension of Fault Zones Bounding the Northern Side of the Caribbean Plate. — 24th Congr. Geol. Intern., Montreal 1972, **3**, 238 – 244, Montreal.
— (1973): Basement Rock Structural Trends in Southern Mexico. — Geol. Soc. Amer., Bull., **84**, 1059 – 1064, Boulder/Col.
— (1976): Tectonic Evolution of Porphyry Copper Mineralization in Panama. — Geol. Soc. Amer., Abstr. with Programs, **8**, no. 6, 954, Boulder/Col.
— (1978): Metallogenesis of the Caribbean region. — J. Geol. Soc. London, **135**, 429 – 441, London.
KESLER, S. E., ASCARRUNZ, K. R. & ANDERSON, T. H. (1972): Exploration Guides for Pb-Zn Mineralization in the Chiantla District, Southern Altos Cuchumatanes, Guatemala. — Transact. VI. Caribb. Geol. Conf., 141 – 145, Caracas.
KESLER, S. E. & ASCARRUNZ, R. (1973): Lead-Zinc Mineralization in Carbonate Rocks, Central Guatemala. — Economic Geol., **68**, 1263 – 1274, Lancaster.
KESLER, S. E., BATESON, J. H., JOSEY, W. L., CRAMER, G. H. & SIMMONS, W. A. (1971): Mesoscopic Structural Homogeneity of Maya Series and Macal Series, Mountain Pine Ridge, British Honduras. — Amer. Assoc. Petrol. Geol., Bull., **55**, 97 – 103, Tulsa, Oklahoma.
KESLER, S. E. & HEATH, S. A. (1970): Structural Trends in the southernmost North American Precambrian, Oaxaca, Mexico. — Geol. Soc. Amer., Bull., **81**, 2471 – 2476, Boulder/Col.
KESLER, S. E., ISSIGONIS, M. J. & VAN LOON, J. C. (1975): An Evaluation of the Use of Halogen and Water Abundances in Efforts to distinguish Mineralized and Barren Intrusive Rocks. J. Geochem. Explor., **4**, 235 – 245, Amsterdam.
KESLER, S. E., JONES, L. M. & WALKER, R. L. (1975): Intrusive Rocks Associated with Porphyry Copper Mineralization in Island Arc Areas. — Economic Geol., **70**, 515 – 526, Lancaster.
KESLER, S. E. & JOSEY, W. L. (1973): Comparison of the Pre-Late Paleozoic Basement Complex in opposite sides of the Cuilco-Chixoy-Polochic Fault Zone in Western Guatemala and Southern Mexico. — Publ. Geol. del ICAITI, **4**, 115 – 122, Guatemala, C.A.
KESLER, S. E., JOSEY, W. L. & COLLINS, E. M. (1970): Basement rocks of Western Nuclear Central America: The Western Chuacús Group, Guatemala. — Geol. Soc. Amer., Bull., **81**, 3307 – 3322, Boulder/Col.
KESLER, S. E., KIENLE, C. F. & BATESON, J. H. (1974): Tectonic Significance of Intrusive Rocks in the Maya Mountains, British Honduras. — Geol. Soc. Amer., Bull., **85**, 549 – 552, Boulder/Col.
KESLER, S. E. & SUTTER, J. F. (1977): Progress Report on radiometric determinations in the Caribbean. — 8th Caribb. Geol. Conf., Abstr., 85 – 86, Amsterdam.
KESLER, S. E., SUTTER, J. F. & ISSIGONIS, M. J. (1976): Tectonic evolution of porphyry copper mineralization in Panama (Abstr.). — Geol. Soc. Amer., Abstr. and Programs, **8**, no 6, 954, Boulder/Col.
KESLER, S. E., SUTTER, J. F., ISSIGONIS, M. J., JONES, L. M., WALKER, R. L. (1977): Evolution of Porphyry Copper Mineralization in Ocean Island Arc: Panama. — Economic Geol. **72**, 1142 – 1153, Lancaster.
KHUDOLEY, K. M. & MEYERHOFF, A. A. (1971): Paleogeography and Geological History of Greater Antilles. — Geol. Soc. Amer. Mem., **129**, 199 p., Boulder/Col.
KILBRIDGE, P., N. (1979): Magmatic evolution of Izalco Volcano and relation to the Santa Ana Complex, El Salvador. — M. A. Thesis, 72 p. Rutgers, State Univ. New Jersey, New Brunswick, N.J.
KILBURG, J. A. (1978): Stratigraphy, Petrology and Origin of some Tertiary volcanic rocks from the Highlands of western Guatemala, Central America. — EOS (Amer. Geophys. Union Transact.) **59**, 4, 400, Washington D.C.
KLING, S. A. (1960): Permian fusulinids from Guatemala. — J. Paleont., **34**, 637 – 655, Menasha, Wisc.

KOCH, A. J. & MCLEAN, H. (1975): Pleistocene Tephra and Ash-Flow Deposits in the Volcanic Highlands of Guatemala. — Geol. Soc. Amer., Bull., **86**, 529 – 541, Boulder/Col.

KOJAN, E. (1958): Geology of the Aluminous Laterite Deposits of the General Valley, San Isidro de el General, Costa Rica. — Alcoa de Costa Rica.

KRUCKOW, T. (1974): Landhebung im Valle Central und Wachstum der Küstenebenen in Costa Rica (Mittelamerika). — Jahrb. Wittheit zu Bremen, **18**, 247 – 263, Bremen.

— (1976 a): Arbeitsziele und erste Arbeitsergebnisse der geologischen Untersuchungen in Costa Rica (Mittelamerika). — Münster Forsch. Geol., Paläont., **38/39**, 219 – 227, Münster (Westf.).

— (1976 b): The evolution and the dependences of the Peninsula of Puntarenas. — Publ. Geol. del ICAITI, **5**, 177 – 179, Guatemala, C.A.

KRUCKOW, T. & GÓMEZ, L. D. (1974): Notes on the palaeocology of the fossil Algae of Costa Rica. I. — Brenesia, **4**, 23 – 29, San José.

KRUSHENSKY, R. D. (1972): Geology of the Istarú Quadrangle, Costa Rica. — Geol. Surv., Bull., **1358**, 46 p., Washington, D.C.

— (1973): Geología Istarú. — Direcc. de Geología, Minas y Petróleo, Inform. Técn. y Notas Geol., **48**, 83 p., San José.

KRUSHENSKY, R. D. & ESCALANTE, G. (1967): Activity of Irazú and Poás volcanoes, Costa Rica, November 1964 – July 1965. — Bull. volcanol., **31**, 75 – 84, Napoli.

KRUSHENSKY, R. D., MALAVASSI, E. & CASTILLO, R. (1976): Geology of Central Costa Rica and its implications in the Geologic History of the Region. — J. Res. U.S. Geol. Surv., **4**, 127 –134, Washington, D.C.

KRUSHENSKY, R. D., SCHMOLL, H. R. & DOBROVOLNY, E. (1974): Geologic constrains on planning for the reconstruction or relocation of Managua, Nicaragua. — Bol. Inform. Asoc. Venezolana de Geología, Minería y Petróleo, **17**, 217 – 224, Caracas.

— — — (1976): The 1972 Managua earthquake: Geologic constrains on planning for the reconstruction or relocation of Managua, Nicaragua. — Transact. VII. Conf. Géol. des Caraïbes, 609 – 614, Saint-François (Guadeloupe).

KUANG, S. J. (1971): Estudio geológico del Pacífico de Nicaragua. — Catastro e Inventario de Recursos Naturales, División de Geología, Inform. Geol., **3**, 1 – 101, Managua, Nicaragua, C.A.

KUENZI, W. D., HORST, O. H. & MCGEHEE, R. V. (1969): Effect of volcanic activity on fluvial-deltaic sedimentation in a modern arc-trench gap, southwestern Guatemala. — Geol. Soc. Amer., Bull., **90**, 827 – 838, Boulder/Col.

KUENZI, W. D., MCGEHEE, R. V. & HORST, O. H. (1975): Effect of volcanic activity on fluvial-deltaic sedimentation in a modern arc-trench gap, southwestern Guatemala. — Geol. Soc. Amer., Abstracts and Programs (USA), **7**, 1154 – 1155, Boulder/Col.

KUNO, H. (1966): Lateral Variation of Basalt Magma Across Continental Margins and Island Arcs. — Canad. Geol. Surv. Pap. No. **66 – 15**, 317 – 336, Ottawa.

KUPFER, D. H. & GODOY, J. (1967): Strike-slip faulting in Guatemala (abstr.). — Amer. Geophys. Union Transact., **48**, 215, Washington, D.C.

KUYPERS, E. & DENYER Ch., P. (1979): Volcanic exhalative manganese deposits of the Nicoya Ophiolite Complex, Costa Rica. — Economic Geol., **74**, 672 – 678, Lancaster.

LADD, J. W. (1976 a): Relative motions of South America with respect to North America and Caribbean tectonics. — Geol. Soc. Amer., Bull., **87**, 969 – 976, Boulder/Col.

— (1967 b): Relative motion between North and South America and Caribbean Tectonics. — Transact. VII. Conf. Géol. des Caraïbes, 63 – 68, Saint-François (Guadeloupe).

LADD, J. W., IBRAHIM, A. K., MCMILLEN, K. J., LATHAM, G. V., VON HUENE, R. E., WATKINS, J. S., MOORE, J. C. & WORZEL, J. L. (1978): Tectonics of the Middle America Trench offshore Guatemala. — Univ. of Texas Marine Sci. Inst. Contrib., **274**, 1 – 9, Galveston Geophys. Lab.

LARSON, R. L. & CHASE, C. G. (1970): Relative velocities of the Pacific, North America and Cocos plates in the middle America Region. — Earth and Planetary Sci. Letters, **7**, 425 – 428, Amsterdam.

LAWRENCE, D. P. (1975): Petrology and structural geology of the Sanarate-El Progreso area, Guatemala. — Ph. D. Thesis, State Univ. of New York, Binghampton.

— (1976): Tectonic implications of the Geochemistry and Petrology of the El Tambor Formation: Probable Oceanic Crust in Central Guatemala. – Geol. Soc. Amer., Abstr. Programs (USA), **8**, No. 6, 973 – 974, Boulder/Col.

LEVY, E. (1970): La metalogénesis en América Central. — Publ. Geol. del ICAITI, **3**, 17 – 57, Guatemala, C.A.

LJUNGGREN, P. (1958 a): Mineralogical examination of black beach sands from "Lago de Izabal", Eastern Guatemala. — Kungl. Fysiogr. Sällskapets i Lund Förhandl., **28**, 133 – 139, Lund.

— (1958 b): The black beach sands of Iztapa on the Pacific coast of Guatemala. — Kungl. Fysiogr. Sällskapets i Lund Förhandl., **28**, 109 – 119, Lund.

LLOYD, J. J. (1963): Tectonic History of the South Central-American Orogen. — Amer. Assoc. Petrol. Geol., Mem., **2**, 88 – 100, Tulsa, Oklahoma.

LLOYD, J. J. & DENGO, G. (1960): Posibilidades Petrolíferas de la Cuenca del Petén, Guatemala. — Bol. Facultad de Ingenería, Univ. de Guatemala, **2**, **1**, 3 – 11, Guatemala, C.A.

LÖSCHNER, R. (1978): Die künstlerische Darstellung Lateinamerikas im 19. Jahrhundert unter dem Einfluß Alexander von Humboldts. — In: Deutsche Künstler in Lateinamerika, Berlin (Ibero-Amerikanisches Inst. Preußischer Kulturbesitz).

LÖTSCHERT, W. & MÄDLER, K. (1975): Die pliopleistozäne Flora aus dem Sisimico-Tal (El Salvador). — Geol. Jb., **B 13**, 97 – 191, Hannover.

LOMNITZ, C. & SCHULZ, R. (1966): The San Salvador Earthquake of May, 3, 1965. — Bull. Seism. Soc. Amer., **56**, 561 – 575, Berkeley.

LONSDALE, P. & KLITGORD, K. D. (1978): Structure and tectonic history of the eastern Panama Basin. — Geol. Soc. Amer., Bull., **89**, 981 – 999, Boulder/Col.

LÓPEZ RAMOS, E. (1969): Geología del Sureste de México y Norte de Guatemala. — Publ. Geol. del ICAITI, **2**, 57 – 67, Guaemala, C.A.

— (1973): Estudio Geológico de la Península de Yucatán. — Bol. Asoc. Mex. de Geol. Petrol., **23**, 23 – 76, México, D.F.

— (1974): Geología General y de México, 3rd Ed., 509 p. — México, D.F.

— (1975): Geological Summary of the Yucatan Peninsula. — In: NAIRN, A. E. M. & STEHLI, F. G. (eds.): The Ocean Basins and Margins, **3**, 257 – 282, New York and London (Plenum Press).

LOWRIE, A. (1978): Buried trench south of the Gulf of Panama. — Geology, **6**, 434 – 436, Boulder/Col.

MACDONALD, H. C. (1969): Geologic evaluation of radar imagery from Darién province, Panama. — Modern Geology, **1**, 1 – 63, New York.

MACDONALD, W. D. (1972): Continental Crust, Crustal Evolution, and the Caribbean. — Geol. Soc. Amer., Mem., **132**, 351 – 363, Boulder/Col.

— (1976 a): Cretaceous-Tertiary evolution of the Caribbean. — Transact. VII. Conf. Géol. des Caraïbes, 69 – 78, Saint François (Guadeloupe).

— (1976 b): The importance of Central America to the tectonic evolution of the Caribbean. — Publ. Geol. del ICAITI, **5**, 23 – 30, Guatemala, C.A.

MADRIGAL G., R. (1977): Evidencias Geomórficas de Movimientos Tectónicos Recientes en el Valle de El General. — Rev. Cienc. Técn., **1**, (1), 97 – 108, Univ. de Costa Rica, San José.

MALAVASSI V., E. (1961): Some Costa Rican larger Foraminiferal localities. — J. Paleont., **35**, 498 – 501, Menasha, Wisc.

— (1977): Explicación de un perfil geológico a través de Costa Rica. — V. Reunión de Geólogos de América Central, 7 p., Managua, Nicaragua.

MALAVASSI V., E. & CHAVES, C. R. (1970): Estudio Geológico Regional de la Zona Atlántico Norte de Costa Rica. — Direcc. de Geol., Minas y Petróleo, Inform. Técn. y Notas Geol., **35**, 1 – 15, Ciudad Universitaria, Costa Rica.

MALAVASSI V., E. & MADRIGAL G., R. (1970): Reconocimiento Geológico de la Zona Norte de Costa Rica. — Direcc. de Geol., Minas y Petróleo, Inform. Técn. y Notas Geol., **38**, 1 – 10, Ciudad Universitaria, Costa Rica.

MALFAIT, B. T. & DINKELMAN, M. G. (1972): Circum-Caribben Tectonic and Igneous Activity and the Evolution of the Caribbean Plate. — Geol. Soc. Amer., Bull., **83**, 251 – 272, Boulder/Col.

MARTIN-KAYE, P. H. A. & WILLIAMS, A. K. (1972): Radargeologic Map of Eastern Nicaragua. — Novena Conf. Interguayanas, Mayo 7 – 14, Ciudad Guayana, Venezuela. — Bol. de Geol., Publ. Espec. No. **6**, 600 – 605.

MARTÍNEZ, M. & VIRAMONTE, J. G. (1971): Geología de la Cordillera de los Marabios. — Catastro e Inventario de Recursos Naturales, Abril 15, Managua, Nicaragua, C.A.

— (1973): Estudio geológico de la Cordillera de los Marabios, Nicaragua. — Publ. Geol. del ICAITI, **4**, 139 – 148, Guatemala, C.A.

MATUMOTO, T. & LATHAM, G. (1973): Aftershocks and Intensity of the Managua Earthquake of 23 December 1972. — Science, **181**, 545 – 547, Washington, D.C.

MATUMOTO, T., LATHAM, G. OHTAKE, M. & UMAÑA, J. (1976): Seismic Studies in Northern Costa Rica (Abstr.). — Transact. Amer. Geophys. Union, **57**, 290, Washington, D.C.

MATUMOTO, T., OHTAKE, M., LATHAM, G. & UMAÑA, J. (1977): Crustal Structure in Southern Central America. — Bull. Seismol. Soc. Amer., **67**, 121 – 135, Berkeley.

MATUMOTO, T. & UMAÑA, J. E. (1975): Informe sobre la Erupción del Volcán Arenal occurida el 17 de Junio de 1975. — Inst. Costarricense de Electricidad, Direcc. de Igenería de Energía, Dept. de Geol., San José.

MCBIRNEY, A. R. (1955): The Origin of the Nejapa Pits near Managua, Nicaragua. — Bull. volcanol., Sér. **II**, **17**, 145 – 154, Napoli.

— (1956): The Nicaraguan Volcano Masaya and its Caldera. — Transact. Amer. Geophys. Union, **37**, 83 – 96, Washington, D.C.

— (1958): Active Volcanoes of Nicaragua and Costa Rica. — In: Catalogue of the Active Volcanoes of the World including Solfatara Fields. Part **VI**, Central America, 107 – 146, Napoli.

— (1963): Geology of a part of the central Guatemalan Cordillera. — Univ. Calif. Publ. geol. Sci., **38**,**4**, 177 – 242, Berkeley and Los Angeles.

— (1964): Notas sobre los Centros Volcánicos Cuaternarios al este de la Depresión Nicaraguense. — Bol. Serv. Geol. Nac., **8**, 91 – 97, Managua.

— (1968): Guatemala: Preliminary Zirkon Ages from Central Cordillera. — Science, **162**, 121 – 122, Washington, D.C.

— (1969 a): Compositional variations in cenozoic calc-alkaline suites of Central America. — Intern. Upper Mantle Project Sci. Rep. **16**, 185 – 189, Oregon.

— (1969 b): Andesitic and Rhyolitic Volcanism of Orogenic Belts. — Geophys. Monographs, **13**, 501 – 507, Washington, D.C.

— (1971): Petrology of the Central American Volcanic Province. — Upper Mantle Project, United States Program, Final Rep., 208 – 209, Juli 1971.

MCBIRNEY, A. R. & BASS, M. N. (1969 a): Geology of Bay Islands, Gulf of Honduras. — Amer. Assoc. Petrol. Geol., Mem., **11**, 229 – 243, Tulsa, Oklahoma.

— — (1969 b): Structural relations of pre-Mesozoic rocks of Northern Central America. — Amer. Assoc. Petrol. Geol., Mem., **11**, 269 – 280, Tulsa, Oklahoma.

MCBIRNEY, A. R. & WEILL, D. F. (1966): Rhyolite Magmas of Central America. — Bull. volcanol., **29**, 435 – 448, Napoli.

MCBIRNEY, A. R. & WILLIAMS, H. (1965): Volcanic History of Nicaragua. — Univ. Calif. Publ. geol. Sci., **55**, 1 – 65, Berkeley and Los Angeles.

MCLEAN, H. (1970): Stratigraphy, Mineralogy, and Distribution of the Supango-Group Pumice Deposits in the volcanic highlands of Guatemala. — Ph. D. Diss., Univ. of Washington.

MELSON, W. G. & SÁENZ, R. (1968): The 1968 eruption of Volcan Arenal: Preliminary summary of field and laboratory studies. — Smithsonian Center of short lived Phenomena, Rep. 7/1968, 35 p., Washington, D.C.

— — (1974): Volume, Energy and Cyclicity of Eruptions of Arenal Volcano, Costa Rica. — Bull. volcanol., **37**, 416 – 437, Napoli.

MÉRIDA, J. (1973): Los Campos Geotérmicos de Panamá. — Direcc. General de Recursos Minerales, Secc. Geotérmica, 10 p., Panamá.

MERINO Y CORONADO, J. (1968): Resumen des las observaciones hechas sobre la reciente erupción del volcán Arenal. — Inst. Geogr. Nac., Informe Semestral, Julio a Diciembre 1968, 13 – 17, San José, C.R.

MERTZMAN, S. A. (1976): A $^{87}Sr/^{86}Sr$ Reconnaissance of the Lake Yojoa Volcanic Field, Honduras. — Publ. Geol. del ICAITI, **5**, 99 – 104, Guatemala, C.A.

MEYER, J. (1964): Stratigraphie der Bimskiese und -aschen des Coatepeque-Vulkans im westlichen El Salvador (Mittelamerika). — N. Jb. Geol. Paläont, Abh., **119**, 215 – 246, Stuttgart.

— (1967 a): Estudios petrográficos de ignimbritas y lavas plio-pleistocénicas de la zona de Ahuachapán, El Salvador, América Central. — Tulane Stud. in Geol., **5**, 167 – 175, New Orleans.

— (1967 b): Geology of the Ahuachapan Area, Western El Salvador, Central America. — Tulane Stud. in Geol., **5**, 195 – 215, New Orleans.

MEYER-ABICH, H. (1956): Los Volcanes Activos de Guatemala y El Salvador (América Central). — An. Serv. Geol. Nac. de El Salvador, Bol., **3**, 1 – 102, San Salvador.

— (1958): Active Volcanoes of Guatemala and El Salvador. — In: Catalogue of the Active Volcanoes of the World including Solfatara Fields. Part VI, Central America, 37 – 105, Napoli.

MEYERHOFF, A. A. (1966): Bartlett fault system: Age and offset. — Third Caribb. Conf. Transact., 1 – 9, Kingston, Jamaica.

MEYERHOFF, A. A. & PINET, P. R. (1973): Diapirlike Features Offshore Honduras: Implications Regarding Tectonic Evolution of Cayman Trough and Central America: Discussion and Reply. — Geol. Soc. Amer., Bull., **84**, 2147 – 2152, Boulder/Col.

MILLS, R. A., HUGH, K. E., FERAY, D. F. & SWOLFES, H. C. (1967): Mesozoic Stratigraphy of Honduras. — Amer. Assoc. Petrol. Geol., Bull., **51**, 1711 – 1786, Tulsa, Oklahoma.

— — — — (1969): Estratigrafía de la Era Mesozoica en Honduras. 83 p. — Honduras (Banco Central).

MILLS, R. A. & HUGH, K. E. (1974): Reconnaissance Geologic Map of Mosquitia Region, Honduras and Nicaraguan Caribbean Coast. — Amer. Assoc. Petrol. Geol., Bull., **58**, 189 – 207, Tulsa, Oklahoma.

MINAKAMI, T. & UTIBORI, S. (1969): The 1968 eruption of Volcano Arenal, Costa Rica. — Earthquake Rec. Inst. Bull., **47**, 769.

MIYAMURA, S. (1975): Recent Crustal Movements in Costa Rica disclosed by Revelling Surveys. — Tectonophysics, **29**, 191 – 198, Amsterdam.

MIYASHIRO, A. (1972): Metamorphism and Related Magmatism in Plate Tectonics. — Amer. J. Sci., **272**, 629 – 656, New Haven, Conn.

MOLNAR, P. & SYKES, L. R. (1969): Tectonics of the Caribbean and Middle America Regions from Focal Mechanism and Seismicity. — Geol. Soc. Amer., Bull., **80**, 1639 – 1684, Boulder/Col.

MONGES CALDERA, J. (1961): Anomalías de la Gravedad al aire libre de Bouguer. — Inst. Geogr. de Costa Rica, Informe Semestral Enero a Junio 1961, San José.

— (1962): Dos Mapas de las Isoanomalías Bouguer de la Gravedad para Centroamérica. — Inst. Geogr. de Costa Rica, Informe Semestral Julio a Diciembre 1961, San José.

MONTERO P., W. (1974): Estratigrafía del Cenozóico del Area Turrucares, Provincia de Alajuela, Costa Rica. — Unpubl. Bachiller Thesis, Univ. de Costa Rica, San Pedro, C.R.

MONTIGNY, R., JAVOY, M. & ALLÈGRE, J. J. (1969): Le problème des andésites. Étude du volcanisme quaternaire du Costa Rica (Amérique centrale) à l'aide des traceurs couples $^{87}Sr/^{86}Sr$ et $^{18}O/^{16}O$. — Bull. Soc. géol. Fr., **(7) XI**, 794 – 799, Paris.

MOODY, J. J. (1975): Central America — Regional Review. — In: FAIRBRIDGE, R. W., The Encyclopedia of World Regional Geology, Part I: Western Hemisphere, 228 – 235, Stroudsburg, Pennsylvania (Dowden, Hutchinson & Ross).

MOORE, G. T. & FAHLQUIST, D. A. (1976): Seismic profile tying Caribbean Sites 153, 151, and 152. — Geol. Soc. Amer., Bull., **87**, 1609 – 1614, Boulder/Col.

MOOSER, F., MEYER-ABICH, H. & MCBIRNEY, A. R. (1958): Catalogue of the Active Volcanoes of the World including Solfatara Fields. Part VI, Central America. 146 p. — Napoli.

MORRIS, A. E. L. (1979): Offshore petroleum potential of Central America. — 4th Latin American Geol. Congress, Abstr., Trinidad & Tobago.

MUEHLBERGER, W. R. (1976): The Honduras Depression. — Publ. Geol. del ICAITI, **5**, 43 – 51, Guatemala, C.A.

MUEHLBERGER, W. R. & RITCHIE, A. W. (1975): Caribbean-American plate boundary in Guatemala and southern Mexico as seen on Skylab IV orbital photography. — Geology, **3**, 232 – 235, Boulder/Col.

MÜLLER-KAHLE, E. (1962): Die Lagerstätte der Grube Montecristo, El Salvador, und ihr geologischer Rahmen. — N. Jb. Geol. Paläont., Abh., **115**, 289 – 334, Stuttgart.

MURATA, K. J. (1964): Notas sobre la Actividad Actual del Volcán Irazú. — Inst. Geogr. de Costa Rica, Informe Semestral Julio a Diciembre 1963, 93 – 104, San José, C.R.

MURATA, K. J., DÓNDOLI, C. & SÁENZ, R. (1966): The 1963 – 65 Eruption of Irazú Volcano, Costa Rica (The period of March 1963 to October 1964). — Bull. volcanol., **29**, 765 – 796, Napoli.

MURRAY, G. E. (1966): Salt Structures of Gulf of Mexico Basin — A Review. — Amer. Assoc. Petrol. Geol., Bull., **50**, 439 – 478, Tulsa, Oklahoma.

— (1976): New oil province, Guatemala and Southeastern Mexico. — 25th Intern. Geol. Congr., **1**, 249 – 250, Sidney.

NAGLE, F., ROSENFELD, J. & STIPP, J. J. (1977): Guatemala, where plates collide. A Reconnaissance Guide to Guatemalan Geology. — 71 p., Miami Geol. Soc.

NAIRN, A. E. M. & STEHLI, F. G. (1975): The Ocean Basins and Margins, **3**, The Gulf of Mexico and the Caribbean. — New York and London (Plenum Press).

NEWCOMB, W. E. (1975): Geology, structure, and metamorphism of the Chuacúas group, Río Hondo Quadrangle and vicinity, Guatemala. — Ph. D. Thesis, State Univ. of New York, Binghamton.

— (1977): Mylonitic and cataclastic rocks from the Motagua Fault Zone, Guatemala. — 8th Caribb. Geol. Conf., Abstr., 141 – 142, Amsterdam.

— (1978): Retrograde cataclastic gneiss north of the Motagua fault zone, East-Central Guatemala. — Geol. en Mijnbouw, **57**, (2), 271 – 276, Amsterdam.

NICOLAUS, H.-J. (1971): Schlußbericht der Deutschen Geologischen Mission in Guatemala (DGMG) 3. 1. 1967 – 31. 3. 1970. — Bundesanstalt f. Bodenforschung, Hannover.

NIELSON, D. R. & STOIBER, R. E. (1973): Relationship of Potassium Content in Andesitic Lavas and Depth to the Seismic Zone. — J. Geophys. Res., **78**, 6887 – 6892, Richmond, Virginia.

PAGE, R. A. (ed.) (1976): Interim report on the Guatemalan Earthquake of 4 February, 1976, and the activity of the U.S. Geological Survey Earthquake Investigation Team. — 26 p., U.S. Dept. of Inter., Geol. Surv., Project Rep. Guatemala Investigations (IR Cu-1), Washington, D.C.

PAGNACCO, P. F. & RADELLI, L. (1962): Note on the geology of the Isles of Providencia and Santa Catalina (Caribbean Sea, Colombia). — Geología Colomb., **3**, 125 – 132, Bogotá.

PAULSEN, S. (1969): Die Gipsvorkommen im Tal des Río Chixoy (Quiché UND Alta Verapaz) und ihre wirtschaftliche Bedeutung. — Bundesanstalt f. Bodenforschung, Deutsche Geol. Mission in Guatemala, Ber. Nr. 3, Guatemala.

PAZ RIVERA, N. (1962): Reconocimiento geológico en la cuenca hidrográfica de los Ríos Coco y Bocay. — Bol. Serv. Geol. Nac., Nicaragua, **6**, 5 – 22, Managua.

PERFIT, M. R. (1977): Petrology and geochemistry of mafic rocks from the Cayman Trench: Evidence for spreading. — Geology, **5**, 105 – 110, Boulder/Col.

PERFIT, M. R. & HEEZEN, B. C. (1978): The geology and evolution of the Cayman Trench. — Geol. Soc. Amer., Bull., **89**, 1155 – 1174, Boulder/Col.

PICHLER, H., STIBANE, F. R. & WEYL, R. (1974): Basischer Magmatismus und Krustenbau im südlichen Mittelamerika, Kolumbien und Ecuador. — N. Jb. Geol. Paläont., Mh., 1974, 102 – 126, Stuttgart.

PICHLER, H. & WEYL, R. (1973): Petrochemical Aspects of Central American Magmatism. — Geol. Rdsch., **62**, 357 – 396, Stuttgart.

— — (1975): Magmatism and Crustal Evolution in Costa Rica (Central America). — Geol. Rdsch., **64**, 457 – 475, Stuttgart.

— — (1976): Quaternary Alkaline Volcanic Rocks in Eastern Mexico and Central America. — Münster. Forsch. Geol. Paläont., **38/39**, 159 – 178, Münster (Westf.).

PIÑEIRO, R. F. & ROMERO, M. S. (1962): Reconocimiento geológico minero de la porción Noroeste de la República de Nicaragua. — Bol. Serv. Geol. Nac. de Nicaragua, **6**, 50 – 91, Managua.

PINET, P. R. (1971): Structural Configuration of the Northwestern Caribbean Plate Boundary. — Geol. Soc. Amer., Bull., **82**, 2027 – 2032, Boulder/Col.

— (1972): Diapirlike Features Offshore Honduras: Implications Regarding Tectonic Evolution of Cayman-Trough and Central America. — Geol. Soc. Amer., Bull., **83**, 1911 – 1922, Boulder/Col.

— (1975): Structural Evolution of the Honduras Continental Margin and the Sea Floor South of the Western Cayman Trough. — Geol. Soc. Amer., Bull., **86**, 830 – 838, Boulder/Col.

PLAFKER, G. (1973 a): Field reconnaissance of the effects of the earthquake of April 13, 1973, near Laguna de Arenal, Costa Rica. — US Geol. Surv. Open-file rep. (200) R290 1881, 17 p., Washington, D.C.

— (1973 b): Field Reconnaissance of the Effects of the Earthquake of April 13, 1973, near Laguna de Arenal, Costa Rica. — Bull. Seismol. Soc. Amer., **63**, 1847 – 1856, Berkeley.

— (1976): Tectonic Aspects of the Guatemala Earthquake of 4 February 12976. — Science, **193**, 1201 – 1208, Lancaster.

PLAFKER, G., SHARP, R. V., BUCKNAM, R. C., BONIS, S. B. & BONILLA, M. G. (1976): Tectonic implications of surface faulting associated with the 4 February 1976 Guatemala Earthquake. — Geol. Soc. Amer., Abstr. Programs (USA), **8**, No. 6, 1051 – 1052, Boulder/Col.

PUSHKAR, P. (1967): Strontium Isotope ratios in volcanic rocks of three Island Arc Areas. — Ann. Progr. Rep. No. COO-689-76 to Res. Div. USAEC, Appendix A-V, Tuscon, Arizona.

— (1968): Strontium Isotope Rations in Volcanic Rocks of Three Island Arc Areas. — J. Geophys. Res., **73**, 2701 – 2714, Richmond, Virginia.

PUSHKAR, P., MCBIRNEY, A. R. (1968): The Isotopic Composition of Strontium in Central American Ignimbrites. — ANN. Progr. Rep. No. COO-689-100 to Res. Div. USAEC, Appendix A-II, Tuscon, Arizona.

PUSHKAR, P., MCBIRNEY, A. R. & KUDO, A. M. (1972): The Isotopic Composition of Strontium in Central American Ignimbrites. — Bull. volcanol., **35**, 265 – 294, Napoli.

PUTNAM, P. C. (1926): The existence of a once homogenous magmamass underlying Central America. — J. Geol. **34**, 807 – 823, Chicago.

QUIRÓS, J. L. (1966): Engineering and economic aspects of the aluminium industry and the potential development of aluminium production in Panamá. — Ph. D. Thesis, 190 p., Univ. of Illinois.

REA, D. K. & MALFAIT, B. T. (1974): Geologic evolution of the northern Nazca plate. — Geology, **2**, 317 – 320, Boulder/Col.

RECCHI, G. (1975): Paleogeografía. — Atlas Nacional de Panamá. Inst. Geogr. Nac. "Tommy Guardia", Panamá.

— (1976): Notas sobre la Geología de Panamá. — Direcc. General de Recursos Minerales, 7 p., Panamá.

REYES A., MUNGUIA, L., LOMNITZ, C. et al. (1977): The February 1976 Guatemala earthquake: an aftershock survey. — Eos, Transact Amer. Geophys. Union, **58,5**, 308, Washington, D.C.

REYNOLDS III, J. H. (1977): Tertiary Volcanic Stratigraphy of Northern Central America. — M. A. Thesis, 89 p., Dartmouth College, Hanover, N.H.

RICHARDS, H. G. (1963): Stratigraphy of the earliest Mesozoic sediments in southeastern Mexico and western Guatemala. — Amer. Assoc. Petrol. Geol., Bull., **47**, 1861 – 1870, Tulsa, Oklahoma.

RICHARDS, H. G. & VINSON, G. L. (1963): Upper Cretaceous and Tertiary Stratigraphy of Guatemala. — Amer. Assoc. Petrol. Geol., Bull., **47**, 702 – 705, Tulsa, Oklahoma.

RIECK, K. (1975): First geological and geochemical investigation of an antimony occurrence near Sta. Cruz Cuchilla in Honduras/Central America. — Erzmetall, **28**, 338 – 39.

RITCHIE, A. W. (1976): Jocotán Fault — Possible Western Extension. — Publ. Geol. del ICAITI, **5**, 52 – 55, Guatemala, C.A.

RITTMANN, A. (1973): Stable Mineral Assemblages of Igneous Rocks. — XVI + 262 p., Berlin-Heidelberg-New York (Springer).

RIVIER, F. (1973): Contribución estratigráfica sobre la geología de la Cuenca de Limón, Zona de Turrialba, Costa Rica. — Publ. Geol. del ICAITI, **4**, 149 – 160, Guatemala, C.A.

ROBERTS, R. J. & IRVING, E. M. (1957): Mineral Deposits of Central America. — Geol. Surv. Bull., **1034**, 205 p., Washington, D.C.

ROBIN, C. & DEMANT, S. (1974): Les trappes de l'Est mexicain: coexistence de séries alcalines et tholeiitiques, caractères différentiels entre le volcanisme des plaines et celui des plateaux. — C.R. Acad. Sci. Paris, **278**, Sér. D, 2413 – 2416, Paris.

ROBIN, C. & TOURNON, J. (1978): Spatial relations of andesitic and alkaline provinces in Mexico and Central America. — Canadian J. Earth Sci., **15**, 1633 – 1641.

RODE, K. (1964): Geologische Erkundung im Lempa-Becken, El Salvador. — Publ. Serv. géol. du Luxembourg, **14**, 311 – 330, Luxembourg.

— (1965): Feldgeologische Notizen über die Becken von Baja Verapaz, Guatemala. — Geol. Rdsch., **54**, 650 – 667, Stuttgart.

— (1975): Das intramontane Lempa-Becken (El Salvador, Mittel-Amerika). — Geol. Jb., **B 13**, 3 – 85, Hannover.

ROMAGNOLI, I. P., CUELLAR, G., JIMÉNEZ, M. & GEZZI, G. 61976): Aspectos Hidrogeológicos del Campo Goetérmico de Ahuachapán, El Salvador. — Proc. 2nd United Nations Sympos. in the Development of Geothermal Resources, **1**, 563 – 574, San Francisco 1975, Washington (U.S. Gov. Print. Off.).

ROPER, P. (1976): Lithologic subdivisions of the Chuacús Group and their structural significance in the Southwestern end of the Sierra de la Minas Range, Guatemala. — Transact. VII. Conf. Géol. des Caraïbes, 589 – 594, Saint-François (Guadeloupe).
— (1978): Stratigraphy of the Chuacús Group in the south side of the Sierra de la Minas Range, Guatemala. — Geol. en Mijnbouw, **57**, 309 – 313, Amsterdam.
ROSE Jr., W. I. (1967): Notes on Fumaroles and recent Activity of Volcán Pacaya. — Excursion Guide Book for Guatemala, Geol., Bull., **4**, 31 – 33, Inst. Geogr. Nac., Guatemala, C.A.
— (1970): The geology of the Santiaguito volcanic dome, Guatemala. — Ph. D. Diss., Dartmouth College, Hanover, N.H.
— (1972): Santiaguito Volcanic Dome, Guatemala. — Geol. Soc. Amer. Bull., **83**, 1413 – 1434, Boulder/Col.
— (1973 a): Notes on the 1902 Eruption of Santa Maria Volcano, Guatemala. — Bull. volcanol., **36**, 29 – 45, Napoli.
— (1973 b): Pattern and Mechanism of Volcanic Activity at the Santiaguito Volcanic Dome, Guatemala. — Bull. volcanol., **37**, 73 – 94, Napoli.
— (1974): Nuée Ardente from Santiaguito Volcano, April 1973. — Bull. volcanol., **37**, 365 – 371, Napoli.
ROSE Jr., W. I., ANDERSON, A. T., BONIS, S. & WOODRUFF, L. G. (1978): The October 1974 basaltic tephra, Fuego volcano, Guatemala: description and history of the magma body. — J. Volcanol. Geotherm. Res., **4**, 3 – 53, Amsterdam.
ROSE Jr., W. I., BONIS, S., STOIBER, R. E., KELLER, M. & BICKFORD, T. (1974): Studies of Volcanic Ash from two Recent Central American Eruptions. — Bull. volcanol., **37**, 338 – 364, Napoli.
ROSE Jr., W. I., GRANT, N. K., HAHN, G. A., LANGE, I. M., POWELL, J. L., EASTER, J. & DEGRAFF, J. M. (1977): The Evolution of Santa Maria Volcano, Guatemala. — J. Geol. **85**, 63 – 87, Chicago.
ROSE Jr., W. I., PEARSON, T. & BONIS, S. (1976/77): Nuée Ardente Eruption from the Foot of a Dacite Lava Flow, Santiaguito, Guatemala. — Bull. volcanol., **40**, 23 – 38, Napoli.
ROSE Jr., W. I. & STOIBER, R. E. (1969): The 1966 eruption of Izalco Volcano. El Salvador. — J. Geophys. Res., **74**, 3119 – 3130, Richmond, Virginia.
— — (1976): Explosive Volcanism in Central America with Special Reference to the 1974 Eruption of Fuego. — Eos (Amer. Geophys. Union Transact.), **57**, 345 – 347, Washington, D.C.
ROSE Jr., W. I., STOIBER, R. E. & BONIS, S. B. (1970): Volcanic Activity at Santiaguito Volcano, Guatemala, June 1968 – August 1969. — Bull. volcanol., **34**, 295 – 307, Napoli.
ROSS, C. A. (1962): Permian Foraminifera from British Honduras. — Palaeontology, **5**, 297 – 306, London and Oxford.
ROSS, D. A. & SHOR, G. G. (1965): Reflection profiles across the Middle America Trench. — J. Geophys. Res., **70**, 5551 – 5571, Richmond, Virginia.
ROWETT, C. L. & WALPER, J. L. (1972): Permian Corals from near Huehuetenango, Guatemala. — Pacific Geol., **5**, 71 – 74, Tokyo.
SÁENZ R., R. (1971 a): El volcán Arenal. — Folleto Guia de la Excursión a la Región de Guanacaste, Costa Rica, 35 – 38, San José, C.R.
— (1971 b): Aparatos volcánicos y fuentes termales de Costa Rica. — Direcc. de Geol., Minas y Petróleo, Inform. Técn. y Notas Geol., **41**, 1 – 13, San José, C.R.
SAITO, T. (1976): Geologic significance of coiling direction in the planktonic foraminifera *Pulleniatina*. — Geology, **4**, 305 – 309, Boulder/Col.
SANDOVAL, M. F. (1966): Geología de una parte de la región noroeste del Valle Central (Hoja de Alajuela). — Direcc. de Geol., Minas y Petróleo, Inform. Técn. y Notas Geol., **21**, Ciudad Universitaria, San Pedro de Montes de Oca, C.R.
— (1969): Posibilidades Mineras de Costa Rica. — Direcc. de Geol., Minas y Petróleo, Inform. Técn. y Notas Geol., **32**, 1 – 11, Ciudad Universitaria, C.R.
— (1971): Geología de una parte de la Región Noreste del Valle Central (Hoja Grecia). — Direcc. de Geol., Minas y Petróleo, Inform. Técn. y Notas Geol., **44**, 1 – 14, San José, C.R.
SAPPER, C. (1897): Über die räumliche Anordnung der mittelamerikanischen Vulcane. — Z. Deutsch. Geol. Ges., **49**, 672 – 682, Berlin.
— (1899): Über Gebirgsbau und Boden des nördlichen Mittelamerika. — Petermanns Mitt., Erg.-H. **127**, Gotha.

— (1913): Die mittelamerikanischen Vulkane. — Petermanns Mitt., Erg.-H. 178, Gotha.
— (1937): Mittelamerika. — Handb. Reg. Geol., **8,4 a,** 160 p., Heidelberg.
SCHMIDT, V. A. & ANDERSON, T. H. (1978): Mesozoic crustal evolution of Middle America and the Caribbean: Geophysical considerations. — Eos, Transact. Amer. Geophys. Union, **59,** 404 – 405, Washington, D.C.
SCHMIDT-EFFING, R. (1975): El primer hallasgo de amonites en América Central Meridional y notas sobre las facies cretácicas en dicha región. — Inst. Geogr. Nac., Informe Semestral Enero a Junio de 1974, 52 – 61, San José, C.R.
— (1976 a): El Liásico Marino de México y su Relación con la Paleogeografía de América Central. — Publ. Geol. del ICAITI, **5,** 22 – 23, Guatemala, C.A.
— (1976 b): Daten zur Entwicklungsgeschichte von Golf, Karibik und Atlantik im Mesozoikum. — Münster. Forsch. Geol. Paläont., **38/39,** 201 – 217, Münster (Westf.).
— (1978): Ozeanische Krustengesteine (Oberjura bis Eozän) in stratigraphischer und fazieller Sicht von Costa Rica. — 6. Geowiss. Lateinamerika Koll., 37 – 39, Stuttgart.
— (1979 a): Alter und Genese des Nicoya-Komplexes, einer ozeanischen Paläokruste (Oberjura bis Eozän) im südlichen Zentralamerika. — Geol. Rdsch. **68**(2), 457 – 494, Stuttgart.
— (1979 b): Geodynamic history of oceanic crust in Southern Central America. — IV. Congr. Latinamericano de Geol., Trinidad and Tobago, Preprint.
SCHMIDT-THOMÉ, M. (1975 a): Das Diatomitvorkommen im Tal des Río Sisimico (El Salvador, Zentralamerika). — Geol. Jb., **B 13,** 87 – 96, Hannover.
— (1975 b): The geology in the San Salvador area (El Salvador, Central America), a basis for city development and planning. — Geol. Jb., **B 13,** 207 – 228, Hannover.
SCHMOLL, H. R., KRUSHENSKY, R. D. & DOBROVOLNY, E. (1975): Geologic Considerations for Redevelopment Planning of Managua, Nicaragua, Following the 1972 Earthquake. — Geol. Surv. Prof. Pap., **914,** 23 p., Washington, D.C.
SCHULZ, R. (1964): Estudio sobre la sismicidad en la región centro-americana. — Bol. Bibliogr. de Geofís. y Oceanogr. Amer., **2,** 135 – 144, México.
— (1965): Explicación para el mapa sísmico de la República de El Salvador. — Bol. Sismol. del Centro de Estudios y Investigaciones Geotécn., **10,** 8 San Salvador.
SCHWARTZ, D. P. (1972): Petrology and Structural Geology along the Motagua Fault Zone, Guatemala. — Transact. VI. Caribb. Geol. Conf., 299, Caracas.
— (1976): Geology of the Zacapa quadrangle and vicinity, Guatemala, Central America. — Ph. D. Thesis, State Univ. of New York, Binghampton.
— (1976): The Motagua Fault Zone, Guatemala: Tertiary and Quaternary Tectonics. — Geol. Soc. Amer., Abstr. Programs (USA), **8,** No. 6, 1092 – 1093.
— (1977): Active faulting along the Caribbean — North American plate boundary in Guatemala. — 8th Caribb. Geol. Conf. Abstr., 180 – 181, Amsterdam.
SCHWARTZ, D. L., PACKER, D. R. & WEAVER, K. D. (1975): New information on offset of the Central American Volcanic Chain near Managua, Nicaragua. — Geol. Soc. Amer., Abstr. and Programs (USA), **7,** 1262, Boulder/Col.
SEEBACH, K. VON (1892): Über die Vulkane Zentralamerikas. — Abh. Kgl. Ges. Wiss., **38,** 251 p., Göttingen.
SEELY, D. R., VAIL, P. R. & WALTON, G. G. (1974): Trench slope Model. — In: BURK, C. A. & DRAKE, C. L. (eds.): The Geology of Continental Margins, 249 – 260. — Berlin-Heidelberg-New York (Springer).
SEIFERT, J. (1977): Fossile Frösche (Diplasiocoela NOBLE 1931) aus einer Kieselgur von El Salvador. — Geol. Jb., **B 23,** 29 – 45, Hannover.
SHEETS, P. D. (1976): Ilopango Volcano and the Maya Protoclassic. — Univ. Mus. Stud., **9,** Southern Illinois Univ., Carbondale.
— (1977): Environmental and Cultural Effects of the Ilopango Eruption. — 42nd Ann. Meet., Soc. Amer. Archaeol., New Orleans.
— (1979): Maya Recovery from Volcanic Disasters. Ilopango and Cerén. — Archaeology, **32,** 32 – 42, Washington.
SHOR, G. G. & FISHER, R. L. (1961): Middle America Trench: Seismic-refraction studies. — Geol. Soc. Amer., Bull., **72,** 721 – 730, Boulder/Col.

SIESSER, W. G. (1967): Stratigraphy and plutonic rocks of the Cuilco quadrangle, Guatemala. — M. S. Thesis, Louisiana State Univ., 87 p.

SILLITOE, R. H. (1972): A plate tectonic model for the origin of porphyry copper deposits. — Economic Geol., **67**, 184 – 197, Lancaster.

SIMONSON, B. M. (1976): Igneous Petrology of the Minas de Oro Quadrangle, Central Honduras. — Publ. Geol. del ICAITI, **5**, 78 – 83, Guatemala, C.A.

STEEN-MCINTYRE, V. (1976): Petrography and particle size analysis of selected tephra samples from Western El Salvador: A preliminary Report. 68 – 77. — In: SHEETS, P. D., 1976.

STEHLI, F. G. & GRANT, R. E. (1970): Permian brachiopods from Huehuetenango, Guatemala. — J. Paleont., **44**, 23 – 36, Menasha, Wisc.

STEWART, R. H. (1975): Panama. — In: FAIRBRIDGE, R. W., The Encyclopedia of Worlds Regional Geology I: Western Hemisphere, 416 – 419, Stroudsburg, Pennsylvania (Dowden, Hutchinson & Ross).

STEWART, R. H. & LOWRIE, A. (1978): Evolving plate boundaries in the Darien of Panama. — Eos, Transact. Geophys. Union, **59**, 377, Washington, D.C.

STIBANE, F. R. (1977): Desarrollo tectónico y sedimentológico en Nicoya, Costa Rica. — 8th Caribb. Geol. Conf., Abstr. 203 – 204, Amsterdam.

STIBANE, F. R., SCHMIDT-EFFING, R. & MADRIGAL, R. (1977): Zur stratigraphisch-tektonischen Entwicklung der Halbinsel Nicoya (Costa Rica) in der Zeit von Oberkreide bis Unter-Tertiät. — Giessener Geol. Schr., **12**, 315 – 358, Giessen.

STOIBER, R. E. & CARR, M. J. (1971): Lithosperic Plates, Benioff Zones, and Volcanoes. — Geol. Soc. Amer., Bull., **82**, 515 – 522, Boulder/Col.

— (1974): Quaternary Volcanic and Tectonic Segmentation of Central America. — Bull. volcanol., **37**, 304 – 325, Napoli.

STOIBER, R. E. &. EBERL, D. (1969): Fumaroles of Guatemala. — Publ. Geol. del ICAITI, **2**, 39 – 43, Guatemala, C.A.

STOIBER, R. E., LEGGET, D. C., JENKINS, T. F., MURRMANN, R. P. & ROSE Jr., W. I. (1971): Organic Compounds in Volcanic Gas from Santiaguito Volcano, Guatemala. — Geol. Soc. Amer., Bull., **82**, 2299 – 2302, Boulder/Col.

STOIBER, R. E. & ROSE Jr., W. I. (1969): Recent Volcanic and Fumarolic Activity at Santiaguito Volcano, Guatemala. — Bull. volcanol., **33**, 476 – 502, Napoli.

— — (1970): The geochemistry of Central American volcanic gas condensates. — Geol. Soc. Amer., Bull., **81**, 2891 – 2912, Boulder/Col.

— — (1973): Sublimates at Volcanic Fumaroles of Cerro Negro Volcano, Nicaragua. — Publ. Geol. del ICAITI, **4**, 63 – 68, Guatemala C.A.

— — (1974 a): Cl, F, and SO_2 in Central American Volcanic Gases. — Bull. volcanol., **37**, 454 – 460, Napoli.

— — (1974 b): Fumarole incrustations at active Central American volcanoes. — Geochim. cosmochim. Acta, **38**, 495 – 516, London.

STOIBER, R. E., ROSE Jr., W. I., LANGE, I. M. & BIRNIE, R. W. (1975): The Cooling of Izalco Volcano (El Salvador) 1964 – 1974. — Geol. Jb., **B 13**, 193 – 205, Hannover.

STRECKEISEN, A. L. (1967): Classification and Nomenclature of Igneous Rocks (Final Report of an Inquiry). — N. Jb. Miner., Abh., **107**, 144 – 240, Stuttgart.

SUTTER, J. F. (1977): K/Ar Ages of Cenozoic volcanic rocks in Northern Central America. — 8th Caribb. Geol. Conf., Abstr., 202 – 206, Amsterdam.

SWAIN, F. M. (1961): Reporte preliminar de los sedimentos del fondo de lago Nicaragua y Managua, Nicaragua. — Bol. Serv. Geol. Nac., Nicaragua, **5**, 11 – 29, Managua.

— (1966): Bottom sediments of Lake Nicaragua and Lake Managua, Western Nicaragua. — J. Sed. Petrol., **36**, 522 – 540, Tulsa, Oklahoma.

TAYLOR, G. (1973): Preliminary report on the stratigraphy of Limón. — Publ. Geol. del ICAITI, **4**, 161 – 166, Guatemala, C.A.

— (1975): The geology of the Limón area of Costa Rica. — Ph. D. Thesis, Luisiana State Univ., Agricult. and Mechan. Coll.

TAYLOR, P. S. & STOIBER, R. E. (1973): Soluble Material on Ash from Active Central American Volcanoes. — Geol. Soc. Amer., Bull., **84**, 1031 – 1042, Boulder/Col.

TERMER, F. (1932): Geologie von Nordwestguatemala. — Z. Ges. Erdkd. Berlin, 240 – 248, Berlin.
— (1936): Zur Geographie der Republik Guatemala. I. Teil: Beiträge zur physischen Geographie von Mittel- und Süd-Guatemala. — Mitt. Geogr. Ges. Hamburg, **44**, 89 – 275, Hamburg.
— (1956): Carlos Sapper-Explorador de Centro América (1866 – 1945). — Anal. Soc. Geogr. Histor. de Guatemala, **29**, 55 – 101, Guatemala.
— (1964): Die Tätigkeit der Vulkane von Guatemala in den Jahren 1960 – 1963. — Petermanns Geogr. Mitt., 4. Quartalsh. 261 – 268, Gotha.
— (1966): Karl Theodor Sapper 1866 – 1945. Leben und Wirken eines deutschen Geographen und Geologen. — Lebensdarstellungen deutscher Naturforscher, **12**, 1 – 89, Deutsch. Akad. Naturforsch. Leopoldina. Leipzig (A. B. Barth).
TERRY, R. A. (1956): A geological reconnaissance of Panamá. — Californ. Acad. Sci., Ocas. Pap., **23**, 1 – 91, San Francisco.
TOURNON, J. (1973): Présence de Basaltes Alcalins Récents au Costa Rica (Amérique Centrale). — Bull. volcanol., **36**, 140 – 147, Napoli.
TRICART, J. (1961): Aperçu sur le Quaternaire du Salvador (Amérique Centrale). — Bull. Soc. Géol Fr., Sér. 7,**3**, 59 – 68, Paris.
TRIPET, J.-P. (1977): Quelques Aspects de la Géographie, de la Géologie et des Karsts du Guatemala. — Cavernes (Suisse), **21**, 3 – 8, 39 – 45.
TROTTER, R. J. (1977): Unravelling a Mayan Mystery. — Science News (USA), **111**, 74, 75, 78.
TRUSHEIM, F. (1976): Zur strukturellen Entwicklung von Wulstfalten-Ketten. — Z. Deutsch. Geol. Ges., **127**, 147 – 181, Hannover.
UCHUPI, E. (1973): Eastern Yucatan Continental Margin and Western Caribbean Tectonics. — Amer. Assoc. Petrol. Geol., Bull., **57**, 1075 – 1085, Tulsa, Oklahoma.
— (1975): Physiography of the Gulf of Mexico and Caribbean Sea. — In: NAIRN, A. E. M. & STEHLI, F. G., The Ocean Basins and Margins, **3**, 1 – 64, New York and London (Plenum Press).
UI, T. (1973): Recent Volcanism in Masaya-Grande Area, Nicaragua. — Bull. volcanol., **36**, 174 – 190, Napoli.
UMAÑA, J. E. (1966): Geología del Sitio de Presa del Proyecto Cachí, Costa Rica. — Publ. Geol. del ICAITI, **1**, 27 – 41, Guatemala, C.A.
VESSEL, R. K. & DAVIES, D. K. (1978): Rates of denudation and sediment transport from the active volcano Fuego, Guatemala. — Geol. Soc. Amer., Abstr. and Programs, **10**, 3, 151, Boulder/Col.
VIKSNE, A., LISTON, T. & SAPP, C. D. (1969): SLR reconnaissance of Panama. — Geophysics, **34**, 54 – 64.
VINIEGRA, O. F. (1971): Age and Evolution of Salt Basins of Southeastern Mexico. — Amer. Assoc. Petrol. Geol., Bull., **55**, 478 – 494, Tulsa, Oklahoma.
VINSON, G. L. (1962): Upper Cretaceous and Tertiary Stratigraphy of Guatemala. — Amer. Assoc. Petrol. Geol., Bull., **46**, 425 – 456, Tulsa, Oklahoma.
VIRAMONTE O., J. G. (1973): Las últimas erupciones en Nicaragua (Período 1968 – 1970). — Publ. Geol. del ICAITI, **4**, 69 – 80, Guatemala, C.A.
VIRAMONTE, O., J. G. & DI SCALA, L. (1970): Summary of the 1968 Eruption of Cerro Negro, Nicaragua. — Bull. volcanol., **34**, 347 – 351, Napoli.
VIRAMONTE O., J. G. & WILLIAMS, R. L. (1973): Estudio preliminar sobre las ignimbritas andesíticas de Nicaragua. — Publ. Geol. del ICAITI, **4**, 171 – 178, Guatemala, C.A.
VIVÓ ESCOTO, J. A. (1978): Los sismos de Guatemala en febrero y marzo de 1976 y su relación con la morfología estructural de América Central. — Anuario de Geogr. Univ. Nac. Auton. de México, **16**, 11 – 44, México, D.F.
WALDRON, H. H. (1967): Debris flow and erosion control problems caused by ash eruptions of Irazú Volcano, Costa Rica. — U.S. Geol. Surv. Bull., **1241**, 37 p., Washington, D.C.
WALPER, J. L. (1960): Geology of Cobán-Purulhá Area, Alta Verapaz, Guatemala. — Amer. Assoc. Petrol. Geol., Bull., **44**, 1273 – 1315, Tulsa, Oklahoma.
WARD, P. L., GIBBS, J., HARLOW, D. & ABURTO Q., A. (1974): Aftershocks of the Managua, Nicaragua, earthquake and the tectonic significance of the Tiscapa fault. — Seismol. Soc. Amer., Bull., **64**, 1017 – 1030, Berkeley.
WEAVER, J. (ed.) (1975): Geology, Geophysics and Resources of the Caribbean. — Report of the IDOE Workshop on Geology and Marine Geophysics of the Caribbean Region and its resources, 15 p., Kingston, Jamaica (Mayaguez, P.R.).

WEICHERT, D. H. (1976): Earthquake Reconnaissance: Guatemala, February 1976. — Geosci. Canada, 3, 208 – 214.
WEYL, R. (1957 a): Beiträge zur Geologie der Cordillera de Talamanca Costa Ricas (Mittelamerika). — N. Jb. Geol. Paläont., Abh., **105**, 123 – 204, Stuttgart.
— (1957 b): Contribución a la Geología de la Cordillera de Talamanca de Costa Rica (Centro América). — 75 p., Inst. Geogr. de Costa Rica, San José, C. R.
— (1961 a): Die Geologie Mittelamerikas. — 226 p. Berlin (Borntraeger).
— (1961 b): Mittelamerikanische Ignimbrite. — N. Jb. Geol. Paläont., Abh., **113**, 23 – 46, Stuttgart.
— (1965): Erdgeschichte und Landschaftsbild in Mittelamerika. — 175 p., Frankfurt/M. (Kramer).
— (1966 a): Tektonik, Magmatismus und Krustenbau in Mittelamerika und Westindien. — Geotekt. Forsch., **23**, 67 – 109, Stuttgart.
— (1966 b): The palaeogeographic development of the Central American — West Indian Region. — Bol. Inform. Assoc. Venezolana de Geología, Minería y Petróleo, **9**, 99 – 120, Caracas.
— (1967): Krustenbau und sialischer Magmatismus. — Geol. Rdsch., **56**, 367 – 372, Stuttgart.
— (1969 a): Magnetitsande der Küste Nicoyas (Costa Rica, Mittelamerika). — N. Jb. Geol. Paläont., Mh., **1969**, 499 – 511, Stuttgart.
— (1969 b): Arenas magnetíticas de la costa de Nicoya (Costa Rica, América Central). — Inst. Geogr. Nac., Informe Semestral Enero a Junio 1969, San José, C.R.
— (1969 c): Magmatische Förderphasen und Gesteinschemismus in Costa Rica (Mittelamerika). — N. Jb. Geol. Paläont., Mh., **1969**, 423 – 446, Stuttgart.
— (1969/1970): Geologische Bilder aus Mittelamerika, 1 – 5. — Natur u. Museum **99**, 415 – 423, 559 – 570; **100**, 120 – 128, 269 – 278, 362 – 370, Frankfurt/M.
— (1971 a): Die Morphologisch-Tektonische Gliederung Costa Ricas (Mittelamerika). — Erdkunde, Arch. f. wiss. Geogr., **25**, 223 – 230, Bonn.
— (1971 b): La Clasificación morfotectónica de Costa Rica (América Central). — Inst. Geogr. Nac., Informe Semestral Julio a Diciembre 1971, 107 – 125, San José, C.R.
— (1971 c): Mittelamerika (Literaturbericht). Mit Beitrag von BOHNENBERGER, O. H. — Zbl. Geol. Paläont., Teil I, 1970, 1003 – 1051, Stuttgart.
— (1974 a): Die paläogeographische Entwicklung Mittelamerikas (Literaturbericht). — Zbl. Geol. Paläont., Teil I, 1973, 432 – 466, Stuttgart.
— (1974 b): El desarrollo paleogeográfico de América Central. — Bol. Assoc. Mexicana de Geólogos Petroleros, **25**, 374 – 424, México, D.F.
— (1978 a): Magmatismus und Metallogenese in Mittelamerika. — Münster Forsch. Geol. Paläont., **44/45**, 43 – 85, Münster (Westf.).
— (1978 b): The Economic Significance of Magmatism and Metallogenesis in Central America. — Natural Resources and Development, **8**, 115 – 143, Tübingen.
WEYL, R. & PICHLER, H. (1973): Petrochemical Aspects of Central American Magmatism. — Publ. Geol. del ICAITI, **4**, 81 – 90, Guatemala, C.A.
— — (1976): Magmatism and Crustal Evolution in Costa Rica (Central América). — Publ. Geol. del ICAITI, **5**, 56 – 70, Guatemala, C.A.
WHITE, L. (1977): Central America: Diverse mineralization provides targets fro exploration. — Engineering and Mining J., November 1977, 159 – 198.
WIESEMANN, G. (1973): Arbeiten der Bundesanstalt für Bodenforschung in der Republik El Salvador in den Jahren 1967 – 1971. — Münster Forsch. Geol. Paläont, **31/32**, 277 – 285, Münster (Westf.).
— (1974): Misión Geológica Alemana en El Salvador 1967 – 1971. Reporte Final. — Bundesanstalt f. Bodenforschung, Hannover.
— (1975): Remarks on the Geologic Structure of the Republic of El Salvador, Central America. — Mitt. Geol.-Paläont. Inst. Univ. Hamburg, **44**, 557 – 574, Hamburg.
— (1977): Mapa Geológico de la República de El Salvador. Introducción — Einleitung — Introduction. — Hannover.
WILLIAMS, H. (1952 a): The great eruption of Coseguina, Nicaragua, in 1835. — Univ. Calif. Publ. geol. Sci., **29**, 21 – 46, Berkeley and Los Angeles.
— (1952 b): Volcanic History of the Meseta Central Occidental, Costa Rica. — Univ. Calif. Publ. geol. Sci., **29** (4), 145 – 180, Berkeley and Los Angeles.
— (1960): Volcanic history of the Guatemalan Highlands. — Univ. Calif. Publ. geol. Sci., **38** (1), 1 – 86, Berkeley and Los Angeles.

WILLIAMS, H. & MCBIRNEY, A. R. (1964): Petrologic and structural contrast of the Quaternary volcanoes of Guatemala. — Bull. volcanol., **27**, 61, Napoli.

— — (1969): Volcanic History of Honduras. — Univ. Calif. Publ. geol. Sci., **85**, 1 – 101, Berkeley and Los Angeles.

WILLIAMS, H., MCBIRNEY, A. R. & DENGO, G. (1964): Geologic Reconnaissance of Southeastern Guatemala. — Univ. Calif. Publ. geol. Sci., **50**, 1 – 56, Berkeley and Los Angeles.

WILLIAMS, H. & MEYER-ABICH, H. (1955): Volcanism in the Southern Part of El Salvador. With Particular Reference to the Collapse Basins of Lakes Coatepeque and Ilopango. — Univ. Calif. Publ. geol. Sci., **32** (1), 1 – 64, Berkeley and Los Angeles.

WILLIAMS, M. D. (1975): Emplacement of Sierra de Santa Cruz, Eastern Guatamala. — Amer. Assoc. Petrol. Geol., Bull., **59**, 1211 – 1216, Tulsa, Oklahoma.

WILMER S., F. (1976): Estudio Geológico relacionado con una Mineralización de Súlfuros en Punta Gorda, Nicoya, Costa Rica. — Unpubl. Thesis, Escuela Centroamer. Geol., Univ. de Costa Rica, San Pedro, C.R.

WILSON, H. H. (1974): Cretacous Sedimentation and Orogeny in Nuclear Central America. — Amer. Assoc. Petrol. Geol., Bull., **58**, 1348 – 1396, Tulsa, Oklahoma.

WILSON, H. H. & HORNE, G. S. (1977): Pre-Cretaceous Rocks of Northwestern Honduras: Basement Terrane in Sierra de Omoa: Discussion. — Amer. Assoc. Petrol. Geol. Bull., **61**, 269 – 273, Tulsa, Oklahoma.

WING, R. S. (1970): Structural analysis from radar imagery, Eastern Panamanian Isthmus. — Ph. D. Thesis, Univ. of Kansas.

WING, R. S. & MACDONALD, H. C. (1973): Radar Geology — Petroleum Exploration Technique. Eastern Panama and Northwestern Colombia. — Amer. Assoc. Petrol. Geol., Bull., **57**, 825 – 840, Tulsa, Oklahoma.

WITHMORE, F. C., Jr. & STEWART, R. H. (1965): Miocene mammals and Central American seaways. — Science, **148**, 180 – 185, Lancaster.

WOODRING, W. P. (1966 a): Estratigrafía Tercariaria de la Zona del Canal y Partes Adaycentes de la República de Panamá. — Publ. Geol. del ICAITI, **1**, 43 – 45, Guatemala, C.A.

— (1966 b): The Panama land bridge as a sea barrier. — Proc. Amer. Philos. Soc., **110**, 425 – 433, Philadelphia.

— (1973): Affinities of Miocene Molluscan Faunas on Pacific side of Central America. — Publ. Geol. del ICAITI, 4, 179 – 187, Guatemala C.A.

— (1974): The Miocene Caribbean Faunal Province and its Subprovinces. — Verh. Naturf. Ges. Basel, **84**, 209 – 213, Basel.

— (1976): Age of El Salto Formation in Nicaragua. — Publ. Geol. del ICAITI, **5**, 18 – 20, Guatemala, C.A.

WOODRING, W. P. & MALAVASSI, E. V. (1961): Miocene foraminifera, molluscs, and a barnacle from the Valle Central, Costa Rica. — J. Paleont., **35**, 489 – 497, Menasha, Wisconsin.

WOODRUFF, L. G., ROSE, W. I., Jr. & RIGOT, W. (1978): Contrasting fractionation patterns for sequential magmas from two calc-alkaline volcanoes in Central America. — Geol. Soc. Amer., Abstr. and Programs, **10**, 3, 154, Boulder/Col.

ZOPPIS BRACCI, L. (1957): Estudio geológico de la región de Palacaguina y de su depósito de Antimonio. — Bol. Serv. Geol. Nac. Nicaragua, **1**, 29 – 34, Managua.

— (1961): Estudio Preliminar de la Geología de las mineralizaciones de Tungsteno y Molibdeno de Mazuelito, Departamento de Nueva Segovia. — Bol. Serv. Geol. Nac. de Nicaragua, **5**, 33 – 51, Managua.

ZOPPIS, BRACCI, L. & GIUDICE, D. DEL (1958 a): Geología de la Costa del Pacífico de Nicaragua. — Bol. Serv. Geol. Nac. de Nicaragua, **2**, 19 – 68, Managua.

— (1958 b): Un reconocimiento geológico del Río Bocay y parte del Río Coco. — Bol. Serv. Geol. Nac. de Nicaragua, **2**, 81 – 112, Managua.

— (1960): Reconocimiento geológico del valle de Punta Gorda. — Bol. Serv. Geol. Nac. de Nicaragua, **4**, 61 – 83, Managua.

ZOPPIS DE SENA, R. (1957): El Yacimiento de Hierro de Monte Carmelo. — Bol. Serv. Geol. Nac. de Nicaragua, **1**, 13 – 27, Managua.

ZUNIGA, I., M. A. (1975): Gravity and Magnetic Survey of the Sula Valley, Honduras, Central America. — Ph. D. Thesis, Univ. of Texas, Austin, Texas.

Author Index

Locality Index

As a guide for the reader, locality names and subjects have, where possible, identified by the following codes:

B = Belize
C = Costa Rica
ES = El Salvador
G = Guatemala
H = Honduras
N = Nicaragua
P = Panama

Subject Index

BEITRÄGE ZUR REGIONALEN GEOLOGIE DER ERDE

Die Geologie Mittelamerikas — Band 1

von Prof. Dr. RICHARD WEYL. Großoktav. 1 Kunstdrucktafel, 61 Textabbildungen, 6 Klapptafeln, 11 Tabellen, XVI, 226 Seiten. 1961. Ganzleinen.
Supplement für die Zeit 1961–1970 (Sonderdruck aus Zentralblatt für Geologie und Paläontologie, Teil I, Heft 7/8, Jahrgang 1970). Von Prof. Dr. RICHARD WEYL. 15×23,5 cm. 2 Tabellen, 49 Seiten. 1970. Geheftet.

Die Geologie von Paraguay — Band 2

von Dr. HANNFRIT PUTZER. Großoktav. 74 Textabbildungen, 2 Ausschlagtafeln, 2 Tafeln Fossilien, 10 Tabellen, 1 mehrfarb. geol. Karte. XII, 184 Seiten. 1962. Ganzleinen.

Geologie von Chile — Band 3

von Prof Dr. WERNER ZEIL. Großoktav. 10 Ausklapptafeln, 43 Textabbildungen und 57 Abbildungen auf Tafeln. XII, 234 Seiten. 1964. Ganzleinen.

Geologie der Antillen — Band 4

von Prof. Dr. RICHARD WEYL. Großoktav. 126 Abbildungen, 48 Photos, 25 Tafeln und 24 Tabellen. VIII, 410 Seiten. 1966. Ganzleinen.

Rocky Mountains — Band 5

Der geologische Aufbau des Kanadischen Felsengebirges von Dr. DIETRICH HANS ROEDER. Großoktav. 108 Abbildungen, davon 16 auf 12 Ausklapptafeln, 1 Kunstdrucktafel und 5 Beilagen. XII, 318 Seiten. 1967. Ganzleinen.

Geologie von Syrien und dem Libanon — Band 6

von Dr. REINHARD WOLFART. Großoktav. 78 Abbildungen, 42 Tabellen, 1 geologische und 1 bodenkundliche Übersichtskarte. XII, 326 Seiten. 1967. Ganzleinen.

Geologie von Jordanien — Band 7

von Dr. FRIEDRICH BENDER. Großoktav. 5 Fossiltafeln, 14 Tabellen, 159 Textabbildungen, 9 Abbildungen und 2 Tabellen auf Ausschlagtafeln sowie 1 geologische Karte. XI, 230 Seiten, 1968. Ganzleinen.
Supplementary edition (translated from the German edition) with minor revisions. IX, 196 pages, 46 figures, 15 tables, 4 folders. 1974. Stiff cover.

Geologie von Ungarn Band 8

Von Dr. LÁSZLÓ TRUNKÓ. Großoktav. Mit 99 Abbildungen und 8 Tabellen im Text, 55 Fotos auf Kunstdruck, 10 Abbildungen und 1 Tabelle auf Ausschlagtafeln und 2 geologischen Übersichten. IX, 257 Seiten. 1969. Ganzleinen.

Geologie von Brasilien Band 9

von Prof. Dr. KARL BEURLEN. Großoktav. Mit 76 Abbildungen und 6 Tabellen im Text, 1 Orientierungskarte, 1 Tabelle, 2 Ausschlagtafeln. VIII, 444 Seiten. 1970. Ganzleinen.

Geology of the South Atlantic Islands Band 10

In English. Großoktav. By Dr. R. C. MITCHELL-THOMÉ. With 106 figures, 95 tables in the text and on 8 folders. X, 367 pages. 1970. Cloth.

Geologie von Ecuador Band 11

von Prof. Dr. WALTHER SAUER. Mit einem Beitrag von Prof. Dr. HANNFRIT PUTZER. Großoktav. Mit 29 Abbildungen, 15 Tabellen, 31 Photos, 4 Ausschlagtafeln und 2 geologischen Karten. IX, 316 Seiten. 1971. Ganzleinen.

Geology of the Middle Atlantic Islands Band 12

In English. Großoktav. By Dr. R. C. MITCHELL-THOMÉ. With 102 figures and 68 tables in the text and on 5 folders. IX, 382 pages. 1976. Cloth.

The Andes · A Geological Review Band 13

In English. Großoktav. By Prof. Dr. W. ZEIL. With 143 figures in the text and on 7 folders. VIII, 260 pages. 1979. Cloth.

Geologie von Afghanistan Band 14

von Dr. R. WOLFART und Dr. H. WITTEKINDT. Großoktav. Mit 76 Abbildungen, 41 Tabellen und 3 Übersichtskarten im Text und auf 16 Beilagen. XVI, 500 Seiten. 1980. Ganzleinen.

Geology of Central America Band 15

Second, completely revised edition. In English. By Prof. Dr. R. WEYL. Großoktav. With 202 figures and 13 tables in the text and on 8 folders. VIII, 371 pages. 1980. Cloth.